D1218377

The Enzymes

VOLUME XIX

MECHANISMS OF CATALYSIS

Third Edition

THE ENZYMES

Edited by

David S. Sigman
Department of Biological Chemistry
and Molecular Biology Institute
University of California
Los Angeles, California

Paul D. Boyer
Department of Chemistry and Biochemistry
and Molecular Biology Institute
University of California
Los Angeles, California

Volume XIX

MECHANISMS OF CATALYSIS

THIRD EDITION

ACADEMIC PRESS, INC.
Harcourt Brace Jovanovich, Publishers
San Diego New York Boston
London Sydney Tokyo Toronto

This book is printed on acid-free paper. ♾

Academic Press, Inc.
San Diego, California 92101

United Kingdom Edition published by
Academic Press Limited
24–28 Oval Road, London NW1 7DX

Library of Congress Cataloging-in-Publication Data

(Revised for vol. 19)

The Enzymes.

Vols. 17- edited byPaul D. Boyer and
Edwin G. Krebs.
Vol. published in: Orlando.
Vol. 19 edited by David S. Sigman, Paul D. Boyer.
Includes bibliographical references and indexes.
1. Enzymes. I. Boyer, Paul D., ed. II. Krebs,
Edwin G., ed. III. Sigman, D. S. [DNLM: 1. Enzymes
QU 135 B791e]
QP601.E523 574.1'925 75-117107
ISBN 0-12-122702-2 (v.2 : alk. paper)
ISBN 0-12-122719-7 (v.19 : alk.paper)

Printed in the United States of America
90 91 92 93 9 8 7 6 5 4 3 2 1

Contents

4. Analysis of Protein Function by Mutagenesis

Kenneth A. Johnson and Stephen J. Benkovic

5. Mechanism-Based (Suicide) Enzyme Inactivation

Mark A. Ator and Paul R. Ortiz de Montellano

6. Site-Specific Modification of Enzyme Sites

Roberta F. Colman

7. Stereochemistry of Enzyme-Catalyzed Reactions at Carbon

Donald J. Creighton and Nunna S. R. K. Murthy

Preface

Most of the volumes in "The Enzymes" treatise present in-depth coverage of a selected area of enzymology. However, two decades ago, Volumes I and II of the Third Edition of this work covered the general properties of enzymes and enzymic catalysis. Although these earlier volumes were well received and served as a useful resource for students and scholars who needed a perspective of the field, much progress has ensued and an up-to-date survey of the status of the field is needed.

The goal of this volume is to present the type of information that has lasting value. This is what graduate students need to make knowledge of enzymes part of their professional competency. Such information also provides the seasoned researcher with a sound basis for experimentation involving the use and understanding of enzymes.

This volume presents coverage by selected authors in areas in which considerable advance has occurred. This includes the understanding of enzyme catalysis as related to binding energies and an updating of kinetic probes of reaction mechanism. The increased understanding of how biological electron transfer occurs is covered. The newer areas of analysis of protein function by mutagenesis, of mechanism-based enzyme inactivation, and of site-specific modification are surveyed. A review of the powerful new methods of stereochemical analyses concludes the volume.

Also recorded in this Preface is another product of the passage of time. This is the last volume for which Paul Boyer will serve as an editor. David Sigman has had the principal responsibility for the preparation of this volume, and is planning subsequent volumes for the series.

David S. Sigman
Paul D. Boyer

1

Binding Energy and Catalysis

DAVID D. HACKNEY

Department of Biological Sciences and
Biophysics and Biochemistry Program
Carnegie Mellon University
Pittsburgh, Pennsylvania 15213

I. Introduction

Enzymic reactions are characterized by two principal properties: high catalytic efficiency and selectivity. Both properties ultimately derive from the binding of substrate molecules at the active site and the subsequent stabilization of the tran-

THE ENZYMES, Vol. XIX

sition state. It is only this binding process, in fact, which distinguishes enzymic catalysis from simple homogeneous chemical catalysis since it is expected that enzymes otherwise obey all the rules of standard chemical reactions. Both types of reactions share common features such as use of general acids and bases to facilitate proton transfers and have predictable variations in reaction rate with experimental parameters. For example, model reactions which involve an increase in charge in the transition state often are found to proceed more rapidly in polar solvents, and enzymes can provide a constellation of hydrogen-bonding and charged residues which are complementary to charges that develop in the transition state. Only recently has it been possible to synthesize nonenzymic models which begin to approach the complexity and specificity of enzymic active sites (*1–4*).

Our knowledge of the influence of binding on catalysis and the related area of induced conformational changes has evolved to a fairly mature state based on many contributions. Experimental studies of ligand binding to enzymes, as well as extensive kinetic investigations (see Chapter 3, this volume), have provided a basis for theoretical generalizations. Especially important have been those studies which allow evaluation of how the total possible binding energy is utilized by the different intermediates that are passed through during enzymic turnover. This has been possible through the development of a number of approaches which allow the determination of the partitioning of individual enzyme intermediates between the forward and reverse directions (or among multiple alternatives in a branched scheme). Transient kinetic experiments have obviously played a critical role since they allow determination of individual rate constants in favorable cases. Equally important has been the analysis of isotopic exchange reactions. Since these reactions result from reversible passage through only a limited subset of the complete reaction scheme, they provide information about the kinetics of that isolated part of the complete scheme. Analysis of the multiple possible isotopic exchange reactions, which result from different overlapping subsets of the complete reaction, often allows determination of most, if not all, of the individual rate constants. Numerous extensions and variations have been developed [see Ref. (*5*) for review], and other techniques, such as kinetic isotope effects, have also proven useful. Detailed free energy diagrams for progress of the net reaction through a series of intermediate states are now available for a number of enzymes and are discussed in detail in Chapter 4 (this volume) with regard to what can be learned from analysis of mutants.

This chapter presents an integrated view of these central concepts in enzymology, based on this collective knowledge. Previous treatments and reviews in the general area can be found in Refs. (*6–11*) and references therein. The specific examples, which are mentioned here, are intended for illustrative purposes only, and no attempt has been made to present an exhaustive compilation of all the possible examples for each concept.

II. Relationship of Binding to Catalysis in Unimolecular Reactions

A. Necessity of Binding to Substrate and Product as Well as Transition State

Catalysis of any kind, in the broadest sense, must derive from a reduction in the energy of the transition state. In the case of enzymic catalysis, the transition state energy is lowered by virtue of its binding interaction with the enzyme. As was pointed out some time ago by Pauling (*12*), this is equivalent to saying that the enzyme binds the transition state tightly. It is sometimes said that this represents a meaningless truism which is merely a restatement of the definition of catalysis. If so, it is an incorrect truism, since tight binding to the transition state is a necessary, but insufficient, condition for catalysis. An enzyme which had to wait for a transition state to be formed before it could bind would be no catalyst at all.

This is illustrated by the reaction coordinate diagram given in Fig. 1A for the change in free energy on conversion of a hypothetical substrate S to product P via a high-energy transition state species, TS. (Note: TS is used to refer to the transition state of the uncatalyzed chemical reaction, regardless of whether the E·TS complex is the highest energy intermediate in a particular enzymic scheme.) According to simple transition state rate theory, the reaction rate will be proportional to $e^{-\Delta G^{\ddagger}/RT}$, where $\Delta G^{\ddagger}$ is the difference in energy between the ground state and the transition state. The nonenzymic reaction proceeds along the solid lines in Fig. 1A, and the transition state energy barrier is given by $\Delta G_1^{\ddagger}$. The same energy barrier, and thus the same rate, results if the transition state for the uncatalyzed reaction must be generated before it can bind to the enzyme and be stabilized (solid line to free transition state followed by interaction with enzyme and conversion to product via dashed lines).

In order to provide a kinetically efficient pathway to the E·TS complex, binding of the substrate must occur at some stage before complete conversion to the free transition state. This is indicated by the dotted line in Fig. 1A, B for conversion of the substrate to the bound transition state without passage through the free transition state. The energy barrier for this process is given by $\Delta G_2^{\ddagger}$, and its smaller value results in a more rapid rate via this pathway. Thus, binding to the substrate, as well as to the transition state, is not merely a consequence of the necessary structural similarity of the two species but is required for catalysis. Microscopic reversibility requires that these considerations apply to the reverse reaction as well, and analogous interactions with the product must also occur.

A general treatment does not require that the detailed shape of the free energy surface for interaction of S with E contain a discrete minimum corresponding to a stable E·S complex. The only requirement is that the enzyme must interact with the substrate before it is fully converted to the transition state. It is usually

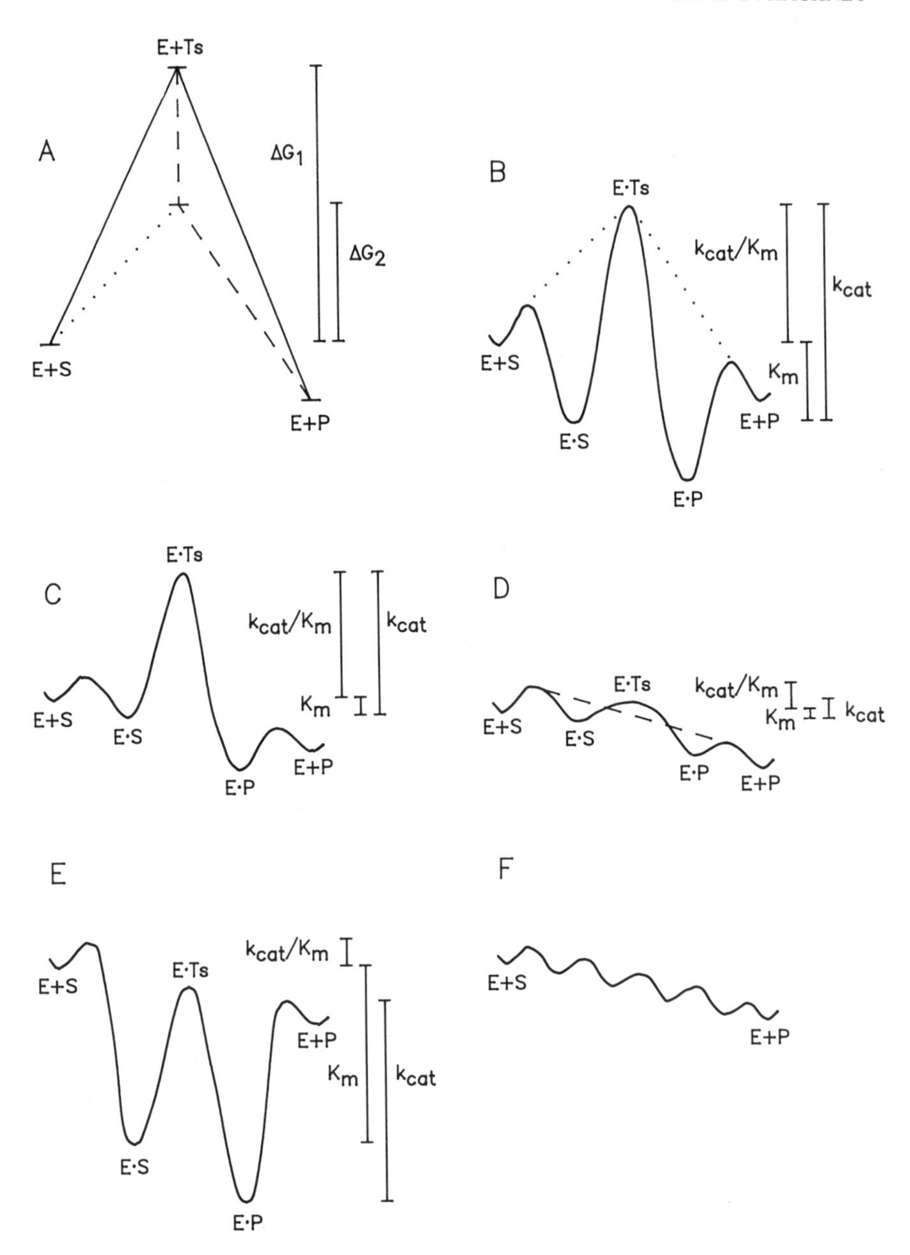

FIG. 1. Free energy profiles for enzymic reactions. For cases B–E, the difference in free energy between two of the states is the predominant factor in determining the values of k_{cat} and K_m for the forward reaction. For purposes of illustration, the energy difference between these two states is indicated with the understanding that exact calculation of k_{cat} and K_m requires consideration of the energy levels of additional states.

observed, however, that enzymic reactions do contain at least one minimum corresponding to a discrete E·S complex, as indicated by the solid line in Fig. 1B. This is reasonable because initial interaction of the enzyme with S is favorable through the common structural features shared by both S and TS, whereas immediate further progress of E·S to E·TS is unlikely, owing to the high-energy barrier for this step. Thus, a minimum in the free energy surface results, and E·S is a discrete species.

Formation of an E·S complex results in the saturation kinetics usually associated with enzymic reactions. At a fixed level of enzyme, the rate initially increases linearly with substrate concentration, but then levels off to approach a limiting maximal rate at saturating substrate level. The maximal rate is given by

$$V_{max} = k_{cat}[E_t]$$

where E_t is the total enzyme concentration, which is equal to the sum of free E plus E·S and any other enzyme species which may occur in a more complex scheme. The value of k_{cat} is determined by the energy levels of the internal states and is limited by how rapidly the E·S complex can produce and release product with regeneration of the free enzyme. Internal states refer to states with some form of the bound substrate, including E·TS and E·P, whereas external states contain free enzyme. It is possible to demonstrate that conformation changes in the free enzyme do occur between release of the product and binding of the substrate [see Britton (*13*)]. These conformational changes could theoretically contribute to k_{cat}, but they are generally not rate limiting under steady-state conditions.

For the simple scheme

$$E + S \underset{k_{-1}}{\overset{k_1}{\rightleftharpoons}} E{\cdot}S \underset{k_{-2}}{\overset{k_2}{\rightleftharpoons}} E{\cdot}TS \underset{k_{-3}}{\overset{k_3}{\rightleftharpoons}} E{\cdot}P \underset{k_{-4}}{\overset{k_4}{\rightleftharpoons}} E + P$$

with the energy levels given in Fig. 1B, k_{cat} is approximately equal to k_2. The S level at 50% V_{max} is designated the K_m and is a measure of the degree of saturation of the enzyme expressed as a dissociation constant. For the above equation with the energy levels of Fig. 1B,

$$K_m = [E][S]/[E{\cdot}S] = (k_{-1} + k_2)/k_1 \cong k_{-1}/k_2$$

under initial rate assumptions for the forward reaction. K_m also influences the rate. For Fig. 1B, $k_4 \gg k_{-3}$ and thus

$$\begin{aligned} \text{Rate} &= k_1[S][E][k_2/(k_{-1} + k_2)] \\ &= (k_{cat}/K_m)[S][E] \end{aligned}$$

where $k_1[S][E]$ is the bimolecular rate for formation of E·S, and $[k_2/(k_{-1} + k_2)]$ is the fraction of E·S which releases ligand as P. At low S concentration, $[E] = [E_t]$, and k_{cat}/K_m is the bimolecular rate constant. For more complex schemes, including consideration of the reverse reaction, these generalizations still apply, but k_{cat} and K_m will be complex functions of the larger number of individual rate constants.

Thus, k_{cat} describes how rapidly the bound states can produce free product and regenerate free enzyme whereas the k_{cat}/K_m ratio describes how well the enzyme operates at low concentration at which the reaction is also limited by the rate of substrate binding. Much of this discussion focuses on the distribution of catalysis between k_{cat} and k_{cat}/K_m, with k_{cat} being more important in some circumstances whereas k_{cat}/K_m is the more relevant parameter in others, such as in consideration of reversibility and substrate selectivity.

B. Necessity of Destabilizing Binding to Substrate and Product

In the case of Fig. 1B the enzyme has equal binding affinity for S, TS, and P as indicated by the equal free energy differences between the bound and free forms of each species. The energy barrier for conversion of S to TS is the same as that for conversion of E·S to E·TS, and k_{cat} equals the rate constant, k_n, for the nonenzymic process. This situation as well does not produce catalysis since the net rate for conversion of S to P is not increased by addition of E as shown by

$$\begin{aligned}
\text{Nonenzymic velocity} &= k_n[\mathrm{S_{total}}] \\
\text{Velocity with enzyme} &= k_n[\mathrm{S}] + (k_n/K_m)[\mathrm{S}][\mathrm{E}] \\
&= k_n[\mathrm{S}] + k_n[\mathrm{ES}] \\
&= k_n([\mathrm{S}] + [\mathrm{ES}]) \\
&= k_n[\mathrm{S_{total}}]
\end{aligned}$$

Catalysis does not occur in spite of the facts that k_{cat}/K_m is greater than k_n for a standard state of 1 *M* and that k_{cat}/K_m can be further increased at will by an additional increase in the amount of uniform binding energy. In this case, k_{cat} is the better measure of catalysis.

In order to produce catalysis, it is also necessary that the binding of S and P be destabilized with respect to the binding of TS, as indicated in Fig. 1C. Now the energy difference between E·S and E·TS is smaller than the difference between S and TS, and k_{cat} is consequently greater than k_n. These considerations can be summarized by a thermodynamic cycle (*11*) which establishes that the increase in k_{cat} over k_n is directly related to the difference in binding energy between S and TS.

The prediction that enzymes must bind their transition states tightly indicates that compounds, which are structurally similar to the transition state, may also bind tightly. Such transition state analogs have been synthesized for a number of specific enzymic reactions. It is generally found that they do, in fact, bind to their respective target enzymes much more strongly than do the substrates. This approach has been limited by the difficulty of designing stable compounds which are good analogs of an unstable transition state and which at best yield a crude approximation of the true transition state structure.

C. Optimized Catalysis

The overall strategy for the enzyme must therefore consist of binding the TS as strongly as possible, while keeping binding to S and P weak. In an extreme limit, this would reduce to the dashed line of Fig. 1D with no discreet E·S or E·P. There is still an energy barrier above the level of free E and S owing to the requirement for E and S to come together from their standard state of 1 *M*. The magnitude of this diffusional barrier is approximately the same for association of any small substrate with a large enzyme and corresponds to a bimolecular rate constant of 10^8–10^9 M^{-1} sec^{-1} (*14*). In the optimized limit, both k_{cat} and K_m would approach infinity but would do so in a fixed ratio, with k_{cat}/K_m remaining equal to the diffusional limit.

In a real situation, discreet E·S and E·P intermediates are likely, as discussed above. The best that can be done is to have the levels of all intermediate species lie as close to the dashed line as possible. This more reasonable situation is indicated by the solid line in Fig. 1D. As long as the energy of all the other states are significantly below the level of the transition state for initial diffusion-controlled binding, then the k_{cat}/K_m ratio will be at its maximum value, and no further increase is possible. The sizes of the deviations from the dashed line will determine the magnitude of k_{cat}. Further decrease in the deviations from the dashed line will result in increased k_{cat} values, but with compensating increases in K_m. Figure 2 illustrates this point for two diffusion-controlled reactions with

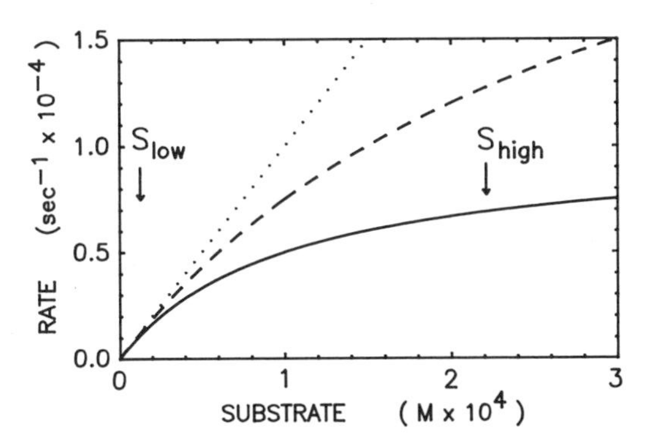

FIG. 2. Influence of substrate level on enzyme perfection. Curves were calculated for k_{cat} and K_m values, respectively, of 1×10^4 sec^{-1} and 1×10^{-4} *M* (solid line); 3×10^4 sec^{-1} and 3×10^{-4} *M* (dashed line); and the limiting case of k_{cat} and K_m both approaching infinity with a ratio of k_{cat}/K_m of 10^8 M^{-1} sec^{-1}.

different values of k_{cat}. At a concentration of S indicated by S_{low}, the maximal rate is obtained when $K_m \gg S_{low}$, and no further increase is possible by increasing k_{cat}. Increasing k_{cat} will, however, increase the rate at higher concentrations of S as indicated at S_{high}. As emphasized by Fersht (*8*), it is advantageous to have K_m exceed the physiological concentration of S so that catalysis will be maximized. It should also be noted that it is advantageous to have $K_m > S$ for reasons of physiological control. If the reverse were true, then the metabolic flux through the catalyzed step would always be at its V_{max} value and could not respond to changes in the concentration of S.

Albery and Knowles (*9*) have referred to the situation in which k_{cat}/K_m is at the diffusion-controlled limit and K_m in excess of the physiological S concentration as a "perfectly evolved" enzyme, even though K_m is still finite. Unqualified perfection should, however, be reserved for the absolutely perfect limiting case of the dashed line of Fig. 1D. Perhaps the above case should rather be referred to as "physiologically perfect," indicating that further improvement in total catalytic capacity will not result in improvement at the physiological S concentration. Even this more limited honor may not be totally justified, since the physiological S level is also subject to evolutionary pressure. Thus, the situation may exist where k_{cat}/K_m is at the diffusional limit and the current S level is slightly below the K_m, appearing optimal by the above criteria, but this situation is not truly optimal. This could arise if it would be beneficial to the cell to increase the physiological S level, but the S level cannot be increased above the K_m without losing proper control sensitivity through that step. Further evolutionary increase in the K_m owing to increased discriminatory binding, however, would allow the physiological substrate level to evolve to a higher value.

D. How Do Enzymes Do It?

The first condition to be met is that the enzyme must bind the TS tightly. Specifically, this must include strong, favorable interaction with the area of the TS where the chemical change occurs. For example, in the reaction of the serine proteases as illustrated in Fig. 3, this involves stabilizing the serine nucleophile

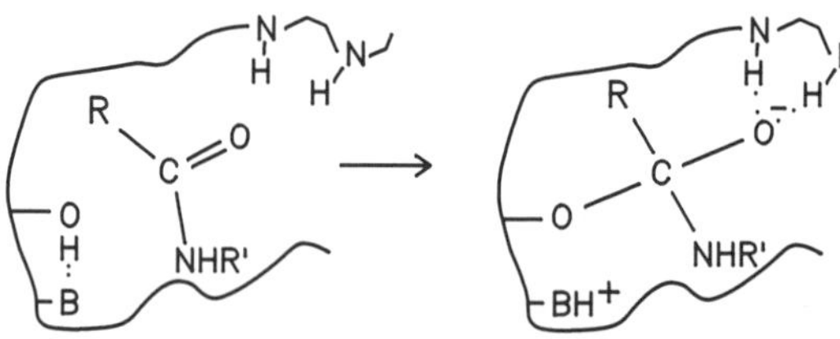

FIG. 3. Mechanism of chymotrypsin.

by providing a general base system to remove the proton; providing appropriately positioned hydrogen bonds to stabilize the developing oxy anion; and later providing a proton source to the departing amine as well as providing a generally favorable microenvironment. In other words, the enzyme uses all the tricks available to model reactions and additionally provides them in an optimal, preorganized array. This not only stabilizes the transition state by means of the favorable interactions, but also provides a major entropic advantage, to be discussed in more detail below.

Beyond this, the enzyme can aid catalysis by using binding interactions with parts of the substrate which are remote from the site of chemical rearrangement. A distinction can be made between uniform contributions, which stabilize all of the bound species equally, and discriminatory contributions, which selectively favor interaction with only a limited set of the bound species. By use of uniform binding, enzymes can lower their K_m values while maintaining k_{cat} unchanged, thus effecting an increase in k_{cat}/K_m as discussed by Albery and Knowles (*9*). This advantage from uniform binding can be increased until the binding of P becomes so strong that its release becomes the new rate-limiting step and k_{cat} is decreased, as indicated in Fig. 1E. In this limit, further increase in uniform binding will decrease k_{cat} and K_m in parallel with no net change in k_{cat}/K_m.

Discriminatory binding takes advantage of both the obvious changes in the part of the substrate undergoing chemical reaction and also the resulting changes in the overall geometry. For the serine proteases, this involves a change from a planar, conjugated amide bond with a sp^2-hybridized carbonyl carbon toward a transition state which is largely tetrahedral at the central carbon and has a negative charge on the former carbonyl oxygen. The enzyme selectively stabilizes the transition state by providing hydrogen bond donors to this developing negative charge and by providing a hydrogen bond acceptor to the serine nucleophile. Additionally, as the substrate moves toward the TS, the angle between the R group on the acyl carbon and the R′ leaving group changes, and it is clear that the enzyme cannot be optimally complementary to both S and TS. If the enzyme is rigidly fixed in a conformation which is complementary to the TS, then the E·S and E·P forms can utilize only a subset of the maximal interactions. The S and P complexes will utilize whichever subset provides the strongest interaction in each case. This is illustrated in Fig. 4A with the interaction of the R group in E·S assumed to be more favorable than the interaction of R′. Conversion of E·S to E·TS allows both groups to simultaneously interact with the enzyme and results in tighter binding. Correspondingly, P can make only a subset of interactions available to the TS. Figure 4 describes the basic principle underlying the use of intrinsic binding energy as developed by Jencks (*10*).

When experiments are conducted in which substrates are modified in such a way as to provide additional binding interaction in one R group, it is often found that the increased favorable interaction is manifest mainly as an increase in k_{cat}

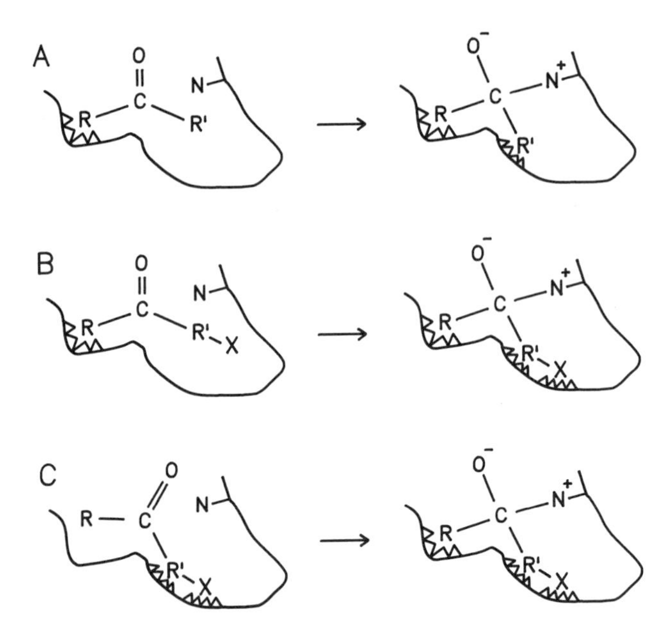

FIG. 4. Binding modes for modified substrates. Zigzag lines indicate regions of strong interaction between the ligand and the enzyme surface.

rather than a decrease in K_m. This result can be explained as illustrated in Fig. 4B. Adding an additional group X onto R′ will not decrease K_m if that group is not able to interact in the E·S complex, but will increase k_{cat} if interaction is possible in the E·TS complex. This process cannot go on without limit, however, since eventually the binding potential of R′–X will become so favorable that the substrate binding mode will switch to favor selective binding of R′–X with a decrease in K_m and no further increase in k_{cat}, as illustrated in Fig. 4C. In many cases, it is anticipated that the simple all or none binding of R versus R′ will not be rigorously true, and an intermediate result will be obtained with changes in both k_{cat} and K_m. Such complexities are a major potential problem in interpretation of the binding and kinetics of modified substrates.

In this regard, it must not be forgotten that the structures of substrates as well as enzymes have been under selective pressure during evolution. It has been pointed out a number of times that the common metabolic intermediates are much larger than needed on the basis of their chemical reactions alone. One reason would be to aid catalysis by providing handles for use as X groups as discussed above. Thus, the phosphate group of many of the glycolytic intermediates is not directly involved in the chemical interconversions, but does provide a nice handle. (Of course, the phosphate group may also serve other purposes such as allowing accumulation of small intermediates inside a cell at high concentration.) The AMP portion of ATP is another example since pyrophosphate is sufficient for all the reactions of ATP involving transfer of the γ-phosphoryl. A particularly good example of the use of handles can be found in the work of Jencks and co-workers on coenzyme A (*15, 16*).

Much attention has been focused on the role of strain, defined as a distortion of the ground state E·S complex toward the geometry characteristic of the TS. Significant distortion of this type is unlikely on theoretical grounds (*17*). The difference in structure between a strained and an unstrained S is expected to be much smaller than the difference in structure between a TS and an unstrained S. Introduction of strain would, therefore, require that large differences in energy be associated with very small differences in the structure of the E·S complex. In general, vibrational movements in proteins are of low energy (i.e., little force is required to deform the enzyme slightly), and a strained E·S could readily relax to an unstrained conformation by small, low-energy movements. The limited experimental evidence on this point also is consistent with little distortion of the E·S complex, although some examples of perturbations have been observed (*18, 19*). The concepts of intrinsic binding energy and discriminatory binding provide a way to obtain the same net result without actual ground state distortion. In effect, the enzyme works by actively favoring the binding of the transition state rather than destabilizing the substrate. Rather than strain, it is more of a seduction in which one subset of possible interactions with the substrate is used as a lure to effect initial binding, while the remaining favorable interactions are held back to induce the transformation to the transition state. In this regard, it is important to note that this transformation is likely to be concerted. That is, when an energized E·S complex starts partial movement in the direction of the transition state, the increased favorable interactions which start to become available help further progress toward the transition state.

A second experimental finding is that the equilibrium between E·S and E·P for an exergonic reaction is often shifted in favor of E·S compared to the equilibrium value for the free species. Partially, this is due to the fact that the ideal situation has all of the intermediates as close to the dashed line of Fig. 1D as possible. This will naturally tend to equalize all the energy levels, both to not let any transition state be too high and to not let any bound intermediate be too low. This is best illustrated by consideration of a multiple step reaction (Fig. 1F) which is optimized by keeping all species as close to the perfect line as possible. This results in the energy levels for the internal states producing two descending series, one series for the ground state complexes and a second series for the transition state complexes. This situation has been referred to by Brenner and co-workers as a "descending staircase" and has been justified on theoretical grounds (*20, 21*) (also, see the discussion by Rees and Farrelly in Chapter 2, this volume, in relation to electron-transfer proteins). This situation may be more meaningfully referred to as a "double descending staircase" in order to emphasize that the separate staircases for the ground state and transition state complexes alternate during their descent.

A different perspective on the relationship of the energy levels for the E·S and E·P complexes is given by the Hammond postulate. This states that for an exergonic reaction the transition state will be more similar to the reactant (S) than the product (P). Thus, an enzyme optimized to bind the TS will tend to be more complementary to S than P since S is more similar to the TS structure. This will result in tighter binding of S with respect to P, which tends to equalize the energy levels of the E·S and E·P states. Albery and Knowles originally postulated that the energy levels of E·S and E·P should be equal, rather than descending. As illustrated above, however, the only requirement is that the energy levels be decreasing. It should be noted that this distinction is only meaningful for irreversible reactions since the energy levels are also equal for a descending staircase model in the case of a reaction at equilibrium. This confusion derives at least in part from the approximations inherent in defining a simple efficiency function for a process which is optimally evolved only in the limit of infinity and whose apparent efficiency is a function of the S concentration.

E. Enzyme Flexibility

1. *Necessity for Flexibility*

The analysis of Fig. 4 assumes that the enzyme remains rigidly fixed in the conformation which is exactly complementary to the structure of the TS. The weak nature of the interactions between enzyme and substrate, which prevents accumulation of strain, however, also allows for some accommodation of the conformation of the enzyme to increase interaction with S or P. The final configuration of the E·S complex depends on the trade-off between the energy cost of deforming the ground state conformation of the enzyme and the energy yield from the favorable new interactions with the ligand which are produced. This will lower the net K_m but will also lower k_{cat}, since the enzyme must return to the original conformation in order to be optimally complementary to the TS. Substrates as well often have considerable conformational flexibility, particularly about single bonds, but usually only one conformation binds to the enzyme.

For optimal catalysis, it is necessary that these conformational changes be rapid. In this regard, it has been speculated that the conformational changes, which occur on conversion of E·S to E·P, may be identical to a vibrational mode of the free enzyme. It is even possible that this corresponds to a major hinge bending mode. Further experimental work is required to test this hypothesis.

Flexibility in the enzyme and its complexes with ligands may be highly localized. For example, in the complex of myosin with ADP and P_i, the bound P_i is apparently free to rotate and scramble the orientation of its four oxygen atoms, whereas essentially no scrambling of the three branch β-phosphoryl oxygens occurs on the neighboring ADP (*22*).

2. *Induced Fit: Specificity*

Discrimination against a substrate analog which is larger than the natural substrate is conceptually easy since the active site can be designed in such a way that the larger analog cannot fit. Discrimination against molecules which are smaller than the desired substrate is more difficult, however, as their access to the active site cannot be blocked. This is a particular problem for enzymes, such as hexokinase, which transfer a phosphoryl group from ATP to a sugar hydroxyl yet must discriminate against water as a nucleophile. Koshland proposed (*23*) that discrimination could be produced if substrate binding were required in order to convert the enzyme from a noncatalytic initial conformation to the conformation complementary to the TS. In this model, part of the binding energy of the large substrate is used to induce the unfavorable conformational change in the enzyme. Thus, water would have its rate greatly reduced because very little of the free enzyme would be in the conformation complementary to the TS, and water would be unable to induce the required conformational change. The sugar substrate, on the other hand, would be able to induce the change and would therefore have the high k_{cat} value characteristic of the enzyme complementary to the TS. The net K_m for the sugar would be increased, however, since part of its binding energy would be required to produce the unfavorable conformational change.

Induced fit provides a competitive advantage to the sugar when the physiological S level is above the K_m and catalysis is controlled by k_{cat}. Evidence for such a conformational change with hexokinase comes from acceleration of ATP hydrolysis by xylose and lyxose (*24*). These sugars lack the terminal hydroxymethyl group of glucose and cannot be phosphorylated, but they still induce a conformational change in the enzyme which, at least partially, aligns the catalytic groups. A large glucose-induced conformational change is also observed by X-ray crystallography. In the E·S complex, the enzyme wraps around the sugar to partially bury it (*25*). Significant conformational changes have also been observed for a number of other enzymes (*26*).

This proposal can be criticized on the grounds that many enzymes operate physiologically in the $K_m > S$ range where induced fit is of reduced importance. For discrimination between two fixed substrates on the basis of size, this is certainly true owing to the trade-off of K_m versus k_{cat}. As long as the concentration of S is significantly below K_m, catalysis is determined by k_{cat}/K_m, which is solely a function of the total energy difference between the E·TS complex and free S and free E in the ground state conformation. Both large and small substrates will have to pay the same energetic price to reverse any conformational change which is required in the ground state enzyme.

There is a more subtle advantage to induced fit, however. Evolution to larger substrates can result in increased uniform binding which lowers the energy difference between E·TS and the free S and E and thus increases k_{cat}/K_m. This provides increased discrimination over smaller substrates without need for induced fit, but this process cannot be continued without limit. Eventually, uniform binding becomes so strong that product release becomes the rate-limiting step (Fig. 1E). Induced fit produces the equivalent of negative uniform binding, and its addition thus allows continued evolution to larger substrates without limitation from rate-limiting product release. In effect, induced fit with its negative uniform binding energy can be used to push an enzyme in the unfavorable condition of Fig. 1E with a large substrate back to the more favorable condition of Fig. 1C without offering a comparable advantage to the smaller substrate.

3. *Induced Fit: Other Advantages*

Induced conformational changes produce a number of other advantages as well. For one, they allow an enzyme to have more side-chain residues in contact with the substrate in the E·S complex than would be possible with a rigid enzyme. Thus, the enzyme can literally wrap around the substrate and surround it with catalytic groups. Those residues which cover the substrate must be able to move out of the way during binding and release steps since, otherwise, the substrate could never get in nor the product get out. A second effect related to induced fit is "substrate synergism" (*27*), observed in reactions where two substrates, A and B, must both bind to their respective subsites in the active site before reaction can occur. It is often found that the binding of A to the enzyme is very different from the binding of A to the E·B complex, and vice versa for binding of B. This effect can in part be due to the mere presence of the other species at the active site, but it is also likely that the first substrate induces conformational changes which alter the other subsite. Discussion of the related area of allosteric and cooperative interactions is presented in Section IV.

III. Beyond Unimolecular Reactions: Entropy

The above discussion of the influence of binding on catalysis has focused on the minimal case of a unimolecular process. In reality, however, no enzymic reaction is actually unimolecular. Most reactions involve two or more substrate or product molecules which undergo chemical rearrangement. Even an apparently unimolecular reaction such as the interconversion of dihydroxyacetone phosphate and glyceraldehyde 3-phosphate catalyzed by triose-phosphate isomerase is not, in fact, unimolecular as it involves catalytic groups. At least one carboxylate acts as a general base, and probably a second group acts to polarize the carbonyl group (*28*). Any realistic model reaction would have to include these groups as buffer components and thus would become bimolecular or termolecular. Additionally, water molecules are also involved in a model reaction through their differential interaction with the transition state versus the substrate. Thus, an evaluation of the factors contributing to the catalytic ability of enzymes requires comparison between a multibodied model reaction in solution and an enzymic reaction of greatly reduced reaction order. Clearly, the E·S complex, with all of the required groups confined and properly aligned, will undergo reaction at a rate which is greatly in excess of the rate for reaction of the free components, which has a large entropic barrier to overcome since all of the reactants must first be brought together.

It has been recognized for some time that this reduction in reaction order resulted in large rate accelerations in model reactions and was likely in itself to provide a major contribution to enzymic catalysis (*29, 30*). Koshland and coworkers proposed an analysis based on dividing the total effect into a proximity component, required to bring the reactants together, and a further factor, required for proper relative alignment of the reactants for optimal reactivity (*31, 32*). It was argued that the proximity factor should be small for formation of a loose complex, but much greater rate accelerations could be obtained if the reactants were to have high angular preference for reaction and if this angular orientation were optimal in the ground state approximated complex. The magnitude of the maximum rate acceleration would depend on the angular size of the "reactive window" for each reactant. This analysis became known as orbital steering, and it was supported by theoretical analyses which indicated that it was consistent with both collision and transition state theory (*33, 34*). Page and Jencks (*35–37*) directly calculated the loss of entropy which occurs on formation of a unimolecular transition state from two free reactants. They also concluded that the total effect could be quite large for a tightly confined transition state, with maximal accelerations of up to 10^8.

In spite of the approximate agreement on the magnitude of the total factor, uncertainty has remained regarding the relationship of these two treatments to each other and how best to assign relative importance to the proximity and orientation components, or even what exactly is meant by these two terms (*37, 38*). The resolution lies in the appreciation that orbital steering as originally defined is analogous to a classic *macroscopic* description of the system whereas a complete analysis requires consideration of entropy and results in a description which has been corrected for quantum mechanical effects arising from the *microscopic* nature of actual chemical reactions.

A. Microscopic versus Macroscopic Systems

1. *The Model System*

The distinction between macroscopic and microscopic systems can be illustrated by simple quantum mechanical calculation of the entropy loss associated with formation of loose and tight complexes from two balls on the scale of small molecules and how this would differ from such calculations performed on macroscopic balls.* Analysis of the microscopic system parallels that originally introduced by Page and Jencks (*35*), except that performing the calculation on balls rather than on specific small molecules allows the general features to be seen more clearly. The calculation was performed in the gas phase† on balls having a molecular weight of 50 and a radius, *r*, of 1.5 Å, which can serve as an approximate model for C_2H_5OH, CH_3CONH_2, CH_3COOH, or other similar small organic molecules. Entropy calculations were performed by standard statistical mechanical methods (*39, 40*) for the association reaction indicated in Fig. 5, and the results are presented in Table I.

For each species there will be three degrees of translational and three degrees of rotational freedom corresponding to 31.3 and 22.9 eu, respectively, in the gas phase† at 298 K and 1 *M* concentration. On formation of a loose complex with an average center-to-center distance, $2a$, of 4 Å, there will be a net loss of three degrees of translational freedom, but this will be largely offset by the new degrees of freedom of the complex. The complex will have rotational freedom about the *Y* and *Z* directions and vibrational freedom along the *X* axis joining the two balls. The rotational freedom of each component about the *X* direction is conserved in the rotational freedom of the complex about the *X* axis and the internal rotational freedom of the two balls relative to each other about the *X* axis. The rotational freedom about the *Y* and *Z* directions is conserved in the four other internal rotations of the complex. The net result‡ is a modest loss of

*For simplicity in discussion it is assumed that the balls are exactly spherical, but in order to avoid problems concerning degeneracy it is more appropriate to consider the balls as slightly nonspherical. The three moments of inertia will be assumed to be sufficiently similar so that they can be approximately equated in the calculations but are understood to not be exactly equal. These calculated differences in entropy are relatively insensitive to the exact mass and size of the balls.

†These calculations are presented for gas phase molecules. It is not feasible to separate out translational and rotational terms in the liquid phase, but approximately the same total entropy is likely to be present. Page and Jencks (*35*) have summarized evidence that no large difference in entropy of the reaction is generally expected between reactions in the gas and liquid phases for many processes although $\Delta S^{\ddagger}$ will generally be somewhat smaller in solution.

‡The contribution of the stretching mode between the two components is difficult to estimate and depends on the tightness of the complex. Considerable freedom should remain if a loose complex is considered analogous to a pair enclosed in the same solvent cage. Page and Jencks (*35*) have summarized arguments that such pseudotranslational entropies can be large, and a value of 4 eu is assigned here as a conservative estimate between 10 eu per degree for the free translation and the minimal value of zero.

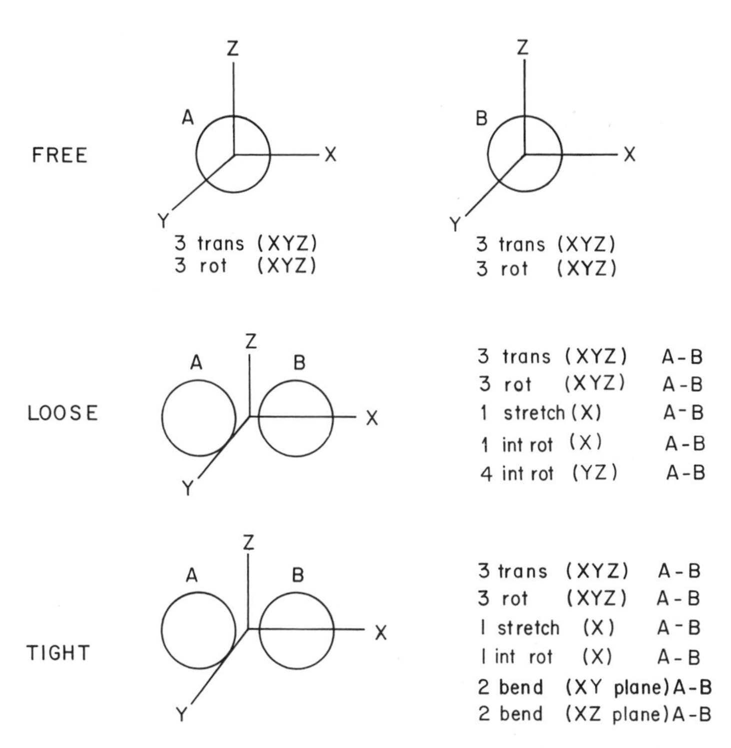

FIG. 5. Model system for bimolecular association.

TABLE I

ENTROPY CALCULATIONS FOR BIMOLECULAR ASSOCIATION[a]

State of association					
Free		Loose		Tight	
Mode	Entropy (eu)	Mode	Entropy (eu)	Mode	Entropy (eu)
$3 \times A_{trans}(XYZ)$	31.3	$3 \times AB_{trans}(XYZ)$	33.4	$3 \times AB_{trans}(XYZ)$	33.4
$3 \times B_{trans}(XYZ)$	31.3	$3 \times AB_{rot}(XYZ)$	28.4	$3 \times AB_{rot}(XYZ)$	28.4
$3 \times A_{rot}(XYZ)$	22.9	$1 \times AB_{stretch}(X)$	~4.0	$1 \times AB_{stretch}(X)$	~0
$3 \times B_{rot}(XYZ)$	22.9	$1 \times AB_{int\ rot}(X)$	7.7	$1 \times AB_{int\ rot}(X)$	7.7
		$2 \times AB_{int\ rot}(Z)$	15.4	$2 \times AB_{bend}(XY)$	~0
		$2 \times AB_{int\ rot}(Y)$	15.4	$2 \times AB_{bend}(XZ)$	~0
Total	108.4		104.3		69.5
ΔS	—		4.1		38.9
ΔΔS	—		—		34.8

[a]Calculated for a standard state of 1 *M* at 289 K in the gas phase for A and B of mass of 50 daltons, radius (r) of 1.5 Å and a center-to-center distance of 4 Å in the associated state. S_{trans} calculated as $R\{5/2 + \ln[V(2\pi mkT)^{3/2}/Nh^3]\}$. S_{rot}A or B calculated as $R\{3/2 + \ln[\pi^{1/2}(8\pi^2IkT/h^2)^{3/2}]\}$ with $I = 2/5(mr^2)$. S_{rot}AB pair calculated as $R\{3/2 + \ln[\pi^{1/2}(8\pi^2I_xkT/h^2)^{1/2} \times (8\pi^2I_ykT/h^2)^{1/2}(8\pi^2I_zkT/h^2)^{1/2}]\}$ with $I_x = 4/5(mr^2)$ and $I_y = I_z = 2[ma^2 + 2/5(mr^2)]$. $S_{int\ rot}$ approximated as $R[1/2 + \ln(8\pi^3I_rkT/h^2)^{1/2}]$ with $I_r = 2/5(mr^2)$.

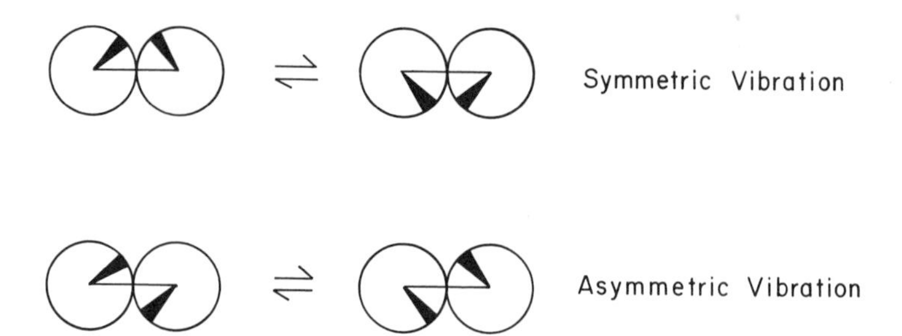

FIG. 6. Bending modes of the bimolecular complex.

only 4.1 eu. Transition state rate theory predicts that the rate of a reaction is proportional to $e^{\Delta S^{\ddagger}/R}$ if $\Delta H^{\ddagger}$ is fixed, which corresponds to an increase of approximately 8-fold in the possible rate acceleration of the preformed complex over a bimolecular reaction. This small net effect is consistent with earlier assignment (*31*) of a small effect to proximity in the formation of a loose complex. The same conclusion has been reached by Dafforn and Koshland (*34*) by analysis of the combination of two bromine atoms which possess no rotational freedom in the free state and by Page (*36*) for a loose association.

Formation of a tight complex along the *X* direction, however, results in a much larger decrease in entropy. The vibrational freedom along the *X* direction will be effectively lost as well as the ability of the individual components to rotate about the *Y* and *Z* directions. The four degrees of internal rotational freedom will be converted to four degrees of bending vibrational freedom, specifically the symmetrical and asymmetrical vibrational modes in both the *XY* and *YZ* planes (Fig. 6). The internal rotational freedom about the *X* direction will not necessarily be impaired on formation of this complex. The loss of these five degrees of internal rotational and vibrational freedom amount to 34.6 eu or a rate acceleration of 4×10^7-fold beyond that already achieved by approximation. The total effect for conversion of the bimolecular species to a tight complex would then be 38.9 eu for this model reaction, in good agreement with the values estimated for similar reactions by Page and Jencks (*35*) with allowance for some internal freedom remaining in the tight complex. In some cases the mode of association may also hinder the rotational freedom of the components in the complex about the *X* direction. For example, such loss results from the double point bonding that develops during cyclopentadiene dimerization (*35*) and would increase the maximum values given in Table I before adjustment for residual freedom in the complex. Reactions with looser transition states will not produce the full effect given in Table I as the new vibrational modes produced will still possess considerable freedom with its associated entropy.

2. *Correlation to Orbital Steering*

The angular dependence of orbital steering corresponds to the loss of the rotational freedom of the reactants in the Y and Z directions in the model of Fig. 5. In the physical model of orbital steering the restraint was considered to be due to the inability of the reaction to occur unless the reactive windows of the two reactants were aligned. One objection to this model is that the orientational factor can theoretically be increased without limit if sufficiently small reactive windows are allowed, whereas an entropy analysis has an upper limit set by the magnitude of the rotational entropy as pointed out by Page and Jencks (*35*). The key to the understanding of this discrepancy is the realization that the physical model proposed originally by orbital steering is a macroscopic one in which it is implicitly assumed that an infinite number of orientational states exist for the rotational freedom of the free reactants. Thus, the total freedom is infinitely divisible into increasingly smaller units as the size of the reactive window is reduced to smaller values, allowing the orientation factor to become infinitely large. This is approximately true of a macroscopic model but is not true of microscopic molecules, which are subject to quantum restrictions that impose a significant energy spacing between the rotational energy levels.

For the model of Fig. 5, the basic energy spacing calculated as $h^2/8\pi^2I$ is 7.4×10^{-17} ergs for a single ball, which is small compared to the value of kT of 4.1×10^{-14} ergs at 298 K but is still finite. The small relative size of the spacing does allow the summation over all the quantum states required for the evaluation of the rotational partition function to be replaced by an integral which can be solved analytically, but it is still large enough that it will limit the total number of quantum states which are reasonably accessible at 298 K. The rotational partition function for each ball will be finite at a value of 2.3×10^4 for the model of Fig. 5, and this will set an upper limit on the rotational freedom and the associated entropy. At the molecular level, freedom and disorder are more closely associated with the number of accessible states, as represented by the entropy, than with the macroscopic concept of angular uncertainty.

It was, of course, appreciated in the initial proposal of orbital steering that reactive windows could not be smaller than the vibrational amplitude of the bond which was being formed in the transition state and that this would place a maximum limit on the rate acceleration which could be obtained. The bond along the X axis will have a zero point energy equal to $\frac{1}{2}\, h\nu$ which can be thought of as a consequence of the uncertainty principle since the energy of the system in the lowest quantum state cannot be zero and thus precisely defined. This energy can also be expressed as $\frac{1}{2}\, ka^2$ where a is the angular displacement from the optimal value and k is the angular force constant. Reasonable estimates of 1000 cm^{-1} for ν and 5×10^{-12} erg/(radian)^2 for k lead to an angular displacement of 0.2 radians or 11.4 degrees. Reactive windows of this magnitude result in rate accelerations of at most 10^5 and not infinity. In a more complete comparison, the effect of the vibrational freedom along the X direction would also have to be included.

From an entropic viewpoint, imposition of even a fairly mild restriction on the vibrational amplitude reduces the entropy associated with that vibration essentially to zero, and little significant additional loss is possible on further restriction. Although a bond with a ν of 1000 cm^{-1} has a substantial vibrational amplitude owing to its zero point energy, it has only 0.1 eu of entropy because only a small number of quantum states are effectively occupied. Further restriction on the magnitude of this vibrational amplitude will further reduce this small residual entropy but will not significantly increase the much larger difference in entropy between these restricted states and that of a free rotation.

This analysis in many ways parallels the theoretical treatment presented by Dafforn and Koshland [see Ref. (*37*)] as an extension and refinement of the initial proposal of orbital steering. They clearly established that the total effect could be large even after allowing for residual internal freedom in the transition state and that a "loose" complex as exemplified by bromine combination could occur with little loss of entropy. The point which is developed more fully here is that the microscopic model used for the statistical mechanical calculations is not the same macroscopic model used in the original definition of orbital steering.

B. Experimental Effective Molarities

1. *Nucleophiles*

It is useful to consider some particular reactions within the framework presented above. It has been argued (*35*) that succinic acid, which does not even appear to have its reactive groups fully juxtaposed, should not exhibit the large effective molarities* of 3×10^5 observed in its cyclization to the anhydride and similar reactions unless juxtaposition per se was the major factor. The two carboxylic acid groups of succinic acid are, however, more constrained than the simple approximated pair of Fig. 5.

This can be illustrated by analysis of the situations presented in Fig. 7 for a complex of two acetic acid molecules and for succinic acid viewed as an approximated dimer of two acetic acids. Using the calculations of Table I, the loose complex of two acetic acids will have lost only 4.1 eu relative to two free acetic acids. The succinic acid, however, is much more constrained. Its three remaining modes of internal rotation can be seen to correspond to the rotation of the two

*Effective molarity is used here to represent the concentration of a free reactant which would be required in an analogous bimolecular reaction to yield the same net rate as observed in a unimolecular reaction.

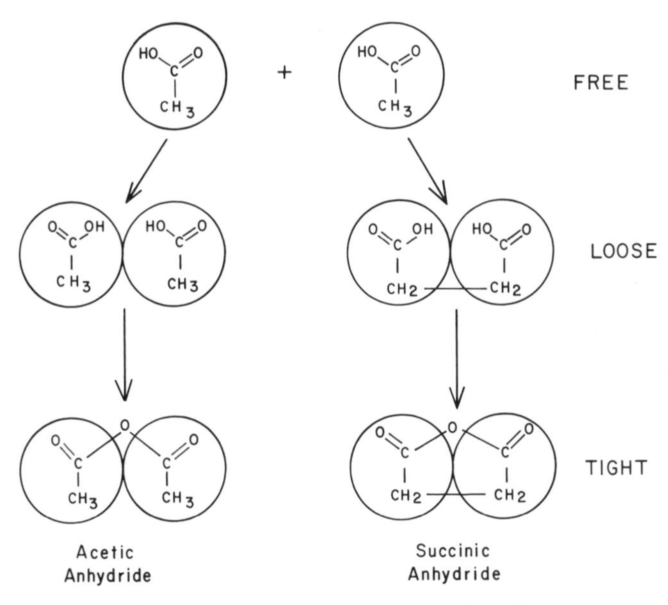

FIG. 7. Model for succinic anhydride formation.

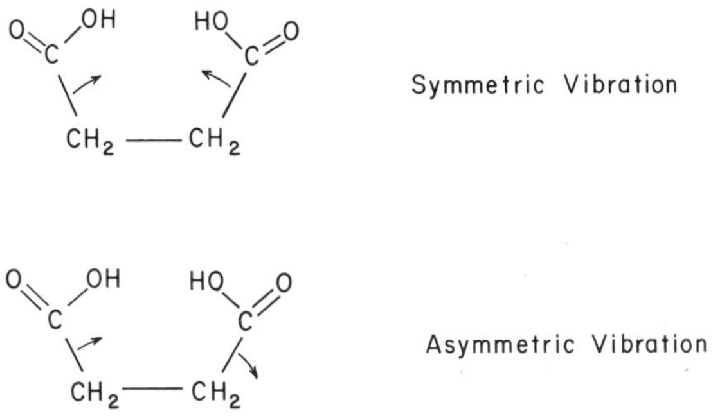

Symmetric Vibration

Asymmetric Vibration

FIG. 8. Bending modes of succinic acid.

acetic acid units about the *Z* direction and the internal rotation about the *X* direction. The rotations of the acetic acids about the *Y* direction have been lost through conversion to the low-entropy symmetrical and asymmetrical bending modes of the system (Fig. 8). The large-amplitude internal vibration along the *X* axis of the loose complex has been converted to the low-entropy vibration of a full single bond between C-2 and C-3. The combined loss of the two degrees of internal rotation and one internal vibration amounts to a loss of 19.4 eu over the approximated loose complex and 23.5 eu over the free acetic acids, corresponding to an effective molarity of over 10^5, in good agreement with the experimental values. Thus, the large effective molarity observed in formation of succinic anhydride can be accounted for by the loss of entropic freedom of succinic acid over that of a loose approximated complex of two acetic acids. The high value of the experimental effective molarity does not nullify the conclusion that the effective molarity for formation of a loose complex should be small.

2. *General Acid and Base Catalysis*

Catalysis by general acids and bases constitutes an excellent model for formation of a loose complex as the hydrogen bonding linking the components is expected to be weak and result in large-amplitude movements. Thus, it is expected that maximum effective molarities for intramolecular general acid or base catalysis should be small, and this is usually observed as summarized by Kirby (*41*). The weak binding does not greatly lower the internal entropy of these complexes, and this provides experimental justification for the conclusion that approximation per se should only contribute a small factor. This result furthermore helps justify the assumption that reactive groups could be juxtaposed without large effect as long as large-amplitude movements were not unduly restricted. It is only when the large-amplitude internal movements become restricted that major effects occur.

3. *Solvation*

The reactants are not the only species which are constrained at the active site of an enzyme. The array of binding surfaces which envelops the substrates in effect replaces much if not most of the solvent shell in which model reactions occur. The restricted mobility of this array of binding surfaces makes possible the types of selective interactions between S and TS of Fig. 3 which result in discriminatory binding. For example, in the case of the "oxy anion hole" of chymotrypsin, two backbone amide protons are prevented from making optimal hydrogen bonds with the carbonyl carbon in the substrate, but they can make optimal hydrogen bonds to the oxy anion which develops in the TS. In addition, however, the restricted mobility of the binding site also provides an entropic advantage. With a model reaction in solution, the development of a negative charge on the oxygen atom results in much tighter confinement of the water molecules which are hydrogen bonded to it with an accompanying loss of entropy. In the enzyme, the stabilizing groups have already lost their entropic freedom in the free enzyme, and no further entropy loss occurs on formation of the E·TS complex. For a single hydrogen bond, this entropic contribution would not be expected to be large, as discussed for general acid and base catalysis above; however, a number of hydrogen bonds are usually observed in an E·TS complex, and other types of solvation can have similar entropic contributions. The total effect of all of these interactions can be significant.

C. Approximation and Orientational Constraint

The orientational requirements of reactions thus produce two main types of effects on conversion of a bimolecular reaction to a unimolecular one. The magnitude of the entropy loss on conversion to a unimolecular reaction results directly from the degree of orientational restriction which must occur in the complex. The more severe the orientational requirements of a reaction, the larger will be the entropy loss on forming the reactive complex and the larger will be the advantage which can be obtained from prior restraint of the reactants in a favorable orientation in the ground state. Severe orientational requirements for optimal reaction can also be a disadvantage, however, in a unimolecular reaction which is not properly aligned. The net rate advantage of prior orientation restraint in any particular orientation will be controlled by the relative magnitudes of these opposing trends.

Figure 9 illustrates these opposing effects of orientational requirements. The discussion up to this point has focused on comparison of an intermolecular reaction and a constrained intramolecular reaction in which favorable approach of the two reactive groups was possible, as illustrated by cases A and B. Case B achieves the full benefit of prior orientational restraint with respect to the intermolecular model A, without any disadvantage owing to improper alignment in the restrained state. Case D also has an advantage over case A in terms of orientational restraint, but it suffers from poor alignment and thus would have a much smaller effective molarity. Case D would exhibit a high effective molarity, however, when compared to the more appropriate intermolecular reaction of case C in which optimal approach of the reactants is sterically blocked. The transition

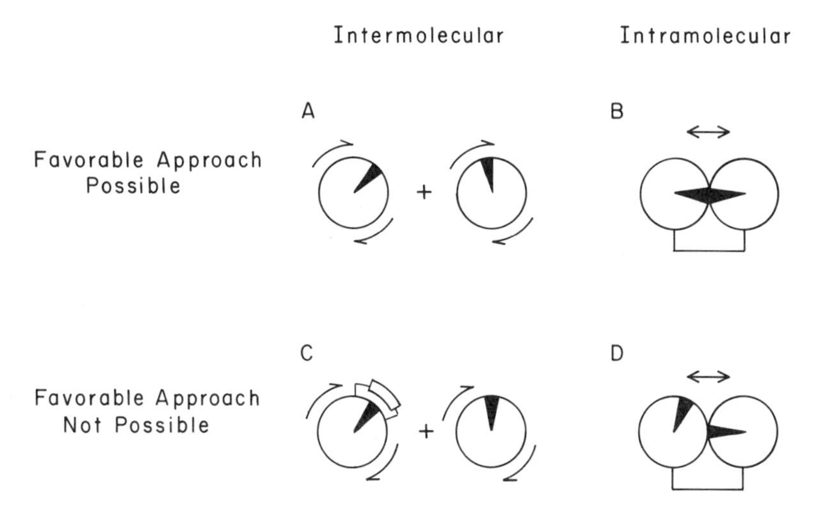

Fig. 9. Comparison of properly aligned and misaligned models.

states of both cases C and D will be destabilized owing to this forced misalignment and will both react more slowly than cases A and B, respectively, but the effective molarity of case D compared to case C will be similar to the effective molarity of case B compared to case A. In such extreme cases, the advantage made possible by orientational restraint results in an entropic advantage of the intramolecular reaction (B versus A or D versus C), whereas the consequences of unfavorable alignment would result in an enthalpic disadvantage (C versus A or D versus B). For reactions in solution, the complicating effects of differences in solvation of the reactants and transition states would tend to obscure such a clear distinction of entropic and enthalpic contributions.

The effect of orientation on the observed effective molarity in a bimolecular reaction is thus more subtle than envisioned in a simple macroscopic model. The exact orientation of the reactive groups need not be confined to a very narrow angular window in order to obtain high effective molarities. The dominant factor is the quantum restrictions which limit the number of accessible states. These considerations allow ready explanation of the fact that numerous examples have not been found in which very low effective molarities are observed in intramolecular reactions where at least a reasonable degree of alignment appears possible. A model evoking very narrow reaction windows would be expected to yield a high frequency of such slowly reacting examples owing to slight misalignment of the reactive groups. The general result of extensive studies which have been carried out, however, is that such slowly reacting examples are rare in the absence of other factors [see review of Kirby (*41*)]. Clearly, the orientation can deviate significantly from the exact optimal value without providing a highly unfavorable situation, in agreement with generally held concepts of the angular freedom of covalent bonds and the partial bonds which often occur in transition states. This does not imply that larger magnitude misorientations would not give rise to greatly decreased rates.

The exact significance of these studies is difficult to determine, however. One major problem is that the relative orientations of the reactive groups are usually not known precisely in the absence of X-ray crystallography or other structural data. This complicates both the orientational analysis and the evaluation of the contribution from strain. This latter problem is particularly acute in most of the confined model systems where strain is often a significant factor. Certainly strain is a major factor in the extremely high and low effective molarities observed in a study of the reactivities of a series of substituted maleamic acids by Kirby and Lancaster (*42*). In this case the range of the effect was very large and involved structures where strain could readily be predicted either in the ground state leading to abnormally high reactivity or in the transition state leading to decreased reactivity. In more subtle cases it may be difficult at best to separate the various effects.

The exact definition and use of the terms proximity and orientation have been a source of confusion and misunderstanding. In a sense, any division of the total effect into component parts is arbitrary, but such division is still useful as a conceptual framework in which to integrate results. The analysis presented here readily lends itself to a division into a small factor for the formation of a loose complex from the free species and a much larger factor which results from further loss in the internal freedom of the loose complex on proceeding to a tight complex. It may be useful to consider these two components as "approximation" required to form a loose complex and "orientational restraint" required to reduce the internal freedom of the loose complex. Much of the translational entropy is not really lost on formation of a loose complex but is transformed into increased rotational and internal freedom of the complex. In this context a distinction between the contributions of translational and rotational entropy is not highly meaningful even in the gas phase, and it becomes even less useful when considering reactions in solution. The exact orientation of the reactive species is not critical as long as the species are not restrained to orientations which are highly unfavorable for reaction. By far the dominant factor is the orientational restraint which will arise from loss of the internal freedom in the loose complex.

Proximity, as used here, is more general than proximity as defined on the basis of "reactive atoms in contact" (*32*). The latter definition is quite adequate for consideration of the simple bimolecular association of Fig. 5 but not for an intramolecular model which has the reactive groups covalently linked. For example, with succinic acid the reactive groups clearly are confined to be in proximity of each other, but the reactive atoms are not constrained to be in contact. The hydroxyl oxygen of one carboxyl group and the carbonyl carbon of the other carboxyl group are free to move apart from each other owing to the rotational freedom about the three C–C bonds. Holding these reactive groups constrained next to each other is not really analogous to similar constraint with two acetic acid molecules, as the added constraint of the covalent links in succinic acid effectively removes most of the freedom present in the two touching acetic acids. It is precisely this loss of freedom owing to the covalent intramolecular links in succinic acid over and above that owing to mere approximation of the reactive groups which accounts for the high effective molarity observed in this case.

D. Significance

Both experimental and theoretical work have indicated that enzymic reactions are favored by the preassociation of the reactants at the active site. Numerous intramolecular reactions have been studied as models for analogous enzymic reactions, particularly the acyl transfer reactions characteristic of the serine proteases, and much recent work has been directed toward design of models which incorporate binding interactions. Simple intramolecular models have been studied which react in water at rates far in excess of their analogous bimolecular reactions and approach rates sufficient to account for catalysis by the serine proteases (*43, 44*). The understanding of the entropic considerations which is presently available has placed this analysis on a firmer theoretical basis. The enzyme, by virtue of its binding of the substrate molecules and the constraint of the catalytic groups which are part of its covalent structure, can convert a reaction which would be intermolecular in solution to an effectively intramolecular reaction at its active site. The entropic advantage of this prior restraint at the active site should be very large for reactions with tight transition states as developed above for nonenzymic model intramolecular reactions.

This effect can be illustrated by consideration of chymotrypsin, a representative of the family of serine proteases. This enzyme can hydrolyze specific peptide substrates with a k_{cat} of 6 sec^{-1} (*45*) [corrected to 25°C using the temperature dependence reported for hydrolysis of *N*-acetylphenylalanine amide (*46*)]. The maximum rates for other substrates and other serine proteases are not likely to exceed this rate by many orders of magnitude. Figure 10 shows an intramolecular acyl transfer reaction which can serve as a model for the rate-limiting step in amide hydrolysis by chymotrypsin. Like the enzymic reaction, this model reaction is subject to general base catalysis, and the rate constant for catalysis by imidazole at 25°C is $2 \times 10^{-4}\ M^{-1}\ sec^{-1}$ (*47*).[*,†] The model reaction incorporates the entropic advantage gained by confining the nucleophile, but it does not include other differences between the enzymic and model reactions such as the increased electron withdrawal of the *N*-acetyl group in the enzymic substrate, which is expected to cause an acceleration of 24-fold compared to the unsubstituted amide of the model compound.[‡] There are also additional entropic advantages in the enzymic case which are not incorporated into the model reaction. On the basis of the above discussion, it can be conservatively expected that the

[*] It is possible that part of the rate acceleration of the model is due to ground state strain between the hydroxyl and carbonyl groups which might be relieved on going to the transition state. The crystal structure of the model compound (*48*) indicates some adjustment of the norbornane backbone to accommodate the substituents; however, the distance between the hydroxyl oxygen and carbonyl carbon is 2.83 Å, which is not significantly in violation of the 3.1 Å optimal distance for van der Waals contact. Morris and Page (*49*) have presented arguments for the lack of any significant strain in the reactions of related compounds, and Huber and Bode (*50*) have summarized evidence indicating that no significant strain is produced by the shorter 2.6 Å distance for the corresponding interaction in the complex of trypsin with pancreatic trypsin inhibitor.

[†] The rate-limiting step for this reaction is likely to be the breakdown of the tetrahedral intermediate, based on the work of Morris and Page (*49*) on a related set of compounds. This does not invalidate the comparison to chymotrypsin since the observed rate for the model therefore represents a lower limit on the rate for formation of the tetrahedral intermediate.

[‡] Inductive correction is based on the difference in basic hydrolysis rate of *N*-acetylphenylalanine methyl ester versus methyl β-phenyl propionate (*51*).

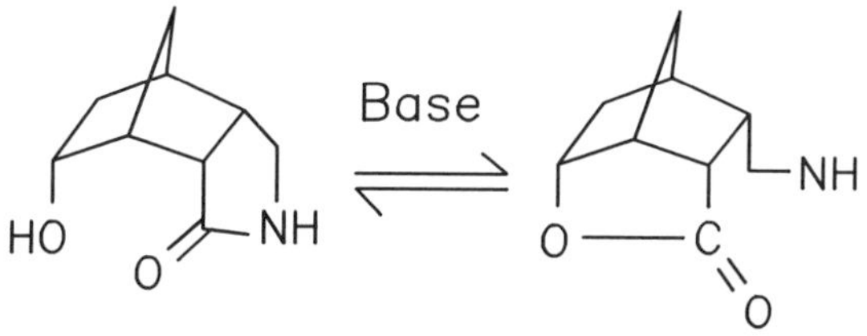

FIG. 10. Intramolecular model for chymotrypsin.

combination of approximation and orientational restraint for the general base in the enzymic reaction will contribute an entropic advantage of at least 10-fold over the free imidazole of the model reaction and that the two hydrogen bonds in the oxy anion hole will contribute 5-fold each. The addition of these inductive and entropic contributions to the bimolecular model reaction results in an adjusted unimolecular rate constant of 1.2 sec^{-1}, which is within an order of magnitude of the enzymic rate.

Obviously, this estimation is only grossly approximate, and much experimental and theoretical work is needed in order to more precisely determine the entropic contribution for chymotrypsin and other enzymes. The striking feature of this analysis, however, is that entropic factors alone can make such a major contribution to the turnover rates of enzymes, even without inclusion of the advantages which the enzyme derives from differential stabilization of the TS versus the S. The detailed quantitative description of all of the factors which contribute to enzymic catalysis for individual enzymes remains a major goal for future work, but it now appears likely that enzymic catalysis can be accounted for on the basis of known chemical principles.

IV. Indirect Uses of Binding

In addition to the direct use of binding energy for catalysis, the binding energy of a molecule can also be used for other processes in an indirect manner. The conformational changes which are induced by binding of ligand at one site are capable of propagating throughout the whole protein, and, in this way, ligand binding can influence the conformation and energetics of the protein at remote sites. These types of interactions fall into several broad classes.

A. CONTROL: ALLOSTERISM AND COOPERATIVITY

Many enzymes with multiple subunits, particularly regulatory enzymes, exhibit complex kinetics produced by the ability of ligand binding to one subunit to induce conformational changes in the other subunits. Classically, two extreme

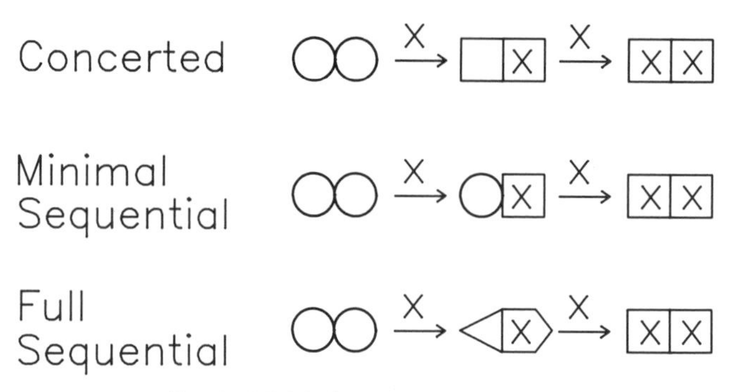

FIG. 11. Models for cooperative interactions.

models have been employed as a conceptual framework for understanding such effects, as illustrated in Fig. 11 for a dimeric enzyme. The concerted model (*52*) requires that the enzyme exist in only two symmetrical conformational states which differ in affinity for the ligand. The minimal version of a sequential model (*53*) allows for the possibility that a hybrid species can be formed on partial ligation. It should be emphasized that these two models are extreme limiting cases which are not likely to relate in detail to any actual enzyme. As emphasized by Koshland (*54*), a meaningful sequential model must allow for a large number of different conformation states. An illustration of such additional complexity is also included in Fig. 11. Now each subunit can respond in a different manner to events at the other subunit. Even this more complex model is hopelessly oversimplified, however, as many other conformation states are also likely. For example, as discussed in Section II, E on induced fit, the free enzyme is likely to be in equilibrium among its various possible conformations even in the absence of ligand. Also, different conformational states are likely for the transition state and product complexes; moreover, the unliganded and fully liganded states do not need to exhibit symmetry.

Hemoglobin is the classic example of cooperative binding and is often modeled in terms of a simple two-state concerted transition. More detailed studies using the techniques of NMR (*55, 56*) and X-ray crystallography (*57*), however, have provided a direct evaluation of the conformations of defined sites on each individual subunit and how they change on litigation. These studies demonstrate that the partially ligated states of hemoglobin do not correspond to either of the limiting concerted or sequential models, but rather to complex interactions of the full sequential type indicated in Fig. 11. Ligand binding to one subunit of glyceraldehyde-3-phosphate dehydrogenase (GPDH) also induces different conformational changes in the other subunits of the tetramer (*58*). Hexokinase provides an even more complex example. Although it exhibits a major conformation change on addition of glucose and is the textbook example for induced fit, the details are complex (*59*). The conformation induced by glucose does not place the glucose close enough to the catalytic nucleotide binding site for phosphotransfer to occur. Some further conformation must occur on formation of the ternary complex which is different from either of the complexes with glucose or nucleotide alone. Even though xylose binding stimulates the ATPase reaction, it is clear that the conformation which it induces is markedly different from the conformation induced by glucose. Furthermore, the enzyme has an asymmetric arrangement of subunits and with only one allosteric nucleotide binding site per dimer, which is located in the subunit interface region.

B. Energy Transductions

Numerous examples of biological energy transduction exist in which two processes are coupled in such a way that energy is conserved. The mechanism of the coupling is readily understandable when it is a direct consequence of the chemical mechanism. A good example of this is the coupling of a redox reaction to a phosphorylation reaction by GPDH (Scheme I). In this case the dehydrogenation reaction produces an unstable thioester intermediate which can be attacked by P_i to yield 3-phosphoglyceroyl phosphate and regenerate the free enzyme. In effect, the enzyme cycles sequentially between the redox and phosphorylation phases because neither partial reaction can be repeated before the enzyme is "reset" by occurrence of the linked process. The general requirement for coupling is that the two reactions be interdigitated in such a way that at least one step of each cycle cannot be performed before some particular step of the other cycle is performed.

The coupling mechanism is less obvious when there is no chemical intertwining of the two processes. In these cases, it is often found that the ligand-binding

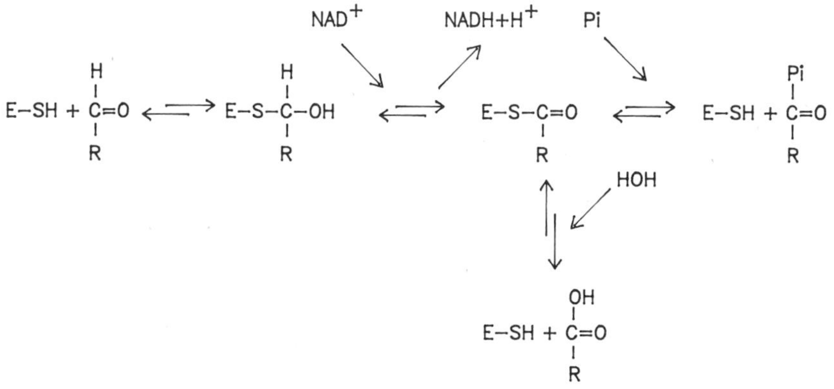

$$\text{E–SH} + \text{H–C(R)=O} \rightleftharpoons \text{E–S–C(H)(R)–OH} \xrightleftharpoons[\text{NAD}^+]{\text{NADH+H}^+} \text{E–S–C(R)=O} \xrightleftharpoons{\text{Pi}} \text{E–SH} + \text{Pi–C(R)=O}$$

$$\text{E–S–C(R)=O} \xrightleftharpoons{\text{HOH}} \text{E–SH} + \text{HO–C(R)=O}$$

Scheme I

reactions provide the linkage through their induced conformation changes. For example, the well-known E1/E2 class of transport ATPases, such as the Na^+,K^+-ATPase, couples transmembrane transport to ATP hydrolysis by requiring that neither partial reaction can be completed in the absence of the occurrence of the other partial reaction. Phosphorylation of the enzyme by ATP can only occur with the binding of Na^+ from one side of the membrane, but hydrolysis of the phosphoenzyme cannot occur until after the conformational change which exposes the ligand to the other side of the membrane and K^+ binding occur. For completion of the cycle, the enzyme must return to the original conformation with a bond K^+. Although this scheme does not involve direct chemical overlap as for GPDH, the overlap is still tangible in the sense that the covalent phosphoenzyme intermediate provides a critical linkage site.

An even more indirect type of coupling is represented by the actomyosin ATPase, which is responsible for movement in muscle, and by the F_1F_0 class of H^+-ATPases, which are responsible for ATP synthesis in oxidative phosphorylation. In both cases, no covalent intermediate has even been detected, and evidence based on stereochemical inversion during hydrolysis (*60, 61*) indicates that none is likely. Now the linkage, by default, is due exclusively to ligand-induced conformational changes. Myosin can exist in two major conformations that are designated the 45 and 90 degree states on the basis of the angle which the myosin head group makes with the long axis of the actin filament. These two states are analogous to the E1/E2 states of the Na^+,K^+-ATPase, and myosin cycles between them during each round of ATP hydrolysis. The basic features of this linkage were originally proposed by Lymn and Taylor (*62*) and involve an antagonism between ATP binding and actin binding which forces the alternation between conformations. ATP binding to the actomyosin complex in the 45 degree state causes the release of the myosin from the actin filament with conversion to the 90 degree state and hydrolysis of the ATP. Further progress is arrested at this stage in the absence of actin because release of the products, ADP and P_i, is very slow. This is analogous to the slow hydrolysis of the Na^+,K^+-ATPase phosphoenzyme in the absence of K^+. Rebinding of myosin to actin completes the cycle, with the induced release of the ADP and P_i coupled to the power stroke which returns the actomyosin complex to the 45 degree state [see Ref. (*63*) for a recent review].

Work on the coupling mechanism of the F_1F_0-ATPase has been hampered by the fact that the proton levels cannot be experimentally altered without introducing the ambiguity associated with binding changes at secondary sites. Considerable progress has been made, however, by studying the effect of the transmembrane proton electrochemical gradient on the partial isotopic exchange reactions accompanying ATP synthesis. These reactions show a number of unusual features which can be accommodated by the "binding change" model of Boyer (*64*) presented in Fig. 12. The F_1 portion has three β subunits which contain the active

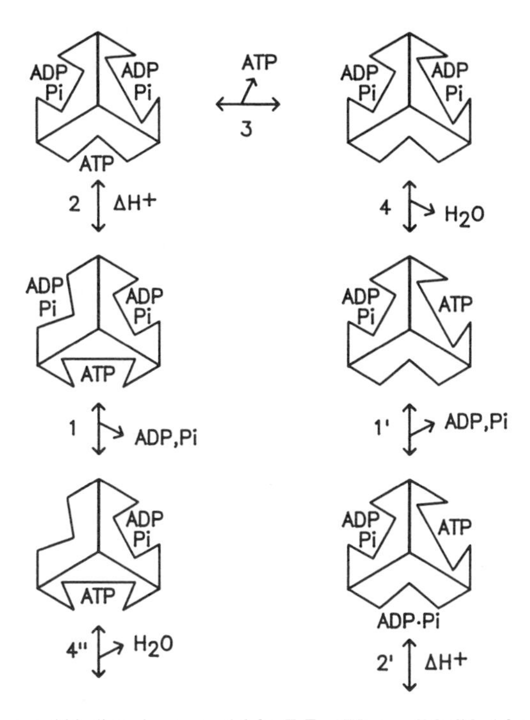

FIG. 12. Proposed binding change model for F_1F_0-ATPases. [Modified from Boyer (*64*).]

sites for ATP binding. The essential features of the model are that a single type of conformational change couples proton transport through the F_0 portion to different changes at each of the three active sites which proceed through net ATP synthesis sequentially. The complete scheme is cyclic, and only one section is indicated in Fig. 12. Each site returns to its starting conformation and ligation state following a complete cycle with synthesis of three ATP molecules. The scheme is reversible, with ATP synthesis corresponding to movement from left to right. Note that the ADP and P_i which bind in Step 1 do not produce the ATP which is released in Step 3. They will not be released as ATP until Step 3″ following complete passage around the circle.

A central question is whether proton transport through the enzyme is directly coupled to the chemical synthesis of the ATP or whether the coupling is indirect. Evidence that the linkage is indirect is provided by the failure of uncouplers (which collapse the proton electrochemical gradient) to block the intermediate $P_i \rightleftharpoons$ HOH oxygen exchange reaction while successfully blocking all of the other exchange reactions which occur with energized vesicles (*65*). The resistance to uncouplers of the intermediate $P_i \rightleftharpoons$ HOH exchange indicates that reversible synthesis of bound ATP continues at the active site in spite of the collapse of the proton electrochemical gradient, which makes net ATP synthesis irreversible. The most likely mechanism for this exchange reaction involves initial hydrolysis of bound ATP to form bound ADP and P_i with one water-derived oxygen incorporated into the P_i; randomization of the oxygens of the bound P_i; resynthesis of bound ATP with one water-derived oxygen (since only one of the four P_i oxygens is derived from water, it will be retained in the ATP three out of four times); rehydrolysis of the ATP to generate P_i with two water-derived oxygens; and continued reversible ATP hydrolysis until release of P_i occurs. In this way, free P_i can be produced which has more than one water-derived oxygen, and the level of incorporation of water-derived oxygen is related to the degree of reversibility of the synthesis of bound ATP. The continuation of this exchange in the absence of a proton electrochemical gradient indicates that energy input is unnecessary for chemical synthesis of bound ATP from bound ADP and P_i in Step 4. In fact, under conditions of low substrate level, as discussed below, even isolated F_1 is capable of reversibly synthesizing bound ATP from bound ADP and P_i, without any possibility of energy input from a proton electrochemical gradient (*66*).

In contrast to the intermediate $P_i \rightleftharpoons$ HOH exchange, many other exchange reactions are inhibited by uncouplers. These other exchanges require either ATP release from the enzyme or ADP and P_i binding to the enzyme in addition to reversible synthesis of bound ATP. Their inhibition indicates that energy input is required for both release of ATP and for competent binding of ADP and P_i. In the absence of a proton electrochemical gradient, bound ATP can still be synthesized from bound ADP and P_i, but ADP and P_i cannot bind in a competent manner and ATP cannot be released. This dual site of energy input is incorporated into the model by having one conformational change in Step 2 be linked both to proton transport and to simultaneous changes at different active sites to energize bound ADP and P_i and to weaken the binding of preformed ATP so that it can be released.

A further prediction of this model is that the extent of the incorporation of water-derived oxygens during ATP synthesis should be influenced by the magnitude of the proton electrochemical gradient. If energy input is limiting, the reaction becomes stalled in the region of Steps 1 and 4″, and many reversals of Step 4 can occur, with extensive incorporation of water-derived oxygens. A large proton electrochemical gradient will result in more rapid release of ATP via Steps 2 and 3 with correspondingly less opportunity for reversal of Step 4″ and less oxygen incorporation. This type of modulation of the exchange reaction by the magnitude of the proton electrochemical gradient has been experimentally observed (*67, 68*). A different type of evidence in support of the model is the direct observation of tightly bound ATP which undergoes "single site catalysis" (*69*).

As emphasized by Jencks (*70, 71*) and others, coupling is a consequence of the kinetics and not directly the thermodynamics of the individual species (more precisely, it is a consequence of the thermodynamics of both the transition states and the ground states and not just a consequence of the thermodynamics of the ground states alone). Thus, there is no need to have the equilibria among the states arranged in any particular pattern; the assignment of the coupling rules via disallowance of certain steps is sufficient to generate coupling. This, however, should not obscure the fact that certain requirements must be met for the process to proceed at a useful rate. A perfectly reasonable thermodynamic scheme is useless to the cell if it only turns over once a year.

C. Catalysis

Glyceraldehyde-3-phosphate dehydrogenase also provides an example of the use of the binding energy of one ligand to drive a separate catalytic process in a cyclic manner. The phosphorolysis of the thioester is in competition with direct hydrolysis by water. The enzyme must strongly discriminate against such attack by water as it results in the decoupling of the redox and phosphorylation processes and the waste of the high-energy thioester. Phosphorolysis of the isolated thioester intermediate, however, is slow, even slower than hydrolysis. This potential disaster is avoided by the binding of NAD^+ to the thioester intermediate. The changes, which are induced by NAD^+ binding, selectively accelerate the attack of P_i in preference to the attack of water (*58*). Thus, part of the binding energy of NAD^+ is not expressed as a reduced K_m for itself, but rather is used to destabilize the complex selectively in the manner which favors attack by P_i. This occurs in spite of the fact that NAD^+ has no direct role in the phosphorolysis reaction and its action must be indirect. The result is to even out the energy profile so that no single step has too high of a transition state energy.

In the case of GPDH, the coupling occurs between reactions which share an overlapping active site and share one common intermediate. Coupling can still occur, however, even when the active sites for the two partial reactions are physically separate. One likely example of this is again provided by the H^+-ATPases which use substrate binding at one site to drive processes at a linked site, as indicated in Fig. 12. Thus, binding of ADP and P_i in Step 1 can be linked to the release of ATP from a different site in Step 3 during ATP synthesis, and ATP binding is linked to release of ADP and P_i during hydrolysis. Evidence for such coupling is of two types. During ATP hydrolysis, extensive incorporation of water-derived oxygens into the remaining ATP is observed owing to sequential binding of ATP, reversible hydrolysis of bound ATP with incorporation of water-derived oxygens, and release of the bound ATP back to the medium. This reaction does not involve medium ADP or P_i directly, yet it is blocked by pyruvate kinase and phosphoenolpyruvate (PEP), which act as an ADP trap (*72*). This inhibition cannot be produced by a simple model, but it is readily accounted for by the model of Fig. 12. ATP can bind in Step 3 and proceed to Step 4″ where incorporation of water-derived oxygens can occur, but this ATP cannot be released since reversal of Step 1 cannot occur in the presence of an ADP trap. A reciprocal inhibition of P_i release by an ATP trap is also observed during ATP synthesis (*73*).

A second type of evidence in favor of the model of Fig. 12 comes from modulation of the extent of isotopic exchange by the substrate concentration. During ATP hydrolysis, limiting levels of ATP should reduce the flux through Step 3, which will stall the reaction in the region of Step 4 with an increase in the amount of reversal of the hydrolysis step and an increase in the incorporation of water-derived oxygens. Such modulation has been observed (*66, 74*) as well as the reciprocal modulation by the ADP and P_i level of the extent of incorporation of water-derived oxygens into ATP during ATP synthesis (*67, 73*). These reciprocal relationships are a consequence of the interdigitation of the steps at the multiple binding sites such that release of product in the direction of either hydrolysis or synthesis cannot occur until substrate is bound to the empty site. This assures that simultaneous alteration of the ligand affinity will occur at all three sites during the conformational changes linked to proton transport.

V. Prospects

The importance of binding interactions for both catalysis and specificity is now well established. Considerable effort is currently being devoted to extending these concepts using the powerful techniques which are available to modify binding sites by genetic and chemical means. Another approach is the continuing development of catalytic antibodies which bind tightly to analogs of transition states. Although beyond the scope of this review, most of these considerations also apply to the binding interactions between nucleic acids, including the role of catalytic RNA. Clearly these investigations into the role of binding will provide additional important information for some time to come.

Acknowledgments

This work was supported in part by the National Institutes of Health.

References

1. Klotz, I. M. (1987). *In* "Enzyme Mechanism" (M. I. Page and A. W. Williams, eds.), pp. 14–34. Royal Society of Chemistry, London.
2. Stoddart, J. F. (1987). *In* "Enzyme Mechanism" (M. I. Page and A. W. Williams, eds.), pp. 35–55. Royal Society of Chemistry, London.
3. Bender, M. L. (1987). *In* "Enzyme Mechanism" (M. I. Page and A. W. .Williams, eds.), pp. 56–66. Royal Society of Chemistry, London.

4. Kirby, A. J. (1987). *In* "Enzyme Mechanism" (M. I. Page and A. W. Williams, eds.), pp. 67–77. Royal Society of Chemistry, London.
5. Boyer, P. D. (1987). *Acc. Chem. Res.* **11,** 218.
6. Page, M. I. (1987). *In* "Enzyme Mechanism" (M. I. Page and A. W. Williams, eds.), pp. 1–13. Royal Society of Chemistry, London.
7. Kraut, J. (1988). *Science* **242,** 533.
8. Fersht, A. (1985). "Enzyme Structure and Mechanism," 2nd Ed. Freeman, New York.
9. Albery, W. J., and Knowles, J. R. (1977). *Angew. Chem. Int. Ed. Engl.* **16,** 285.
10. Jencks, W. P. (1975). *Adv. Enzymol.* **43,** 219.
11. Wolfenden, R. (1972). *Acc. Chem. Res.* **5,** 10.
12. Pauling, L. (1946). *Chem. Eng. News* **24,** 1375.
13. Britton, H. G. (1973). *BJ* **133,** 255.
14. Hammes, G. G., and Eigen, M. (1963). *Adv. Enzymol.* **25,** 1.
15. Moore, S. A., and Jencks, W. P. (1982). *JBC* **257,** 10893.
16. Fierke, C. A., and Jencks, W. P. (1986). *JBC* **261,** 7603.
17. Levitt, M. (1974). *In* "Peptides, Polypeptides and Proteins" (E. R. Blout, F. A. Bovey, M. Goodman, and N. Lotan, eds.), pp. 99–113. Wiley, New York.
18. Robillard, G., Shaw, E., and Shulman, R. G. (1974). *PNAS* **71,** 2623.
19. Argade, P. V., Gerke, G. K., Weber, J. P., and Peticolas, W. L. (1984). *Biochemistry* **23,** 299.
20. Stackhouse, J., Nambiar, K. P., Burbaum, J. J., Stauffer, D. M., and Benner, S. A. (1985). *J. Am. Chem. Soc.* **107,** 2757.
21. Ellington, A. D., and Benner, S. A. (1987). *J. Theor. Biol.* **127,** 491.
22. Dale, M. P., and Hackney, D. D. (1987). *Biochemistry* **26,** 8365.
23. Koshland, D. E., Jr. (1959). "The Enzymes," 2nd Ed., Vol. 1, p. 98.
24. DelaFuente, G., Lagunas, R., and Sols, A. (1970). *EJB* **16,** 226.
25. Bennett, W. S., and Steitz, T. A. (1978). *PNAS* **75,** 4848.
26. Bennett, W. S., and Huber, R. (1984). *Crit. Rev. Biochem.* **15,** 291.
27. Bridger, W. A., Millen, W. A., and Boyer, P. D. (1968). *Biochemistry* **7,** 3608.
28. Knowles, J. R., and Albery, W. J. (1977). *Acc. Chem. Res.* **4,** 105.
29. Koshland, D. E., Jr. (1960). *Adv. Enzymol.* **22,** 83.
30. Westheimer, F. H. (1962). *Adv. Enzymol.* **24,** 455.
31. Koshland, D. E., Jr. (1962). *J. Theor. Biol.* **2,** 75.
32. Storm, D. R., and Koshland, D. E., Jr. (1970). *PNAS* **66,** 445.
33. Dafforn, A., and Koshland, D. E., Jr. (1971). *Bioorg. Chem.* **1,** 129.
34. Dafforn, A., and Koshland, D. E., Jr. (1971). *PNAS* **68,** 2463.
35. Page, M. I., and Jencks, W. P. (1971). *PNAS* **68,** 1678.
36. Page, M. I. (1973). *Chem. Soc. Rev.* **2,** 295.
37. Dafforn, A., and Koshland, D. E., Jr. (1973). *BBRC* **52,** 779.
38. Jencks, W. P., and Page, M. I. (1974). *BBRC* **57,** 887.
39. Barrow, G. M. (1966). "Physical Chemistry," pp. 254–258. McGraw-Hill, New York.
40. Davidson, N. (1962). "Statistical Mechanics," pp. 169–200. McGraw-Hill, New York.
41. Kirby, A. J. (1980). *Adv. Phys. Org. Chem.* **17,** 183.
42. Kirby, A. J., and Lancaster, P. W. (1972). *J. Chem. Soc., Perkin Trans. 2,* p. 1206.
43. Fife, T. H., and Benjamin, B. M. (1974). *Chem. Commun.*, p. 525.
44. Choing, K. N. G., Lewis, S. D., and Schafer, J. A. (1975). *J. Am. Chem. Soc.* **97,** 418.
45. Bauer, C. A., Thompson, R. C., and Blout, E. R. (1976). *Biochemistry* **15,** 1296.
46. Himoe, A., Brandt, K. G., and Hess, G. P. (1967). *JBC* **242,** 3963.
47. Hackney, D. D., and Koshland, D. E., Jr. (1974). Unpublished observations.
48. Olson, A. J., Templeton, S. H., and Templeton, L. K. (1977). *Acta Crystallogr., Sect. B: Struct. Sci.* **33,** 2266.
49. Morris, J. J., and Page, M. I. (1980). *J. Chem. Soc., Perkin Trans. 2,* p. 679.

50. Huber, R., and Bode, W. (1978). *Acc. Chem. Res.* **11,** 114.
51. Silver, M. S., Stoddard, M., Sone, T., and Matta, M. F. (1970). *J. Am. Chem. Soc.* **92,** 3151.
52. Monod, J., Wyman, J., and Changeux, J. (1965). *JMB* **12,** 88.
53. Koshland, D. E., Jr., Nemethy, G., and Filmer, D. (1966). *Biochemistry* **5,** 265.
54. Koshland, D. E., Jr. (1970). "The Enzymes," 3rd Ed., Vol. 1, p. 341.
55. Viggiano, G., and Ho, C. (1979). *PNAS* **76,** 3673.
56. Miura, S., and Ho, C. (1982). *Biochemistry* **24,** 6280.
57. Liddington, R., Zygmunt, D., Dodson, G., and Harris, D. (1988). *Nature (London)* **331,** 725.
58. Byers, L. D., and Koshland, D. E., Jr. (1975). *Biochemistry* **14,** 3661.
59. Steitz, T. A., Shoham, M., and Bennett, W. S. (1981). *Philos. Trans. R. Soc. London B:* **293,** 43.
60. Webb, M. R., and Trentham, D. R. (1980). *JBC* **255,** 8629.
61. Webb, M. R., Grubmeyer, C., Penefsky, H. S., and Trentham, D. R. (1980). *JBC* **255,** 11637.
62. Lymn, R. W., and Taylor, E. W. (1971). *Biochemistry* **10,** 4617.
63. Hibberd, M. G., and Trentham, D. R. (1986). *Annu. Rev. Biophys. Biophys. Chem.* **15,** 119.
64. Boyer, P. D. (1989). *FASEB J.* **3,** 2164.
65. Rosing, J., Kaylar, C., and Boyer, P. D. (1977). *JBC* **252,** 2478.
66. Hutton, R. L., and Boyer, P. D. (1979). *JBC* **254,** 9990.
67. Hackney, D. D. (1984). *Curr. Top. Cell. Regul.* **24,** 379.
68. Stroop, S. D., and Boyer, P. D. (1987). *Biochemistry* **26,** 1479.
69. Penefsky, H. (1988). *In* "The Ion Pumps: Structure, Function and Regulation" (W. D. Stein, ed.), pp. 261–268. Alan R. Liss, New York.
70. Jencks, W. P. (1980). *Adv. Enzymol.* **51,** 75.
71. Jencks, W. P. (1982). *In* "From Cyclotrons to Cytochromes" (N. O. Kaplan and A. Robinson, eds.), pp. 485–508. Academic Press, New York.
72. Kaylar, C., Rosing, J., and Boyer, P. D. (1977). *JBC* **252,** 2486.
73. Hackney, D. D., and Boyer, P. D. (1978). *JBC* **253,** 3164.
74. Russo, J. A., Lamos, C. M., and Mitchell, R. A. (1978). *Biochemistry* **17,** 473.

2

Biological Electron Transfer

DOUGLAS C. REES[*] • DAVID FARRELLY[†]

[*] *Division of Chemistry and Chemical Engineering*
California Institute of Technology
Pasadena, California 91125

[†] *Department of Chemistry and Biochemistry*
University of California, Los Angeles
Los Angeles, California 90024

THE ENZYMES, Vol. XIX

Introduction

Many fundamental processes in biochemistry and chemistry involve electron transfer reactions. Consequently, an intense, multidisciplinary effort, involving biologists, biochemists, chemists, and physicists, has arisen to understand the mechanistic details of electron transfer reactions. The convergence of these diverse fields into a single area has fueled an explosive growth of experiments and ideas. An unfortunate consequence of the increasing level of detail and analysis of electron transfer reactions, however, is the increasing difficulty of appreciating developments in different areas. This is perhaps most true between the biological and physical extremes of the field of biological electron transfer, where the relevant vocabulary and concepts occupy nearly orthogonal spaces. Our objective in this chapter is to help to bridge this gap, by providing a conceptual framework for orienting biochemists and molecular biologists toward the exciting developments in the "physical chemistry" aspects of biological electron transfer.

One might ask, "Why should biochemists and molecular biologists be interested in the details of electron transfer reactions?" Briefly, one answer to this question is simply that electron transfer reactions occupy a position of central importance to the functioning of biological systems, especially in the areas of bioenergetics, biosynthesis, and cellular regulation. A molecular appreciation of these processes requires familiarity with electron transfer reactions. Furthermore, the field of electron transfer stands to benefit from the active involvement of biochemists and molecular biologists. At present, there are many theories, general ideas, and experimental approaches to study electron transfer reactions. Considerable progress has been achieved in the characterization of biological electron transfer reactions, but many fundamental questions remain open: What determines the rate of electron transfer? Can these rates be predicted and rationally altered? How is specificity between the correct donor and acceptor proteins determined? More extensive interest in electron transfer reactions is required, especially on systems of central biological significance that can be manipulated by genetic and biochemical methods. Biochemists and molecular biologists are uniquely suited to address these problems, and so are in a position to make critical contributions to the detailed understanding of one of the most fundamental classes of chemical reactions.

This review is organized to cover the basic features of simple electron transfer reactions. The first three sections develop background material on the thermodynamics, kinetics, and microscopic theory of electron transfer reactions. More general, semiquantitative treatments of these topics are presented, with the objective of introducing the conceptual approaches used to characterize electron transfer processes. The fourth section describes experimental studies on two electron transfer systems, selected from both physiological and nonphysiological processes. These studies provide a representative survey of the current status of theory and experiment in understanding biological electron transfer reactions. The concluding section outlines future directions for study, modification and design, and biotechnological applications of electron transfer proteins.

Given the extensive literature on electron transfer reactions, we have not attempted to provide an exhaustive and historically complete collection of references. In keeping with the general nature of the review, we have tried to cite representative references which serve to illustrate particular points. A number of excellent reviews on electron transfer theory and biological electron transfer have recently appeared (*1–9*); more detailed information on many of the topics may be found in these sources.

I. Thermodynamics of Electron Transfer Reactions

The thermodynamic parameter of central importance to the characterization of electron transfer processes is the reduction potential, E°. The reduction potential provides a measure of the tendency of an oxidized molecule to become reduced. Values of E° are typically expressed in units of either volts (V) or millivolts (mV). For example, the textbook E° value of cytochrome *c* (at 298 K) is about +250 mV (*10*). It is unlikely, however, that an actual laboratory measurement of the E° of cytochrome *c* would give this value. Reduction potentials (in common with all thermodynamic quantities) are constant only for the specific conditions under which they were determined. Changes in the protein and solution environment may alter E° from the values measured under different experimental conditions. These variations provide an opportunity to explore how E° values of a redox group can be modulated by the protein and surrounding solution.

The sensitivity of E° values to details of the protein and solution environment may be illustrated by the effect of pH on E°. The pH dependences of E° for two *c* cytochromes isolated from cow (*11*) and from the purple bacterium *Rhodobacter capsulatus* (*12*) are shown in Fig. 1. Two major points emerge from Fig. 1: (1) although the two cytochromes share a common fold and sequence similarity (*13*), the E° values differ by over 100 mV at low pH, and (2) the E° values are pH dependent, with E° decreasing with increasing pH. Further complexities are revealed if the ionic strength dependence of E° is examined, as illustrated by cytochrome *c*-552 from *Euglena* (*14*) in Fig. 2. Depending on pH, both the magnitude and the sign of the dependence of E° on ionic strength may vary. To understand the origins of these effects, a general framework is presented in this section to assist in the analysis and rationalization of the influence of environmental (protein and solution) factors on E° values.

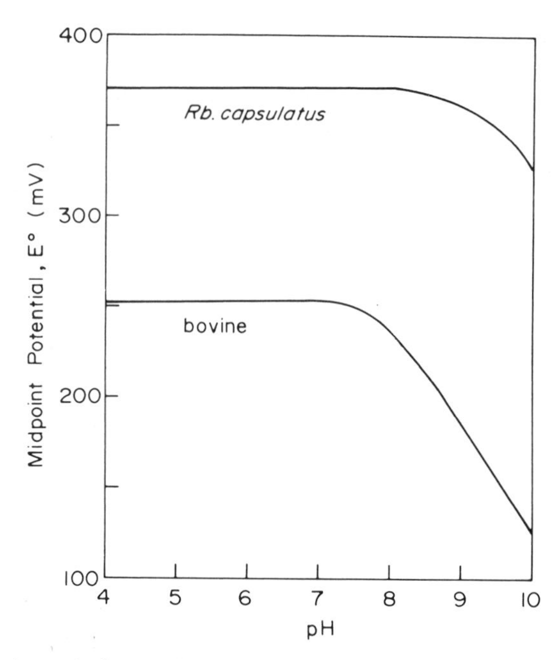

FIG. 1. Dependence of E° on pH for bovine and *Rb. capsulatus* *c* cytochromes. [Adapted from Refs. (11) and (12).]

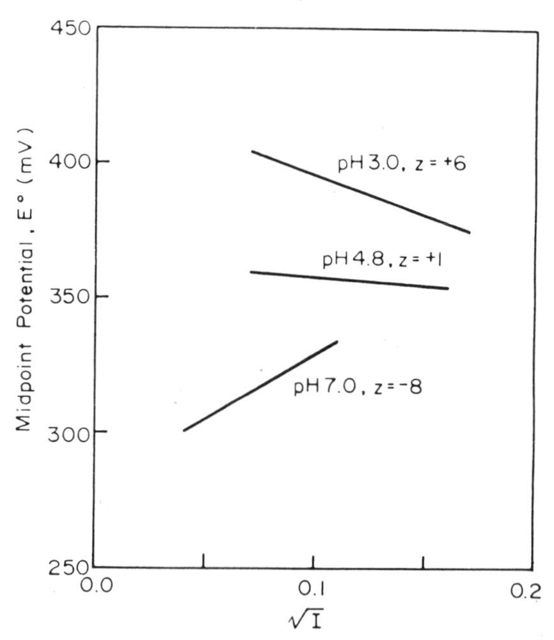

FIG. 2. Dependence of E° values on ionic strength, I, for *Euglena* cytochrome *c*-552 measured at pH 3.0, 4.8, and 7.0, where the net charge, z, on the protein is positive, nearly zero, and negative, respectively. [Adapted from Ref. (14).]

A. Properties of Reduction Potentials

Qualitatively, the higher the value of E°, the greater is the ability of a molecule to accept electrons (*15*). While *absolute* values of E° cannot be determined, the *difference* in E° values between different molecules can be measured experimentally. Let A and B represent two different molecules, with the oxidized and reduced forms of these molecules indicated by the subscripts "o" and "r," respectively. The individual "half-cell" reduction reactions may be written

$$A_o + e^- \rightarrow A_r; \qquad E^\circ_A \tag{1a}$$

$$B_o + e^- \rightarrow B_r; \qquad E^\circ_B \tag{1b}$$

where E°_A and E°_B are the reduction potentials for the two reactions. For the coupled reaction

$$A_o + B_r \rightarrow A_r + B_o; \qquad \Delta E^\circ \tag{2}$$

the measured potential difference is given by $\Delta E^\circ = E^\circ_A - E^\circ_B$. When the value of one standard half-cell potential is assigned by convention, values for all other half-cell reactions may be determined by comparison to the standard reaction. For this purpose, the reference hydrogen half-cell

$$2H^+ + 2e^- \rightarrow H_2; \qquad E^\circ = 0 \text{ mV} \tag{3}$$

is assigned the value of $E^\circ = 0$ mV at pH 0, 298K $[H^+] = 1\ M$, and $[H_2] =$ 1 atm pressure. The value of E°_A for any other half-cell may be measured (at least in principle) from Eq. (2), using the measured potential difference against the hydrogen half-cell in Reaction (1b).

The reduction potential difference, ΔE°, is related to both the equilibrium constant, K_{eq}, and the standard free energy change for the reaction, ΔG°. K_{eq} can be derived from the chemical activities of the various species at equilibrium:

$$K_{eq} = \frac{(a_{A_r})(a_{B_o})}{(a_{A_o})(a_{B_r})} \tag{4}$$

where a_i is the activity of the ith chemical species. ΔE° may be obtained from K_{eq} and ΔG° through the standard thermodynamic relationships:

$$\Delta E^\circ = -\frac{\Delta G^\circ}{nF} = \frac{RT}{nF} \ln K_{eq} \tag{5}$$

where n is the number of moles of electrons transferred in the redox reaction, F is the Faraday constant (which has the value 23.06 cal/mV or 96.48 joules/mV), R is the gas constant (1.987 cal/K or 8.314 joules/K), and T is the absolute temperature. The change in reduction potential, ΔE, when the components are not at unit activity is obtained from the Nernst equation:

$$\Delta E = \Delta E^\circ - \frac{RT}{nF} \ln\left[\frac{(a_{A_r})(a_{B_o})}{(a_{A_o})(a_{B_r})}\right] \tag{6}$$

By convention in biochemical systems, standard state conditions are defined in the following fashion: all components are at unit activity (1 M), except for H^+ (10^{-7} M = pH 7) and gases (1 atm pressure). Activities are related to concentrations, c, by the activity coefficient, γ:

$$a = \gamma c \tag{7}$$

Although it is common to equate activities to concentrations (i.e., assign $\gamma = 1$), there are many situations in biochemical systems (especially when polyelectrolytes such as proteins are involved) for which this is a poor approximation. Some of the consequences of this nonideal behavior for electron transfer reactions are described later.

Approximate limits on the range of observed half-cell potentials in biochemical systems may be set from the following considerations. From Eqs. (3) and (6), the hydrogen cell potential at pH 7 is determined to be -416 mV. This sets an approximate lower limit on E° for biochemical reductants, since reductants with lower potential in aqueous solution would reduce protons to hydrogen. (Note: Reducing centers protected from exposure to water by protein can have lower potentials, however.) An upper limit on reduction potentials is set by the oxygen half-cell reaction:

$$O_2 + 4H^+ + 4e^- \rightarrow 2H_2O \tag{8}$$

which has a potential of $+816$ mV at pH 7. Species with higher reduction potentials could oxidize water to oxygen under appropriate conditions, which does occur in the oxygen-evolving reactions of plant photosynthetic systems. Between these limits, the sequence of carriers in a biochemical electron transfer chain will generally exhibit increasing reduction potentials on moving along the chain to the terminal electron acceptor. It must be recognized, however, that the actual thermodynamically favored direction of electron transfer will depend on the particular concentrations of the reduced and oxidized species [Eq. (6)], and this direction might vary from that predicted by the standard potential differences [Eq. (5)].

B. Protein and Solution Influences on E°

Although ΔE° is related to the equilibrium constant K_{eq}, it should not be imagined that either quantity is actually constant; rather each is defined for a particular set of experimental conditions. Variations in environmental parameters such as temperature, pressure, pH, buffer type, and ionic strength can lead to significant differences in reduction potentials. Furthermore, the same cofactor (such as a heme) can exhibit a range of reduction potentials when associated with different proteins. Factors influencing E° may be somewhat arbitrarily divided into three categories: (1) the intrinsic properties of the redox group, (2) effects of the protein, and (3) effects of the surrounding solution. As with most aspects of protein energetics, the observed E° value reflects the contributions of many significant, yet compensatory, factors. Since this severely complicates quantitative, *ab initio* estimates of E°, the following discussion of protein and solution influences on E° is restricted to a more qualitative treatment that serves to illustrate the types of effects that may be observed. More detailed reviews of the factors that determine the E° of redox groups have appeared (*4, 6, 16–19*).

A useful approach for characterizing the influence of various factors on E° is to consider the relative effects on the stability of the oxidized and reduced states. Interactions that preferentially stabilize the reduced state (or preferentially destabilize the oxidized state) will increase E°, whereas the converse situation will be reflected by a decrease in E°. Formally, if the reduced and oxidized forms of a redox cofactor have association constants for binding the protein of K_r and K_o, respectively, then the observed midpoint potential, E°, is related to the intrinsic potential of the free cofactor, E_i°, by the expression

$$E^\circ = E_i^\circ + \frac{RT}{nF} \ln\left(\frac{K_r}{K_o}\right) \tag{9}$$

Clearly, the magnitudes of K_r and K_o are of less importance (assuming they are high enough to ensure cofactor binding to the protein) than the ratio K_r/K_o in determining E°.

1. *Redox Groups*

Nature exhibits considerable diversity in the selection of redox cofactors for electron transfer proteins. Redox active groups include nonprotein cofactors (single metals, inorganic clusters, organometallic species, and organic groups) as well as amino acid side chains (such as cysteine and tyrosine). A representative sample of various redox groups found in electron transfer proteins is presented in Table I (*20–29*). These groups have been selected from proteins of known three-dimensional structure for which details of the protein environment are available. The list is by no means comprehensive, and it excludes many interesting and important cofactors that have been identified in electron transfer proteins.

An important characteristic of a redox group is the E° value, and E° values span the range from about -400 to $+400$ mV. Most cofactors exhibit either $n = 1$ or $n = 2$ behavior exclusively, although some groups (most notably flavins and quinones) may function as both one- and two-electron carriers. Even a casual perusal of redox cofactors emphasizes the multiplicity of groups exhibiting simi-

TABLE I

ELECTROCHEMICAL PROPERTIES OF SELECTED ELECTRON TRANSFER PROTEINS[a]

Protein	Cofactor	Metal ligands	$E°$ (mV)	Couple	Ref.
Cytochrome *c*	Heme	His, Met	+260	$Fe^{3+/2+}$	*13*
Cytochrome b_{562}	Heme	His, Met	+185	$Fe^{3+/2+}$	*6*
Cytochrome *c′*	Heme	His	~+100	$Fe^{3+/2+}$	*20*
Metmyoglobin	Heme	His, water	+60	$Fe^{3+/2+}$	*21*
Cytochrome b_5	Heme	2 His	+20	$Fe^{3+/2+}$	*6*
Cytochrome-*c* peroxidase	Heme	His, water	−194	$Fe^{3+/2+}$	*22*
Cytochrome *P*-450	Heme	Cys^-, water	−300	$Fe^{3+/2+}$	*22*
Catalase	Heme	Tyr^-	<−420	$Fe^{3+/2+}$	*23*
Plastocyanin, azurin	Cu	2 His, Met, Cys^-	+350	$Cu^{2+/1+}$	*24, 25*
HiPIP	Fe_4S_4	4 Cys^-	+300	$[(Fe_4S_4)(SR)_4]^{1-/2-}$	*26*
Ferredoxin	Fe_4S_4	4 Cys^-	−400	$[(Fe_4S_4)(SR)_4]^{2-/3-}$	*26*
Flavodoxin	Flavin		−150	Oxidized/semiquinone	*27*
			−450	Semiquinone/reduced	
Reaction centers	$(Bchl)_2$		−800	$(Bchl)_2^{+/*}$	*28*
			+450	$(Bchl)_2^{+/0}$	
	Bphe		−600	$Bphe^{0/-}$	
	Quinone		−50	$Q^{0/-}$	
Thioredoxin	Disulfide		−260	$(RS)_2$/2 HSR ($n = 2$)	*29*

[a] Approximate $E°$ values are listed; significant changes may be observed for proteins from different sources and with variations in solution conditions. High-potential iron protein, HiPIP; bacteriochlorophyll, Bchl; bacteriopheophytin, Bphe; quinone, Q.

lar electrochemical properties. Low-potential electron transfer proteins have been isolated that contain iron–sulfur clusters, flavins, NADH, and/or porphyrin cofactors. In several cases, interchangeable pairs of electron transfer proteins performing the same physiological function [such as plastocyanin and cytochrome *c*-552 in algae (*30*) or ferredoxin and flavodoxins in nitrogen-fixing bacteria (*31*)], have been identified. Presumably, environmental availability of metals and biosynthetic capabilities for more complex cofactors contribute to the selection of the actual redox group used by a particular protein. The rationale behind the choice of cofactor for a specific reaction usually remains obscure, although the question is mechanistically and evolutionarily important. It appears, just as in the case of proteolytic enzymes, that there are many ways of designing a protein to carry out a given electron transfer reaction.

2. *Protein Influences on* E°

Interactions with a protein may profoundly change the E° value of an associated redox group from the solution value. The chemical nature of the ligands, the detailed coordination geometry, and noncovalent interactions between the protein and redox group all contribute to the observed E° value.

a. *Chemical Nature of Ligands.* Negatively charged ligands stabilize the more positively charged oxidized state of the redox group, and consequently lower E°. As examples, substitution of a glutamic acid carboxylate group for the water ligand to the heme in myoglobin changes E° by about -200 mV (*21*). Low-potential heme proteins, like cytochrome *P*-450 and catalase, have anionic ligands (cysteine thiolate and tyrosinate, respectively), in contrast to the neutral ligands observed in higher potential heme proteins such as cytochrome b_5 and cytochrome *c* (Table I). More generally, electron-donating ligands stabilize the oxidized state and consequently lower E°, whereas electron-acceptor ligands stabilize the more electron-rich reduced state and increase E°. Model studies indicate that variations in E° of over 150 mV may be associated with changes in metal ligation (*32*).

b. *Coordination Geometry.* When the preferred coordination geometry is different between the oxidized and reduced forms, the protein may alter the relative stability of the two states, and hence E°, by imposing a particular coordination geometry on the redox center. A striking example of this behavior is found in E° values of Cu^{2+}/Cu^{+} couples. Reduced Cu^{+} displays tetrahedral coordination, while oxidized Cu^{2+} prefers square planar coordination. The high redox potentials ($+300$ to $+400$ mV) of copper proteins like plastocyanin and azurin reflect the distorted tetrahedral environment of the copper in these proteins (*24, 25, 33*). By binding the metal center in a coordination environment favoring the reduced form, a protein is able to destabilize the oxidized state of the metal, thereby raising $E°$.

From the few cases where the structures of apo- and metallosubstituted forms of a protein have been crystallographically determined (*34–36*), it appears that the metal binding site is primarily determined by the protein, rather than *vice versa* (i.e., structural organization of the protein by the metal). Thus, the protein is able to adopt and maintain a metal coordination environment that may preferentially stabilize one oxidation state relative to the other. An interesting exception to this generalization may apply to large redox cofactors covalently coordinated to the protein. Recently, the replacement of a cysteine, liganded to an iron–sulfur cluster, by alanine in a ferredoxin from *Azotobacter vinelandii* has been described (*37*). This ligand replacement is accompanied by a structural adjustment that permits cluster coordination by a previously nonliganding cysteine. In this case, the extensive interactions between the protein and a relatively large, rigid cofactor that is covalently coordinated to the protein appear to enable the cofactor to play a dominant role in shaping the coordination environment.

In addition to the influence on $E°$ values through the coordination environment, $E°$ may be modulated by groups outside of the first coordination sphere. One example of this has been proposed to explain protein control of the redox behavior in Fe_4S_4 clusters. Three oxidation states are available to this cluster:

$$[Fe_4S_4](SR)_4^{-} \leftrightarrow [Fe_4S_4](SR)_4^{2-} \leftrightarrow [Fe_4S_4](SR)_4^{3-} \quad (10)$$

The net charge of the cluster includes the contribution of the four cysteine thiolate groups (SR) used to covalently coordinate the cluster. Fe_4S_4 clusters have been found in high-potential iron protein (HiPIP) and ferredoxins. Although the clusters have similar coordination geometries in both proteins, the electrochemical properties are quite different. The $E°$ values for HiPIP and ferredoxin are approximately +300 and −400 mV, respectively. According to the three-state hypothesis (*26*), the two proteins actually undergo electron exchange between different pairs of states. HiPIP utilizes the more oxidized pair $\{[Fe_4S_4](SR)_4^{-}, [Fe_4S_4](SR)_4^{2-}\}$, whereas ferredoxin utilizes the more reduced pair $\{[Fe_4S_4](SR)_4^{2-}, [Fe_4S_4](SR)_4^{3-}\}$. Although the two proteins share structural similarities, they have developed the capability of stabilizing different oxidation states with the same coordinating ligands. One distinguishing feature between the two proteins is the greater number of hydrogen bonds to the cluster sulfur atoms of the type (NH---S) in ferredoxin relative to HiPIP (*38*). By donating more hydrogen bonds (with a partial positively charged hydrogen) to the cluster, ferredoxin can apparently stabilize the more electron-rich fully reduced form of the cluster, compared to the cluster in HiPIP.

A second example illustrates how hydrogen bonding interactions may stabilize the more oxidized form of a redox cofactor. The porphyrin iron in heme proteins is often liganded by a histidine imidazole side chain. The imidazole is frequently hydrogen bonded to a second protein group (*39*), such as a peptide bond carbonyl oxygen [in globins and cytochrome *c* (*39*)] or an aspartate carboxyl group [in cytochrome-*c* peroxidase (*22*)]. By supplying a hydrogen bond acceptor (with a partial negative charge) to a heme ligand, the protein may better stabilize the more positively charged oxidized form of the porphyrin. The lower $E°$ of cytochrome-*c* peroxidase compared to the globins has been partially attributed (*22*) to the increased electron-donating ability of the carboxyl group hydrogen bonded to the histidine in the former, relative to the electron-donating ability of the corresponding peptide carbonyl oxygen in the latter.

These two examples illustrate the potential of hydrogen-bonding interactions involving noncoordinating groups to influence reduction potentials. Donation of hydrogen bonds to the redox cofactor tends to preferentially stabilize the more electron-rich reduced form, thus raising $E°$. In contrast, the presence of hydrogen bond acceptors preferentially stabilizes the more positively charged oxidized form of the cofactor, thus lowering $E°$. Consequently, depending on the specifics of the hydrogen-bonding arrangement, the protein may tune the midpoint potential in either direction.

Additional, structurally subtle mechanisms have been proposed for control of $E°$ by protein ligation. Among these are variation in cytochrome *c* redox potential associated with changes in the methionine sulfur–iron bond length (*19*) and methionine sulfur chirality (*40*); changes in $E°$ associated with differences in the angular orientation between coordinating ligands and redox cofactors (*4*); the greater basicity of the *syn* lone pair electrons of carboxyl groups relative to the *anti* lone pair (*41, 42*), suggesting that the $E°$ for *syn*-coordinated metals will be lower than for *anti*-coordinated metals; the greater basicity of the Nε2 nitrogen of histidine relative to the Nδ1 nitrogen (*43*), suggesting that $E°$ for Nε2-coordinated metals will be lower than for Nδ1-coordinated metals; and $E°$ changes that may be associated with ligand-induced spin state changes (*18, 19*). The existence of these effects indicates that detailed structural and electronic information about electron transfer proteins will be essential for the quantitative understanding of $E°$ values.

c. *Noncovalent Interactions.* The reduction potentials for different proteins with the same redox groups and protein ligands may vary significantly, as illustrated in Fig. 1 for different *c* cytochromes. These differences reflect in part the influence of noncovalent interactions between the protein and redox cofactor in controlling $E°$ values. The major effects in this category are steric and electrostatic interactions.

i. Steric effects. Since the oxidized and reduced forms of the redox groups are distinct chemical species, the detailed structures and properties of the two forms necessarily differ. If the binding site in the protein for the redox group exhibits a higher affinity for one of the oxidation states, then an accompanying shift in E° will be generated in accordance with Eq. (9). This behavior applies not only to ligation interactions between the protein and the redox group but also to more general types of steric interactions. A rigid binding site sterically complementary to one oxidation state will generate large changes in E° when a significant structural difference exists between oxidized and reduced forms of the redox cofactor. An important example of this effect has been observed with flavin cofactors. Whereas the isoalloxazine ring system of the flavin is planar in the oxidized and semiquinone states, the fully reduced flavin ring displays significant nonplanarity in solution (*44*). Protein–cofactor interactions may be designed to exploit this conformational change to selectively stabilize different oxidation states, with a resulting change in E° value for the flavin. The low E° value between the semiquinone and fully reduced forms of the protein-bound flavin in flavodoxin has been proposed (*45*) to reflect the planar conformation of the cofactor in both oxidation states, with the consequent destabilization of the reduced form. In addition, by altering E° values between the three oxidation states of flavin, protein interactions control the ability of the flavin to function as either a one- or two-electron carrier (*46*).

ii. Electrostatic interactions. The reduction potential reflects the free energy required to move an electron to an oxidized redox group from an infinite initial separation. Since the electron is negatively charged, the redox potential will be influenced by electrostatic interactions between the protein and electron. The simplest model of the electrostatic interaction energy, U, between an electron and a group on the protein of valence charge z is given by Coulomb's law:

$$U = -\frac{14435z}{\varepsilon_{\mathrm{eff}} r} \quad \mathrm{mV} \tag{11}$$

where $\varepsilon_{\mathrm{eff}}$ is the effective dielectric constant and r is the separation distance (in Å) between the charged groups. Experimental estimates of $\varepsilon_{\mathrm{eff}}$ are in the range of 20–50 (*47, 48*). Since the dielectric properties of the protein–solvent system are complex, there is no simple relationship between $\varepsilon_{\mathrm{eff}}$ and macroscopically measured dielectric constants (*49, 50*). Therefore, no detailed microscopic significance can be attached to $\varepsilon_{\mathrm{eff}}$. Equation (11) is best considered as a parametrization of electrostatic interactions; more detailed approaches are needed to provide an explicit structural treatment of electrostatic interactions in proteins.

In terms of this simple model, there are three electrostatic mechanisms based on variations in z, r, and $\varepsilon_{\mathrm{eff}}$ by which proteins may influence E°:

(1) Charge (z) effects: All other factors being equivalent (admittedly an important qualification), increasing the number of positively charged groups on a protein will lead to increasing stabilization of the more negatively charged re-

duced form of the redox group, thus resulting in higher values of E°. A simple estimate of the change in E° associated with a change in charge of a protein has been given (*51*). If the protein is modeled as a sphere of radius R (in Å), with net charge Q uniformly distributed over the protein surface, and immersed in a medium of dielectric constant ε, then the change ΔE° in the redox potential associated with a change in charge ΔQ is given by

$$\Delta E^\circ = \frac{14435}{\varepsilon R}\Delta Q \qquad \text{mV} \tag{12}$$

With $\varepsilon = 80$ and $R = 15$ Å (typical of small proteins), a change in ΔQ of $+1$ will result in a potential shift ΔE° of 12 mV. Experimentally, the ΔE° value associated with a $+1$ change in charge has been determined to be about 15–20 mV from chemical modification studies that neutralize specific lysine residues on the surface of cytochrome *c* (*52*). Given the simplistic nature of the electrostatic model used to estimate ΔE°, the agreement between the two values is surprisingly good. Similar results have been obtained from numerical analyses of more detailed electrostatic models (*53–55*). The relative insensitivity of these energetic estimates to the precise electrostatic model is a consequence of the high value for $\varepsilon_{\mathrm{eff}}$ (owing to the surrounding aqueous solution) which is associated with electrostatic interactions involving surface charges (*47, 48*).

A survey of small, soluble electron transfer proteins indicates a tendency for low-potential proteins to be negatively charged, whereas high-potential proteins tend to be positively charged (*51*). From the above considerations, this suggests that electrostatic modulation of E° by charged residues has been utilized by electron transfer proteins. It should be emphasized, however, that there are certainly exceptions to this trend (*56*), which is to be expected given the range of factors that influence E°.

(2) Distance (r) effects: Protein charges positioned close to redox groups should have a more significant impact on E° than more distant charges. An illustration of this behavior is found in the globin family. Replacement of a buried valine near the heme iron in myoglobin by an aspartate residue decreases E° by 200 mV (*21*). This effect is about 10 times greater than the change in E° observed for the modification of single surface charges.

(3) Dielectric ($\varepsilon_{\mathrm{eff}}$) effects: According to simple electrostatic models, the transfer of a charged group from a high to low dielectric medium is energetically unfavorable. This process becomes increasingly unfavorable as the absolute magnitude of the charge on the redox group increases. Since the interior of a protein is relatively apolar compared to water, Kassner (*57, 58*) proposed that these effects could significantly alter the E° of a protein-bound cofactor relative to the E° value of the free cofactor in water. Because the oxidized and reduced forms of the cofactors have different charges, immersion of the two states in an apolar region will have different energetic consequences, which will show up as

a shift in reduction potential. Qualitatively, the direction of the shift in E° will reflect the greater destabilization of the more highly charged form of the cofactor. For heme proteins, the formal charge on the porphyrin changes from $+1$ to 0 on reduction. Consequently, the oxidized form will be electrostatically destabilized in the protein relative to the reduced form, which is manifested by an increased midpoint potential of the porphyrin in the protein relative to water [again, with all other factors (ligands, etc.) remaining the same]. The converse statement applies to proteins where the absolute charge of the redox group increases on reduction, such as ferredoxins. In model studies, Kassner (*57*) demonstrated that E° shifts of at least 300 mV could be generated by changing the polarity of the cofactor environment.

The polarity of the cofactor environment may be influenced not only by the general hydrophobicity of nearby amino acid side chains, but also by the interaction of the redox cofactor with dipoles from peptide bonds and amino acid side chains (*59*), as well as with the solvent (*59, 60*). Despite the complexity of these interactions, promising methods have been developed (*59*) for the theoretical modeling of these dielectric contributions to E°.

iii. Solvent effects. The properties of proteins, including reduction potential, are intimately influenced by interactions with the surrounding solvent. Changes in the temperature, pH, and ion composition of the solution environment may significantly alter E°.

d. *Temperature.* In general, E° values are a function of temperature, as seen from the standard thermodynamic relationship

$$\frac{d\Delta E^\circ}{dT} = \frac{\Delta S^\circ}{F} \tag{13}$$

where ΔS° is the standard entropy change for the redox reaction. For cytochrome *c*, ΔS° is approximately -40 cal/K (*61*), which means that E° decreases by about 1.5 mV/K. Additional changes in E° can result from the temperature dependence of the protein structure. Changes in protein structure associated with temperature changes can also alter E° through the types of effects discussed previously. Besides temperature, variations in other intensive thermodynamic parameters (such as pressure) may also lead to changes in E°, although these have been less well studied than temperature.

e. *pH Effects.* Reduction potentials are also functions of the pH of the solution, since (1) the conformation of many proteins is pH dependent and (2) the ionization state of amino acid side chains varies with pH. Alterations in protein structure accompanying changes in pH can influence E° through the mechanisms previously described. The decreased E° observed in many *c* cytochromes at high pH (Fig. 1) has been attributed to a conformational change (the "alkaline transition") in which the methionine ligand is no longer coordinated to the heme iron (*13*).

The pH-dependent ionization of amino acid residues will alter the electrostatic environment at the redox cofactor. These changes will perturb E° in a fashion analogous to that associated with chemical or genetic changes to charged residues. Removal of protons will decrease the net charge of the protein, thus stabilizing the more positively charged oxidized form of the protein. Typically, E° will decrease with increasing pH (Fig. 1).

The pH-dependence of E° provides information on the ionization behavior of residues interacting with the redox center. A general treatment of the coupling between E° and pH has been presented by Clark (*62*). The main features of this interaction may be appreciated from a simple scheme in which only a protonated form of the oxidized protein can be reduced:

$$\begin{array}{ccc} A_oH & \overset{+e^-}{\rightleftharpoons} & A_rH; \qquad E^\circ \\ {\scriptstyle +H^+} \upharpoonleft\downharpoonright {\scriptstyle K} & & \\ A_o & & \end{array} \tag{14}$$

This mechanism could represent, for example, the interaction of a protonated carboxyl group with a redox center. On deprotonation, the resulting carboxylate group might stabilize the oxidized form of the protein so strongly that it becomes very difficult to reduce. The apparent midpoint potential for this scheme, E°_{app}, may be determined from the potential at which the total concentration of oxidized species ($[A_o] + [A_oH]$) equals $[A_rH]$. The pH dependence of E°_{app} is given by

$$E^\circ_{app} = E^\circ + \frac{RT}{nF} \ln\left(\frac{[H^+]}{[H^+] + K}\right) \tag{15}$$

Consequently, E°_{app} will decrease with increasing pH. The pK of the ionizing group can be determined from the pH dependence of E°_{app}. The pK values of E°_{app} for the bovine and *Rb. capsulatus c* cytochromes illustrated in Fig. 1 are 7.8 and 9.4, respectively (*11*, *12*). It should be emphasized, however, that the situation for real proteins may be more complex than represented in Eq. (14), which can complicate the interpretation of experimental pH–E°_{app} profiles.

Not only does the change in ionization state of a residue influence the electrochemical behavior of the redox group, but the converse situation also applies. The change in oxidation state of the redox group will perturb the electrostatic environment at the ionizing group, thus altering the titration behavior. For these electrostatic reasons, it is usually easier to remove a proton from a basic group in the oxidized (more positively charged) form. Consequently, the pK of a group in the oxidized protein is generally lower than the pK of the same group in the reduced protein. The magnitude of the difference in pK values between oxidation states reflects the extent of interaction between the ionizing and redox groups. Theoretical estimates of the pK shift of a titratable group accompanying oxidation state change in an electron transfer protein are in reasonable agreement with experimental observations (*63*).

f. *Ion Effects.* The presence of ions in the solution surrounding the protein may have important effects on the electrochemical properties of the associated redox groups. Specific binding of ions to the protein may occur. If the binding affinity for ions of the protein varies between oxidation states (oxidation-state linked ion binding), ion binding results in a stabilization of the oxidation state favoring ion binding. The measured $E°$ value will shift to reflect this event. An interesting example of the influence of ion binding on $E°$ has been described for cytochrome *c*. As might be expected, anions such as chloride and phosphate bind more tightly to the oxidized (more positively charged) form of horse cytochrome *c* (*64, 65*). As a result, the $E°$ values measured in the presence of these anions are lower than those observed in the presence of a nonbinding ion system, such as Tris–cacodylate. Surprisingly, the opposite behavior is observed for bovine and tuna *c* cytochromes, where the reduced form binds chloride and phosphate more tightly than the oxidized form (*65*). The difference in ion binding behavior between these different forms of cytochrome *c*, in spite of the overall high degree of sequence similarity, illustrates the exquisite sensitivity of ion binding to the details of the protein structure.

In addition to specific ion binding effects, midpoint potentials are sensitive to more general, ionic strength-dependent Debye–Hückel type effects (*61, 64*). On the basis of qualitative considerations, the oxidation state of the protein with the greatest absolute charge is electrostatically less favored than the other oxidation state. In the presence of ions in the solution, however, a complementary atmosphere of counterions exist around the protein, which reduces the difference in electrostatic energy of the two oxidation states. Consequently, $E°$ values will shift in the direction of the more highly charged form of the redox protein with increasing ionic strength. For proteins with a net positive charge, $E°$ should decrease with increasing ionic strength, whereas the opposite behavior is expected for negatively charged proteins. An illustration of this behavior is provided in Fig. 2 for cytochrome *c*-552 (*14*). Since the isoelectric point of cytochrome *c*-552 is about 5, the net charges on the protein should be positive, nearly zero, and negative at pH 3.0, 4.8, and 7.0, respectively. The slope of the $E°$ dependence on ionic strength at these different pH values is consistent with this qualitative analysis.

A simple Debye–Hückel model (*15*) may be employed to provide a first approximation to the dependence of $E°$ on ionic strength. From Eq. (6), the measured potential E is related to the midpoint potential and the activities of the oxidized and reduced species:

$$E = E^\circ - \frac{RT}{nF}\ln\left(\frac{a_r}{a_o}\right)$$

$$E = E^\circ - \frac{RT}{nF}\ln\left(\frac{\gamma_r}{\gamma_o}\right) - \frac{RT}{nF}\ln\left(\frac{c_r}{c_o}\right) \quad (16)$$

$$E = E^\circ_{app} - \frac{RT}{nF}\ln\left(\frac{c_r}{c_o}\right)$$

E°_{app} is the apparent midpoint potential obtained by measuring the concentration ratio of reduced to oxidized forms as a function of E.

In the simplest Debye–Hückel model, the activity coefficient γ_i is given by

$$\ln \gamma_i = -\frac{z_i^2 e^2}{2\varepsilon k_B T}\kappa \quad (17)$$

where

$$\kappa^2 = \frac{8\pi N_A e^2}{1000\varepsilon k_B T} I \quad (18)$$

contains the ionic strength (I) dependence of γ_i. In these expressions, z_i is the charge on the ith species, e is the elementary charge, N_A is Avogadro's number, and k_B is Boltzmann's constant.

If the oxidized and reduced forms have charges z and $z - 1$, respectively (assuming $n = 1$ and neglecting oxidation state linked ion binding, etc.), then

$$E^\circ_{app} = E^\circ - \frac{RT}{nF}\frac{(2z - 1)e^2\kappa}{2\varepsilon k_B T}$$
$$E^\circ_{app} = E^\circ - \alpha(2z - 1)I^{1/2} \quad (19)$$

and the variation in E° with $I^{1/2}$ is proportional to $-(2z - 1)$, in agreement with the qualitative behavior just described. In aqueous solutions at 298 K, $\alpha \cong 30$ mV, with I in moles/liter.

More detailed models for the ionic strength dependence of E° have been developed that include higher order terms in the Debye–Hückel expression for γ and the actual positions of the charges in the proteins (*66, 67*). Considering the complexity of proteins and the relative simplicity of the models, remarkable agreement has been achieved with the observed ionic strength-dependent effects.

C. Manipulation of E° Values

A summary of the general consequences of various modifications on the E° values of electron transfer proteins is presented in Table II. At present, quanti-

TABLE II

MODIFICATION TO REDUCTION POTENTIALS OF ELECTRON TRANSFER PROTEINS

Modification	Comments
Cofactor replacement	E° shift depends on intrinsic E° of redox group and protein–cofactor interactions
Ligand replacement	Replacement with anionic or more basic ligands will decrease E°; shift may exceed 150 mV
Hydrogen bonding	Hydrogen bond donor to redox group will increase E°; hydrogen bond acceptor to redox group will decrease E°
Electrostatic interactions	
Charge modification	Increasing positive charge on protein will increase E° by approximately 10–20 mV/charge for surface charges; larger effects may be observed for buried charges
Polarity	More polar environment favors more highly charged oxidation state
Ionic strength	Increasing ionic strength favors more highly charged oxidation state
pH	Increasing pH typically decreases E°

tative, *a priori*, estimates of the midpoint potential shifts associated with various modifications are not yet feasible, with the possible exception of some electrostatic effects. The sensitivity of E° values to subtle structural details is illustrated by the observation that reconstitution of apocytochrome b_5 with hemin yields two species that differ in E° by 27 mV (*68*). Advances in the understanding of protein energetics, development of new computational algorithms for evaluating these energies, and the availability of powerful methodologies for synthesizing molecules, however, provide grounds for optimism that rational design of redox proteins with specified electrochemical properties will be achieved in the near future.

Three general approaches may be used to alter the redox properties of a protein: (1) modification of existing electron transfer proteins, (2) introduction of a new redox center into an existing protein, and (3) design of novel redox proteins. All approaches have been implemented at some level and are briefly discussed in this section.

1. *Modification of Existing Electron Transfer Proteins*

Approaches to modification of either existing redox groups and/or the surrounding protein environment may be divided into categories equivalent to those utilized in discussing protein influences on midpoint potentials: (a) Replacement of cofactors with modified groups having different intrinsic E° values. Representative examples of cofactor replacement have been reported for proteins containing porphyrins (*69, 70*), flavins (*71, 72*), and quinones (*73, 74*). (b) Replacement of protein ligands by groups that have different affinities for reduced and oxidized forms of the redox center. Replacement of the water ligand to the heme iron in myoglobin by a glutamic acid residue lowers $E°$ by 200 mV (*21*). Substitutions of different amino acids for the histidine ligand to the heme iron have been reported for cytochrome b_5 (*75*) and cytochrome *c* (*76*). (c) Modification of the general protein environment to differentially alter the stabilities of the different redox states. This could be accomplished by replacing or introducing charged amino acids (by chemical modification or genetic methods) or by changing the polarity of the cofactor environment (perhaps by modification of the solvent exposure). (d) Modification of the solution environment (by changing pH, ionic strength, buffer components, or solvent) to alter the stabilities of the different redox states.

2. *Introduction of New Redox Centers into Existing Proteins*

It is possible to introduce new redox groups into proteins using chemical and/or biosynthetic methods. Attachment of ruthenium to specific histidine residues on various electron transfer proteins has provided a method for the characterization of intramolecular electron transfer processes in proteins (*77, 78*). Studies of intramolecular electron transfer in ruthenated proteins are described in more detail in Section IV.

Genetic methods provide a powerful approach to the biosynthetic introduction of redox groups [including cysteine residues as well as "unnatural" redox active amino acids (*79*)] into proteins. As an example, a disulfide bridge inserted across the active site of T4 lysozyme has been used to create a redox mechanism for regulating enzyme activity (*80*). Oxidation of the cysteines to form the disulfide closes the active site region, whereas reduction exposes the active site and restores catalytic activity.

The combination of chemical and genetic methods has great potential for introducing redox groups into proteins. Genetic engineering of histidine residues into different positions in yeast cytochrome *c* enables attachment of ruthenium to specific sites on the protein (*81*). Related studies have been used to introduce metal binding sites into proteins, which may be useful, for example, in preparing heavy atom derivatives for crystallographic projects (*82–84*). Semisynthetic redox proteins containing both biologically and chemically synthesized protein components have also been created (*85*).

3. *Design of Novel Electron Transfer Proteins*

Recent developments in molecular biology and protein synthesis methods have created two important approaches for obtaining new proteins with specific properties. In the first approach, antibodies with specifically designed ligand binding properties may be created by raising monoclonal antibodies to selectively designed antigens. If the antigens are analogs of the transition state of a reaction, then antibodies with specific enzymic properties may be obtained. The realization of "catalytic antibodies" has opened the door to an era of designed proteins having specific predetermined properties (*86, 87*). The relevance of this approach for creating tailor-made electron transfer proteins has been emphasized in a recent report describing the production of metalloporphyrin-binding antibodies (*88*). Extension to the creation of new electron transfer antibodies based on either porphyrins or other redox groups may be readily envisioned.

The second approach embodies perhaps the most exciting frontier in present-day protein chemistry, namely, the possibility of designing new proteins of defined structure. Novel proteins based on α-helical bundles have already been synthesized and characterized (*89–91*). This structural motif is of special interest since the cofactors of several classes of electron transfer proteins are associated with helical bundles. Although designed electron-transfer proteins apparently have yet to be created, the extension of this approach to the development of novel redox proteins seems inevitable. The successful construction of designed proteins with prespecified electrochemical properties will provide the ultimate test of our understanding of the principles of protein folding and redox control.

II. Kinetics of Electron Transfer Reactions

Although thermodynamic considerations are fundamental in establishing the direction and equilibrium position of an electron transfer reaction, they provide no direct insights into the kinetics or mechanism of the reaction. The kinetics of electron transfer reactions are of central importance for biological function. A key aspect of biological electron transfer is that not only must electrons be transferred along the correct sequence of carriers, but the transfer must be specific, implying that the redox proteins discriminate between a variety of other thermodynamically suitable, although biologically incorrect, donors and acceptors. In essence, electron transfer proteins are designed to provide low kinetic barriers for transfer to correct acceptors, while simultaneously maintaining high kinetic barriers against transfer to incorrect acceptors. General structural considerations that contribute to this specificity in electron transfer reactions, thereby avoiding intolerable "short circuits," are discussed in this section.

A. General Kinetic Considerations

For an electron transfer reaction of the type represented by Eq. (2),

$$A_o + B_r \rightarrow A_r + B_o$$

the overall kinetics of the reaction will depend not only on the rate of the actual electron transfer step, but also on the association and dissociation kinetics of the reactants and products. A simple reaction mechanism including these steps would have the form (*92*)

$$A_o + B_r \underset{k_{-1}}{\overset{k_1}{\rightleftharpoons}} A_oB_r \underset{k_{-2}}{\overset{k_2}{\rightleftharpoons}} A_rB_o \underset{k_{-3}}{\overset{k_3}{\rightleftharpoons}} A_r + B_o \tag{20}$$

The three basic steps in this mechanism are (1) reactant association, (2) electron transfer, and (3) product dissociation. Equilibrium constants associated with these three steps may be defined $K_1 = k_1/k_{-1}$, $K_2 = k_2/k_{-2}$, and $K_3 = k_3/k_{-3}$.

The general kinetic expression for this reaction mechanism is quite involved, but, clearly, the overall velocity will depend on the detailed kinetics of all three steps. If the back electron transfer rate is neglected (i.e., $k_{-3}[A_r][B_o] \cong 0$), and the steady-state approximation is applied to the two central complexes, then the reaction velocity, v, may be derived:

$$\begin{aligned} v &= \frac{d[A_r]}{dt} \\ &= k_{obs}\,[A_o][B_r] \\ &= \frac{k_1k_2k_3}{(k_{-1}k_{-2} + k_{-1}k_3 + k_2k_3)}\,[A_o][B_r] \end{aligned} \tag{21}$$

Various limiting forms for v may apply under special conditions. If the electron transfer steps are rate limiting (k_2 and k_{-2} small), then

$$v = k_2K_1[A_o][B_r] \qquad \text{(electron transfer step rate limiting)} \tag{22}$$

If the substrate association step is rate limiting, then

$$v = k_1[A_o][B_r] \qquad \text{(substrate association rate limiting)} \tag{23}$$

If the first two steps are at equilibrium, and Step 3 is rate limiting, then

$$v = k_3K_2K_1[A_o][B_r] \qquad \text{(product dissociation rate limiting)} \tag{24}$$

The observed rate constant, k_{obs}, never equals the rate constant k_2 for forward electron transfer in these simple models. The presence of multiple steps in the electron transfer mechanism [keeping in mind that Eq. (20) represents a minimal scheme for an electron transfer reaction] emphasizes the difficulties in extracting k_2 values from measurements of k_{obs} under steady-state conditions. Rapid kinetic studies provide a more powerful approach for separating the actual kinetics of electron transfer from the association and dissociation steps, but the analysis may still be complex. Owing to difficulties associated with bimolecular kinetics, many recent studies of electron transfer have emphasized unimolecular processes. Physiologically, however, the bimolecular processes can be of considerable importance for the overall electron transfer kinetics.

Mechanism (20) provides a general framework for characterizing the origin of donor and acceptor specificity in biological electron transfer reactions. In sequential order, discrimination between transfer to correct and incorrect electron acceptors may occur at Steps 1, 2, and/or 3.

Step 1: Reactant association. The surfaces of electron transfer proteins have extensive regions available for contact between donor and acceptor proteins. Even in proteins where the redox group has some solvent exposure, only a small fraction of the total surface area of the protein consists of these groups. In heme proteins surveyed by Stellwagen (*60*), for example, the porphyrin contributes less than 3% of the total protein surface area. Consequently, protein–protein contacts will dominate the energetics of complex formation. In an analogous fashion to enzyme–substrate interactions, complementary structural interactions between interacting surfaces provide a mechanism for discriminating between correct and incorrect redox partners. Complementary interactions between the correct redox partners will stabilize the complex, increasing the concentration of the species A_oB_r actually competent for electron transfer.

Step 2: Electron transfer. Complementary interactions between the correct donor and acceptor proteins in the A_oB_r complex should orient the two redox groups in a favorable fashion for electron transfer, thus increasing k_2. Less favorable orientations between incorrect donor and acceptor proteins in the complex will presumably have decreased k_2 values.

Step 3: Product dissociation. The two proteins must dissociate after electron transfer if they are to become available for other reactions, and, hence, the complex should dissociate rapidly. Too tight association of the product species will decrease the overall rate of the reaction.

The most favorable mechanism for achieving specificity in electron transfer reactions would appear to operate at the reactant association stage. Discrimination or proofreading for incorrect redox partners at Steps 2 and 3 may also occur, but this suffers the disadvantage that at least some fraction of the donor and acceptor will be tied up in nonproductive complexes, thus decreasing the concentration of proteins available for productive transfer.

B. Structure of Electron Transfer Complex

The structural details of the donor–acceptor complexes are of great interest given the central role these species play in the electron transfer mechanism. In the absence of specific structural information from crystallographic or NMR studies, a variety of chemical, genetic, and computer graphics approaches have been used to develop possible models. The initial impetus behind this work came from observations that the kinetics of many electron transfer reactions [in particular those involving cytochrome *c* (*93*)] depend strongly on ionic strength. The effects of ionic strength on rate appeared too large to reflect changes in the rate constant for the actual electron transfer step, k_2. Consequently, ionic strength effects were attributed to changes in association constants between donor and acceptor proteins. The decrease in electron transfer rates with increasing ionic strength indicated that the complex was stabilized by complementary electrostatic interactions (salt bridges).

In a pioneering study, Salemme (*94*) used computer modeling methods to "dock" together cytochrome *c* and cytochrome b_5, whose structures had been independently determined by crystallographic methods. Positively charged residues surround the exposed heme surface in cytochrome *c*, whereas the heme in cytochrome b_5 is ringed by acidic groups. Salemme was able to find a set of complementary salt bridges that could stabilize the complex, with the two heme groups held in a nearly coplanar fashion. The distance of closest approach between cofactor atoms in this model was 8 Å. Subsequent studies (*95–97*) have supported both the general features as well as many of the specific details of this structural model for the cytochrome *c*–cytochrome b_5 complex.

Following the initial work of Salemme, models for other electron transfer complexes have been proposed for cytochrome *c*–cytochrome-*c* peroxidase (*98*), cytochrome *c*–flavodoxin (*99*), cytochrome b_5–hemoglobin (*100*), cytochrome b_5–myoglobin (*101*), cytochrome *c*–photosynthetic reaction center (*102, 103*), and ferredoxin–cytochrome c_3 (*104*). The same general approach of identifying potential salt bridges that could stabilize a complex with approximately coplanar cofactor rings has been utilized in many of these studies.

Are these models accurate representations of electron transfer complexes? Certainly, the general features of the models (identification of the interaction regions, utilization of ionic interactions to stabilize the complex) as well as many of the specific features (involvement of particular residues in the interaction region) appear to be correct. The usefulness of these hypothetical models in the design and interpretation of experiments has been significant. It now appears unlikely, however, that only a single, unique structure of the donor–acceptor complex exists in every case. Rather, the interaction between two electron transfer proteins may generate a collection of molecular species. A dramatic illustration of this was described in the crystallographic analysis of the complex between cytochrome *c* and cytochrome-*c* peroxidase (*105*). Although both proteins were present in the crystal, only cytochrome-*c* peroxidase was sufficiently well ordered to be observed in the electron density map. The cytochrome *c* was apparently orientationally disordered, and it could not be located in this study. This suggests that cytochrome *c* and cytochrome-*c* peroxidase can interact in a variety of orientations, rather than forming a single, well-defined molecular complex. Evidence supporting a diversity of structures for donor–acceptor complexes has also been obtained from solution studies (*106, 107*), studies of the effects of replacing amino acid residues in the interaction region (*108*), and computational simulations (*109, 110*). Decreased electron transfer rates in cases where the complex may be trapped in a single conformation [such as low ionic strength (*106*) or by chemical cross-linking (*107*)] suggest that dynamic interconversion between different structures may be essential for rapid electron transfer. Computational studies (*109, 110*) indicate that the energy requirements needed to interconvert between the different forms are quite modest (on the order of several kcal/mol).

C. $E°$ Values in Electron Transfer Complex

The environment of the redox cofactor will change on association of the acceptor and donor proteins in the electron transfer complex. Conformational changes in the two proteins may be associated with complex formation, and the proximity of the second protein creates differences from the solution environment of the isolated molecule. Since $E°$ depends significantly on the environment, the redox potential difference in the complex, $\Delta E = (RT/nF) \ln K_2$, may vary from the potential difference, $\Delta E° = (RT/nF) \ln K_{eq}$, determined from the electrochemical properties of the isolated proteins. This effect is analogous to the difference in equilibrium constants between substrates and products free in solution, and when bound to the enzyme (*111*). Changes in $E°$ associated with complex formation have been reported in several systems: adrenodoxin–adrenodoxin reductase (*112*), cytochrome *c*–cytochrome-*c* oxidase (*113*), cytochrome *c*–cytochrome-*c* peroxidase (*114*), cytochrome *c*–plastocyanin (*107, 115*), and cytochrome *c*–flavodoxin (*116*). The changes in $\Delta E°$ on complex formation in these examples range between 20 and 100 mV. Many of these complexes are believed to be stabilized by complementary salt bridges between the interacting proteins. The direction of the $\Delta E°$ shift may often be rationalized on electrostatic grounds. The $E°$ of a redox protein associating with a positively charged protein will increase, while the $E°$ of a redox protein associating with a negatively charged protein will often decrease. A recent example of this behavior has been reported for the interaction between the positively charged cytochrome *c* and the negatively charged plastocyanin (*107*). In the complex between these two proteins, the $E°$ values of cytochrome *c* and plastocyanin change by -11 and $+25$ mV, respectively, for an overall shift of $+36$ mV.

Shifts in $\Delta E°$ on complex formation reflect a difference between the internal equilibrium constant K_2 and the overall equilibrium constant K_{eq}. Attention has focused on the relationship between K_2 and K_{eq} in connection with the optimal free energy profile for an enzyme-catalyzed reaction (*117, 118*). In a sequential reaction mechanism, the greatest flux of material occurs along pathways lacking both large kinetic barriers and highly stable intermediates (*119*). In terms of a free energy diagram, this situation requires that no significant energetic valleys or hills exist between the initial and final states, that is, the free energies of all intermediates should decrease smoothly between reactant and product species (Fig. 3).

Albery and Knowles (*117, 118*) suggested for the most efficient enzymes that "the kinetically significant transition state will lie between kinetically significant intermediates of equal free energy." In terms of Mechanism (20), this hypothesis suggests that $K_2 = 1$ for an evolutionarily optimal electron transfer reaction, independent of the value of K_{eq}. Later theoretical studies of this problem (*120, 121*) have modified this conclusion to the more accurate statement that K_2 is bounded by $1 < K_2 < K_{eq}$ in an optimal reaction [i.e., the free energy profile for the reaction resembles a "descending staircase" (*120*)]. Experimental studies on a variety of systems are generally consistent with the concept that internal equilibrium constants are closer to unity than are the overall equilibrium constants (*120*).

This behavior has important implications for the free energy profiles of the most rapid electron transfer reactions (*51*). The fastest rates should occur when the free energy change for electron transfer in the complex approaches zero. Equivalently, the reduction potential difference between donor and acceptor

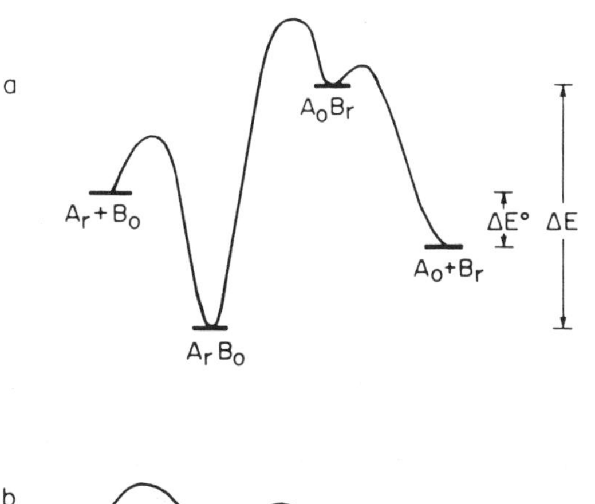

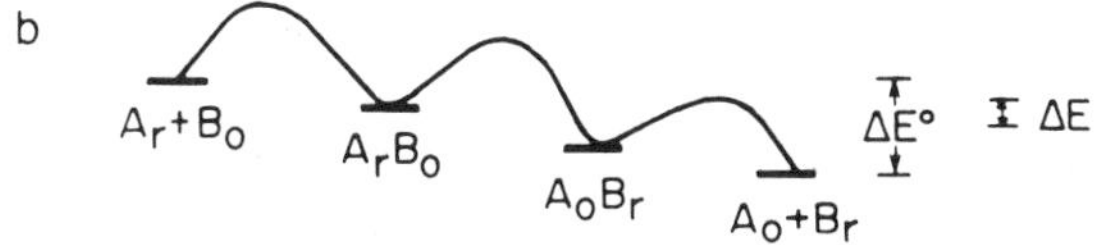

FIG. 3. Free energy profile for electron transfer reactions. (a) Inefficient profile, with large activation energies and stable intermediates along the pathway. (b) Efficient profile, with low activation barriers separating intermediates. Values of $\Delta E°$ and ΔE, the reduction potential difference between free and bound A and B species, respectively, are indicated.

should approach zero in the complex. Typically, this could involve some combination of increasing E° for the donor and decreasing E° for the acceptor. Since E° values are influenced by a variety of factors, many mechanisms could be utilized to minimize the potential difference between redox groups in the complex.

One proposed mechanism for shifting the reduction potentials involves changing the electrostatic environment of the redox groups in the complex (*51*). As previously discussed, many electron transfer complexes are believed to be at least partially stabilized by salt bridges. If the region around the acceptor cofactor is more positively charged than the complementary region around the donor group, electrostatic effects will shift the donor and acceptor potentials in the required direction. The positively charged acceptor will increase the E° of the donor, while the more negatively charged donor will decrease the E° of the acceptor, with the net result that ΔE° in the complex will be closer to zero than for the isolated protein. The observation that high-potential proteins tend to be more positively charged than low-potential proteins (*51*) is consistent with these proposals, although they certainly are not sufficient to establish them. Detailed kinetic analyses of electron transfer between physiological redox partners and nonphysiological redox partners will be necessary to characterize with confidence possible differences in free energy profiles of these systems. It should be emphasized, however, that many physiological electron transfer reactions may be "fast enough" (relative to the rates of other metabolic reactions) so that there is no evolutionary pressure to maximize the rates of these reactions (*2*).

D. Kinetic Complications: Gating

If multiple forms of the donor–acceptor complex exist in rapid equilibrium, then the apparent kinetics of the overall electron transfer reaction are essentially decoupled from the kinetics of interconversion. However, significant effects on the overall kinetics may be observed if the species are interconverted at rates comparable or slower than electron transfer (*122, 123*). In particular, the interconversion mechanism may provide a "gate" which sequentially enhances and depresses the rate of electron transfer. The kinetic consequences of conformational changes have long been appreciated in biochemical systems (allosteric enzymes and proline cis–trans isomerization in protein folding are two of many possible examples). Therefore, it is not surprising that interconversion of conformational states (conformational gating) could also be significant for some electron transfer reactions.

A specific example serves to illustrate the general properties of gated processes. Four-coordinate copper prefers different ligand geometries in the two oxidation states: Cu^{2+} generally exhibits a square planar coordination, while Cu^{+} ligands are generally tetrahedral. If interconversion between these conformations is slower than electron transfer, then gating effects on the kinetics of electron transfer may be observed. Following Bernardo *et al.* (*124*), a general electron transfer scheme may be defined:

$$\begin{array}{ccccccc} \mathrm{O} & \mathrm{Cu^{2+}L} & \leftrightarrow & \mathrm{Cu^{+}L} & \mathrm{P}; & E_{\mathrm{A}}^{\circ} \\ & \upharpoonleft\!\downharpoonright & & \upharpoonleft\!\downharpoonright & & \\ \mathrm{Q} & \mathrm{Cu^{2+}L} & \leftrightarrow & \mathrm{Cu^{+}L} & \mathrm{R}; & E_{\mathrm{B}}^{\circ} \end{array} \tag{25}$$

where L represents the metal ligands, and O, P, Q, and R denote various species. For simplicity, only intramolecular processes are represented, so that if this were a protein–protein electron transfer reaction, the association and dissociation steps would be neglected. O and P have the oxidized state geometry, while Q and R adopt the reduced state geometry. Therefore, O and R represent the stable conformers, while Q and P represent metastable conformers that have similar geometries to the other oxidiation state. The following equilibrium constants may be defined:

$$\begin{aligned} K_{\mathrm{OQ}} &= [\mathrm{Q}]/[\mathrm{O}] \\ K_{\mathrm{PR}} &= [\mathrm{R}]/[\mathrm{P}] \end{aligned} \tag{26}$$

so that K_{OQ} and K_{PR} reflect the ability of each oxidation state to adopt the reduced conformation. E_{A}° and E_{B}° correspond to the midpoint potentials for the reduction of O to P and Q to R, respectively.

Depending on the values of the rate constants, different electrochemical behavior may be observed under different experimental conditions. If $K_{\mathrm{OQ}} \cong 0$ (state O preferred), and states P and R interconvert very slowly, then reduction of O will occur with an apparent midpoint potential E_{A}°. If, however, the reverse experiment is performed (oxidation of R when K_{PR} is very large, and O and Q interconvert very slowly), an apparent midpoint potential of E_{B}° will be measured. Under conditions of rapid equilibrium between all states, the measured midpoint potential E° is given by

$$E^{\circ} = E_{\mathrm{A}}^{\circ} + \frac{RT}{nF}\ln\left(\frac{1 + K_{\mathrm{PR}}}{1 + K_{\mathrm{OQ}}}\right) \tag{27}$$

Significant differences in these potentials have been measured (*124*) in model copper compounds ($E_{\mathrm{A}}^{\circ} < 350$ mV, $E_{\mathrm{B}}^{\circ} = 630$ mV, and $E^{\circ} = 440$ mV).

Conformational gating processes may also lead to different mechanisms for oxidation and reduction reactions. Reduction of O to R can follow two alternate pathways: O → P → R and O → Q → R. Direct conversion of O → R is unlikely, as Hoffman and Ratner (*122*) have demonstrated that the activation energy for the concerted O to R reaction is higher than for the two sequential reactions. It is possible that the reduction of O to R and the oxidation of R to O could follow different pathways, depending on the rate constants for the various steps. Thus, the presence of multiple, slowly interconverted conformational states can influence both the kinetics and mechanism of electron transfer. Model studies demonstrate that gating occurs even in simple systems, so that the opportunities for conformational gating are especially abundant in biological electron transfer processes.

III. Microscopic Theory of Electron Transfer

The structural and electronic complexity of electron transfer proteins and reaction centers is an indicator of the high degree of control which is exercised in nature over this fundamentally important process. In many ways electron transfer in proteins is akin to the operation of a semiconductor device; the job of the protein is to transfer charge in one direction efficiently and with a high level of specificity. Artificial semiconductors used, for example, in solid–liquid interface solar cells are constructed to optimize efficiency, and, on the whole, are better at it than are electron transfer proteins. Because efficiency is often equated to a high throughput of electrons, there is some tendency to assume that electron transfer in living organisms is optimized simply to maximize the rate of electron transfer. In fact, part of the story of why electron transfer proteins exhibit the complexity they do is likely that their physical and electronic structures are fine tuned to provide a rate of electron transfer consistent with the need to maintain overall biological control of the numerous interrelated processes occurring in the living system. If the protein simply "went-for-broke," the danger of short circuits caused by insufficient specificity could be disastrous for the organism. The theoretical and experimental challenge is thus to paint a more global picture which explains how protein structure and function are related to each other and to the overall needs of the biological system.

The experiment which triggered much of the theoretical interest in electron transfer in biological systems was that of DeVault and Chance (*125*), who studied the temperature dependence of the cytochrome *c* (Cyt *c*) reduction of the primary donor $(\mathrm{Bchl})_2^+$ (bacteriochlorophyll) in the photosynthetic bacterium *Chromatium vinosum*. The reaction is

$$\mathrm{Cyt}\ c^{2+} + (\mathrm{Bchl})_2^+ \rightarrow \mathrm{Cyt}\ c^{3+} + (\mathrm{Bchl})_2 \tag{28}$$

and, it was observed to be essentially temperature independent in the range 4–100 K but dependent on temperature above about 130 K. The observation of temperature independence of a rate immediately suggests a tunneling mechanism rather than an activated process (note that lack of temperature dependence does not in general per force imply tunneling; ΔH might be zero, for example). Tunneling is a purely quantal phenomenon in which the wavefunction associated with some state of a particle penetrates into classically forbidden regions (e.g., potential barriers), and it ultimately can lead to physical passage of the particle through the forbidden region of space. The experiment on *C. vinosum* convinced chemists that long-range (tens of angstroms) transfer of electrons was not only possible but that, perhaps surprisingly, it seemed to occur via a classically forbidden process (forbidden does not necessarily mean slow). Assuming that the experiment of DeVault and Chance does imply some form of tunneling, it might seem that the *electron* is tunneling. In fact, the correct conclusion is that *nuclear* rather than electronic tunneling is occurring (of course, the electron is not thereby precluded from tunneling itself). In order to understand that it is necessary to delve a bit more deeply into some of the quantum mechanical aspects of tunneling and the theory of adiabatic and nonadiabatic transitions.

The most widespread and successful theory of electron transfer processes in nonbiological as well as biological systems, and which has fueled a tremendous amount of experimental and theoretical activity, is due to R. A. Marcus (*1, 2, 7–9, 126–128*). This theory was originally applied to electron transfer in solution and inorganic systems, where it continues to play an important role. Particularly during the 1980s, however, it was extensively applied to electron transfer in proteins. Although important extensions to Marcus theory have been made by Sutin and others (*1, 8*), for the purposes of this review the term Marcus theory will be taken to include all such modifications, and no special attempt will be made to trace the historical development of the theory. The general predictions of Marcus theory have been demonstrated in a number of biologically important cases (as well as in nonbiological settings), and the theory appears to provide an excellent picture of electron transfer in cases where the electron is most likely tunneling directly through space between D(onor) and A(cceptor) (*1, 2, 129–131*). Alternative mechanisms to that described by Marcus theory have also been proposed which emphasize electronic aspects of the problem (quantum electronic, or QE, theories) and which make some predictions similar to those of Marcus theory regarding, for example, the distance dependence of electron transfer (*1, 2, 7–9*). Based on experimental observations of electron transfer in model problems, it is hard to determine definitively which, if any, of the various theories provide the most appropriate descriptions of biological electron transfer. However, because the two classes of theory in many ways provide complementary rather than competing viewpoints, it is perhaps best that they be thought of as emphasizing different aspects of the same problem. It seems likely that elements of each will be necessary to describe how biological electron transfer occurs.

The rest of this section is organized as follows: a fairly straightforward account of modern Marcus theory is given in Section III,A which emphasizes the distinction between electron and nuclear tunneling. The superexchange mechanism and other QE theories are considered in Section III,B, and the analogy between them and certain solid state problems are considered. It is worth reemphasizing that, consistent with the overall objective of this chapter, this section is meant to be an introduction to the major theories of electron transfer which are currently in use: readers interested in more detailed aspects of the theory can do no better than to consult the reviews of Marcus and Sutin (*1*), DeVault (*7*), Guarr and McLendon (*9*), and McLendon (*2*).

A. Marcus Theory

Before considering Marcus theory itself, it is worth discussing some of the basic quantum mechanical and semiclassic ideas of tunneling. For the sake of concreteness a definite physical system will be considered, the H_2^+ molecular ion (*132*). At some fixed internuclear separation (i.e., the Born–Oppenheimer approximation has been made, which assumes that the fast moving electrons adjust instantaneously to changes in the positions of the nuclei, just like flies buzzing around an elephant) the potential experienced by the electron is shown in Fig. 4. Neglecting tunneling, each energy level supports two states localized in the left and right wells, Ψ_L and Ψ_R, respectively, and each energy level is thus 2-fold degenerate. The double well is symmetric, and consequently states with the electron localized at the left nucleus are equal in energy to those corresponding to the electron being localized at the right nucleus. Tunneling can be viewed as a perturbation of the Hamiltonian, and its inclusion leads to lifting of the degeneracy and a consequent splitting of the energy level into two energies (E^+ and E^-) symmetrically distributed about their center of gravity. The splitting between the states is mainly determined by the frequency of electron tunneling, while the "band gaps" (i.e., the spacings between the lowest and first

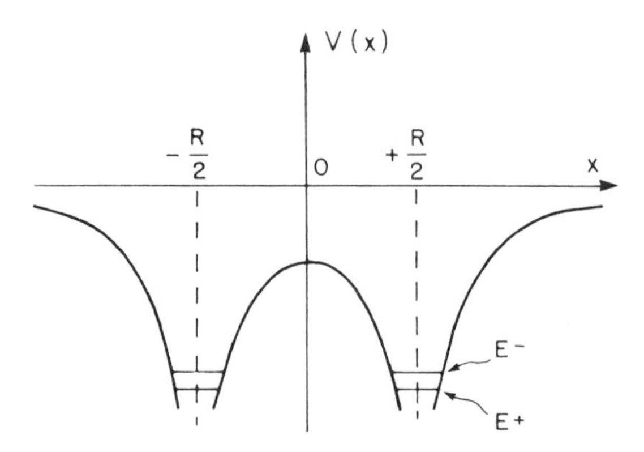

Fig. 4. The potential experienced by an electron in the hydrogen molecular ion at some fixed internuclear separation. In the absence of tunneling, the energy levels E^+ and E^- would collapse into a single doubly degenerate level, $E°$. Tunneling lifts the degeneracy, and the splitting of the levels is determined by the tunneling probability.

excited "bands," etc.) are largely determined by the overall shape of the potential. It is apparent that the propensity of the *electron* to tunnel is a strong function of the particular *nuclear* configuration the ion finds itself in; separation of the nuclei to infinity leads to an exponential drop in tunneling probability. Additionally, transitions between bands (important for some QE theories) are also a strong function of the nuclear configuration and dynamics. If the symmetry between the two wells were broken (e.g., HD^+), then the electron tunneling probability would be depressed substantially compared to H_2^+. This can be understood on noting that energy must be conserved during a tunneling event.

Semiclassically, the tunneling probability T is given by the expression (*133*)

$$T = \text{const.} \exp\left(-\frac{2\pi}{h} \int_a^b dq \sqrt{|2m[E - V(q)]|} \right) \tag{29}$$

where the (constant) energy E occurs explicitly, a and b are the classic turning points (where the momentum of the particle is zero), and q is the electronic coordinate. From a perturbation theory point of view, the energy of the electron must match up with one of the quantized energies of the well being tunneled into. If not, then the probability of tunneling is reduced dramatically (in actuality, the true quantum states are distributed over both wells and some tunneling always occurs). Optimum tunneling therefore occurs for symmetric wells. This simple picture lies at the root of Marcus theory and provides a mechanism by which changes in *nuclear* configuration are able to modify *electron* tunneling rates.

Consider the situation in which D and A are sufficiently well separated so that their mutual electronic interaction is small. The nuclear potential energy surfaces of reactants and products for the reaction

$$D + A \rightarrow D^+ + A^- \tag{30}$$

may then be represented schematically as in Fig. 5a. This picture merits some explanation; the number of vibrational (nuclear) degrees of freedom in a protein is enormous, and thus the potential energy surfaces are functions of many thousands of variables. In addition, solvent interactions can (and likely do) destroy the simple picture of overlapping parabolas shown in Fig. 5a. Nevertheless, the qualitative message of Fig. 5a remains valid: (1) the equilibrium nuclear configurations for R(eactants) and P(roducts) are different, and (2) reaction will occur only at nuclear configurations close to or at the point of intersection of the two curves. The second point is contingent on operation of the Franck–Condon principle, namely, that during the actual electron transfer process the instantaneous positions and momenta of the (relevant) nuclei are unchanged. Because the intersection does not correspond to equilibrium configurations of either R or P, concerted nuclear motion is necessary to allow the reactants to reach the intersection point. Note that at the point X there is a degeneracy which is lifted if

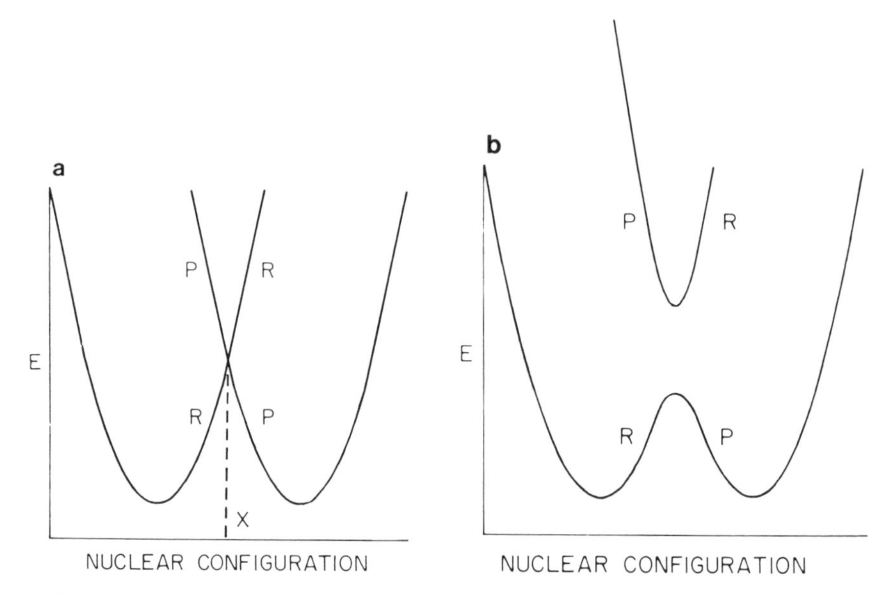

FIG. 5. (a) Schematic reaction coordinate diagram. R is the reactant curve and P corresponds to products. At X the potentials are degenerate or almost degenerate. This corresponds to the nonadiabatic case. In (b) the coupling is sufficiently large that the degeneracy at X is lifted substantially. This corresponds to an adiabatic reaction. Note that nuclear tunneling can occur through the barrier in the lower potential energy surface.

electronic interactions between reactants and products are accounted for, in an analogous way to the lifting of the degeneracy which occurs in H_2^+. The splitting of the states at the degeneracy point is $2H_{RP}$, where H_{RP} is the electronic matrix element describing the interaction. If the electronic coupling is large enough, then an "adiabatic" reaction occurs (Fig. 5b). Otherwise, reactive trajectories will pass through the crossing region with a smaller probability of reaction (nonadiabatic case) since they can jump the gap and stay on the R curve.

The curves in Fig. 5 correspond to nuclear motion and say nothing directly about how the electron itself transfers. As noted, transfer of the electron must conserve energy, and this is the key to nuclear control of electron tunneling. Figure 6, which is based on that appearing in the review by Marcus and Sutin (*1*), illustrates how changes in nuclear configuration can enhance or suppress electron tunneling; the electron can tunnel with appreciable probability only when energy is conserved. The semiclassic expression [Eq. (29)] suggests that the transmission probability for electron tunneling, $\kappa(r)$, decays exponentially with distance with the form

$$\kappa(r) \cong \exp(-\beta r) \tag{31}$$

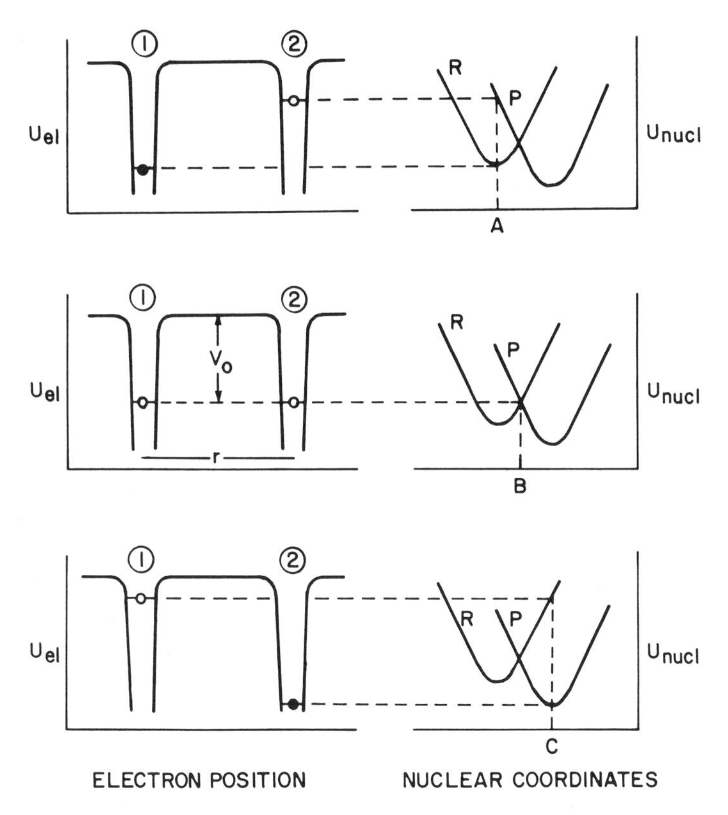

FIG. 6. Electronic potential energy curves (left) at three nuclear configurations. Effective electron tunneling can only occur in case at center when the energy of the electron in both wells is degenerate. [Adapted from Ref. (*1*).]

The overall rate constant, however, will be determined not so much by the rate of electron tunneling, but by the probability of the nuclei reaching the crossing point owing to thermal fluctuations. This rate constant has been obtained by Marcus and is given by (1)

$$k = \kappa A\sigma^2 \exp(-\Delta G^*/RT) \tag{32}$$

where $A\sigma^2$ has dimensions of collision frequency, κ has been averaged over r, and ΔG^* is the free energy of activation. The quantity κ represents the probability that a system in the transition state will pass to products. In a nonadiabatic reaction the system can pass through the transition state (the crossing point in Fig. 5a) many times without reacting and thus $\kappa < 1$, whereas in the adiabatic case $\kappa \cong 1$ (*127*).

An alternative way by which the nuclei can end up on the P surface is to tunnel through the barrier depicted in the lower surface in Fig. 5b. At high temperatures the contribution of nuclear tunneling to the reaction rate will be small, but it becomes relatively more important as the temperature is reduced. On the other hand, temperature does not directly affect the transmission probability of the electron through the barrier. Therefore, the temperature independence of the rates measured by DeVault and Chance (*125*) in *C. vinosum* at low temperatures is indicative of nuclear rather than electronic tunneling.

Inclusion of quantum effects on the nuclear dynamics can be accomplished by using Fermi's Golden Rule (*134*), which is really a manifestation of first-order time-dependent perturbation theory and conservation of energy during a transition. In this level of refinement, and formally allowing for the inclusion of solvent effects, the rate is given by

$$k = \frac{2\pi}{\hbar}|H_{\mathrm{RP}}|^2\mathrm{FCWD} \tag{33}$$

where H_{RP} is the electronic matrix element, and FCWD is the so-called Franck–Condon weighted density of states which describes the mixing of R (vibrational/solvational) states with corresponding P states, together with appropriate weighting by Boltzmann factors. The tunneling matrix element H_{RP} for a single particle tunneling through a barrier between symmetric wells is an exponential function of the distance between wells (i.e., between D and A):

$$H_{\mathrm{RP}} \cong \exp(-\beta r_{\mathrm{DA}}) \tag{34}$$

An important general prediction of Marcus theory is thus that transfer rates should also decrease approximately exponentially with distance. It is worth pointing out that H_{RP} itself depends slightly on temperature since temperature affects the nuclear configuration and thus the effective potential seen by the electron. Such higher order effects are neglected because the overall treatment is essentially first-order perturbation theory. Some QE mechanisms also predict exponential dependence with distance. Further discussion of this point is held over to the next section.

Another prediction of Marcus theory which has been succinctly described in the review by McLendon (*2*) is the phenomenon of "exothermic rate restriction." The free energy of activation in Eq. (32), ΔG^*, can be thought of as containing contributions from two terms: the free energy of activation, ΔG^0, and the reorganization energy, λ. According to Marcus (*1*), ΔG^* is given explicitly by

$$\Delta G^* = (\lambda + \Delta G^0)^2/4\lambda \tag{35}$$

and the reorganization energy may be thought of as the difference in energy between P at the equilibrium configuration of R and P at its own equilibrium geometry. The reorganization energy is commonly divided into two parts, one of which accounts for the inner shell of atoms whereas the other is concerned with solvent reorientation, but this distinction will not be made here. The main point is that the overall activation energy depends on the competition between λ and ΔG^0. If a set of reactions has the same value of λ, then the rate (at a given distance) should vary with ΔG^0 in such a way that it is a maximum when $\Delta G^0 = -\lambda$ and should decrease as the reaction becomes more exothermic. The quantity $-\Delta G^0$ is often referred to as the energy gap, and the effect as the energy gap law. The dependence of rate on ΔG^0 has been demonstrated in pulse radiolysis experiments by Miller *et al.* (*131*) in studies of nonbiological molecules.

An important aspect of some physiological electron transfer processes is the slowness of the back reaction in comparison to the forward reaction. The near perfect efficiency of *in vivo* photosynthesis is a direct consequence of the unidirectionality of the electron transfer process. In principle, all of the theoretical arguments just presented could be inverted to explain why the reverse rate might be expected to be of comparable magnitude to the forward reaction. However, the energy gap law leads one to the conclusion that the forward rate could dominate if the "inverted" region holds sway; assuming that for the forward reaction $\Delta G^0 = -\lambda$ and that λ is the same in both directions, then the forward rate should dominate since now $-\Delta G^0 > -\lambda$ for the reverse process. Another possible explanation attributes the difference between forward and back reactions to changes in H_{RP} because the electronic wavefunctions involved in the forward and reverse reactions are different. Several other explanations have been proposed [see the review by Guarr and McLendon (*9*)], although none is completely satisfactory. This is discussed further when the reaction center is considered in more detail in Section IV.

B. Quantum Electronic Theories of Electron Transfer

Marcus theory approaches the problem of electron transfer from a classic or semiclassic viewpoint in which vibrational nuclear motions are emphasized over electronic structure. It is important to recognize that the theory says little about the specifics of how electrons break through the barrier separating D and A. If the picture of a single electron tunneling through a barrier were correct then the matrix element H_{RP} would decay exponentially with distance, but this assumption is not central to Marcus theory itself. As pointed out by Marcus and Sutin (*1*) the term "electron tunneling" is probably not representative of the actual physical process by which the electron transfers, since it is likely that more than one electron is involved in charge separation. In principle, Eq. (33) could still be used to describe electron transfer provided a better (many electron) approach to the calculation of H_{RP} were employed.

At the outset of the discussion of QE theories, it is necessary to determine when quantum electronic effects are expected to be important. Recall that the splitting between the nuclear potential energy surfaces at the degeneracy point X in Fig. 5a depends on the size of the coupling matrix element H_{RP}. It should also be remembered that Eq. (33) is a result of perturbation theory and therefore has a restricted range of applicability; in fact, the assumptions made in deriving Fermi's Golden Rule put one "between a rock and a hard place" since it applies provided that the perturbation inducing transitions is "small but not too small!" (*134*). If H_{RP} is "large but not too large" then the splitting of the nuclear potential energy curves at the point X in Fig. 5b will be sufficiently big that the reaction can proceed essentially adiabatically, and classic Marcus theory will be a good approximation (i.e., $\kappa \cong 1$). If H_{RP} is too large then there will be substantial electronic overlap between R and P, and perturbation theory itself will break down. In the context of the basic framework of assuming that electron transfer is in some sense a forbidden process (small transition probabilities), electronic effects on the transfer rate will only be important when (seemingly paradoxically) H_{RP} is too small, that is, the reaction is appreciably nonadiabatic. Herein lies the difficulty in determining which (if any) theory accurately describes electron transfer in proteins. Most predict an exponentially small transfer rate but differ in the exact form of the rate law. To establish the precise functional dependence it would be necessary to perform distance dependence studies of transfer rates over a substantial range of distances while keeping as many other factors as possible constant. This is especially difficult to do in protein systems, and consequently the results of such studies can be somewhat ambiguous. However, pulse radiolysis studies of the distance dependence of transfer rates have been possible for homologous series of compounds in which D and A are separated by varying numbers of fairly rigid spacer molecules, and these studies have established fairly well-defined exponential decay laws (*131*). These data can thus be used to compare with theoretical predictions. Details of this kind of study have been reviewed elsewhere (*135*) and, since these systems are not directly relevant to physiological electron transfer, are not considered further.

The most widely referred to of the QE theories is superexchange, which, although having a specific meaning related to antiferromagnetic (spin) interactions between electrons, is sometimes used to refer to any medium-assisted electron transfer via the medium between D and A. This theory has, like Marcus theory, to some extent been carried over to biological problems from inorganic chemistry where it has been used to explain forbidden electron transfer in transition metal compounds. To understand this mechanism as well as some of the other QE theories, it might help to review what a "band" in solid state parlance is. Consider the potential for H_2^+ repeated an infinite number of times, as shown schematically in Fig. 7 (*132*). Because the electron has equal probability of being localized close to any of the lattice sites, its probability distribution is delocalized over the whole crystal. A suitable approximation might be to construct single electron wavefunctions for the crystal lattice and then proceed to fill

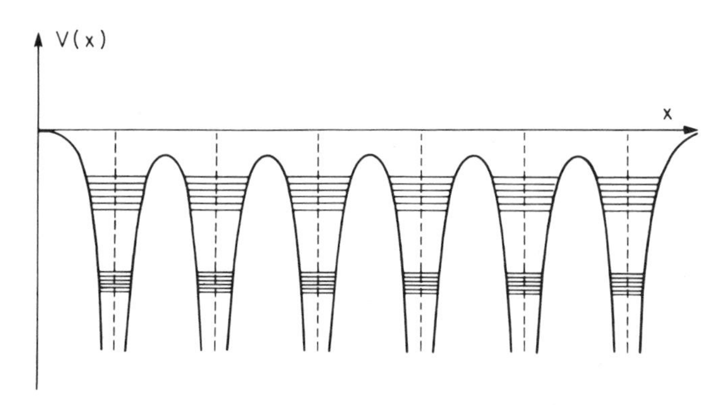

FIG. 7. Energy levels for several repetitions of the potential in Fig. 4. Two bands are shown. [Adapted from Ref. (*132*).]

up these states with the actual number of electrons. Electron–electron interactions could then be thought of as a perturbation on this zeroth order picture. The pairs of almost degenerate levels of Fig. 4 then widen into bands containing many levels, depending on the actual number of lattice points in the crystal (see Fig. 7). As the size of the crystal approaches infinity, the spacing between levels in each band goes to zero, and within a band essentially all values of energy are possible. The spacings between bands ("band gaps") will remain (in general, although there may be overlap), and the electronic properties of the material will be determined largely by the spacings and distributions of bands and the number of electrons in a band. As a considerably simplified statement, if the highest energy band is full, conduction will not occur (the band is a valence band) and the material will be an insulator, whereas if it is less than full it can conduct electricity. In a semiconductor electrons from the valence band are promoted (thermally, for example) into a conduction band, and conduction is then possible. The holes or electron vacancies left in the valence band behave as positively charged particles, and transport of holes is an alternative mechanism to transport of electrons in order to achieve charge separation.

The tendency to form bands and thereby the possibility of conduction depends on the balance between the tendency of an electron to delocalize and repulsion between electrons which come too close to each other. Petrov (*136*) has discussed the idea of conduction and valence bands in proteins as well as the possibility of either electron transfer through conduction bands or hole transfer through valence bands, both processes being made possible energetically by impurities in an analogous way to the situation in certain semiconductors. In this approach it is supposed that the protein is a quasi-periodic crystal (because of the repeating peptide groups), while the impurities are provided by the donor and acceptor groups. Direct conduction is not proposed, but rather a resonant exci-

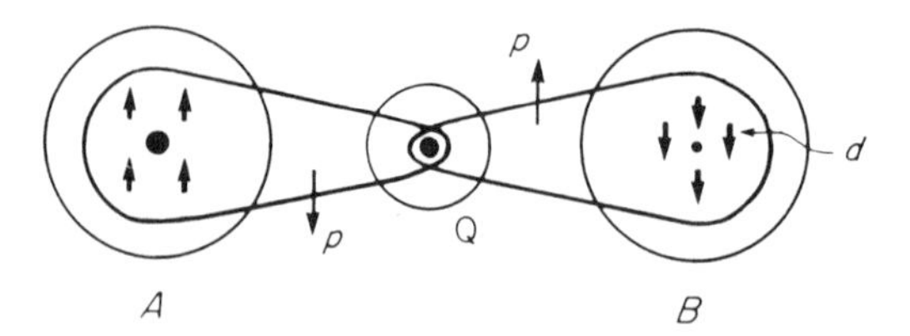

FIG. 8. Schematic of superexchange between two transition metal ions (A and B) via a diamagnetic group, Q, having two antiparallel p electrons.

tation mechanism which predicts an exponential dependence on distance. This idea does not seem to have been widely pursued, partly because it is difficult to quantify the predictions of the theory, but further analysis is probably warranted. An alternative, but in some ways similar mechanism, is provided by superexchange itself (*2*).

Because the term "superexchange" is used so widely, it is worth discussing in a little detail what superexchange is, as first defined by Kramers, Anderson, and others (*137*). Consider a situation in which A and B are transition metal ions separated by a "lattice" of diamagnetic groups (i.e., no permanent magnetic moment), as shown in Fig. 8, where the lattice consists of a single diamagnetic atom. There is clearly a direct interaction between the two ions, but this is too long range to effect efficient electron transfer if A and B are well separated. Assume that A has + spin, B − spin, and the diamagnetic ion Q has a pair of antiparallel electrons. The + spin atom A can then couple to an electron of − spin of the diamagnetic ion Q, and similarly the + spin electron of Q can couple to the − spin atom B. Thus, A and B are coupled to each other through the diamagnetic medium separating them. The difficulty in treating the situation is that A and B both couple strongly to Q but the overall superexchange effect is exponentially small.

Anderson received the Nobel Prize partly for his work on superexchange; his idea, which was truly inspired, was to redefine the zero-order Hamiltonian ($\mathcal{H}_0$) in terms of (quasi-spin) states which took into account the couplings of the transition metals with the diamagnetic group almost exactly, and couplings between transition metal ions could then be treated as small perturbations. In other words, redefining the zero-order problem, superexchange became a small and treatable perturbation. In this approach solutions to the redefined Hamiltonian $\mathcal{H}_0$, included ionized wavefunctions extended over both transition metal ions. However, these states are higher energy than the ground state and thus not permanently occupied. It might seem that conservation of energy would prevent any occupation of such states. This is not the case, and "virtual" transitions to such states are possible and lead to charge separation. This can be understood by using one of two equivalent lines of reasoning: (1) The ionized states are themselves approximations to the true states of the system, since their construction relies on a number of assumptions. Therefore, these wavefunctions are states of an approximate Hamiltonian, $\mathcal{H}_1$. The true Hamiltonian may be written, $\mathcal{H} = \mathcal{H}_1 + \mathcal{H}'$, where $\mathcal{H}'$ is a perturbation which contains all neglected effects. Since the virtual states are not true eigenstates of the Hamiltonian, the perturbation can induce transitions with some small probability into these states, leading to charge separation. Presumably, whatever instigates the electron transfer process in the first place also perturbs the system sufficiently that such energy nonconserving transitions can occur. (2) The time–energy uncertainty principle allows small non-energy-conserving transitions into virtual states. Some thought reveals that this is really a shorthand way of making the same point as (1) above! After all, if the system were in a true eigenstate, then no time evolution could occur.

The idea of superexchange has been extended beyond its original applications, by McConnell (*138*) and others, to electron transfer in purely organic systems via virtual solvent states. Importantly, the distance dependence of superexchange type processes is also exponential, although the specific dependence on system parameters is different (*2*). The idea that intervening molecular orbitals can mediate electron transfer between D and A is also the basis of 'through bond' electron transfer mechanisms (*139, 140*). The basic idea is similar to superexchange, namely, that the donor and acceptor wavefunctions, which would overlap very weakly with each other because of their separation, have enough overlap with intervening quasi-periodic protein orbitals (*139, 140*) to allow electron transfer to occur. This approach has been developed by Hopfield, and colleagues (*139*), who point out that, since the arrangement of intervening groups can be quite sensitive to temperature, the standard picture of a temperature-independent tunneling mechanism might not be appropriate. They also postulate a mechanism in which "through bond" and "through space" tunneling both play a role in the actual process of electron transfer (*139*).

As a final point, note that almost all theories use some form of perturbation theory to discuss charge separation. As with Anderson's model of superexchange, the challenge is to come up with the correct zero-order Hamiltonian. Since the perturbation ("tunneling") is small, if larger effects than the tunneling are left out of $\mathcal{H}_0$, then perturbation theory might not converge! Most of the various theories are simply using a different $\mathcal{H}_0$; the trick is to combine as much of the known biochemistry into $\mathcal{H}_0$ as possible prior to treating the perturbation.

IV. Examples of Electron Transfer Reactions

As we have seen, the rate of electron transfer depends on both nuclear and electronic factors. The most experimentally accessible approaches to characterizing the dependence of the electron transfer rate on these factors are to (1) vary the thermodynamic driving force for the reaction and (2) alter the separation distance and the nature of the medium (both protein and solvent) between and around the donor and acceptor groups. The successful manipulation of electron transfer systems where these parameters may be specifically altered often requires considerable chemical, biochemical, and genetic ingenuity. As has been amply demonstrated in recent years, however, this effort is being rewarded with important insights into the molecular details of the mechanism of biological electron transfer.

The thermodynamic driving force for an electron transfer reaction may be changed by altering the donor and/or acceptor species. This is usually accomplished using chemical methods to either attach or replace redox groups on a protein. If all other factors remain constant, this approach provides a method for evaluating the nuclear factor $\exp\{-[(\Delta G^\circ + \lambda)^2/4\lambda RT]\}$ in the Marcus rate expression. By appropriate analysis of the results, the magnitude of the reorganization energy, λ, may be obtained. As values for λ may be theoretically (computationally) derived, comparison of the experimental and theoretical results can permit a molecular understanding of the various energetic contributions to the reorganization energy.

With appropriate chemical and genetic methods, the relative positions of the same pair of donor and acceptor groups can be varied. In favorable cases, this permits an experimental determination of the distance (r) dependence of the electron transfer rate. This dependence is generally assumed to be of the exponential form $e^{-\beta r}$. The value of β is related to the electronic coupling between donor and acceptor groups and is of great interest for comparison with values derived for different experimental systems (both biological and nonbiological). Additionally, the value of β may provide information on the mechanism or pathway of electron transfer. This latter question can be explored in more detail using both chemical and genetic (site-directed mutagenesis) techniques to vary the nature of the medium between the donor and acceptor groups. In principle, it seems possible to design experimental systems that could distinguish between through space and superexchange mechanisms of electron transfer (assuming only one of these possibilities is operative).

The overall kinetics of bimolecular electron transfer reactions are complicated by the reactant association and product dissociation steps. To eliminate the contributions of these events to the reaction kinetics, many recent studies have concentrated on intramolecular electron transfer processes, where both the donor and acceptor groups are present in the same molecule. These systems provide the opportunity to measure directly the rate of electron transfer in the absence of competing bimolecular processes. For this reason, the examples in this section are concerned only with intramolecular processes.

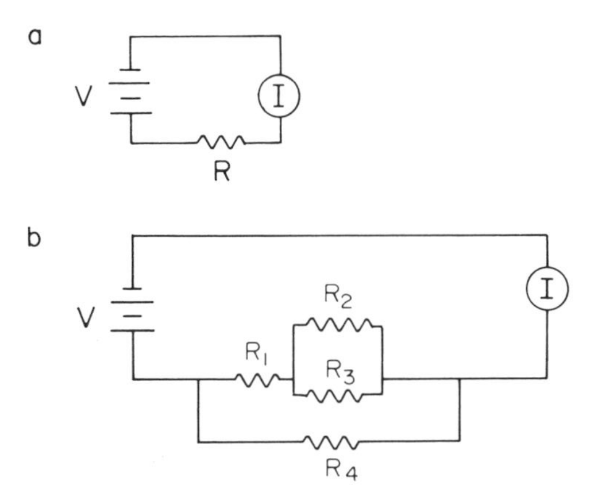

FIG. 9. Simple electrical circuits illustrating the relationship between driving force, V, current flow, I, and resistance, R. (a) Simple series circuit. (b) More complex circuit with multiple resistances R_i connected in series and parallel fashion.

In many ways, analysis of the factors influencing biological electron transfer is analogous to the study of electric conduction in macroscopic systems. The properties of conducting systems may be described by Ohm's law:

$$I = V/R \tag{36}$$

where I is the current (rate of electron flow), V is the voltage drop (driving force), and the resistance R describes medium effects. In the circuits depicted in Fig. 9, measurement of the current I as a function of V permits the resistance R of the intervening material to be defined. Alternatively, the variation in I as a function of R at fixed V may be measured. In the simple circuit of Fig. 9a, I and R will be inversely proportional. In more complex schemes, involving perhaps either series or parallel arrangements of multiple resistances (Fig. 9b), more complicated dependences of I on the R_i would be observed. In principle, from a "shotgun" approach to measuring the current after changing the various resistances, a complete analysis of the circuit could be achieved. One could hope that an analogous dissection of medium effects on biological electron transfer rates will permit an equivalent elucidation of the types of pathways utilized for this process.

Two intramolecular electron transfer systems of current interest, ruthenated proteins and photosynthetic reaction centers, are now discussed. These examples serve to illustrate the types of experimental approaches and information that may be obtained concerning the details of electron transfer processes in biological systems.

A. Ruthenated Proteins

Covalent attachment of ruthenium ions to proteins (ruthenation) provides a powerful approach for introducing a second redox active group into an electron transfer protein (*77, 78*). Under suitable conditions, intramolecular electron transfer may be monitored between Ru and the intrinsic redox group. Ruthenated proteins display a number of advantages for the study of intramolecular electron transfer reactions (*77, 78*):

1. Conditions of ruthenation have been developed that result in the specific modification of single, surface-exposed histidine residues. Typically, the modification reaction results in the coordination of an $Ru(NH_3)_5{}^{3+}$ group to a histidine. Since there are usually few histidines on the surface of a protein, it is possible to prepare singly ruthenated protein species. If the modified histidine is identified in the protein sequence, and if the three-dimensional structure of the protein is known, then the positions and separation of the Ru and intrinsic redox group may be determined.
2. Both Ru^{3+} and Ru^{2+} are relatively substitution inert, so that the Ru will remain attached to the protein during the experimental work.
3. The electrochemical properties of the attached Ru are well suited to studying electron transfer processes in proteins. The $E°$ value for the $Ru(NH_3)_5$His group attached to a protein is about 80 mV. This value can be varied by replacing an NH_3 ligand with different groups, providing a convenient mechanism for altering the driving force of the electron transfer reaction. Since these $E°$ values occupy the middle of the biologically observed range, the modifications are useful for studying a wide range of electron transfer proteins.
4. Ruthenation appears to occur without perturbing the structure of the protein, except for perhaps local changes in the vicinity of the modified histidine. The electrochemical properties of the protein are only slightly altered by ruthenation. The $E°$ value for the intrinsic redox cofactor of a protein typically increases by about 20 mV, consistent with the addition of a positively charged group to the protein.
5. Many clever methods have been developed for generating the thermodynamically unfavored form of the one-electron reduced protein. Details of these methods, which are based on a variety of photochemical and pulse radiolysis techniques, may be found in the original references. The kinetics of intramolecular electron transfer may then be followed by monitoring the rate of approach to the thermodynamically favored state.

Selectively modified and characterized ruthenated proteins have been prepared for cytochrome *c* (*141–144*), cytochrome *c*-551 (*145*), azurin (*146, 147*), plastocyanin (*147*), HiPIP (*148*), and myoglobin (*77, 149–154*). Intramolecular electron transfer rates, *k*, measured in these systems are listed in Table III along

TABLE III

ELECTRON TRANSFER IN RUTHENATED PROTEINS[a]

Protein	Donor	Acceptor	$\Delta E°$ (mV)	d (Å)	k (sec^{-1})	Ref.
Cytochrome c	a_5Ru(His-33)	FeP	180	12	30	*141*
					53	*142*
	a_5Ru(His-33)	ZnP*	360		240	*143*
	a_4(isn)Ru(His-33)	$ZnP\cdot^+$	660		2.0×10^5	*144*
	ZnP*	a_5Ru(His-33)	700		7.7×10^5	*143*
	a_4(py)Ru(His-33)	$ZnP\cdot^+$	740		3.5×10^5	*144*
	ZnP*	a_4(py)Ru(His-33)	970		3.3×10^6	*144*
	a_5Ru(His-33)	$ZnP\cdot^+$	1010		1.6×10^6	*143*
	ZnP*	a_4(isn)Ru(His-33)	1050		2.9×10^6	*144*
Cytochrome c_{551}	a_5Ru(His-47)	FeP	200	8	20	*145*
Azurin	a_5Ru(His-83)	Cu	280	12	1.9	*146*
					2.5	*147*
Plastocyanin	a_5Ru(His-59)	Cu	300	12	0.03	*147*
HiPIP	a_5Ru(His-42)	Fe_4S_4	270	8	18	*148*
Myoglobin	FeP	a_5Ru(His-42)	20	13	0.04	*149*
		a_4(py)Ru(His-42)	275		2.5	*150*
	H_2P*	a_5Ru(His-42)	530		760	*151*
	PdP*		700		9.1×10^3	*152*
	PtP*		730		1.2×10^4	*151*
	MgPDE*		~800		3.2×10^4	*153*
	CdP*		850		6.3×10^4	*151*
	MgP*		870		5.7×10^4	*153*
	ZnP*		880		7.0×10^4	*154*
	PdP*	a_4(py)Ru(His-42)	980		9.0×10^4	*152*
	MgPDE*	a_5Ru(His-81)	~820	19	48	*153*
	MgP*		890		82	*153*
	ZnP*		900		86	*154*
	MgPDE*	a_5Ru(His-116)	~820	20	49	*153*
	MgP*		890		69	*153*
	ZnP*		900		89	*154*
	MgPDE*	a_5Ru(His-12)	~810	22	39	*153*
	MgP*		880		67	*153*
	ZnP*		890		101	*154*

[a] Rate constants, k, are for electron transfer from the donor to acceptor groups in ruthenated electron transfer proteins. Ruthenium ligands are abbreviated as follows: NH_3, a; isonicotinamide, isn; pyridine, py. MP are metallosubstituted porphyrins (M = Fe, Zn, Pd, Pt, Mg, Cd, and H). MP* and $MP\cdot^+$ are the photoexcited and oxidized radical cation forms of MP, respectively. PDE is porphyrin diester. d is the closest edge-to-edge distance between cofactor atoms.

with the driving force and the Ru–cofactor separation distances. Measured rates span the range from about 0 to over 10^6 sec^{-1}. The driving forces for these reactions have been varied between about 0 to over 1000 mV, by combinations of changing the Ru ligand, changing the protein, and reconstituting heme proteins with different metalloporphyrins. The largest driving forces have been obtained by replacing the heme with various metallosubstituted porphyrins (abbreviated MP, where M = Fe, Zn, Mg, Pd, Pt, H, Cd). Photochemical excitation of the substituted porphyrins creates a strongly reducing species (*69, 70*) (designated MP*), which can then reduce Ru^{3+} to Ru^{2+}. In several cases, the rate of the back electron transfer from Ru^{2+} to $MP^{\cdot+}$ (the radical cation generated after oxidation of MP*) has also been measured. Several important observations concerning intramolecular electron transfer in these systems have been described.

(1) *Distance dependence of k.* By measuring k as a function of donor–acceptor separation distance in a given protein, the distance dependence of k may be studied. Myoglobin has four surface histidines, located at varying distances from the heme, that have been selectively ruthenated, and the kinetics of intramolecular electron transfer between the Ru sites and the heme measured. Assuming factors such as $\Delta G°$ and λ remain constant between these different species, the k values should reflect the distance dependence of the electron transfer rates. In order to obtain measurable rates for electron transfer from the more distant Ru sites to the heme, a large driving force for the reaction was generated using photochemical excitation of metallosubstituted porphyrins in myoglobin. Measurements of k for electron transfer from photoexcited Zn- and Mg-porphyrins to the four different Ru sites in myoglobin are consistent with an exponential dependence of k on separation distance, r, of the form $e^{-\beta r}$ (*151*). The β value from this analysis is determined to be about 0.8 $Å^{-1}$. This value is similar to results obtained from nonbiological model systems, which are typically in the range of 0.85–1.2 $Å^{-1}$. Given the exponential dependence of k on r, however, the consequences of these different β values are distinguishable. The variation of β in these systems may reflect differences in the details of electron transfer between proteins and model systems.

An unresolved question in these studies concerns the appropriate measurement of the donor–acceptor separation distance. The distances in Table III represent estimates of the closest edge–edge distance between the Ru–His group and the porphyrin ring atoms. The corresponding metal–metal distances are some 6 Å longer. These distances implicitly assume a through space type electron transfer mechanism which is sensitive to the straight line distance between redox groups. If a through bond mechanism is operative, however, then the appropriate distance would be the length of the shortest covalently bonded path between cofactors. In general, this distance will be much longer than that measured from the straight line path (*151*). Furthermore, the two distance measures (through bond and through space) will not generally be proportional to one another. An additional complication is the possibility (likelihood?) of having through bond transfer in combination with "jumps" across nonbonded regions (*151*). The distinction between these various possibilities is of critical importance, given the theoretical expectation of the increased efficiency of transfer via the through bond mechanism. Resolution of this issue is a current goal of work in this field. Using site-directed mutagenesis techniques to position histidines (and hence Ru atoms) at variable through bond and through space distances from the heme provides a promising approach to address this question (*81*).

Although electron transfer rates within myoglobin appear to follow an exponential dependence on distance, the derived rate expression is not directly transferable to other electron transfer proteins. A particularly striking comparison is between the *c* cytochromes and the copper proteins plastocyanin and azurin. Intramolecular electron transfer rates are at least 10–100 times slower in the copper proteins compared to the *c* cytochromes, even though the distances and driving forces for the reactions are comparable. The origin of this behavior is unclear, but it does suggest caution in the quantitative transfer of rate expressions between different systems.

(2) *Driving force dependence.* In the ruthenated systems studied to date, an increase in the driving force (ΔE°) between a given pair of redox sites results in an increased electron transfer rate (Table III). In terms of Marcus theory, this suggests that in all cases $-\Delta G^\circ$ is less than the reorganization energy λ needed to achieve the transition state, that is, the "inverted region" past which the rate decreases with increasing $-\Delta G^\circ$ has not yet been reached. Values of λ may be obtained from the variation in k with ΔG° (again, assuming all other parameters remain constant) and are observed to be in the range of about 1.5–2 eV (about 30–50 kcal/mol). The magnitude of λ is sensitive to the reorganization energies of both the solvent and the protein about the redox groups. Theoretical and model studies indicate that the contribution of the solvent terms to λ increases with increasing solvent polarity (since the larger dipole moments of polar solvent molecules respond more strongly to the change in charge distribution associated with electron transfer). The relatively large values of λ observed for intramolecular electron transfer in these ruthenated systems almost certainly reflect a substantial contribution from the reorganization of water about the redox sites.

Additional insight into the various energetic contributions to λ may be derived from the temperature dependence of k. Within the Marcus framework, it is possible to extract the enthalpy of reorganization of the protein from the observed activation enthalpy. The reorganization enthalpies for cytochrome *c* and azurin have an upper limit of about 0.3 eV (7 kcal/mol), whereas the value for myoglobin is about 0.9 eV (20 kcal/mol) (*77*). The experimental reorganization energy for cytochrome *c* is consistent with a theoretical prediction of this value obtained by Warshel and Churg (*155, 156*). As physiological electron transfer proteins, cytochrome *c* and azurin might be expected to exhibit small reorganization energies as part of a structural mechanism to increase electron transfer rates. In contrast, myoglobin is not primarily an electron transfer protein. The changes in spin state and ligation accompanying oxidation–reduction of myoglobin should contribute to a large reorganization energy for this process.

With appropriate selection of the driving force, it has been possible to measure electron transfer rates in both directions between a porphyrin and attached Ru. The forward and reverse electron transfer rates between Ru attached to His-48 of myoglobin and the heme (for which $\Delta G^\circ \cong 0$ kcal/mol) are equal within experimental error. Rapid electron transfer rates ($k \cong 10^6$ sec^{-1}) have been measured in both directions between Ru and the Zn-porphyrin in ZnP-substituted cytochrome *c*, with driving forces exceeding 700 mV in each direction (*143, 144*). These experimental results suggest that structural rearrangements are not rate limiting. Consequently, conformational gating does not appear to contribute to the kinetics of electron transfer in these simple systems.

B. Photosynthetic Reaction Centers

Reaction centers (RCs) are integral membrane protein complexes that carry out light-mediated electron transfer reactions from a donor to a series of acceptors in the initial steps of photosynthesis. A landmark achievement in the study of intramolecular electron transfer reactions was the crystallization (*157*) and structure determination (*158–160*) of the RC from the purple bacterium *Rhodopseudomonas (Rps.) viridis* by Michel and colleagues. With the subsequent structure determination of the homologous RC from *Rhodobacter (Rb.) sphaeroides* (*161–170*), RCs presently provide the best characterized examples of an intramolecular electron transfer system. The convergence of structural, spectroscopic, chemical, genetic, and theoretical approaches has created an unprecedented opportunity to dissect RC function at the atomic level. More detailed accounts of these exciting developments may be found in recent symposia proceedings and reviews (*171–177*).

Reaction centers in photosynthetic bacteria typically contain three membrane-bound subunits (L, M, and H), and the following cofactors: four bacteriochlorophyll (Bchl or B), two bacteriopheophytin (Bphe or Φ), two quinones (Q), and one Fe atom (*28, 178*). The sequence of electron transfer steps along the various cofactors has been established largely by spectroscopic methods. The primary donor, D, which initially absorbs light (creating the excited state D*) is a dimer of Bchl molecules [also designated $(Bchl)_2$ or P]. Electron transfer proceeds from D* to an intermediate acceptor (a Bphe molecule), to a primary acceptor, Q_A, and finally to the secondary acceptor Q_B. After these initial events, the RC

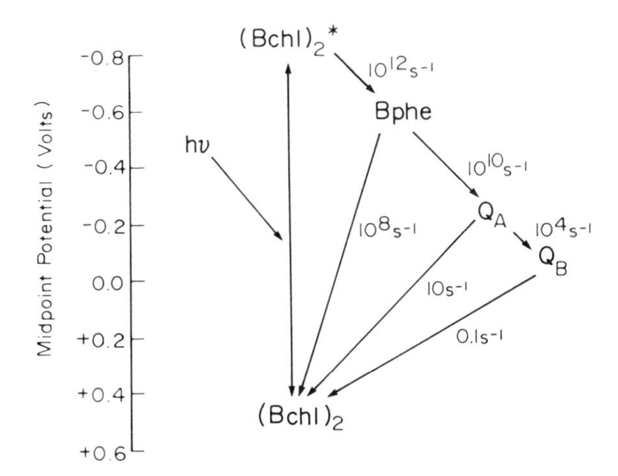

FIG. 10. Electron transfer rate constants (to nearest order of magnitude) measured in bacterial reaction centers in the forward and reverse directions. Approximate reduction potentials are indicated on the vertical axis.

exists in the charge separated state $D^+Q_B^-$. A critical aspect of the electron transfer scheme is that the rates of competing back-reaction steps ("short circuits") are much slower than the rates of the forward reactions (*173*). Consequently, the overall quantum efficiency of the reaction is essentially unity (i.e., perfect) (*179*). The kinetics and energetics of these reactions are illustrated in Fig. 10.

The crystallographic work on the RC has provided a structural framework for interpreting the kinetic measurements. Eleven transmembrane α-helices from the protein subunits serve as a scaffold to organize the cofactors. The cofactors are arranged in two connected branches, A and B (termed L and M in the *Rps. viridis* RC structure), extending from one side of the membrane-spanning region of the RC to the other (Fig. 11). Along each branch, cofactors appear in the sequence $(Bchl)_2$, Bchl, Bphe, Q, and Fe. The $(Bchl)_2$ and Fe groups are common to both branches. The branch association for the Bchl, Bphe, and Q groups is indicated by an A or B subscript. A striking aspect of the RC is the approximate symmetry of the structure, where both the two cofactor branches and the L and M subunits are closely related by a 2-fold rotation operation. The RC structure provided confirmation of two central features of the electron transfer scheme that had been deduced spectroscopically: (a) the identity of the primary donor as a dimer of two Bchl molecules and (b) the sequence of electron transfer steps in the order $(Bchl)_2$ to Bphe to Q. A surprising aspect of the RC structure was the existence of two cofactor branches of similar structures, although spectroscopic

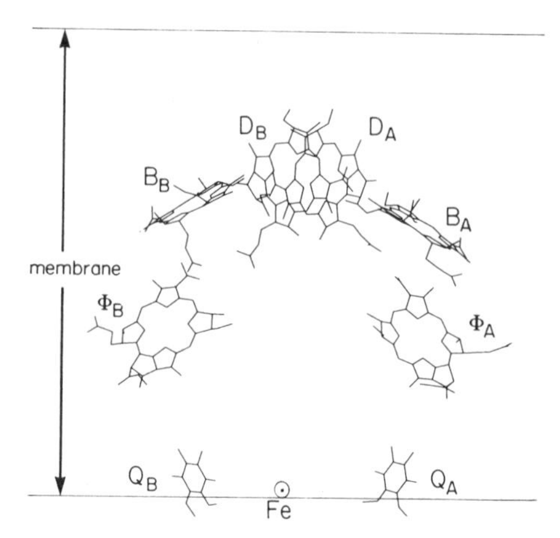

FIG. 11. Orientation of the cofactors in the bacterial reaction center. The approximate position of the membrane–water interface is indicated. Phytyl and isoprenoid tails have been truncated for clarity. [Modified from Ref. (*177*).]

evidence indicated the existence of only one photochemically active branch (the A branch).

Rate constants of the individual electron transfer reactions are listed in Table IV, along with estimates of the driving force and the separation distance between cofactors. The rates of the first two steps ($D^* \rightarrow \Phi_A$ and $\Phi_A^- \rightarrow Q_A$) are particularly impressive relative to the rates observed for comparable distances (10 Å) and driving forces (~0.5 V) in ruthenated proteins. This suggests that more favorable nuclear and/or electronic factors have been achieved for the electron transfer reactions in the RC.

Marcus theory predicts that the nuclear factor in the electron transfer rate expression will be maximal when $-\Delta G° = \lambda$. Under these conditions, the electron transfer process will be temperature independent. The first two electron transfer steps in the RC approximately exhibit this behavior. The rate of the $D^* \rightarrow \Phi_A$ step actually increases slightly (by a factor of 2–4) as the temperature decreases from 300 to 8 K (*180*). Consequently, λ may be estimated to be in the range 0.3 to 0.5 V (~7–10 kcal/mol) from the $\Delta E°$ values for these two steps (Table IV). These values are approximately 4 times smaller than those observed for ruthenated proteins discussed previously. Sequestration of the redox groups in a membrane-bound protein complex, away from aqueous solution, may serve to decrease the value of λ by minimizing the reorganization energy of a highly polar solvent.

The electronic overlap between donor and acceptor groups may be enhanced either by increasing the coupling between the groups (by a superexchange

TABLE IV

ELECTRON TRANSFER RATES IN BACTERIAL PHOTOSYNTHETIC REACTION CENTERS[a]

Step	$\Delta E°$ (mV)	*d*–ring (Å)	*d*–all (Å)	k (sec^{-1})
Forward reactions				
$D^* \rightarrow \Phi_A$	300	10	4	10^{12}
$\Phi_A^- \rightarrow Q_A$	500	10	4	10^{10}
$Q_A^- \rightarrow Q_B$	100	16	13	10^4
Back reactions				
$\Phi_A^- \rightarrow D^+$	1100	10	4	10^8
$Q_A^- \rightarrow D^+$	600	23	4	10^1
$Q_B^- \rightarrow D^+$	500	23	4	10^{-1}
Other relevant distances				
$D \rightarrow B_A$		6	4	
$B_A \rightarrow \Phi_A$		5	4	

[a] Approximate values for the driving force, $\Delta E°$, and electron transfer rate constants, k (to the nearest order of magnitude), are obtained from Refs. (*28, 173, 176, 178*). *d*–ring and *d*–all are the closest distance between cofactor ring atoms and the closest distance between any cofactor atoms, including the phytyl and isoprenoid tails. Distances were measured from the *Rb. sphaeroides* RC structure (*166*).

mechanism) or by positioning them closer together. Both types of mechanisms may be involved in the initial RC electron transfer steps. The primary electron transfer step has attracted considerable attention in this regard since that rate is approaching the theoretical upper limit of about 10^{13} sec^{-1} (*1*). Without an unreasonably small value for β, it is difficult to reconcile a rate of 10^{12} sec^{-1} for electron transfer with a 10 Å cofactor edge–edge separation distance. The location of the monomer Bchl, B_A, bridging D and Φ_A suggests the possible involvement of this group in the electron transfer reaction. Two functions for B_A have been proposed. First, B_A serves as a transient intermediate in a $D^* \rightarrow B_A \rightarrow \Phi_A$ electron transfer pathway (*181*). Femtosecond spectroscopic studies (*180*) of the initial electron transfer step have not detected the presence of this intermediate, however. If B_A^- does exist, the rate of electron transfer from B_A^- to Φ_A cannot exceed 10 fsec. Second, B_A may function in a superexchange mechanism to enhance the D to Φ_A coupling interaction (*182, 183*). Experimental and theoretical work suggest that the D and B groups exhibit significant electronic coupling, consistent with some role for B in the $D^* \rightarrow \Phi_A$ electron transfer step.

An alternative mechanism for enhancing the coupling between D and Φ_A involves van der Waals contacts between Φ_A and the phytyl tail of D. This could provide a direct coupling mechanism between D and Φ_A responsible for the rapid electron transfer rate, although this possibility has not been explored as extensively as the role of B_A.

Similar considerations apply to the $\Phi_A \rightarrow Q_A$ and $Q_A \rightarrow Q_B$ electron transfer steps. Theoretical calculations (*184*) suggest that a tryptophan side chain (residue 252 of the *Rb. sphaeroides* M subunit) in contact with both Φ_A and Q_A may serve a superexchange function in this electron transfer reaction. Direct contact between the phytyl and isoprenoid tails of Φ_A and Q_A occurs, again providing the possibility of electron transfer actually occurring over a short distance. A superexchange role has been proposed for the Fe atom bridging Q_A and Q_B (*185*). The role of the Fe atom in the RC function is also of general interest, since it does not undergo oxidation–reduction, and it may be substituted with a variety of divalent metals with little effect on the Q_A to Q_B electron transfer rate (*186*).

In contrast to the rapidity of the initial steps, two other types of reactions in the RC are of interest because they are relatively slow.

(1) *Recombination reactions.* The unit quantum efficiency arises from the much greater electron transfer rates in the forward direction, compared to the rate of recombination of the charge-separated species to regenerate D. The origins of these rate differences are unclear, but several types of explanations have been proposed. (a) The large driving forces, $\Delta E°$, associated with some of the reactions (most notably Φ_A^- to D^+) may exceed λ. Such reactions would be in the Marcus "inverted region," where the rate decreases with increasing driving force. (b) Gating mechanisms may exist, where conformational changes accompanying the charge separation make recombination less favorable. (c) Once the electron proceeds past Φ_A, the increasing distance of the electron from D^+ implies that the electronic overlap between successive cofactors is larger than for the corresponding recombination reaction, resulting in faster forward electron transfer rates.

(2) *Preferential electron transfer along the A branch.* The rate of electron transfer is at least 20 times faster along the A branch than the B branch, in spite of the structural similarities of the two pathways [reviewed in Ref. (*187*)]. Undoubtedly, deviations from exact symmetry and differences in the protein environment provided by the nonidentical L and M subunits are responsible for the preferred use of the A pathway. A theoretical study (*187*) of the contribution of some of these factors to the electron transfer rate in the *Rps. viridis* RC yielded a rate ratio of 45 between the two branches, consistent with the experimental observations.

One of the most promising approaches for developing and testing models for electron transfer in the RC is the use of site-directed mutagenesis techniques to replace specific residues with other amino acids. These studies can serve to identify amino acid side chains that may be involved in electron transfer pathways, although it is more difficult to assess the role of main chain atoms in electron transfer by these methods. The usual strategy has been to replace conserved residues or residues in contact with cofactors by different amino acids, and then study the consequences on photosynthetic function both *in vivo* and *in vitro*. Residues are typically substituted with other amino acids of similar size but which have different polarity and/or hydrogen-bonding capabilities to minimize general structural disruptions to the RC. Other approaches are possible, however, such as complete substitution of every amino acid at a single position or exhaustive mutagenesis of a given region of the protein. Technical advances are extremely rapid in this area, so that ever more creative techniques are being developed which will increase the power of these methods.

The position of conserved residues in homologous RCs provide a guide for directing the mutational analysis. The underlying philosophy is that conserved residues must be structurally and/or functionally important. Among the RCs from four purple photosynthetic bacteria, 48% (20/42) of residues in contact with D, B, or Φ cofactor rings are conserved in all four species, all five Fe ligands are conserved, and 77% (10/13) of residues in contact with the Q rings are conserved (*168*). For comparison, 35% of residues in the membrane-spanning α helices are conserved in all four species. Of the 60 conserved residues in contact with the cofactors, 19 contain potential hydrogen-bonding groups (Table V). These groups are of special interest in mutational analysis, since hydrogen bonds form specific, directional interactions that might influence the stability of the cofactors in different states, thereby altering the spectroscopic and/or kinetic properties of the RC.

Nearly half of the 19 potential hydrogen-bonding residues have been replaced by different amino acids as of June, 1989. The results of these substitutions are summarized in Table V (188–193). The availability of suitable genetic systems has restricted this work to *Rb. capsulatus* and *Rb. sphaeroides*. Although the detailed spectroscopic and kinetic properties of the RCs may be altered by these modifications, very few residues appear to be absolutely essential for maintaining the RC function. A striking example of this is the replacement of the histidine residue ligating the Mg atom of the D_B group by either isoleucine or phenylalanine (*192, 193*). In response to this modification, the $(Bchl)_2$ dimer is replaced by a heterodimer containing both Bchl and Bphe. Surprisingly, this RC is photosynthetically competent, even though the electronic properties of the primary donor have been significantly changed as a consequence of the amino acid substitution.

The mutational analyses suggest that RC function is, in general, rather robust to changes that apparently would significantly alter ligation and hydrogen-bonding interactions to different cofactors (although, certainly specific exceptions to this general statement exist). The same conclusion could perhaps have been predicted from the similar electron transfer rates exhibited by RCs from different purple photosynthetic bacteria, even though over 50% of the residues in contact with D, B, and Φ cofactor rings are not conserved. The results so far

TABLE V

CONSERVED AMINO ACID RESIDUES IN CONTACT WITH COFACTOR RINGS IN REACTION CENTERS[a]

Conserved position	Cofactor	Initial residue	Photosynthetic ability of modified RCs		Comments	Ref.
			Active	Inactive		
L104	Φ_A	Glu	Gln, Leu	Lys		*188*
L152	B_A	His	Thr			*189*
L162	P	Tyr	Val	Lys		*189*
L168	B_B/D_A	His				
L172	D_A	His	Gln			*189*
L190	Fe, Q_B	His				
L212	Q_B	Glu	Gln		Reduced Q_B protonation rate	*190*
L223	Q_B	Ser	Pro		Reduced Q_B binding	*191*
L230	Fe	His				
M129	Φ_B	Trp				
M182	B_B	His				
M202	D_2	His	Gln, Leu, Phe		Leu or Phe gives hetero-dimer special pair that remains active	*192, 193*
M210	D_B, B_A, Φ_A	Tyr				
M219	Fe, Q_A	His				
M222	Q_A	Thr				
M234	Fe	Glu				
M252	Q_A	Trp		Val		*189*
M259	Q_A	Asn				
M266	Fe	His				

[a]From four species of purple photosynthetic bacteria. Only residues with side chains capable of participating in hydrogen-bonding interactions are included (not every residue is observed to form hydrogen bonds in the RC crystal structure, however). Positions are designated by subunit (L/M) and sequence number in the *Rb. sphaeroides* RC. The identification of modified RCs as active/inactive is difficult, owing to differences between various *in vivo* and *in vitro* assays and possible complications arising from perturbations in RC assembly and interaction with other membrane components. The original references should be consulted for more detailed analyses of the consequences of residue substitutions.

obtained emphasize the need to combine the mutational analysis with structural, kinetic, and theoretical work to achieve the eventual goal of understanding the function of specific amino acids and cofactors in electron transfer.

C. Summary

The specific electron transfer characteristics of ruthenated proteins and photosynthetic RCs reflect relevant features of biological electron transfer systems in general:

1. For similar driving forces, a single rate–distance expression does not necessarily describe the behavior of different electron transfer systems. When all other factors are kept approximately constant, an exponential dependence of rate on distance is observed in ruthenated myoglobins. The nature of the intervening medium between donor and acceptor groups can significantly influence the rates of electron transfer observed in different systems, however. The presence of appropriate cofactors, amino acid side chains, and phytyl or isoprenoid tails between donors and acceptors appears to contribute to a substantial acceleration of the rate of electron transfer in RCs, relative to ruthenated proteins.
2. Rates are sensitive to the magnitude of the reorganization energy associated with a particular electron transfer step. The rapid rates of the initial electron transfer steps in RCs are associated with low values of the reorganization energy (about 0.3–0.5 eV). In contrast, the reorganization energies of ruthenated proteins are large (about 2 eV). As a consequence, the rates of electron transfer in these latter systems may be enhanced by increasing $\Delta E°$. The reorganization enthalpies of different proteins vary considerably, and they most probably reflect functionally important properties of the protein.
3. Conformational gating does not appear to be involved in electron transfer reactions in ruthenated systems, but it may influence the rates of the back reactions in RCs.
4. The use of site-directed mutagenesis techniques should be of great value in defining the pathways of electron transfer in proteins. Observations on the RC system suggest that electron transfer reactions appear rather robust to changes in conserved residues, in that very few residues which interact with the various cofactors are uniquely required for activity.

V. Future Prospects

As an outgrowth of the rapid development of experimental and theoretical approaches to biological electron transfer, significant progress in the 1990s may be expected in the following areas.

Thermodynamics of electron transfer. It should become possible to specifically alter reduction potentials by suitable modification of proteins and/or redox groups. The shifts in $E°$ arising from these modifications should also be accurately predictable from theoretical calculations. Reasonable agreement between observation and theory has been already reported for the effects of certain electrostatic modifications on $E°$ (*49, 53, 54, 59*). The more difficult problem of *ab initio* calculations of $E°$ values will take longer to accomplish, although encouraging results have been described for quinones (*194*).

Theoretical predictions of the $E°$ shifts accompanying redox protein modification are in many respects analogous to estimation of the effects of protein modification on the pK_a values of ionizable groups. A systematic comparison between experimental and theoretical pK shifts in subtilisin (*195–197*) demonstrated considerable success in the ability to modify pK values in a rational manner. Similar results should be obtainable for manipulation of reduction potentials.

Kinetics of electron transfer. The central problem in investigations of the kinetics of electron transfer is to establish the mechanism and pathways of electron transfer. Is the through space mechanism relevant to this process, or have specific electron conduction pathways (through bond or superexchange) been designed into electron transfer proteins? Once this question is resolved, then a quantitative understanding of electron transfer rates may be feasible.

Do well-defined electron transfer pathways exist in proteins? As numerical simulations by Kuki and Wolynes have illustrated (*198*), pathways can perhaps be defined only in some average fashion, since the electron is not a point particle quantum mechanically. The ambiguity in defining single pathways and the multiplicity of electron transfer mechanisms guarantee that establishing the number and nature of electron transfer pathways will be difficult. The most promising approaches to these problems are to use site-directed mutagenesis methods to replace different amino acids between (or around) the donor and acceptor groups and monitor the influences of these changes on the electron transfer rate. Since amino acid substitutions may also have longer range effects on the overall protein conformation, the interpretation of these results must be made carefully.

An instructive parallel may be drawn between the identification and characterization of electron transfer pathways and the identification and characterization of protein folding pathways (*199–202*). In the simplest cases, both processes are intramolecular and involve a transition between an initial and final state. An advantage to understanding electron transfer pathways is that both states are structurally ordered, whereas protein folding converts a "disordered" initial state into a structurally ordered final state. In both situations, a number of mechanisms for this transition have been envisioned, ranging from a few dominant pathways to many, different pathways. The challenge in both cases is to develop experimental approaches that will define and enumerate the number of pathways.

Recent work has concentrated on the contribution of individual residues to a pathway, using site-directed mutagenesis to replace specific amino acids. An interesting example of this approach to study protein folding is presented in Ref. (*202*). The main difficulty in these projects is the interpretation of experimental results: it may be quite challenging to separate the specific effects of an amino acid replacement on a process from the effect of a more general structural perturbation. For example, if the change of an amino acid alters the rate of electron transfer (or protein folding), is this a consequence of a direct involvement in a particular pathway, or does it reflect a change in structure that disrupts the process? It is clear that considerable experimental and theoretical ingenuity will be necessary to define the pathways of electron transfer (and protein folding) at the atomic level. Given the central importance of these problems to biochemistry and molecular biology, however, the challenge is certainly worth taking.

Specificity. Discrimination between correct and incorrect redox partners appears to be mediated by complementary structural interactions that can be satisfied by a correct pair of proteins. Important areas of investigation related to this process are the energetics of association and dissociation (partners need to interact specifically, but not so tightly that they are unable to dissociate after electron transfer), the structure and dynamics of the complex between redox partners, and how the complex facilitates the actual electron transfer. These questions are central to the biologically relevant question of how the correct electron transfer reactions occur in the presence of other thermodynamically suitable, but physiologically incorrect, donor and acceptor proteins.

Design of novel electron transfer proteins. Design of novel electron transfer proteins is a rather long-range goal, but it represents the ultimate test of an understanding of both protein structure and electron transfer mechanisms. The ability to design novel electron transfer molecules with specific properties will open up a vast expanse of scientific and biotechnological applications: (a) Design of electron transfer proteins with new redox properties. The potential of these developments is illustrated by the conversion of cytochrome b_5 to a peroxidase after alteration of the heme axial ligands by site-directed mutagenesis (*75*). The construction of tailor-made oxidation–reduction enzymes may be envisioned. (b) Development of more efficient solar cells. The nearly perfect quantum efficiency of photosynthetic RCs is achieved at the expense of losing more than 60% of the energy available in the photon initially absorbed (*176*). It might be possible to modify RCs (or develop analogous systems) that permit a more efficient energy conversion of the incident radiation. (c) Design of molecular electronic devices. The size of electronic devices is limited by the dimensions of the individual elements. Electron transfer proteins provide a mechanism for generating organized electronic circuits of molecular size. The potential of these devices has been illustrated by a recent proposal (*203*) for molecular shift registers (central to computational devices) based on electron transfer reactions.

By the very nature of the process, biological electron transfer has commanded the interest of molecular biologists, biochemists, as well as chemists and physicists. Rapid and exciting progress in experimental and theoretical areas promises developments in the understanding and design of electron transfer processes that will significantly impact on a variety of disciplines. The involvement of molecular biologists and biochemists is critical to the eventual realization of these goals.

Acknowledgments

Discussions with J. P. Allen, H. B. Gray, J. B. Howard, B. T. Hsu, H. Komiya, R. A. Marcus, and J. D. Morrison are gratefully acknowledged. This work was supported in part by the National Institutes of Health and the National Science Foundation. DCR is an A. P. Sloan research fellow.

References

1. Marcus, R. A., and Sutin, N. (1985). *BBA* **811,** 265.
2. McLendon, G. (1988). *Acc. Chem. Res.* **21,** 160.
3. Gray, H. B., and Malmstrom, B. G. (1989). *Biochemistry* **28,** 7499.
4. Cusanovich, M. A., Meyer, T. E., and Tollin, G. (1988). *Adv. Inorg. Biochem.* **7,** 37.
5. Tollin, G., Meyer, T. E., and Cusanovich, M. A. (1986). *BBA* **853,** 29.
6. Mathews, F. S. (1985). *Prog. Biophys. Mol. Biol.* **45,** 1.
7. DeVault, D. (1980). *Q. Rev. Biophys.* **13,** 387.
8. Sutin, N. (1983). *Prog. Inorg. Chem.* **30,** 441.
9. Guarr, T., and McLendon, G. (1985). *Coord. Chem. Rev.* **68,** 1.
10. Lehninger, A. (1982). "Principles of Biochemistry," 2nd Ed., p. 474. Worth, New York.
11. Rodkey, F. L., and Ball, E. G. (1950). *JBC* **182,** 17.
12. Pettigrew, G. W., Meyer, T. E., Bartsch, R. G., and Kamen, M. D. (1975). *BBA* **430,** 197.
13. Dickerson, R. E., and Timkovich, R. (1976). "The Enzymes," 3rd. Ed., Vol. 11, pp. 397–547.
14. Schejter, A., Aviram, I., and Goldkorn, T. (1982). *In* "Electron Transport and Oxygen Utilization" (C. Ho, ed.), pp. 95–99. Elsevier, New York.
15. Eisenberg, D., and Crothers, D. (1979). "Physical Chemistry with Applications to the Life Sciences," Chap. 9. Benjamin-Cummings, Menlo Park, California.
16. Perrin, D. D. (1959). *Rev. Pure Appl. Chem.* **9,** 257.
17. Gutmann, V. (1973). *Struct. Bonding (Berlin)* **15,** 141.
18. Moore, G. R., and Williams, R. J. P. (1976). *Coord. Chem. Rev.* **18,** 125.
19. Moore, G. R., and Williams, R. J. P. (1977). *FEBS Lett.* **79,** 229.
20. Weber, P. C., Howard, A., Xuong, N. H., and Salemme, F. R. (1981). *JMB* **153,** 399.
21. Varadarajan, R., Zewert, T. E., Gray, H. B., and Boxer, S. G. (1989). *Science* **243,** 69.
22. Poulos, T. L. (1988). *Adv. Inorg. Biochem.* **7,** 1.
23. Murthy, M. R. N., Reid, III, T. J., Sicignano, A., Tanaka, N., and Rossmann, M. G. (1981). *JMB* **152,** 465.
24. Colman, P. M., Freeman, H. C., Guss, J. M., Murata, M., Norris, V. A., Ramshaw, J. A. M., and Venkatappa, M. P. (1978). *Nature (London)* **272,** 319.
25. Adman, E. T., and Jensen, L. H. (1981). *Isr. J. Chem.* **21,** 8.

26. Carter, C. W., Jr., Kraut, J., Freer, S. T., Alden, R. A., Sieker, L. C., Adman, E., and Jensen, L. H. (1972). *PNAS* **69,** 3526.
27. Mayhew, S. G., and Ludwig, M. L. (1975). "The Enzymes," 3rd Ed., Vol. 12, pp. 57–118.
28. Okamura, M. Y., Feher, G., and Nelson, N. (1982). *In* "Photosynthesis" (Govindjee, ed.), Vol. 1, pp. 195–272. Academic Press, New York.
29. Holmgren, A., Söderberg, B.-O., Eklund, H., and Brändén, C.-I. (1975). *PNAS* **72,** 2305.
30. Wood, P. M. (1978). *EJB* **87,** 9.
31. Yoch, D. C., and Valentine, R. C. (1972). *Annu. Rev. Microbiol.* **26,** 139.
32. Marchon, J.-C., Mashiko, T., and Reed, C. A. (1982). *In* "Electron Transport and Oxygen Utilization" (C. Ho, ed.), pp. 67–72. Elsevier, New York.
33. Guss, J. M., Harrowell, P. R., Murata, M., Norris, V. A., and Freeman, H. C. (1986). *JMB* **192,** 361.
34. Garrett, T. P. J., Clingeleffer, D. J., Guss, J. M., Rogers, S. J., and Freeman, H. C. (1984). *JBC* **259,** 2822.
35. Rees, D. C., Howard, J. B., Chakrabarti, P., Yeates, T., Hsu, B. T., Hardman, K. D., and Lipscomb, W. N. (1986). *In* "Zinc Enzymes" (I. Bertini, C. Luchinat, W. Maret, and M. Zeppezauer, eds.), pp. 155–166. Birkhauser, Boston.
36. Schneider, G., Cedergren-Zeppezauer, E., Knight, S., Eklund, H., and Zeppezauer, M. (1985). *Biochemistry* **24,** 7503.
37. Martin, A., Burgess, B. K., Stout, C. D., Cash, V., Dean, D. R., Jensen, G. M., and Stephens, P. J. (1989). *PNAS* **87,** 598.
38. Adman, E., Watenpaugh, K. D., and Jensen, L. H. (1975). *PNAS* **72,** 4854.
39. Valentine, J. S., Sheridan, R. P., Allen, L. C., and Kahn, P. C. (1979). *PNAS* **76,** 1009.
40. Senn, H., and Wüthrich, K. (1985). *Q. Rev. Biophys.* **18,** 111.
41. Gandour, R. D. (1981). *Bioorg. Chem.* **10,** 169.
42. Carrell, C. J., Carrell, H. L., Erlebacher, J., and Glusker, J. P. (1988). *J. Am. Chem. Soc.* **110,** 8651.
43. Reynolds, W. F., Peat, I. R., Freedman, M. H., and Lyerla, J. R., Jr. (1973). *J. Am. Chem. Soc.* **95,** 328.
44. Tauscher, L., Ghisla, S., and Hemmerich, P. (1973). *Helv. Chim. Acta* **56,** 630.
45. Burnett, R. M., Darling, G. D., Kendall, D. S., LeQuesne, M. E., Mayhew, S. G., Smith, W. W., and Ludwig, M. L. (1974). *JBC* **249,** 4383.
46. Hemmerich, P., Massey, V., Michel, H., and Schug, C. (1982). *Struct. Bonding (Berlin)* **48,** 93.
47. Rees, D. C. (1980). *JMB* **141,** 323.
48. Mehler, E. L., and Eichele, G. (1984). *Biochemistry* **23,** 3887.
49. Warshel, A., and Russell, S. T. (1984). *Q. Rev. Biophys.* **17,** 283.
50. Warshel, A. (1987). *Nature (London)* **330,** 15.
51. Rees, D. C. (1985). *PNAS* **82,** 3082.
52. Smith, H. T., Staudenmayer, N., and Millett, F. (1977). *Biochemistry* **16,** 4971.
53. Rogers, N. K., Moore, G. R., and Sternberg, M. J. E. (1985). *JMB* **182,** 613.
54. Rogers, N. K. (1986). *Prog. Biophys. Mol. Biol.* **48,** 37.
55. Rogers, N. K., and Moore, G. R. (1988). *FEBS Lett.* **228,** 69.
56. Moore, G. R., Pettigrew, G. W., and Rogers, N. K. (1986). *PNAS* **83,** 4998.
57. Kassner, R. J. (1972). *PNAS* **69,** 2263.
58. Kassner, R. J. (1973). *J. Am. Chem. Soc.* **95,** 2674.
59. Churg, A. K., and Warshel, A. (1986). *Biochemistry* **25,** 1675.
60. Stellwagen, E. (1978). *Nature (London)* **275,** 73.
61. George, P., Hanania, G. I. H., and Eaton, W. A. (1966). *In* "Hemes and Hemoproteins" (B. Chance, R. W. Estabrook, and T. Yonetani, eds.), pp. 267–270. Academic Press, New York.

62. Clark, W. M. (1960). "Oxidation–Reduction Potentials of Organic Systems." Williams & Wilkins, Baltimore, Maryland.
63. Bashford, D., Karplus, M., and Canters, G. W. (1988). *JMB* **203,** 507.
64. Margalit, R., and Schejter, A. (1973). *EJB* **32,** 500.
65. Gopal, D., Wilson, G. S., Earl, R. A., and Cusanovich, M. A. (1988). *JBC* **263,** 11652.
66. Koppenol, W. H., Vroonland, C. A. J., and Braams, R. (1978). *BBA* **503,** 499.
67. Koppenol, W. H., and Margoliash, E. (1982). *JBC* **257,** 4426.
68. Walker, F. A., Emrick, D., Rivera, J. E., Hanquet, B. J., and Buttlaire, D. H. (1988). *J. Am. Chem. Soc.* **110,** 6234.
69. McGourty, J. L., Blough, N. V., and Hoffmann, B. M. (1983). *J. Am. Chem. Soc.* **105,** 4470.
70. Simolo, K. P., McLendon, G. L., Mauk, M. R., and Mauk, A. G. (1984). *J. Am. Chem. Soc.* **106,** 5012.
71. Massey, V., and Hemmerich, P. (1980). *Biochem. Rev.* **8,** 246.
72. Simondsen, R. P., and Tollin, G. (1983). *Biochemistry* **22,** 3008.
73. Okamura, M. Y., Isaacson, R. A., and Feher, G. (1975). *PNAS* **72,** 3491.
74. Gunner, M. R., Braun, B. S., Bruce, J. M., and Dutton, P. L. (1985). *In* "Antennas and Reaction Centers of Photosynthetic Bacteria" (M. E. Michel-Beyerle, ed.), pp. 298–304. Springer-Verlag, Berlin.
75. Sligar, S. G., Egeberg, K. D., Sage, J. T., Morikis, D., and Champion, P. M. (1987). *J. Am. Chem. Soc.* **109,** 7896.
76. Sorrell, T. N., Martin, P. K., and Bowden, E. F. (1989). *J. Am. Chem. Soc.* **111,** 766.
77. Mayo, S. L., Ellis, W. R., Crutchley, R. J., and Gray, H. B. (1986). *Science* **233,** 948.
78. Sykes, A. G. (1988). *Chem. Br.* **24,** 551.
79. Noren, C. J., Anthony-Cahill, S. J., Griffith, M. C., and Schultz, P. G. (1989). *Science* **244,** 182.
80. Matsumura, M., and Matthews, B. W. (1989). *Science* **243,** 792.
81. Bowler, B. E., Meade, T. J., Mayo, S. L., Richards, J. H., and Gray, H. B. (1989). *J. Am. Chem. Soc.* **111,** 8757.
82. Brandhuber, B. J., Boone, T., Kenney, W. C., and McKay, D. B. (1987). *JBC* **262,** 12306.
83. Parker, M. W., Pattus, F., Tucker, A. D., and Tsernoglou, D. (1989). *Nature (London)* **337,** 93.
84. Stock, A. M., Mottonen, J. M., Stock, J. B., and Schutt, C. E. (1989). *Nature (London)* **337,** 745.
85. Kaiser, E. T., and Lawrence, D. S. (1984). *Science* **226,** 505.
86. Tramontano, A., Janda, K. D., and Lerner, R. A. (1986). *Science* **234,** 1566.
87. Pollack, S. J., Jacobs, J. W., and Schultz, P. G. (1986). *Science* **234,** 1570.
88. Schwabacher, A. W., Weinhouse, M. I., Auditor, M.-T. M., and Lerner, R. A. (1989). *J. Am. Chem. Soc.* **111,** 2344.
89. Eisenberg, D., Wilcox, W., Eshita, S. M., Pryciak, P. M., Ho, S. P., and DeGrado, W. F. (1986). *Proteins* **1,** 16.
90. Regan, L., and DeGrado, W. F. (1988). *Science* **241,** 976.
91. DeGrado, W. F., Wasserman, Z. R., and Lear, J. D. (1989). *Science* **243,** 622.
92. Wherland, S., and Gray, H. B. (1977). *In* "Biological Aspects of Inorganic Chemistry" (W. R. Cullen, D. Dolphin, and B. R. James, eds.), pp. 289–368. Wiley, New York.
93. Margoliash, E., and Bosshard, H. R. (1983). *TIBS* **8,** 316.
94. Salemme, F. R. (1976). *JMB* **102,** 563.
95. Stonehuerner, J., Williams, J. B., and Millet, F. (1979). *Biochemistry* **18,** 5422.
96. Mauk, M. R., Mauk, A. G., Weber, P. C., and Matthew, J. B. (1986). *Biochemistry* **25,** 7085.
97. Rodgers, K. K., Pochapsky, T. C., and Sligar, S. G. (1988). *Science* **240,** 1657.
98. Poulos, T. L., and Kraut, J. (1980). *JBC* **255,** 10322.

99. Simondsen, R. P., Weber, P. C., Salemme, F. R., and Tollin, G. (1982). *Biochemistry* **21,** 6366.
100. Poulos, T. L., and Mauk, A. G. (1983). *JBC* **258,** 7369.
101. Livingston, D. J., McLachlan, S. J., LaMar, G. N., and Brown, W. D. (1985). *JBC* **260,** 15699.
102. Allen, J. P., Feher, G., Yeates, T. O., Komiya, H., and Rees, D. C. (1987). *PNAS* **84,** 6162.
103. Tiede, D. M., Budil, D. E., Tang, J., El-Kabbani, O., Norris, J. R., Chang, C.-H., and Schiffer, M. (1988). *In* "The Photosynthetic Bacterial Reaction Center" (J. Breton and A. Vermeglio, eds.), pp. 13–20. Plenum, New York.
104. Cambillau, C., Frey, M., Mosse, J., Guerlesquin, F., and Bruschi, M. (1988). *Proteins* **4,** 63.
105. Poulos, T. L., Sheriff, S., and Howard, A. J. (1987). *JBC* **262,** 13881.
106. Hazzard, J. T., Moench, S. J., Erman, J. E., Satterlee, J. D., and Tollin, G. (1988). *Biochemistry* **27,** 2002.
107. Peerey, L. M., and Kostic, N. M. (1989). *Biochemistry* **28,** 1861.
108. Hazzard, J. T., McLendon, G., Cusanovich, M. A., Das, G., Sherman, F., and Tollin, G. (1988). *Biochemistry* **27,** 4445.
109. Wendoloski, J. J., Matthew, J. B., Weber, P. C., and Salemme, F. R. (1987). *Science* **238,** 794.
110. Northrup, S. H., Boles, J. O., and Reynolds, J. C. L. (1988). *Science* **241,** 67.
111. Fehrst, A. (1985). "Enzyme Structure and Mechanism," 2nd Ed. Freeman, New York.
112. Lambeth, J. D., Seybert, D. W., and Kamin, H. (1979). *JBC* **254,** 7255.
113. Brooks, S. P. J., and Nicholls, P. (1982). *BBA* **680,** 33.
114. Pettigrew, G. W., and Seilman, S. (1982). *BJ* **201,** 9.
115. Geren, L. M., Stonehuerner, J., Davis, D. J., and Millett, F. (1983). *BBA* **724,** 62.
116. Hazzard, J. T., Cusanovich, M. A., Tainer, J. A., Getzoff, E. D., and Tollin, G. (1986). *Biochemistry* **25,** 3318.
117. Albery, W. J., and Knowles, J. R. (1976). *Biochemistry* **15,** 5631.
118. Knowles, J. R., and Albery, W. J. (1977). *Acc. Chem. Res.* **10,** 105.
119. Atkinson, D. E. (1977). "Cellular Energy Metabolism and Its Regulation." Academic Press, New York.
120. Ellington, A. D., and Benner, S. A. (1987). *J. Theor. Biol.* **127,** 491.
121. Yeates, T. O. (1987). Unpublished results.
122. Hoffman, B. M., and Ratner, M. A. (1987). *J. Am. Chem. Soc.* **109,** 6237; Hoffman, B. M., and Ratner, M. A. (1988). *J. Am. Chem. Soc.* **110,** 8267.
123. McLendon, G., Pardue, K., and Bak, P. (1987). *J. Am. Chem. Soc.* **109,** 7540.
124. Bernardo, M. M., Robandt, P. V., Schroeder, R. R., and Rorabacher, D. B. (1989). *J. Am. Chem. Soc.* **111,** 1224.
125. DeVault, D., and Chance, B. (1966). *BJ* **6,** 825.
126. Marcus, R. A. (1957). *J. Chem. Phys.* **26,** 867.
127. Marcus, R. A. (1964). *Annu. Rev. Phys. Chem.* **15,** 155.
128. Marcus, R. A. (1982). *Faraday Discuss. Chem. Soc.* **74,** 7.
129. Miller, J. R. (1975). *Science* **189,** 221.
130. Guarr, T., McGuire, M., Strauch, S., and McLendon, G. (1983). *J. Am. Chem. Soc.* **105,** 616.
131. Miller, J. R., Closs, G. L., and Calcaterra, L. T. (1984). *J. Am. Chem. Soc.* **106,** 3047.
132. Cohen-Tannoudji, C., Diu, B., and Laloë, F. (1977). "Quantum Mechanics," 2nd Ed., pp. 1156–1168. Wiley (Interscience), New York.
133. Landau, L. D., and Lifshitz, E. M. (1977). "Quantum Mechanics," 3rd Ed., pp. 164–194. Pergamon, Oxford.
134. Landau, L. D., and Lifshitz, E. M. (1977). "Quantum Mechanics," 3rd Ed., pp. 1299–1302. Pergamon, Oxford.

135. Endicott, J. F. (1988). *Acc. Chem. Res.* **21,** 59.
136. Petrov, E. G. (1979). *Int. J. Quantum Chem.* **16,** 133.
137. Anderson, P. W. (1959). *Phys. Rev.* **115,** 2.
138. McConnell, H. M. (1961). *J. Chem. Phys.* **35,** 508.
139. Beratan, D. N., Onuchic, J. N., and Hopfield, J. J. (1987). *J. Chem. Phys.* **86,** 4488.
140. Larsson, S. (1988). *Chem. Scripta* **28A,** 15.
141. Nocera, D. G., Winkler, J. R., Yocom, K. M., Bordignon, E., and Gray, H. B. (1984). *J. Am. Chem. Soc.* **106,** 5145.
142. Isied, S. S., Kuehn, C., and Worosila, G. (1984). *J. Am. Chem. Soc.* **106,** 1722.
143. Elias, H., Chou, M. H., and Winkler, J. R. (1988). *J. Am. Chem. Soc.* **110,** 429.
144. Meade, T. J., Gray, H. B., and Winkler, J. R. (1989). *J. Am. Chem. Soc.* **111,** 4353.
145. Osvath, P., Salmon, G. A., and Sykes, A. G. (1988). *J. Am. Chem. Soc.* **110,** 7114.
146. Kostic, N. M., and Margalit, R., Che, C.-M., and Gray, H. B. (1983). *J. Am. Chem. Soc.* **105,** 7765.
147. Jackman, M. P., McGinnis, J., Powls, R., Salmon, G. A., and Sykes, A. G. (1988). *J. Am. Chem. Soc.* **110,** 5880.
148. Jackman, M. P., Lim, M.-C., Sykes, A. G., and Salmon, G. A. (1988). *J. Chem. Soc. Dalton Trans.*, p. 2843.
149. Crutchley, R. J., Ellis, W. R., and Gray, H. B. (1985). *J. Am. Chem. Soc.* **107,** 5002.
150. Lieber, C. M., Karas, J. L., and Gray, H. B. (1987). *J. Am. Chem. Soc.* **109,** 3778.
151. Cowan, J. A., Upmacis, R. K., Beratan, D. N., Onuchic, J. N., and Gray, H. B. (1988). *Ann. N.Y. Acad. Sci.* **550,** 58.
152. Karas, J. L., Lieber, C. M., and Gray, H. B. (1988). *J. Am. Chem. Soc.* **110,** 599.
153. Cowan, J. A., and Gray, H. B. (1988). *Chem. Scripta* **28A,** 21.
154. Axup, A. W., Albin, M., Mayo, S. L., Crutchley, R. J., and Gray, H. B. (1988). *J. Am. Chem. Soc.* **110,** 435.
155. Churg, A. K., Weiss, R. M., Warshel, A., and Takano, T. (1983). *J. Phys. Chem.* **87,** 1683.
156. Warshel, A., and Churg, A. K. (1983). *JMB* **168,** 693.
157. Michel, H. (1982). *JMB* **158,** 567.
158. Deisenhofer, J., Epp, O., Miki, K., Huber, R., and Michel, H. (1984). *JMB* **180,** 385.
159. Deisenhofer, J., Epp, O., Miki, K., Huber, R., and Michel, H. (1985). *Nature (London)* **318,** 618.
160. Michel, H., Epp, O., and Deisenhofer, J. (1986). *EMBO J.* **5,** 2445.
161. Allen, J. P., and Feher, G. (1984). *PNAS* **81,** 4795.
162. Allen, J. P., Feher, G., Yeates, T. O., Rees, D. C., Deisenhofer, J., Michel, H., and Huber, R. (1986). *PNAS* **83,** 8589.
163. Allen, J. P., Feher, G., Yeates, T. O., Komiya, H., and Rees, D. C. (1987). *PNAS* **84,** 5730.
164. Allen, J. P., Feher, G., Yeates, T. O., Komiya, H., and Rees, D. C. (1987). *PNAS* **84,** 6162.
165. Yeates, T. O., Komiya, H., Rees, D. C., Allen, J. P., and Feher, G. (1987). *PNAS* **84,** 6438.
166. Yeates, T. O., Komiya, H., Chirino, A., Rees, D. C., Allen, J. P., and Feher, G. (1988). *PNAS* **85,** 7993.
167. Allen, J. P., Feher, G., Yeates, T. O., Komiya, H., and Rees, D. C. (1988). *PNAS* **85,** 8487.
168. Komiya, H., Yeates, T. O., Rees, D. C., Allen, J. P., and Feher, G. (1988). *PNAS* **85,** 9012.
169. Chang, C.-H., Schiffer, M., Tiede, D., Smith, U., and Norris, J. (1985). *JMB* **186,** 201.
170. Chang, C.-H., Tiede, D., Tang, J., Smith, U., Norris, J., and Schiffer, M. (1986). *FEBS Lett.* **205,** 82.
171. Michel-Beyerle, M. E. (ed.) (1985). "Antennas and Reaction Centers of Photosynthetic Bacteria." Springer-Verlag, Berlin.
172. Biggens, J. (ed.) (1987). "Progress in Photosynthesis Research." Martinus Nijhoff, Dordrecht.
173. Kirmaier, C., and Holten, D. (1987). *Photosynth. Res.* **13,** 225.

174. Breton, J., and Vermeglio, A. (eds.) (1988). "The Photosynthetic Bacterial Reaction Center." Plenum, New York.
175. Deisenhofer, J., and Michel, H. (1989). *EMBOJ.* **8,** 2149.
176. Feher, G., Allen, J. P., Okamura, M. Y., and Rees, D. C. (1989). *Nature (London)* **339,** 111.
177. Rees, D. C., Komiya, H., Yeates, T. O., Allen, J. P., and Feher, G. (1989). *Annu. Rev. Biochem.* **58,** 607.
178. Feher, G., and Okamura, M. Y. (1978). *In* "The Photosynthetic Bacteria" (R. K. Clayton and W. R. Sistrom, eds.), pp. 349–386. Plenum, New York.
179. Wraight, C. A., and Clayton, R. K. (1973). *BBA* **333,** 246.
180. Fleming, G. R., Martin, J. L., and Breton, J. (1988). *Nature (London)* **333,** 190.
181. Marcus, R. A. (1987). *Chem. Phys. Lett.* **133,** 471.
182. Bixon, M., Jortner, J., Michel-Beyerle, M. E., Ogrodnik, A., and Lersch, W. (1987). *Chem. Phys. Lett.* **140,** 626.
183. Creighton, S., Hwang, J.-K., Warshel, A., Parson, W. W., and Norris, J. (1988). *Biochemistry* **27,** 774.
184. Plato, M., Michel-Beyerle, M. E., Bixon, J., and Jortner, J. (1989). *FEBS Lett.* **249,** 70.
185. Okamura, M. Y., and Feher, G. (1989). *BJ* **55,** 221a.
186. Debus, R. J., Feher, G., and Okamura, M. Y. (1986). *Biochemistry* **25,** 2276.
187. Michel-Beyerle, M. E., Plato, M., Deisenhofer, J., Michel, H., Bixon, M., and Jortner, J. (1988). *BBA* **932,** 52.
188. Bylina, E. J., Kirmaier, C., McDowell, L., Holten, D., and Youvan, D. C. (1988). *Nature (London)* **336,** 182.
189. Bylina, E. J., Jovine, R., and Youvan, D. C. (1988). *In* "The Photosynthetic Bacterial Reaction Center" (J. Breton and A. Vermeglio, eds.), pp. 113–118. Plenum, New York.
190. Paddock, M. L., Rongey, S. H., Feher, G., and Okamura, M. Y. (1989). *PNAS* **86,** 6602.
191. Paddock, M. L., Rongey, S. H., Abresch, E. C., Feher, G., and Okamura, M. Y. (1988). *Photosynth. Res.* **17,** 75.
192. Bylina, E. J., and Youvan, D. C. (1988). *PNAS* **85,** 7226.
193. Kirmaier, C., Holten, D., Bylina, E. J., and Youvan, D. C. (1988). *PNAS* **85,** 7562.
194. Reynolds, C. A., King, P. M., and Richards, W. G. (1988). *Nature (London)* **334,** 80.
195. Sternberg, M. J. E., Hayes, F. R. F., Russell, A. J., Thomas, P. G., and Fersht, A. R. (1987). *Nature (London)* **330,** 86.
196. Russell, A. J., Thomas, P. G., and Fersht, A. R. (1987). *JMB* **193,** 803.
197. Russell, A. J., and Fersht, A. R. (1987). *Nature (London)* **328,** 496.
198. Kuki, A., and Wolynes, P. G. (1987). *Science* **236,** 1647.
199. Creighton, T. E. (1985). *J. Phys. Chem.* **89,** 2452.
200. Harrison, S. C., and Durbin, R. (1985). *PNAS* **82,** 4028.
201. Kim, P. S., and Baldwin, R. L. (1982). *Annu. Rev. Biochem.* **51,** 459.
202. Goldenberg, D. P., Frieden, R. W., Haack, J. A., and Morrison, T. B. (1989). *Nature (London)* **338,** 127.
203. Hopfield, J. J., Onuchic, J. N., and Beratan, D. N. (1988). *Science* **241,** 817.

3

Steady-State Kinetics

W. W. CLELAND
Enzyme Institute
University of Wisconsin-Madison
Madison, Wisconsin 53705

I. Introduction

With the new tool of site-directed mutagenesis available to the enzymologist, it becomes possible to change putative catalytic groups on enzymes and thus test their proposed roles in the reaction. Once mutants are made and purified, however, it is necessary to characterize their catalytic properties, and this requires fairly detailed kinetic analysis. Thus, it is more than ever important for the enzymologist to have a good working knowledge of enzyme kinetic methods.

THE ENZYMES, Vol. XIX

It was 20 years ago that this author wrote a chapter on steady-state kinetics for Volume II of this treatise (*1*). Most of what is in that chapter is still valid and useful today, so we do not repeat it here, but rather discuss the new methods developed during the intervening period. We assume that the reader has read the previous chapter in Volume II. The reader may also benefit from other general articles on enzyme kinetics (*2, 3*).

There are four stages in the kinetic analysis of enzyme mechanisms. First, one determines the kinetic mechanism, which defines the order of combination of substrates and release of products. Most of the methods for doing this were described in Volume II, but a few new ones, such as the use of isotope effects, have been developed and are described here. This is fundamentally qualitative information.

Second, one determines the relative rates of the various steps in the kinetic mechanism; this is quantitative information. This includes determining whether the substrates are sticky (i.e., react to give products after combining with the enzyme faster than they are released back off of the enzyme), as well as determining which are the slow steps that limit V and V/K. A number of new methods have been developed in this area, and these are described in Section VI.

Third, one determines the chemical mechanism (i.e., a description of what happens in the terms used by a physical organic chemist). The type of acid–base chemistry involved and whether a reaction is concerted or stepwise (and, in this case, the nature of the intermediate) are involved here. The kinetic methods for determining chemical mechanism are almost all entirely new, and we spend a major part of this chapter in a description of the use of pH profiles and isotope effects to deduce the chemical mechanism. This can only be done after the kinetic mechanism and relative rates of the steps in it are known, however, so Steps 1 and 2 must be undertaken first.

The fourth stage of analysis is determination of transition state structures. Clearly, this requires all of the information from Steps 1–3, but in favorable cases we can use isotope effects in the same fashion as the physical organic chemist to get information on transition state structure. The use of structure–function relationships is not so practical, since specificity problems usually distort the results and make interpretation difficult. We shall describe the methods now available for determining intrinsic isotope effects on bond-breaking steps so that this approach can be applied.

II. Nomenclature and Notation

We use the nomenclature and notation which this author introduced in 1963 (*4*). If the reader is not familiar with it, he or she should consult Volume II or read the original article (*4*).

III. Basic Theory

Steady-state kinetics applies whenever the concentration of the substrate is well above that of the enzyme, so that the rate of change of substrate concentration greatly exceeds the rate of change of the concentration of any enzyme form. The resulting equations for velocity are the ratio of polynomials in reactant concentrations, and if only one substrate concentration is varied, the velocity is usually given by

$$v = VA/(K + A) \quad (1)$$

where v is velocity, V is the maximum velocity at infinite substrate concentration, A is substrate concentration, and K is the Michaelis constant. The Michaelis constant measures affinity in the steady state (as opposed to at equilibrium), but it is not an independent kinetic constant. The independent kinetic constants are V and V/K [the apparent first-order rate constant at low substrate level where $v = (V/K)A$]. V and V/K vary independently with pH, ionic strength, temperature, and the concentrations of reactants, activators, and inhibitors. The Michaelis constant is simply the ratio of V and V/K, and it can be greater than, equal to, or less than the dissociation constant of the substrate.

Equation (1) is usually inverted for plotting and analysis, since this gives a straight line with slope K/V (the reciprocal of V/K) and vertical intercept $1/V$ (the reciprocal of V) when $1/v$ is plotted versus $1/A$:

$$1/v = (K/V)(1/A) + (1/V) \quad (2)$$

In analyzing initial velocity or inhibition patterns, one considers separately the effects on the slopes of such reciprocal plots (which represent effects on V/K) and on the intercepts (which represent effects on V). Similarly, one considers isotope effects on V or V/K separately, and one plots the logarithms of V or V/K versus pH for pH profiles.

Steady-state kinetic studies are usually carried out by measuring the initial velocities in separate reaction mixtures containing different concentrations of substrates, activators, or inhibitors. There are several reasons for this. First, the concentrations of reactants are changing as the reaction proceeds, and in many cases developing product inhibition as well as substrate depletion changes the rate. The integrated rate equations for a reaction involving more than one substrate are quite complex, and it is not an easy task to try to extract multiple kinetic constants from time course data. Second, the enzyme may die as the reaction proceeds (although the presence of inert protein such as serum albumin prevents the surface denaturation that often causes this); this is particularly a problem at extremes of pH. Third, the pH may change if protons are taken up or released during reaction. Finally, statistical analysis of time course data is a real problem. The points on a time course of a reaction are highly correlated, and thus a least-squares fit to the integrated rate equation gives a misleadingly low standard error for the parameters. In practice one needs to run several time courses at different reactant or inhibitor levels in order to determine all of the kinetic constants, and no one has yet solved the problem of how to extract meaningful standard errors for the kinetic constants from such analysis.

Initial velocity measurements, by contrast, are made on separate reaction mixtures, and thus are all independent measurements. This simplifies statistical analysis, and Fortran programs are available for fitting most rate equations that will be encountered in steady-state enzymic reactions (*5*). A word of caution: Do not use a pocket calculator to fit a straight line to kinetic data fitting Eq. (2). A fit in reciprocal form must be weighted with either the second powers of the velocities (if errors in velocities are proportional to velocities) or the fourth powers (if errors in velocities are constant). The available computer programs make such weighted fits to Eq. (2), or similar equations in reciprocal form, and then fit the velocities (if errors are constant) to the equation in nonreciprocal form by iterative least-squares methods (*6*). If errors in velocities are proportional to velocities, the final fits are made in the log form [i.e., $\log v = \log[V\mathrm{A}/(K + \mathrm{A})]$ for Eq. (1), for example].

IV. Derivation of Rate Equations

At the time that the previous chapter in Volume II was written, the method of King and Altman (*7*) was the method of choice for deriving steady-state rate equations for enzymic reactions, and this is still true for any mechanism involving branched reaction pathways. The best description of this method may be found in Mahler and Cordes (*8*). A useful advance was made in 1975 with the introduction of the net rate constant method (*9*), and because it is the simplest method to use for any nonbranched mechanism, as well as for equations for isotopic exchange, positional isotopic exchange, isotope partitioning, etc., we shall present it here.

The method involves replacing the mechanism being considered, such as the following:

$$\mathrm{E} \underset{k_2}{\overset{k_1\mathrm{A}}{\rightleftharpoons}} \mathrm{EA} \underset{k_4}{\overset{k_3}{\rightleftharpoons}} \mathrm{EPQ} \underset{k_6\mathrm{P}}{\overset{k_5}{\rightleftharpoons}} \mathrm{EQ} \xrightarrow{k_7} \mathrm{E + Q} \quad (3)$$

by a corresponding one in which each step is irreversible, but the flux through each step and the distribution of enzyme forms are the same as for Mechanism (3):

$$\mathrm{E} \xrightarrow{k_1'} \mathrm{EA} \xrightarrow{k_3'} \mathrm{EPQ} \xrightarrow{k_5'} \mathrm{EQ} \xrightarrow{k_7'} \mathrm{E + Q} \quad (4)$$

In Mechanism (4), k_1', k_3', k_5', and k_7' are thus "net rate constants" which will produce the same rate and distribution of enzyme forms as in Mechanism (3). It is easy to demonstrate that the rate equation is then

$$v = E_t/(1/k_1' + 1/k_3' + 1/k_5' + 1/k_7') \tag{5}$$

To determine the net rate constants for use in Mechanism (4), one starts with an irreversible step such as the release of Q in Mechanism (3). For such a step the net rate constant equals the real one:

$$k_7' = k_7 \tag{6}$$

One next moves to the left and writes

$$k_5' = k_5[k_7/(k_6\mathrm{P} + k_7)] \tag{7}$$

That is, the net rate constant will be the real forward rate constant times the fraction of EQ that reacts forward as opposed to undergoing reversal to EPQ. Moving to the left again,

$$k_3' = k_3[k_5'/(k_4 + k_5')] = k_3k_5k_7/(k_4k_6\mathrm{P} + k_4k_7 + k_5k_7) \tag{8}$$

The net rate constant equals the true one times the partition ratio for the EPQ complex. Moving again to the left,

$$\begin{aligned} k_1' &= k_1\mathrm{A}[k_3'/(k_2 + k_3')] \\ &= k_1k_3k_5k_7\mathrm{A}/(k_2k_4k_6\mathrm{P} + k_2k_4k_7 + k_2k_5k_7 + k_3k_5k_7) \end{aligned} \tag{9}$$

Note that when a reactant adds, it is the product of the rate constant and the concentration factor that is used. The overall rate equation is now obtained by substituting Eqs. (6)–(9) into Eq. (5).

If one does not want the entire rate equation, but only V or V/K, these can be written down directly. Since A is infinite at saturating A, k_1' will be infinite, and thus will not contribute to Eq. (5). V is then given by E_t divided by the sum of the reciprocals of k_3', k_5', and k_7. To calculate V/K, we remember that the concentration of A is extrapolated to near zero, so that k_1' is very much smaller than the other net rate constants and thus will totally dominate the rate equation, which now is given by $v = (V/K)\mathrm{A}$. Since only the reciprocal of k_1' is important in Eq. (5), V/K is given by $k_1'E_t/\mathrm{A}$.

For mechanisms involving random addition of substrates, the King–Altman method gives squared terms in numerator and denominator of the rate equations, which are messy and difficult to work with. The method of Cha (*10*) treats each random segment as if it were in rapid equilibrium, and this simplifies the rate equation. The fact that data fit such a simplified equation does not prove that the mechanism is a rapid equilibrium one (see the rules in Section V,A,2 below) but does facilitate initial velocity analysis.

The rate equations for multisite Ping-Pong mechanisms require a special approach when reaction at one or more sites is Ping-Pong. The interested reader can consult the article by Cleland which describes how to derive such equations (*11*).

V. Determination of Kinetic Mechanism

A. Initial Velocity Studies

1. *Crossover Point Analysis*

The initial velocity patterns for two-substrate cases were dealt with in Volume II, but one quantitative analysis of such patterns has been developed since then. A sequential mechanism with two substrates has an initial velocity rate equation:

$$v = V_1\mathrm{AB}/(K_{ia}K_b + K_a\mathrm{B} + K_b\mathrm{A} + \mathrm{AB}) \tag{10}$$

which plots in reciprocal form ($1/v$ vs. 1/A or 1/B) as a series of lines intersecting to the left of the vertical axis. The vertical coordinate of the crossover (Xover) point (which is the same whether 1/A or 1/B is plotted on the horizontal axis) is

$$1/v_{\mathrm{Xover}} = (1/V_1)(1 - K_a/K_{ia}) \tag{11}$$

In Eqs. (10) and (11), K_{ia} and K_a are the dissociation constant from EA and the Michaelis constant (with B = ∞) for substrate A, which is the first one to add in an ordered mechanism, or either one in a random mechanism.

An equation similar to Eq. (10) applies in the reverse direction, with P and Q replacing B and A. The vertical coordinate of the crossover point in this direction is given by

$$1/v_{\mathrm{Xover}} = (1/V_2)(1 - K_q/K_{iq}) \tag{12}$$

where K_{iq} and K_q are the dissociation constant from EQ and the Michaelis constant (P = ∞) for Q, the last product released in an ordered mechanism, or either one in a random mechanism. V_1 and V_2 are maximum velocities in the forward and reverse directions.

As long as V_1 and V_2 are measured at equal enzyme concentrations (or corrected to this), we can calculate a parameter R:

$$R = \frac{\Sigma\text{vertical coordinates of crossover points}}{1/V_1 + 1/V_2} \tag{13}$$

This parameter tells where the rate-limiting step is in the direction with the slower maximum velocity. In an ordered mechanism, R can only vary from zero to one, with 0 corresponding to a Theorell–Chance mechanism where second product release is totally rate limiting and catalysis and first product release are so fast that there is a very low steady-state level of ternary complexes. In fact, in a simple Theorell–Chance mechanism, the crossover point in Eq. (11) is given by $1/V_1 - 1/V_2$ in the forward direction and that in Eq. (12) by $1/V_2 - 1/V_1$ in the reverse direction, and failure of the data to fit this pattern is very good evidence against a Theorell–Chance mechanism and in favor of a random mechanism with dead-end EAP and EBQ complexes (these mechanisms give the same product inhibition patterns). In a random mechanism, R can have any value less than unity, including negative ones.

An R value of 1.0 indicates an equilibrium ordered mechanism in which catalysis and release of the first product totally limit V. In the general case of an ordered mechanism, R is the proportion of enzyme in central complexes in the steady state in the slower direction, and $R/(1 - R)$ is thus the ratio of central complex concentration to the concentration of the E–second product complex, and also the ratio of the rate constant for second product release and the rate constant for catalysis and first product release. Application of this method to glycerokinase gave an R value of 5/6, and, thus, in the slow back reaction, release of glycerol (the second product) was 5 times faster than catalysis and release of MgATP (*12*).

2. *Initial Velocity Patterns for Terreactant Mechanisms*

The general form for the rate equation when there are three substrates (as long as reciprocal plots are always linear) is

$$v = VABC/[\text{constant} + (\text{coef. A})A + (\text{coef. B})B + (\text{coef. C})C + K_aBC + K_bAC + K_cAB + ABC] \quad (14)$$

where V is the maximum velocity, and K_a, K_b, and K_c are Michaelis constants for A, B, and C. The definitions of the coefficients of the A, B, and C terms in the denominator depend on the mechanism.

In Ping-Pong mechanisms the constant term is absent, whereas in sequential mechanisms it is always present. Terreactant Ping-Pong mechanisms were discussed in Volume II, but only a few of the possible sequential cases were discussed. Viola and Cleland have described all of the possible sequential terreactant mechanisms (*13*), and we shall describe them briefly.

The denominator of Eq. (14) for a sequential mechanism will normally contain the constant and ABC terms, and at least one single concentration term (A, B, or C) and one double concentration term (AB, AC, or BC). Further, at least one double concentration term must contain a letter present as a single concentration term (i.e., A and AB terms are possible, but one cannot have only A and BC terms). This is because the single and double concentration terms (when only one is present) correspond to the complexes that can form, and there must be a path by which substrates add sequentially to reach the EABC complex.

Viola and Cleland formulated four rules for sequential terreactant initial velocity patterns (*13*): (1) The number of terms in the denominator of Eq. (14) depends on whether the first two reactants to add do so in steady-state fashion (k_{off} is less than, or does not greatly exceed, V/E_t, the turnover number) or in rapid equilibrium ($k_{off} \gg V/E_t$), but it does not matter whether the last substrate to add does so in steady-state or rapid equilibrium fashion. (2) When reactants add in rapid equilibrium fashion, some steps connecting the various complexes can be missing (i.e., go at negligible rate) as long as all complexes present can still be interconverted by some rapid equilibrium pathway. (3) Where a substrate adds in steady-state fashion, it always has a finite Michaelis constant. Conversely, the lack of a K_aBC, K_bAC, or K_cAB term indicates: (a) rapid equilibrium addition of the corresponding substrate when it adds in first or second position and (b) absence of the corresponding complex (i.e., lack of a K_aBC term means no EBC complex). When a Michaelis constant is finite, however, you may have either steady-state or rapid equilibrium addition of the corresponding substrate, and the ternary complex may be absent if addition of the substrate is in steady state. (4) The lack of an A, B, or C term in the denominator of Eq. (14) indicates absence of the corresponding binary complex in the mechanism, and thus an obligatory order of addition for at least one substrate (i.e., it cannot add until one of the others does, and then it prevents the other one from leaving the enzyme). Presence of an A, B, or C term does not require that the corresponding binary complex form, however, if the mechanism is a steady-state one.

Note that these rules correctly predict that in a two-substrate case the only sequential mechanism in which a term is missing from the denominator is an ordered one in which the first substrate addition is in rapid equilibrium. Rapid equilibrium binding in a random mechanism does not change the initial velocity rate equation (Rule 1), since both substrates can add in the second position. These rules can easily be generalized for cases with four or more substrates.

We describe the possible sequential terreactant mechanisms below; the reader should consult Viola and Cleland (*13*) for a more complete discussion.

1. All terms in the denominator of Eq. (14) present. This mechanism entails fully random addition, or random steady-state addition of two substrates, with ordered addition of the third. A number of enzymes have fully random mechanisms.
2. One single concentration term missing. The missing term is the B term, and B cannot add until A is present. C adds either third (fully ordered mechanism) or randomly with respect to A and B. Glutamate dehydrogenase ($NADP^+$) has ordered addition of NADPH, α-ketoglutarate, and ammonia (*14*), whereas the oxidation of mevaldate by $NADP^+$ in the presence of coenzyme A (CoA) by hydroxymethylglutaryl-CoA reductase (NADPH) shows ordered addition of CoA and mevaldate and random addition of $NADP^+$ (*15*).

3. Only one single concentration term present, but all three Michaelis constants have finite values. The term present is the A term, and the mechanism involves steady-state addition of A, followed by random addition of B and C. Citrate cleavage enzyme [ATP citrate (*pro*-3*S*)-lyase] has this mechanism (*16*).

4. All single concentration terms present, but one Michaelis constant is zero. The addition of the substrate whose K_m is zero must be in rapid equilibrium, and it must add first or second. No mechanisms of this type are known. See Viola and Cleland (*13*) for the possible mechanisms, which are very unrealistic.

5. All single concentration terms are present, but two Michaelis constants are zero. The two substrates whose K_m values are zero (A and B) must add randomly and in rapid equilibrium, with C adding third and also forming a dead-end EC complex. This is an unrealistic mechanism.

6. One single concentration term and one double concentration term (sharing the same letter) missing. One possibility is a fully ordered mechanism with addition of A in rapid equilibrium and addition of B in steady state. The missing terms are the B and K_aBC ones. Another possibility is the equilibrium ordered addition of A and B in that order, with C adding randomly (K_b is not zero because B can add last). A third possibility is that A and B add randomly, with addition of A in rapid equilibrium but addition of B in steady state, and with C adding only to the EAB complex. In this case, C and K_aBC terms are missing. Biotin carboxylase has this mechanism in the presence of high free Mg^{2+} (A is bicarbonate, B MgATP, and C biotin) (*17*).

7. One single concentration term missing, and the corresponding Michaelis constant equal to zero (B and K_bAC terms missing). This is an unlikely ordered mechanism in which A adds in steady state, B adds in rapid equilibrium, and C forms a dead-end EC complex.

8. Only one single concentration term present, and corresponding Michaelis constant equal to zero (B, C, and K_aBC terms missing). Rapid equilibrium addition of A is followed by random addition of B and C.

9. Only one single concentration term present, and another Michaelis constant equal to zero (B, C, K_bAC terms missing). This is an ordered mechanism with addition of A in steady state and B in rapid equilibrium, and examples include glutamate dehydrogenase with α-ketovalerate or α-ketobutyrate as substrates (*14*).

10. One single concentration term missing and two Michaelis constants equal to zero, with one corresponding to the missing single concentration term (B, K_aBC, and K_bAC terms missing). This unlikely mechanism entails equilibrium ordered addition of A and B, with C adding third, and also forming an EC dead-end complex.

11. One single concentration term missing, and the two noncorresponding Michaelis constants equal to zero (C, K_aBC, and K_bAC terms missing). The mechanism proceeds by rapid equilibrium random addition of A and B, with C

adding only to EAB. Biotin carboxylase should show this pattern if the level of free Mg^{2+} were lowered enough for MgATP addition to be in rapid equilibrium (*17*).

12. Only one single concentration term present and one finite Michaelis constant (A and K_cAB terms present). This mechanism requires equilibrium ordered addition of A and B, followed by C.

To determine which terms are present or absent in Eq. (14), one can run initial velocity patterns for two of the substrates at high or low levels of the third (*13*), but the best procedure is to measure initial velocities as a function of all three substrate concentrations and fit the overall set of data (obtained on the same day) to Eq. (14). If the coefficient of any denominator term comes out negative or not significantly different from zero, the data should be fitted to the equation with this term left out. If the values of the other kinetic constants do not change appreciably and the residual least square does not increase, the omission of the term is justified. For example, with initial velocity data for glutamate dehydrogenase, the coefficient of the [α-ketoglutarate] term in the denominator was 0.014 $\pm$ 0.015, but no other term had a standard error as large as half its value (*14*). When this term was left out, the residual least square dropped from 0.0488 to 0.0486 (the drop comes from increasing the degrees of freedom by one), demonstrating the absence of the [α-ketoglutarate] term and the ordered nature of the reaction.

Not many terreactant mechanisms have had full initial velocity patterns determined, but because of the many possible mechanisms that can be distinguished, this type of kinetic study gives much information, and should be more widely used. In particular, the altered patterns when a slow alternate substrate (or altered mutant enzyme) are used can give much information on the kinetic mechanism.

3. *Haldane Relationships*

Haldane relationships are relationships between the equilibrium constant and the various kinetic constants defined for a given mechanism. They exist because we define for each mechanism more kinetic parameters than there are independently determinable parameters. They are of two types, kinetic and thermodynamic, and every mechanism has at least one of each. Thermodynamic Haldanes consist of the cross product of reciprocal dissociation constants for the substrates and dissociation constants for the products (i.e., the product of equilibrium constants for each step in the mechanism). For mechanisms with at least three substrates, the Cleland notation defines dissociation constants as K_i values (i.e., K_{ia}, K_{ib}, K_{ic}, etc.), but for Ordered Uni Bi and Bi Bi mechanisms, Cleland defined the dissociation constants of the inner substrates differently (*4*). The dissociation constants for A and Q were K_{ia} and K_{iq}, but that for B was

$K_{ib}K_aV_2/(K_{ia}V_1)$, whereas that for P was $K_{ip}K_qV_1/(K_{iq}V_2)$. The thermodynamic Haldane for Ordered Uni Bi is thus

$$K_{eq} = V_1K_{ip}K_q/(V_2K_{ia}) \quad (15)$$

whereas that for Ordered Bi Bi is

$$K_{eq} = (V_1/V_2)^2K_{ip}K_q/(K_aK_{ib}) \quad (16)$$

For Ordered Ter Bi, the thermodynamic Haldane is

$$K_{eq} = K_{ip}K_{iq}/(K_{ia}K_{ib}K_{ic}) \quad (17)$$

The kinetic Haldane is the ratio of the apparent rate constants in forward and reverse directions when reactant concentrations are very low. With only one substrate, the apparent rate constant is the V/K value, but with more than one substrate the V/K value for the last substrate to add is multiplied by the reciprocal dissociation constants of all substrates that have previously added to the enzyme. The kinetic Haldane relationships for several sequential mechanisms are given as

Uni Uni: $$K_{eq} = V_1K_p/(V_2K_a) \quad (18)$$

Ordered Uni Bi: $$K_{eq} = V_1K_pK_{iq}/(V_2K_a) \quad (19)$$

Ordered Bi Bi: $$K_{eq} = V_1K_pK_{iq}/(V_2K_{ia}K_b) \quad (20)$$

Ordered Ter Bi: $$K_{eq} = V_1K_pK_{iq}/(V_2K_{ia}K_{ib}K_c) \quad (21)$$

Ping-Pong mechanisms have separate thermodynamic and kinetic Haldanes for each half-reaction, and the overall Haldanes are the product of those for each half-reaction. Since one can combine two thermodynamic or two kinetic Haldanes, or one of each, there are four possible Haldanes for a Ping-Pong mechanism with two half-reactions, although only two of these will have V_1/V_2 to the first power (the other two have this ratio to the second or zero power).

Haldanes are useful for checking the consistency of experimental kinetic constants with the measured equilibrium constant. The Haldanes for a number of mechanisms are given by Cleland (*18*).

4. *Alternate Substrate Studies*

Although the kinetics of the physiological substrates for an enzyme are always of interest, much useful information can be obtained about the mechanism by using alternate substrates. Substrates that react at less than 10% the rate of the normal ones will generally show rapid equilibrium binding and not be sticky.

This may change the initial velocity pattern, as in the case of fructose-6-sulfate, which is a slow substrate for phosphofructokinase. This substrate has lost sufficient affinity for the enzyme that it binds only when MgATP is present, and thus the mechanism changes from a random one with both substrates sticky with fructose 6-phosphate (fructose-6-P) to an equilibrium ordered one with MgATP adding first (*19*). Slow alternate substrates give cleaner and more easily interpreted pH profiles, and isotope effects are often (but not always) more fully expressed (see Sections VII,A and VII,B below).

Alternate substrates with a more rigid shape than the normal ones are very useful in defining the shape of enzyme active sites and the precise nature of the substrate. For example, fructokinase and phosphofructokinase were both shown to be specific for the β-furanose isomers of fructose or fructose-6-P by the use of stable cyclic ether analogs. Whereas sugars mutarotate between various ring forms (and rapidly, in the case of fructose), cyclic ethers lacking the anomeric hydroxyl group of course cannot. 2,5-Anhydromannitol was an excellent substrate for fructokinase, and 2,5-anhydroglucitol was phosphorylated on carbon 6, rather than carbon 1, showing that the hydroxymethyl group that was phosphorylated had to have the same configuration as C-1 in β-fructofuranose (*20*). Cyclic ethers with six-membered rings similar to those in the pyranose anomers of fructose were neither substrates nor inhibitors. With phosphofructokinase, 2,5-anhydromannitol-6-P was a substrate, whereas 2,5-anhydroglucitol-6-P was not, again demonstrating β specificity for fructose-6-P (*21*).

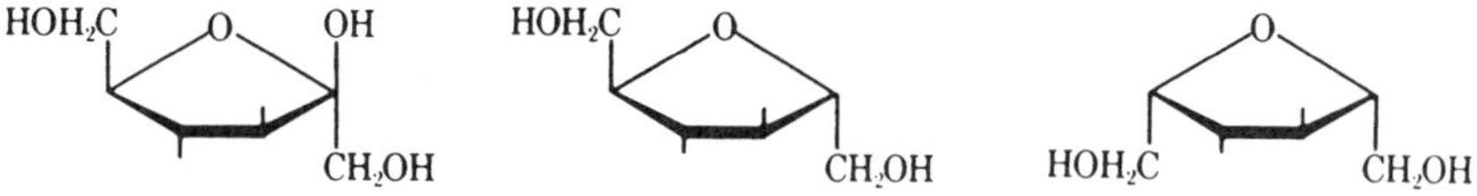

β-Fructofuranose 2,5-Anhydromannitol 2,5-Anhydroglucitol

MgATP exists as a rapidly equilibrating complex [dissociating 7000 times per second (*22*)], with full coordination to the β- and γ-phosphates and approximately 50% coordination at the α-phosphate. The bidentate isomers can exist as two screw-sense isomers:

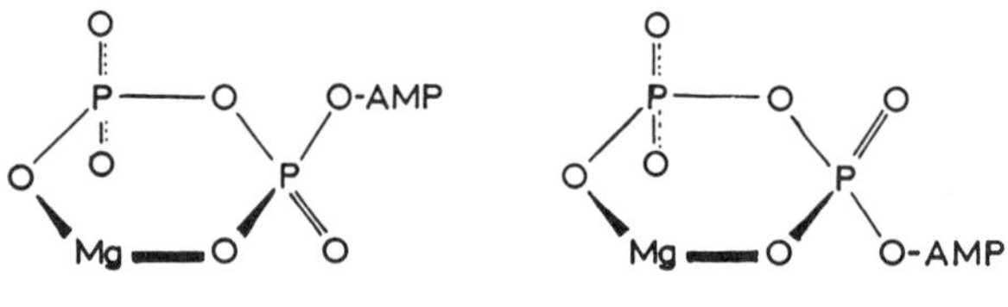

Λ isomer Δ isomer

but one cannot tell directly which is the substrate for a given enzyme. Two methods have been used for this analysis. One involves the use of Cr(III),

Co(III), or Rh(III), which form inert complexes with nucleotides that exchange ligands on the time scale of days or weeks (especially at low temperatures). It is possible, for example, to separate the Λ and Δ isomers of CrATP and use them as substrates in single turnover experiments with various enzymes (*23, 24*). When the enzyme catalyzes multiple turnovers, the developing circular dichroic (CD) spectrum as one isomer is converted to a product without a CD spectrum can determine the screw-sense specificity. Thus, hexokinase and glycerokinase use the Λ isomer of CrATP as a substrate, and pyruvate kinase and myokinase (adenylate kinase) use the Δ isomer (*25*). The absolute configurations of the ADP and ATP complexes of these metal ions are now known and have been correlated with the CD spectra (*26–30*).

An alternate method for determining the screw-sense specificity is to use chirally sulfur-substituted nucleotides together with Mg^{2+} [which prefers to coordinate oxygen rather than sulfur by a ratio of 31,000 (*22*)] and a metal ion such as Cd^{2+} [which prefers sulfur over oxygen by a factor of 60 (*22*)]. The preference for one isomer of ATPβS over the other as the metal ion is switched defines the screw-sense specificity, since the screw sense of the majority of the complex in solution will differ with Mg^{2+} or Cd^{2+}. This method requires that the sulfur-substituted nucleotide be very free from ATP (which often reacts much faster than the sulfur analog) and that V/K values, as well as V values, be compared. The reader should consult Refs. (*31* and *32*) for more details.

B. Inhibition Studies

Inhibition studies are a very useful tool in determining kinetic mechanisms, and they were discussed in detail in Volume II of this treatise. We shall thus only discuss those topics where new theory has been developed since 1970.

1. *Inhibition in Multisite Ping-Pong Mechanisms*

The only multisite Ping-Pong mechanism known in 1970 was that of transcarboxylase (methylmalonyl-CoA carboxyltransferase) (*33*), but a number have been identified since then, including not only reactions in which biotin, lipoic acid, and 4-phosphopantetheine are carriers between active sites, but also reactions where oxidation and reduction of a group on the enzyme occur at different sites [e.g., glutamate synthase (*34*)].

Three rules can be formulated to predict product or dead-end inhibition patterns in such mechanisms. (1) When an inhibitor occupies the same portion of an active site as the variable substrate, the inhibition is competitive. (2) When an inhibitor combines in a different portion of the same active site as the variable substrate, the inhibition is noncompetitive. Thus, with pyruvate carboxylase, MgADP and P_i are both noncompetitive inhibitors versus bicarbonate, although in accordance with Rule 1 they are competitive versus MgATP. (3) When an inhibitor combines at a different active site than the variable substrate, it is noncompetitive only if (a) it can affect the ratio between the forms of the carrier that interact at the site where the variable substrate combines, and (b) reaction at this site is random sequential and not Ping-Pong. (a) If reaction at the site where the variable substrate combines is Ping-Pong or (b) if the inhibitor cannot by itself reconvert the form of the carrier produced by the variable substrate back to the form that reacts with the substrate, then the inhibition is uncompetitive. Dead-end inhibitors in this category are always uncompetitive, whereas product inhibitors may be either noncompetitive or uncompetitive. For example, with pyruvate carboxylase, neither MgADP nor P_i by themselves can convert carboxybiotin back to free biotin, and thus they are uncompetitive versus pyruvate (*35*). At high pyruvate concentrations, these inhibitors keep MgATP off of the enzyme and slow down the carboxylation of biotin, but at low pyruvate the carboxylation of biotin will come to equilibrium (even if more slowly in the presence of the inhibitors), and thus the rate of reaction with pyruvate is not affected. Interestingly, if MgADP and P_i are used together as inhibitors, they should be noncompetitive, since they can decarboxylate carboxybiotin when present together.

The results of having a Ping-Pong site reaction are shown by the product inhibition patterns of NADH with pyruvate dehydrogenase (*36*). NADH is uncompetitive versus pyruvate because, even though it can reduce lipoate and make the oxidized form less available for reaction, it cannot prevent pyruvate from reaction with thiamin-PP to form hydroxyethylthiamin-PP, and this process (which limits the rate at low pyruvate concentrations) will go on at the same rate regardless of how much lipoate is reduced by NADH. Acetyl-CoA is also uncompetitive versus pyruvate, but it would be even if the reaction of pyruvate with lipoate were not Ping-Pong, since acetyl-CoA cannot convert acetyl lipoate back to oxidized lipoate, but rather only slow down the formation of reduced lipoate from acetyl lipoate. At low pyruvate concentration, all of the lipoate will be in the oxidized form regardless of the acetyl-CoA level.

Fatty-acid synthase has a seven-site Ping-Pong mechanism, with three of the active sites (the two transacetylases and the condensation site) having Ping-Pong mechanisms, and the two dehydrogenases, the dehydrase, and the palmitylpantetheine hydrolase site having sequential mechanisms (*37*). CoA is competitive versus malonyl-CoA or acetyl-CoA and uncompetitive versus NADPH, whereas $NADP^+$ is competitive versus NADPH and uncompetitive versus either thioester substrate.

2. *Alternate Substrate Inhibition*

The technique of alternate substrate inhibition, which was pioneered by Fromm (*38*), involves the use of alternate substrates (that produce alternate products) as inhibitors of the formation of the specific product from the variable substrate. The mechanism thus contains branched reaction pathways, and in certain cases one will observe nonlinear reciprocal plots or unusual dependence of the inhibition constants on the concentration of the nonvaried substrate that are highly diagnostic for mechanism. The method has been little used, but it is potentially quite powerful.

a. *Ping-Pong Mechanism.* Let A and B be substrates that produce P and Q as products, respectively, and C be an alternate substrate for A. If the formation of P is measured, C will be competitive versus A and noncompetitive versus B. For the competitive inhibition,

$$K_{is} = K_c/(1 + K_{b(c)}/B) \tag{22}$$

where $K_{b(c)}$ is the Michaelis constant for B with C as a substrate, and thus K_{is} varies from K_c (the Michaelis constant for C) at high B to zero as B approaches zero. It is easy to see why K_{is} is equal to K_c at infinite B, since K_c measures the affinity of the enzyme for C in the steady state under just these conditions, but the approach to zero as B approaches zero concentration is trickier to understand. At low B most of the enzyme is in the F form that reacts with B, and the level of C required to keep it there (and thus not in the E form that can react with A) gets lower and lower as B is decreased.

When B is varied, the inhibition constants for the noncompetitive inhibition of C are

$$K_{is} = K_c K_{b(a)} A/(K_{b(c)} K_a) \tag{23}$$

$$K_{ii} = K_c(1 + A/K_a) \tag{24}$$

where $K_{b(a)}$ is the Michaelis constant for B with A as substrate, and K_a is the Michaelis constant for A. Because of the competition between A and C it is expected that both K_{is} and K_{ii} will increase linearly with A, and the fact that K_{ii} equals K_c at low A is expected. The approach of K_{is} to zero as A approaches zero is unusual, however, and reflects the same situation discussed above. At low B where the enzyme is almost all F, it takes only a very low C level to keep things that way so that no E exists for A to react with. This unique variation of K_{is} values in both competitive and noncompetitive patterns will occur only in a Ping-Pong mechanism and thus is diagnostic for such a mechanism.

b. *Sequential Mechanisms.* In a random sequential mechanism, an alternate substrate is competitive versus the substrate it replaces, and noncompetitive versus the other one. If C replaces A, the competitive inhibition constant is

$$K_{is} = K_c(1 + K_{ib}/B)/(1 + K_{b(c)}/B) \tag{25}$$

where K_{ib} and $K_{b(c)}$ are the dissociation constant of B from EB and the Michaelis constant for B with C as the other substrate. Note the contrast with Eq. (22) above. For the noncompetitive inhibition of C versus B,

$$K_{is} = K_{ic}(1 + A/K_{ia}) \tag{26}$$

where K_{ic} and K_{ia} are dissociation constants of C from EC and A from EA, and K_{ii} is given by Eq. (24). The inhibition constants are linear functions of A, as expected, but K_{is} has a finite value at low levels of A [compare Eqs. (23) and (26)].

In an ordered sequential mechanism where C is an alternate substrate for the second substrate, B, one observes competitive inhibition by C versus B and noncompetitive inhibition versus A. K_{is} for the competitive inhibition is given by an equation analogous to Eq. (25) containing K_{ia}, $K_{a(c)}$, and A in place of K_{ib}, $K_{b(c)}$, and B. For the noncompetitive inhibition, K_{is} is given by an equation analogous to Eq. (26), with $K_{ia}K_c/K_{a(c)}$, $K_{ia}K_b/K_{a(b)}$, and B replacing K_{ic}, K_{ia}, and A, and K_{ii} is given by an equation analogous to Eq. (24), with B and K_b replacing A and K_a. The patterns thus do not differ from those seen for a random mechanism.

When C is an alternate substrate for A, the first substrate to add in an ordered mechanism, however, more complex patterns are observed. C is competitive versus A, with K_{is} given by an equation similar to Eq. (25) with $K_{ic}K_{b(c)}/K_c$ replacing K_{ib} in the numerator. When B is varied, however, the reciprocal plots are not linear (except fortuitously), and are 2/1 functions:

$$1/v = [a + b(1/B) + c(1/B)^2]/[1 + d(1/B)]/V \tag{27}$$

where

$$a = (1 + K_a/A)\{1 + C/[K_c(1 + A/K_a)]\} \tag{28}$$

$$b = K_{b(a)}(1 + K_{ia}/A) + d(1 + K_a/A) \tag{29}$$

$$+ K_aC(K_{b(c)} + K_{ia}K_{b(a)}/K_a)/(K_cA)$$

$$c = dK_{b(a)}(1 + K_{ia}/A)\{1 + C/[K_{ic}(1 + A/K_{ia})]\} \tag{30}$$

$$d = K_{ic}K_{b(c)}/K_c \tag{31}$$

Two ratios determine the shape of these reciprocal plots: (1) $(K_{ic}K_{b(c)}/K_c)/(K_{ia}K_{b(a)}/K_a)$, which is the ratio of the apparent Michaelis constants for B with low levels of either C or A as the other substrate, and (2) K_{ic}/K_c, the ratio of the dissociation constant of C from EC and its Michaelis constant with B as the second substrate. If either ratio is unity, reciprocal plots are linear in the presence of C, and the inhibition constants for the noncompetitive inhibition of C are given by Eqs. (26) and (24). When one ratio is greater than unity and the other less than 1.0, reciprocal plots are concave downward with C present. When both ratios are less than 1.0, reciprocal plots are concave up in the presence of C, but they cannot have a minimum. When both ratios are greater than unity, reciprocal plots in the presence of C are also concave upward but can have a minimum at high enough C values (i.e., the reciprocal plot shows partial substrate inhibition) if

$$K_{ic}/K_c > [1 + K_{ia}K_{b(a)}/(K_aK_{b(c)})] \tag{32}$$

These 2/1 reciprocal plots will have linear asymptotes at low levels of B, and the slopes and intercepts of these asymptotes will be linear functions of C, so that the inhibition will appear noncompetitive. The inhibition constant for the asymptote slopes will be given by Eq. (26), but that for the asymptote intercepts will be

$$K_{ii} = K_{ic}(1 + A/K_a)/[1 + K_{ia}K_{b(a)}(1 - K_c/K_{ic})/(K_aK_{b(c)})] \tag{33}$$

which can have a negative value if

$$K_c/K_{ic} > [1 + K_aK_{b(c)}/(K_{ia}K_{b(a)})] \tag{34}$$

so that the asymptote intercept will *decrease* with increasing C and become negative at high enough C values. This can occur only if the two ratios given above are less than unity and the reciprocal plots in the presence of C are concave upward without a minimum. Note that if Eq. (34) is an equality, the asymptote intercept will not vary with C even though the true intercepts increase linearly with C and the reciprocal plots are concave upward. When both ratios are above unity, Eq. (33) gives only positive values for K_{ii} even though the reciprocal plots are also concave up and may have a minimum. The asymptote intercept thus increases with C, although more slowly than the true intercept.

When Rudolph and Fromm used thionicotinamide adenine dinucleotide (thio-DPN) as an alternate substrate for NAD^+ and varied the concentration of ethanol with liver alcohol dehydrogenase [following the reaction at 342 nm, the isosbestic point for thio-DPN and reduced thio-DPN (thio-DPNH)], they saw what appeared to be concave upward reciprocal plots with partial substrate inhibition in the presence of thio-DPN (*38*). However, the asymptote intercepts appeared to decrease with increased thio-DPN concentration, which is not what the above equations predict for a case where a minimum is present in the curve. There must have been other interactions that caused the substrate inhibition by ethanol in the presence of thio-DPN.

3. *Use of Dead-End Inhibitors to Determine Rate-Limiting Steps*

The use of dead-end inhibitors to deduce kinetic mechanisms was covered in Volume II, but two points deserve further mention.

a. *Order versus Synergism.* Inhibition patterns for dead-end inhibitors have long been used to identify ordered kinetic mechanisms. Thus, an inhibitor competitive versus A will be noncompetitive versus B, whereas an inhibitor that, like the second substrate B, can only add to the EA complex is competitive versus B and uncompetitive versus A. Care must be taken in interpretation, however, since "order" may simply be a reflection of a high degree of synergism in binding. For example, glycerokinase appears to have an ordered kinetic mechanism with glycerol adding before MgATP, since CrATP is competitive versus MgATP and uncompetitive versus glycerol (*12*). However, the enzyme does act as an ATPase, with a V value 6×10^{-5} that with glycerol and a Michaelis constant for MgATP of 0.54 mM at pH 8 (vs. 15 μM for the glycerokinase activity). The ATPase activity does appear to result from the same active site, since it is suppressed by high levels of (R)-1-mercaptopropanediol, which is very slowly phosphorylated on sulfur at 10^{-6} the rate with glycerol. Thus, MgATP does form a binary complex with the enzyme, although the prior presence of glycerol increases affinity by a factor of 36 (*39*).

Glycerokinase will also phosphorylate (S)-1-mercaptopropanediol on the 3-hydroxyl group with a rate 0.035 that for glycerol, but now the K_m and K_i for MgATP are nearly the same as the K_m observed for the ATPase reaction; that is, there is no synergism in the binding of the substrates. The enzyme also phosphorylates the unprotonated forms of both (S)- and (R)-1-aminopropanediol, the former on nitrogen (at 0.004 the rate with glycerol) and the latter on the 3-hydroxyl group (V value 0.007 that with glycerol), but in both cases the initial velocity pattern appeared to be equilibrium ordered, with MgATP adding first! The K_i value for MgATP was the same as the K_m for the ATPase reaction (~2 mM at pH 9.4), while the K_m values for the amino analogs were 0.28 mM for the (S) isomer and 8 mM for the (R) isomer (*39*).

An equilibrium ordered pattern implies that the Michaelis constant for the first substrate is zero and the dissociation constant of the second substrate is infinity (i.e., it does not form a binary complex with the enzyme but only combines with EA). It is clear that what is happening here is that the binding of the amino analogs and MgATP is highly synergistic, as is the binding of glycerol and MgATP (but not the binding of MgATP and the thiol analogs), and that MgATP binds more loosely than glycerol, but more tightly than the amino analogs. When substrate binding is synergistic by at least an order of magnitude, the more tightly bound reactant appears to bind first in an ordered fashion. An equilibrium ordered pattern is seen with a slow substrate, whereas a normal intersecting initial velocity pattern is seen with a fast substrate like glycerol which is sticky and does not add in rapid equilibrium fashion. Glycerokinase thus has a random mechanism, but with synergistic binding of glycerol and its amino analogs, but not of the thiol analog. Many apparently ordered mechanisms may be of this type, but in other cases the binding site for the second substrate may not exist until the first is present, either to create a portion of the site physically or to induce a conformation change that produces it.

b. *Determination of Rate-Limiting Steps.* When a dead-end inhibitor combines with an EA complex in an ordered mechanism, it is competitive versus B, and the inhibition constant is the dissociation constant for EA, as long as saturating A is used. When the reaction is run backward, the inhibition will appear uncompetitive versus either P or Q, and, if the nonvaried substrate is saturating, the size of the K_{ii} value will indicate the rate-limiting step for the maximum velocity in the reverse direction. Thus,

$$k_2/(V_2/E_t) = K_{ii}/K_{is} \tag{35}$$

where k_2 is the rate constant for release of A from the EA complex, V_2/E_t is the turnover number in the reverse direction, K_{is} is the dissociation constant of the inhibitor from EA at saturating A, and K_{ii} is the inhibition constant for A versus either P or Q at saturating levels of the other.

Because V_2/E_t is the reciprocal of the sum of reciprocals of k_2 and k_4 (the rate constant for catalysis and release of B from the enzyme), K_{ii}/K_{is} equals $(1 + k_2/k_4)$, and

$$k_2/k_4 = K_{ii}/K_{is} - 1 \tag{36}$$

With glycerokinase, CrATP was competitive versus MgATP ($K_{is} = 0.5$ mM) and uncompetitive versus MgADP at saturating glycerol 3-P ($K_{ii} = 1.4$ mM), giving a ratio of 1.8 ± 0.6 for k_2/k_4 (*12*). This value came out 5 ± 2 from analysis of crossover points of the initial velocity patterns (see Section V,A,1 above), but in any case it appears that the rate of glycerol release from E–glycerol is greater than the rate of transphosphorylation and release of MgATP. Glycerol is released 11–23 times slower than the turnover number in the *forward* direction, however, since V_1 is so much greater than V_2, and thus glycerol is a quite sticky substrate.

A dead-end inhibitor will often have affinity for both EA and EQ in an ordered mechanism and thus gives noncompetitive patterns in both directions. As long as it only causes inhibition by combining with EA and EQ, however, one can simply compare the K_{is} value in one direction with the K_{ii} value in the other as done above. Since one has data in both directions of the reaction, one learns the relative rates of the steps in both directions. This is a powerful technique that should be applied more often.

4. *Multiple Inhibition*

The multiple inhibition method where initial velocity is measured at constant substrate levels but in the presence of various levels of two competitive inhibitors (I and J) was described in Volume II. Such experiments tell whether the competitive inhibitors are mutually exclusive (parallel pattern of $1/v$ vs. I at different J levels), or, where both inhibitors can be on the enzyme at the same time, how synergistic or antisynergistic their combination is. The proper equation to use for such a case is

$$1/v = 1/V + [K/(V\text{A})][1 + \text{I}/K_\text{i} + \text{J}/K_\text{j} + \text{IJ}/(\alpha K_\text{i}K_\text{j})] \quad (37)$$

where K_i and K_j are dissociation constants of I or J, and α is the interaction coefficient ($\alpha = \infty$, >1, 1.0, or <1 for mutually exclusive, antisynergistic, independent, or synergistic binding). If the substrate concentration is held constant, the horizontal coordinate of the crossover point when $1/v$ is plotted versus I is $-\alpha K_\text{i}$, whereas if $1/v$ is plotted versus J it is $-\alpha K_\text{j}$. With the values of K_i and K_j from separate competitive inhibition experiments, α is easily determined. Alternatively, the substrate as well as the two inhibitor concentrations can be varied and the data fitted to Eq. (37) or to the corresponding nonreciprocal form to determine all of the parameters directly from one experiment. Cook *et al.* (*40*) did this with creatine kinase, where the substrate was creatine phosphate and the two inhibitors creatine and nitrate (MgADP was constant at a saturating level). These patterns were determined over the pH range of 5–10 and showed that, whereas the K_i for creatine increased below a pK of 6, that for nitrate was pH independent, as was the value of α of 0.014. The two inhibitors are highly synergistic in forming the E–MgADP–creatine–nitrate complex, which mimics the transition state for the phosphoryl transfer.

This method is not limited to competitive inhibitors, but in the general case tests for the presence in the denominator of the rate equation of a term containing both inhibitor concentrations (i.e., a term containing IJ as a factor with or without other reactant concentrations multiplied by it). In such cases it is more convenient to use the following equation:

$$1/v = (1/v_\text{o})[1 + \text{I}/K_\text{i} + \text{J}/K_\text{j} + \text{IJ}/(\beta K_\text{i}K_\text{j})] \quad (38)$$

where v_o is the velocity in the absence of inhibitors. K_i and K_j are now only apparent dissociation constants, and β a measure of the degree of interaction of the two inhibitors. This approach was used with isocitrate dehydrogenase ($NADP^+$) with $NADP^+$ as a product inhibitor and substrate inhibition levels of α-ketoglutarate to show that the substrate inhibition by α-ketoglutarate was not caused by a dead-end E–$NADP^+$–α-ketoglutarate complex (β was ∞) (*41*).

This method was also used with alanine dehydrogenase with NADH as a product inhibitor and substrate inhibition levels of L-alanine (*42*). At first glance the observed β value of 4.9 suggests that a dead-end E–NADH–alanine complex caused the substrate inhibition; in an ordered mechanism, however, β could not exceed 1.0 unless alanine had as great or greater affinity for E–NADH than its apparent Michaelis constant at low NAD^+, and, with NAD^+ present at its K_m (as was the case), the value of β could not have exceeded 0.6. The equation for the apparent β value in an ordered mechanism like this one where the two inhibitors are Q (product) and substrate inhibition levels of B is as follows:

$$\text{app } \beta = [1 + K_{ia}K_b/(K_aK_{Ib})][V_1K_q/(V_2K_{iq})]/(1 + K_a/A) \quad (39)$$

where $K_{ia}K_b/K_a$ is the apparent Michaelis constant of B at very low A, K_{Ib} is the dissociation constant of B from EQ, and K_a is the Michaelis constant for A. The expression $V_1K_q/(V_2K_{iq})$ is the proportion of enzyme that is in EQ at saturation with substrate, and it has a maximum value of 1.0.

Grimshaw and Cleland (*42*) suggested that the high value of β resulted from alanine combination with (1) the initial E–NAD^+ complex before it had a chance to isomerize to the E–NAD^{+*} complex that combines productively with alanine, (2) a central complex, or (3) E–NADH–pyruvate. These interactions would contribute to the I/K_i term in Eq. (38) (I is alanine), but not to the $IJ/(\beta K_iK_j)$ term. The first possibility was thought more likely (but note that alanine combination would have to *prevent* the isomerization!).

5. *Induced Substrate Inhibition*

Inhibition by substrates that also act as dead-end inhibitors was discussed in Volume II. In an ordered mechanism, however, addition of an inhibitor that mimics the first substrate may permit binding of the second substrate. In such a situation the presence of the inhibitor will induce substrate inhibition by the second substrate, as long as the first substrate addition is not in rapid equilibrium:

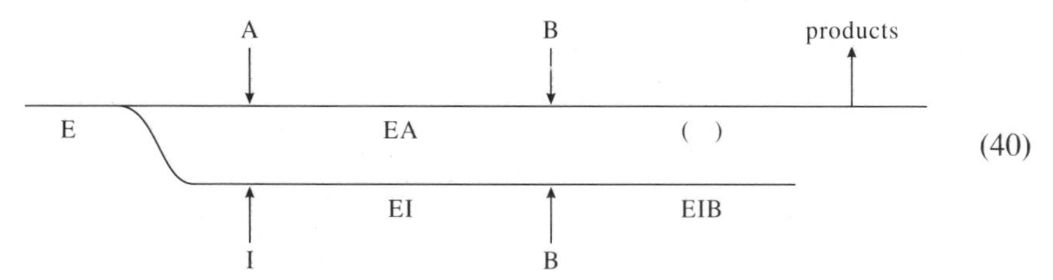

(40)

The rate equation is

$$v = VAB/\{(K_{ia}K_b + K_aB)[1 + (I/K_i)(1 + B/K_{Ib})] + K_bA + AB\} \quad (41)$$

where K_i and K_{Ib} are dissociation constants of I from EI and B from EIB, and the other kinetic constants have their usual meaning for an ordered mechanism. The substrate inhibition by B occurs only in the presence of I, and it is competitive when A is the varied substrate at a fixed level of I.

Few cases of induced substrate inhibition have been reported, partly because few mechanisms are truly ordered (the presence of B must keep I from dissociating in order to see the effect), and partly because the inhibitor may fail to induce the conformation change that permits addition of the second substrate. Bromodeoxyuridine 5′-monophosphate (BrdUMP) does induce substrate inhibition of thymidylate synthase by methylene tetrahydrofolate as the result of the abortive covalent complex of enzyme, BrdUMP, and methylene THFA that forms (*43*).

Substrate inhibition could also be induced in a three-substrate ordered mechanism by an inhibitor mimicking the first or second substrates to add to the enzyme. An inhibitor combining in place of A would induce substrate inhibition by B that was competitive versus A and uncompetitive versus C, if B combines with EI and prevents release of I. If C could then combine with the EIB complex, substrate inhibition by C would also be induced that was competitive versus A and uncompetitive versus B. An inhibitor that combines in place of B would induce substrate inhibition by C that was competitive versus B and uncompetitive versus A if C formed an EAIC complex from which I could not dissociate. No such cases are known, but such inhibition has not often been looked for. With glutamate dehydrogenase, oxalylglycine, which is an excellent mimic of α-ketoglutarate (it has the same dissociation constant), failed to induce substrate inhibition by ammonia, suggesting that ammonia will only combine if it can react with the carbonyl carbon, which is possible with α-ketoglutarate but not with oxalylglycine (*14*).

In a related phenomenon, enhanced substrate inhibition by α-ketoglutarate of saccharopine dehydrogenase (NAD^+, L-lysine-forming) is seen at high levels of lysine (*44*). The mechanism is ordered, with NADH, α-ketoglutarate, and lysine adding in that order and saccharopine released before NAD^+. α-ketoglutarate combines weakly with E–NAD^+, but subsequent addition of lysine traps α-ketoglutarate on the enzyme and greatly enhances the substrate inhibition, which is uncompetitive versus NADH and lysine, as expected for the above mechanism.

Induced substrate inhibition has also been seen in random mechanisms where there is strong synergism in binding of substrates, but the substrate inhibition is partial in this case. With yeast hexokinase, lyxose induces partial substrate inhibition by MgATP that is competitive versus glucose (*45*). The dissociation constant of MgATP is approximately 5 m*M* in the absence of sugars, but about 0.1 m*M* in the presence of glucose or lyxose. By inducing much tighter binding of lyxose at levels that are saturating for the normal reaction, MgATP enhances the inhibition by lyxose and causes substrate inhibition. Lyxose can dissociate slowly from the E–lyxose–MgATP complex, however, so that the substrate inhibition is only partial.

6. *Product Inhibition and Oversaturation with Proline Racemase*

A racemase is a difficult enzyme with which to study product inhibition, since $K_{eq} = 1$ and, if one puts in enough product to inhibit the enzyme, one is near equilibrium or the reaction goes backward. In their epic studies on proline racemase, Knowles and co-workers solved this problem by using optical rotation to follow the time course of the reaction to equilibrium (*46*). At levels of D- or L-proline up to 16 m*M*, the initial velocity increased as one would expect and showed saturation, with K_m values of approximately 3 m*M*. The time courses behaved in the expected manner, with the half-times for approach to equilibrium divided by initial substrate concentrations decreasing with increasing substrate concentration and reaching a limiting value at saturation.

When initial concentrations of proline from 16 to 190 m*M* were used, however, a different pattern was observed. Although the initial velocities were the same, the half-times for approach to equilibrium divided by the initial substrate concentrations *increased* with increasing substrate concentrations. This phenomenon, which was called "oversaturation," results from developing noncompetitive product inhibition during the reaction. If the developing product inhibition were competitive, the normalized time courses would be identical for all saturating substrate levels.

Noncompetitive product inhibition in a racemase indicates an Iso mechanism in which the form of free enzyme reacting with the L substrate is different from that reacting with the D substrate, and the interconversion between these two forms is kinetically important. The most likely explanation for the oversaturation phenomenon is that it occurs when so much enzyme is tied up by substrate and product that the unimolecular conversion of E′ (the form combining with product) back to E (the form combining with substrate) becomes rate limiting. This is the case with proline racemase, and this occurs at 125 m*M* reactants (*46*). Although a number of enzymes may have mechanisms where E and E′ conversion has to occur, this will never be a major rate-limiting step at physiological reactant levels. With proline racemase, V/E_t is 2000 sec^{-1} in either direction, whereas the E to E′ step has a rate of 10^5 sec^{-1}.

A less likely mechanism that produces oversaturation, but one which does apply to carbonic anhydrase, is one in which initial saturation with the substrate ties up enough enzyme for free enzyme isomerization to become rate limiting (i.e., E′ to E conversion limits *V*, rather than ES to E′ conversion) (*46*). Oversaturation then occurs at still higher substrate concentrations where there is more EA than free E, and EA conversion via E′ to E becomes rate limiting. With carbonic anhydrase, the basic reaction can be written

$$\text{E–Zn–OH}^- + CO_2 \rightleftharpoons \text{E–Zn–O—CO}_2\text{H}^- \xrightleftharpoons[]{H_2O \quad HCO_3^-} \text{E–Zn–OH}_2 \qquad (42)$$

The "isomerization" of E–Zn–OH_2 (the form that reacts with bicarbonate) back to E–Zn–OH^- (the form that reacts with CO_2) is rate limiting for *V*, as shown by the D_2O solvent isotope effect on *V* but not on *V*/*K* or the rate of exchange between CO_2 and bicarbonate (*47*). Carbonic anhydrase has, in fact, a Ping-Pong mechanism in which reaction of buffer with E–Zn–OH_2 to give E–Zn–OH^- and protonated buffer is the second half-reaction. Thus, small anions such as I^-, SCN^-, or acetate are uncompetitive versus CO_2 because they replace water in $E\text{–}Zn\text{–}OH_2$ but cannot displace hydroxide from $E\text{–}Zn\text{–}OH^-$. Reference (*47*) is an excellent review on the mechanism of carbonic anhydrase.

C. Isotopic Exchange

This topic was dealt with in Volume II, but we shall summarize more recent developments.

1. *Ping-Pong Mechanisms*

A reaction that has a Ping-Pong mechanism will always be at equilibrium when only the reactants in one partial reaction are present, so one is free to vary the concentration of any reactant and determine the resulting initial velocity of isotopic exchange. Thus, in a Ping-Pong mechanism with two substrates and two products, one can measure A–P exchange in the absence of B and Q, and vice versa. The pattern when reciprocal exchange velocity is plotted versus 1/A at different P levels is a parallel one, with the apparent Michaelis constants for A and P being the dissociation constants of these reactants from the central complex (the reaction is at equilibrium, of course). The maximum exchange velocity is V_1K_{ia}/K_a or V_2K_{ip}/K_p. The exchange between B and Q determines K_{ib} and K_{iq} and a maximum exchange velocity of V_1K_{ib}/K_b or V_2K_{iq}/K_q. These maximum exchange velocities are related to the maximum chemical velocities by

$$1/V^*_{A\text{–}P} + 1/V^*_{B\text{–}Q} = 1/V_1 + 1/V_2 \tag{43}$$

where $V^*_{A\text{–}P}$ is the maximum exchange velocity between A and P, and $V^*_{B\text{–}Q}$ is the same for the B–Q exchange. This relationship must hold for a Ping-Pong mechanism, and the failure of it in the case of acetate kinase (the MgATP–MgADP exchange is fast enough, but the acetate–acetyl-P exchange is too slow and cannot be detected if charcoal is used to remove traces of nucleotides from the enzyme) provided strong evidence that the mechanism was in fact sequential (*48*).

When the partial reaction of a Ping-Pong mechanism involves three reactants,

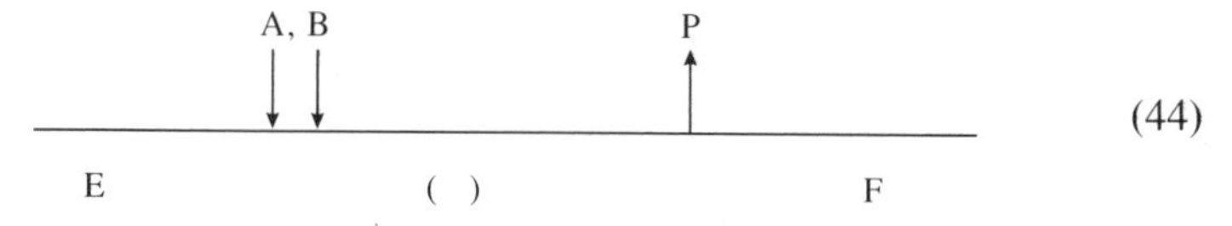

(44)

it is simple to determine the order of addition of A and B by measuring the isotope exchange pattern of P with A or B as a function of the concentrations of A and B with P held constant. If P exchanges with A and addition of A and B occurs in that order, the A–B exchange pattern is intersecting with uncompetitive substrate inhibition by B. If P exchanges with B and the order of addition is A followed by B, the A–B exchange pattern is equilibrium ordered. If combination of A and B is random, an intersecting pattern is seen with no substrate inhibition. The technique was first used with asparagine synthetase (aspartate–ammonia ligase), where MgATP and aspartate add randomly (*49*).

The technique must be used with caution with multisite Ping-Pong mechanisms. With an enzyme like pyruvate carboxylase, for example, bicarbonate will show uncompetitive substrate inhibition versus MgATP of exchanges between MgATP and either P_i or MgADP, since bicarbonate can keep carboxybiotin out of the site and prevent it from reacting with MgADP and P_i.

2. *Sequential Mechanisms*

Isotopic exchange in sequential mechanisms is carried out either at equilibrium with all reactants present or by measuring the transfer of label from a product back into a substrate while the reaction is proceeding. The latter approach is an excellent way to determine the relative rates of release of products, since reverse transfer of label can only occur if products are released slowly enough that sufficient E–product complex is present in the steady state to react with the labeled product and reform the central complex. This method was originally used to show ordered release of glucose before P_i with glucose-6-phosphatase, and with ordered release of products the ratio of exchange back from P into substrate to the forward chemical reaction is P/K_{is}, where K_{is} is the slope inhibition constant for product inhibition by P (*50*).

5-Aminolevulinate synthase catalyzes the reaction of succinyl-CoA with an enzyme-bound imine of pyridoxal phosphate and glycine to give CO_2, 5-aminolevulinate (ALA), and CoA. Although the initial velocity pattern looks parallel, label from $^{14}CO_2$ is found in glycine only when succinyl-CoA is present, thus ruling out a Ping-Pong mechanism. The amount of back exchange is doubled by addition of CoA, tripled by the addition of ALA, and increased 5 times by the presence of both CoA and ALA (*51*). Thus CO_2 is released first, with CoA and ALA released in random and partly rate-limiting fashion, so that the E–CoA–ALA complex is present in the steady state. ALA is released slightly more rapidly than CoA from this complex, but CoA release from the ternary complex is significant, since the complex level can be doubled by adding CoA.

The major use of isotope exchange studies of sequential mechanisms has been to test for order by measuring exchange at equilibrium. In a bireactant mechanism, one can measure exchange between like substrates and products, and thus one has A–Q and B–P exchanges to study. The exchange involving the group transferred during the reaction (A–P or B–Q exchange) is often affected by primary isotope effects (especially for dehydrogenases!), and is not normally studied. One usually varies one substrate and one product concentration in constant ratio in order to stay at equilibrium (although in theory one could vary all reactants at once, or raise one substrate and lower the other). Any substrate and product pair can be varied, but the A–Q and B–P pairs are usually used, since raising A and P or B and Q together often leads to abortive complexes with resulting substrate inhibition of the exchange.

In a random mechanism, varying A and Q, or B and P, together will give linear reciprocal plots for both A–Q and B–Q exchanges, but the exchange rates will differ for the two exchanges unless the catalytic conversion is totally rate limiting (i.e., the mechanism is a rapid equilibrium one in both directions). Varying A and P, or B and Q, will then identify the possible dead-end complexes. On the basis of such exchange studies, creatine kinase at pH 8 has a rapid equilibrium random mechanism with both E–MgADP–creatine and E–MgATP–creatine phosphate dead-end complexes (*52*). This pattern should not be observed at neutral pH, where creatine-P is sticky (*40*), and its slow release should make the creatine–creatine-P exchange slower than the MgATP–MgADP exchange. Hexokinase (*53*), galactokinase (*54*), and fructokinase (*55*) all show faster MgATP–MgADP exchange than sugar–sugar phosphate exchange, presumably because sugar release from the enzyme is a slow step in these random mechanisms.

In an ordered mechanism, reciprocal plots for A–Q and B–P exchanges when A and Q are varied in constant ratio at constant B and P levels should be linear and intersect on the horizontal axis, with the B–P exchange faster than A–Q exchange. The exact analytical geometry is given in Ref. (*2*). When B and P are varied in constant ratio, the reciprocal plots for A–Q and B–P exchanges are parallel and the B–P one is linear. The A–Q exchange will, however, show total substrate inhibition as the result of B and P keeping all of the enzyme as central complexes so that A and Q cannot dissociate from the enzyme. These experiments are sensitive tests for order, since any finite rate of dissociation of A and Q from the central complexes will change the pattern to one of only partial substrate inhibition. This is the case for liver alcohol dehydrogenase with cyclohexanol and cyclohexanone as reactants (*56*). The randomness also makes the apparent K_m values for A–Q and B–P exchanges different when A and Q are varied together, with that for the A–Q exchange being higher than that for the B–P exchange, since it now represents dissociation from central complexes rather than from binary EA and EQ complexes.

These patterns also provide a sensitive test for the presence of central complexes. A completely Theorell–Chance mechanism for liver alcohol dehydrogenase was ruled out by observation of a finite maximum exchange velocity for the cyclohexanol–cyclohexanone exchange when the concentrations of these two reactants were varied together (*56*). In a Theorell–Chance mechanism where EA and B react to give P and EQ without formation of central complexes, the reciprocal plot of the isotopic exchange would have to go through the origin. The vertical intercept of this reciprocal plot is independent of the A and Q levels, and it equals the sum of the vertical coordinates of the crossover points in initial velocity patterns in forward and reverse directions. It thus can be used to calculate the parameter R described above in Section V,A,1.

When catalysis is much faster than reactant release in a random mechanism, drastically different apparent K_m values can be observed in isotope exchange studies. For example, with isocitrate dehydrogenase, $NADP^+$, NADPH, and isocitrate are all very sticky and dissociate slowly from the enzyme. CO_2–isocitrate exchange was normally faster than $NADP^+$–NADPH exchange; however, when isocitrate and α-ketoglutarate were varied together, the apparent K_m for the $NADP^+$–NADPH exchange was so much lower than for the CO_2–isocitrate one (40-fold) that the reciprocal plots would cross at low enough isocitrate and α-ketoglutarate levels (*57*). The low apparent K_m for $NADP^+$–NADPH exchange reflects the low degree of saturation with CO_2 and isocitrate needed to keep catalysis faster than nucleotide release. In contrast, half-saturation with isocitrate is needed to give half of the maximum rate for CO_2–isocitrate exchange. Similarly, the apparent K_m when $NADP^+$ and NADPH were varied together was 15-fold lower for the CO_2–isocitrate exchange than for the nucleotide exchange because only a low degree of saturation with nucleotides was needed to keep catalysis faster than isocitrate release.

A number of isotope exchange studies were carried out in the 1960s and 1970s, but the laborious nature of the experiments and the complexity of the equations have kept the method from extensive use. Recently, however, Wedler has begun to use this approach in parallel with simulation studies to study the mechanisms of enzymes like aspartate transcarbamylase (*58, 59*).

3. *Positional Isotopic Exchange*

The positional isotopic exchange (PIX) technique involves observation of exchange of label between two positions in a reactant. The method was first used by Midelfort and Rose in 1976 to follow β–γ bridge to β-nonbridge transfer of ^{18}O in ATP incubated with glutamine synthetase (*60*):

$$\text{AMP—O—}\overset{\displaystyle \text{O}}{\underset{\displaystyle \text{O}}{\text{P}}}\text{—}\bullet\text{—}\overset{\displaystyle \bullet}{\underset{\displaystyle \bullet}{\text{P}}}\text{—}\bullet \longrightarrow \text{AMP—O—}\overset{\displaystyle \bullet}{\underset{\displaystyle \text{O}}{\text{P}}}\text{—O—}\overset{\displaystyle \bullet}{\underset{\displaystyle \bullet}{\text{P}}}\text{—}\bullet \qquad (45)$$

and has since been used to measure such exchange in phosphate, carboxyl, and guanidinium groups. The observation of PIX shows that (a) intermediates are formed on the enzyme reversibly and (b) rotation of the labeled group is possible before reversal or product release in order to produce the exchange. Conversely, the failure to observe PIX does not show the lack of intermediates, since restricted rotation will prevent PIX. For example, no biotin-containing enzyme has been shown to catalyze PIX, even under conditions where the transfer of ^{18}O from bicarbonate to phosphate shows clearly that a carboxyphosphate intermediate is formed. When it is seen, however, PIX can be a great help in deciding the relative rates of steps in the mechanism, since it directly measures the partition ratio of a key intermediate complex. The interested reader should consult a recent review by the prime practitioners of the art (*61*).

The progress of PIX can be followed either by mass spectrometry or by NMR, since ^{18}O causes small upfield changes in ^{31}P and ^{13}C chemical shifts which are readily resolved in modern NMR spectrometers. Because the shift is dependent on bond order, a nonbridge oxygen that has partial double bond character shows a greater upfield shift than a bridging one that is necessarily single bonded. The NMR method requires more material, but the mass spectrometry method requires chemical processing of the reaction mixture prior to analysis.

Since PIX is generally carried out at the same time as the chemical reaction, the level of the substrate in which PIX is being observed is constantly decreasing. The initial rate of PIX in this case is given by

$$v_{PIX} = X(A_o/t)[\ln(1 - F)]/[\ln(1 - X)] \qquad (46)$$

where A_o is the initial concentration of labeled substrate, X is the fraction of substrate converted at time t, and F is the fractional approach to the final equilibrium value for positional isotopic exchange. Thus, for the situation in Eq. (45), the final equilibrium position has two-thirds of the β-phosphate with nonbridge ^{18}O and one-third with bridge ^{18}O. Each γ-phosphate has three nonbridge ^{18}O attached at all times, but at final equilibrium, one-third will also have a bridging ^{18}O attached while two-thirds will not. The PIX can be measured by following either (or both) the β- or γ-phosphate signals, since each differently labeled species will give separate NMR peaks. The reader should consult Ref. (*61*) for a detailed discussion of this methodology.

PIX is essentially a partitioning experiment, as can be seen by considering the following mechanism:

$$E \underset{k_2}{\overset{k_1 A}{\rightleftharpoons}} EA \underset{k_4}{\overset{k_3}{\rightleftharpoons}} EX \xrightarrow{k_5} E + \text{products} \qquad (47)$$

where A is the substrate in which we are observing the PIX, EX is the intermediate in which rotation of the labeled group is possible, and the interconversion of EA and EX involves the chemical transfer reaction. Although the rate equations for both PIX and the velocity of the chemical reaction have the usual denominator expressing the distribution of the enzyme among its possible forms, the denominators will be the same, and thus the ratio of PIX and chemical reaction will be given solely by the partitioning of EX. This is k_5 in the forward direction, and $k_4[k_2/(k_2 + k_3)]$ in the reverse direction. Thus,

$$v_{PIX}/v_{chem\ rxn} = k_2k_4/[k_5(k_2 + k_3)] \qquad (48)$$

It is clear that any decrease in k_5 enhances the ratio of PIX to chemical reaction, whereas if k_5 is too large no PIX is observed. It is thus surprising that with biotin carboxylase no PIX was observed when MgATP and bicarbonate were present in the absence of biotin (*62*). The slow bicarbonate-dependent ATPase under these conditions shows that carboxyphosphate is forming and breaking down, but k_5 is not large. Rotation of the β-phosphate is clearly not occurring, possibly because the two metal ions present (a Mg^{2+} ion is necessary in addition to the one coordinated to ATP) immobilize the phosphate completely.

Where k_5 is too large to see PIX, the addition of products may alter the partitioning so that it can be observed. This was first done with argininosuccinate lyase (*63*); arginine and fumarate are the products, and the PIX was measured by exchange of ^{15}N from the bridge to nonbridge position of the guanidinium group of argininosuccinate (^{15}N NMR was used). If EX in Mechanism (47) can dissociate either product, and we add P,

$$\begin{array}{lll} E \xrightarrow{k_5} EP + Q & & \\ \quad \underset{k_{10}P}{\overset{k_9}{\searrow\!\!\nwarrow}} & & \\ & EQ \xrightarrow{k_{11}} E + Q & \end{array} \qquad (49)$$

the net rate constant for forward reaction of EX is no longer just k_5 but is now $k_5 + k_9k_{11}/(k_{11} + k_{10}P)$, so that the ratio of PIX to chemical reaction becomes

$$v_{PIX}/v_{chem\ rxn} = k_2k_4/\{(k_2 + k_3)[k_5 + k_9k_{11}/(k_{11} + k_{10}P)]\} \qquad (50)$$

If $k_9 = 0$ (i.e., Q dissociates before P), $v_{PIX}/v_{chem\ rxn}$ will be independent of P, and is given by Eq. (48). If $k_5 = 0$ (P is released first), $v_{PIX}/v_{chem\ rxn}$ will be a linear function of P, with a value of $k_2k_4/[k_9(k_2 + k_3)]$ at zero P and a horizontal intercept of k_{11}/k_{10}. If k_5 and k_9 are both finite (random release of P and Q), $v_{PIX}/v_{chem\ rxn}$ will be a hyperbolic function of P, with a value of $k_2k_4/[(k_2 + k_3)(k_5 + k_9)]$ at zero P and a value given by Eq. (48) at infinite P.

This latter pattern was observed for argininosuccinate lyase (*63*), with no PIX (<0.15 for $v_{PIX}/v_{chem\ rxn}$) observed at low fumarate but a value of 1.8 for $v_{PIX}/v_{chem\ rxn}$ seen at infinite fumarate. Raushel and Garrard estimated that fumarate was released more than 6 times faster than V_2/E_t, whereas arginine was released at at least one-half the value of V_2/E_t.

When the substrate in which PIX is being observed is the first one to add in an ordered mechanism, a high level of the second substrate will prevent PIX. This is readily seen in Mechanism (51):

$$E \underset{k_2}{\overset{k_1A}{\rightleftharpoons}} EA \underset{k_4}{\overset{k_3B}{\rightleftharpoons}} EAB \underset{k_6}{\overset{k_5}{\rightleftharpoons}} EPQ \xrightarrow{k_7} EQ \xrightarrow{k_9} E \qquad (51)$$

where PIX measures the partitioning of EPQ. The net rate constant for release of A from EPQ is $k_2k_4k_6/[k_2(k_4 + k_5) + k_3k_5B]$, and thus

$$v_{PIX}/v_{chem\ rxn} = k_2k_4k_6/\{k_7[k_2(k_4 + k_5) + k_3k_5B]\} \qquad (52)$$

If the mechanism is truly ordered, $v_{chem\ rxn}/v_{PIX}$ will plot linearly versus B; in a random mechanism, however, this plot would be a hyperbola, and $v_{PIX}/v_{chem\ rxn}$ would have a finite value at infinite B.

Further quantitative analysis is possible. Thus,

$$k_2/(V_1/E_t) = (1 + \text{intercept})/[(\text{slope})K_b] \tag{53}$$

where "slope" is of the plot of $v_{chem\ rxn}/v_{PIX}$ versus B, "intercept" is for the same plot, and K_b is the Michaelis constant for B in Mechanism (51). This method thus gives an exact value for $k_2/(V_1/E_t)$ to compare with one derived by the isotope partitioning method (see Section VI,A, below).

The following relationship also holds:

$$k_7/(V_2/E_t) = k_7/k_2 + k_7/k_4 + \text{intercept} \tag{54}$$

where "intercept" is the same as in Eq. (53), and the rate constants are for Mechanism (51). Although this equation gives a lower limit on k_7, k_4 is not known. If the same type of PIX experiments are carried out in the reverse direction, one determines k_9 from an equation similar to Eq. (53) and obtains an equation similar to Eq. (54) that also contains k_4 and k_7 as unknowns. Simultaneous solution of this equation and Eq. (54) permits determination of all four rate constants from PIX experiments in forward and reverse directions.

This method has been applied to UDPglucose pyrophosphorylase (UTP–glucose-1-phosphate uridylyltransferase), where MgUTP adds before glucose-1-P and MgPP is released before UDPG. The PIX in UTP was shown to be linearly suppressed by glucose-1-P as predicted by Eq. (52) in reciprocal form, and likewise the PIX in UDPG was suppressed linearly by PP_i. Hester and Raushel were able to combine the results from PIX experiments plus the other kinetic constants to calculate all ten rate constants in Mechanism (51), including k_8 and k_{10} (*64*).

In a Ping-Pong mechanism,

$$E \underset{k_2}{\overset{k_1 A}{\rightleftharpoons}} EA \underset{k_4}{\overset{k_3}{\rightleftharpoons}} FP \underset{k_6 P}{\overset{k_5}{\rightleftharpoons}} F \tag{55}$$

where one wishes to observe PIX in substrate A in the absence of the second substrate, one must add excess P to the system to reconvert F to E, since, in the absence of P, all of the enzyme is converted to F in the first turnover and no further catalysis is possible. The presence of excess P serves another function, since the initial rate of formation of unlabeled A from P represents the rate at which F is initially formed from E and labeled A. The high level of unlabeled P also dilutes out P* formed from A* and prevents apparent PIX resulting from reaction of P* that has dissociated from the enzyme. Thus, it is possible by comparing the PIX rate to the rate of formation of unlabeled A to calculate $v_{PIX}/v_{chem\ rxn}$ from Eq. (48). The corrections for the level of P* that has dissociated and recombined to give A and apparent PIX are discussed by Hester and Raushel (*65*), who also showed that, in this system,

$$k_5/(V_2/E_t) \geqq v_{\text{chem rxn}}/v_{\text{PIX}} \tag{56}$$

They measured PIX (β-nonbridge to β-bridge ^{18}O) in UDPglucose in the presence of galactose-1-P uridylyltransferase (UTP–hexose-1-phosphate uridylyltransferase) and a 4- to 12-fold excess of unlabeled glucose-1-P and found that glucose-1-P was released at least 3.4 times faster than V_2/E_t. This type of study should be very useful in determining the relative rates of reactant release in Ping-Pong mechanisms.

4. *Countertransport or Tracer Perturbation Method*

The countertransport or tracer perturbation method, introduced by Britton and Clarke (*66*), involves a test for obligate free enzyme isomerization (i.e., an Iso mechanism). The method was used by Knowles and co-workers to prove the Iso mechanism for proline racemase and show that V is limited by EA to E′P, rather than by E′ to E, conversion (*67*).

In the experiment, 33 mM labeled DL-proline and 183 mM unlabeled L-proline were incubated with proline racemase and the percent label in D- and L-proline monitored as L-proline was converted to D-proline and the reaction approached chemical and isotopic equilibrium. The percent label in L-proline rose to 65%, then gradually dropped to 50%, showing that there was countertransport of label from D- to L-proline induced by the high flux from L- to D-proline. This countertransport results from the high level of E′ formed by the reaction E + A → P + E′, which increases the rate of the reverse reaction, P* + E′ → E + A*. The fractional excess of label in A at any point during the reaction is given by

$$(A^* - P^*)/(A_o^* + P_o^*) = [(A - P)/(A_o + P_o)]\{1 - [(A - P)/(A_o - P_o)]^x\} \tag{57}$$

where asterisks indicate labeled species, and the subscript "o" indicates initial values. $A_o^* = P_o^*$, but $A_o \gg P_o$, because of the large excess of unlabeled A added at the start of reaction. Note that the fractional excess in A is zero both at the beginning of the reaction, where $(A - P) = (A_o - P_o)$, and also at the end, where A = P. The point of maximum fractional excess label in A occurs at

$$(A - P) = (A_o - P_o)/(1 + x)^{1/x} \tag{58}$$

In Eqs. (57) and (58), x is the ratio of $(A_o + P_o)$ to the concentration at which the rate-limiting step switches from the conversion of EA to E′ + P to the conversion of E′ to E. This point, which was approximately 120 mM for proline racemase (*67*), indicates a mechanism where EA to E′ + P conversion limits V and the E′ to E conversion only limits the rate in the oversaturated region. This is a very important experiment, since it clearly distinguishes between the two possible mechanisms for proline racemase [the other being one in which E′ to E conversion limits V, in which case x in Eqs. (57) and (58) would be the ratio of $(A_o + P_o)$ and the average K_m for the substrates].

The fractional excess of label in A at the point given by Eq. (58) will be

$$[(A^* - P^*)/(A_o^* + P_o^*)]_{max} = [(A_o - P_o)/(A_o + P_o)]x/(1 + x)^{(1+1/x)} \quad (59)$$

Thus, the countertransport experiment requires that $(A_o + P_o)$ be high enough for the exponent x in Eq. (57) to be at least 0.3, if the maximum fractional excess label is to be approximately 0.1. With proline racemase, x was 1.6, and the maximum excess label was 0.30. It is also necessary that $A_o \gg P_o$ in order to observe the maximum possible countertransport (a value of 17 was used with proline racemase).

Interestingly, proline racemase showed no countertransport in ammonium bicarbonate buffer, as the buffer ions catalyze the E′ to E conversion so that it no longer can be made rate limiting (*67*). The E′ to E conversion involves the deprotonation of a water (or small buffer) molecule in the active site to give hydroxide (or deprotonated buffer) by one base, and the subsequent protonation of hydroxide (or buffer) by the other base. Thus, one of the bases on the enzyme (which are sulfhydryl groups) must be protonated and the other unprotonated for activity, and which one is protonated determines whether L- or D-proline is adsorbed (*68*).

VI. Stickiness and Methods for Measurement

A. Isotope Partitioning Method

A sticky substrate is one that reacts to give products as fast or faster than it dissociates from the enzyme, and the ratio of these two rates can be defined as the stickiness ratio, S_r. The concept of stickiness was introduced (although not by that name; the present author will take the blame for that) by Rose *et al.* in 1974 with their development of the isotope partitioning method (*69*). As the name implies, this is a partitioning method that involves the fate of an enzyme complex during a single turnover. In a small volume one incubates enzyme with a radioactive substrate (A) that forms a binary complex, trying to get as much of the enzyme tied up as possible. One then dilutes this solution into a rapidly mixing solution of much greater volume that contains (1) a high concentration of unlabeled A and (2) a variable level of the other substrate, B. After several seconds one stops the reaction (usually with acid, but any method is fine as long as it is fast) and analyzes for labeled product.

In the control experiment, the labeled substrate is in the large solution, and only the enzyme is added in the small volume. The control measures the amount of highly diluted substrate converted to product during the several-second-long incubation. The difference between experiment and control is thus the product formed during the first turnover. To improve the accuracy of the method, the dilution of labeled substrate (ratio of volumes times ratio of concentrations in the two solutions) should be as high as possible, and the time of incubation should be as short and uniform as possible. A stopped-flow apparatus is not needed; the method works well simply by using a syringe to add the small volume to a rapidly stirred solution in a beaker. One must make sure that no second substrate is present in the enzyme, or reaction will take place with the EA* complex before it is added to the large solution; this can be controlled by adding the labeled EA* complex to a solution of acid and determining whether any labeled product has been formed.

The system involved here for a random mechanism is

$$\begin{array}{ccccc} EA^* & \underset{k_4}{\overset{k_3B}{\rightleftharpoons}} & EA^*B & \xrightarrow{k_5} & E + \text{product}^* \\ \downarrow{\scriptstyle k_2} & & \downarrow{\scriptstyle k_7} & & \\ E + A^* & & EB + A^* & & \end{array} \tag{60}$$

where an asterisk indicates the label. Any A* that dissociates will be diluted by the high level of unlabeled A present, and thus we are only concerned with the partitioning of EA* in the first turnover as a function of the level of B present in the large solution.

The ratio k_5/k_7 in Scheme (60) is the stickiness ratio for A, and is given by

$$S_r = k_5/k_7 = 1/\{E_t/[\text{app } P^*_{max}(1 + K_{ia}/A)] - 1\} \tag{61}$$

where K_{ia} is the dissociation constant of A, E_t is the concentration of active sites, and app P^*_{max} is the level of labeled product formed at infinite concentration of B. app P^*_{max} is determined from a reciprocal plot of $1/P^*$ versus $1/B$.

Note that A in Eq. (61) is the level of *free* A in the original incubation, and that if E_t approaches K_{ia} this will be less than the total amount of A added. In practice this is not a serious problem, since the $(1 + K_{ia}/A)$ term is a correction for not tying up all of the enzyme as EA*, and if K_{ia} is less than E_t it is easy to add enough A so that almost all of the enzyme is in the EA* form. When the binding is loose and one cannot get enough enzyme in solution to convert most of it to EA*, this correction term is important, but A will be nearly equal to total A added in this case.

Note also in Eq. (61) that it is k_5, and not V_1/E_t that is compared to k_7 (k_5 includes catalysis and release of the first product). It may be that a substrate has a low stickiness ratio and yet is still released more slowly than V_1/E_t if k_5 is much greater than the rate constant for second product release.

The rate with which A dissociates from EA relative to V_1/E_t (this *is* a comparison with the turnover number) is then given by

$$(K'/K_b)E_t/[\text{app } P^*_{max}(1 + K_{ia}/A)] \geqq k_2/(V_1/E_t) \geqq K'/K_b \qquad (62)$$

where K' is the apparent Michaelis constant from the reciprocal plot of 1/P* versus 1/B, K_b is the Michaelis constant of B in the chemical reaction, and the other parameters are the same as in Eq. (61). If the mechanism is ordered [i.e., k_7 is zero in Mechanism (60)], S_r from Eq. (61) is infinite, and Eq. (62) gives an exact solution for $k_2/(V_1/E_t)$. In a random case, $k_2/(V_1/E_t)$ equals the left-hand expression in Eq. (62) if B is not sticky (i.e., cannot be trapped in isotope partitioning studies) and equals the right-hand expression if B is quite sticky. If both A and B can be trapped and their stickiness determined,

$$k_2/(V_1/E_t) = (K'/K_b)(1 + S_{ra})(1 + S_{rb})/(S_{ra}S_{rb} + S_{ra} + S_{rb}) \qquad (63)$$

where S_{ra} and S_{rb} are stickiness ratios from Eq. (61) for A and B.

In Eqs. (61) and (62), one must use a value for K_{ia} from other experiments, and this is often the most poorly known kinetic constant. Where it is not possible to tie up a high percentage of the enzyme as EA*, one can run the experiments with different levels of A* as well as B, and then the equation for the amount of labeled product will be

$$P^* = P^*_{max}AB/[(K_{ia} + A)(K' + B)] \qquad (64)$$

Because the denominator is factored, this equation is particularly easy to fit, and this is one of the better ways to get accurate dissociation constants, especially if the substrates are at all sticky so that correct values are not obtained from initial velocity patterns. Equation (61) becomes

$$S_r = 1/[E_t/P^*_{max} - 1] \qquad (65)$$

and the left term in Eq. (62) is $(K'/K_b)(E_t/P^*_{max})$. This method has so far only been used with yeast hexokinase (*70*).

In a Ping-Pong mechanism, the EA* complex will react to give F and P* to the extent determined by K_{eq} for the half-reaction, and, after dilution with unlabeled A, this reaction will not reverse because of dilution of P*. Thus, the level of P* will be independent of the concentration of B in the diluting solution, and the zero value of K' establishes a Ping-Pong mechanism. This would be a very useful technique where one has a parallel initial velocity pattern and K_{eq} is high enough for A–P isotopic exchange not to be readily observed. K' would be zero in a sequential mechanism only if k_2 truly were far less than V_1/E_t.

B. Variation of *V/K* with Viscosity

Determining stickiness on the basis of variation of V/K with viscosity depends on the fact that diffusion is inversely proportional to the microviscosity in solution. Thus, molecules like glycerol or sucrose raise the microviscosity, and 30% sucrose will produce a relative viscosity of approximately 3 and reduce diffusion-limited steps by this factor. Polymers like Ficoll increase the macroviscosity, but have little effect on diffusion because they do not affect the microviscosity.

In the following mechanism,

$$\mathrm{E + A} \underset{k_2}{\overset{k_1}{\rightleftharpoons}} \mathrm{EA} \xrightarrow{k_3} \mathrm{E + products} \tag{66}$$

we presume that k_1 and k_2 will be affected by viscosity and k_3 will not. Thus, if A is not sticky ($k_2 > k_3$), neither V, K, nor V/K will be affected by viscosity, since the effect on k_1 and k_2 will be equal. If the substrate is sticky ($k_3 > k_2$), V/K is equal to k_1E_t, and will be inversely proportional to viscosity. In the general case,

$$1/(V/K) = \eta_{rel}/(k_1E_t) + (k_2/k_3)/(k_1E_t) \tag{67}$$

The horizontal intercept of a plot of $1/(V/K)$ versus η_{rel} is thus k_2/k_3, the reciprocal of S_r, the stickiness ratio, while the slope is $1/(k_1E_t)$, from which k_1 can be determined if E_t is known.

In order to compare data for different substrates, one can plot the ratio of V/K at $\eta_{rel} = 1$ to V/K at different η_{rel} values versus η_{rel}:

$$(V/K)_{\eta_{rel}=1}/(V/K)_{\eta_{rel}} = \eta_{rel}S_r/(S_r + 1) + 1/(S_r + 1) \tag{68}$$

The plot will have a slope of unity if S_r is very large but zero for a nonsticky substrate. In the general case, S_r is (slope)/(1 − slope).

This method was used successfully by Kirsch and colleagues with chymotrypsin (*71*) and β-lactamase (*72*), but it has failed in many cases. It seems to work best for simple hydrolytic enzymes where product release is not rate limiting. Among the problems one can encounter are the following: (1) If product release is rate limiting for V/K, this will make V/K viscosity dependent even if the substrate is not sticky. (2) If the viscosogen has any affinity for the active site it will act as an inhibitor. (3) The viscosogen may have an effect on the conformation changes that precede or follow catalysis, and thus on k_3 in Mechanism (52). For example, Kurz *et al.* observed an effect of sucrose on the V/K for adenosine with adenosine deaminase (*73*), although comparison of the pH profiles for V/K of adenosine and the pK_i of a competitive inhibitor clearly shows that adenosine is not sticky (*74, 75*).

Similarly, with malic enzyme, sucrose or glycerol caused a decrease in the K_m for malate and a slight increase in V (*76*), with the result that V/K was greatly increased (by a factor of 2 with sucrose and a factor of 5 with glycerol). These changes decreased the ^{13}C isotope effect on decarboxylation, presumably as the result of changes in the commitments in the system (see Section VII,B below).

Thus, the viscosity method must be used with care, and it is essential that slow, nonsticky substrates used as controls not show pronounced sensitivity to viscosity if valid results are to be obtained. The method does, however, give information on the stickiness of the last substrate to add to the enzyme in an ordered mechanism. Initial velocity patterns are not sensitive to the stickiness of the last substrate to add (see Section V,A,2 above), and the isotope partitioning method does not work for this substrate. Analysis of pH profiles and the pH variation of isotope effects will determine stickiness, however (see Section VII,A,4 below).

VII. Determination of Chemical Mechanism

The methods described above or in the chapter in Volume II serve primarily to deduce the kinetic mechanism and to determine the relative rates of the steps in it. The two methods to be described in this section provide information on kinetic mechanisms and the relative rates of the steps in them, but they are also capable of giving direct evidence on the nature of the chemical mechanism and, in the case of isotope effects, on transition state structures. They are currently being used actively to determine enzyme mechanisms and are among the most important tools available to enzymologists. Particularly when coupled with the use of slow alternate substrates and enzymes modified by site-directed mutagenesis to make catalysis more rate limiting, these are very powerful methods indeed.

A. pH PROFILES

It is often the case that the substrate has a required protonation state for binding and/or catalysis, and, similarly, catalytic groups in the active site usually have required protonation states as well. Thus, when the substrate or a key group on the enzyme ionizes in the accessible pH range, there will be a variation of the kinetic parameters with pH. Such pH variation often provides a clue to the acid–base catalysis, and thus the chemical mechanism of the reaction. The interested reader may find other articles on pH profiles useful (*2, 3, 77*).

The p*K* values of the substrate are readily determined by titration or change in NMR chemical shift with pH, and the use of an alternate substrate with an altered p*K* will confirm the assignment. Identification of the group in the active site is more difficult, although the observed p*K* values can usually be correlated with groups seen in an X-ray structure and the assignments confirmed by site-directed mutagenesis. Acid–base catalysis is usually worth approximately 10^5 in rate acceleration, so deleting a key group causes a drop in rate of this order of magnitude and may also change the entire shape of the pH profile if protons or hydroxide ions have to replace the group on the enzyme to bring about the reaction.

The parameters that are plotted versus pH are (1) $\log(V/K)$ for each substrate, (2) $\log(V)$, (3) pK_i (logarithm to the base 10 of the reciprocal of the dissociation constant) for a competitive inhibitor or a substrate not adding last to the enzyme, and (4) pK_i or pK_m for metal ion activators. It is particularly important to consider the pH variation of V/K and V, the two independent kinetic constants, and not simply to determine the rate at some arbitrary concentration of each substrate. The Michaelis constant is merely the ratio of V and V/K, so its pH profile is a combination of effects on V and V/K. Although we shall discuss the shapes of pH profiles, the reader should remember that graphical plotting is for a preliminary look at the data, and that the data must be fitted to the appropriate rate equation by the least-squares method to obtain reliable estimates of kinetic parameters, pK values, and their standard errors (*5*). Because pH profiles commonly show decreases of a factor of 10 per pH unit over portions of the pH range, the fits are always made in the log form [i.e., $\log(V)$, $\log(V/K)$, or pK_i versus pH].

1. *pK_i Profile for a Competitive Inhibitor*

As noted above, pH profiles are plotted as log–log plots, and they consist of linear segments with whole number slopes connected by curved segments 2 pH units wide. We consider first pK_i profiles for a competitive inhibitor (or substrate that adds prior to the last substrate). This profile shows only the pK values of the inhibitor (or substrate) or of groups on the enzyme that are important for binding. When only the deprotonated inhibitor will bind and no groups on the enzyme ionize, the pK_i profile will be flat from 1 pH unit above the pK of the inhibitor to the high-pH end of the profile, and it will have a slope of 1 at low pH:

$$\text{Apparent } pK_i = pK_i - \log(1 + H/K) \tag{69}$$

where H is $[H^+]$, K_i is the dissociation constant of the inhibitor from EI, and K is the acid dissociation constant of the inhibitor. The curved portion of the profile will extend from 1 pH unit below the pK to 1 pH unit above it, and the asymptotes to the two portions of the curve will cross at the pK. The same pK_i profile will be seen if the inhibitor does not have any pK values in the accessible pH range, but binding is prevented by protonation of a group on the enzyme.

The great advantages of pK_i profiles are that (1) the pK values are seen at their correct values because the inhibition constant is a parameter determined at equilibrium (i.e., extrapolated to zero level of the substrate) and (2) the profile shows only the requirements for binding and not for catalysis.

It often happens that protonation of a group causes a change in the strength of binding but does not totally prevent it. In this case the pK_i profile begins to drop at the pK of the group in free inhibitor or enzyme, but levels out at a new plateau value reflecting the pK_i for the less favorable protonation state:

$$\text{app } pK_i = pK_{i\,\text{high pH}} - \log\{(1 + H/K)/[1 + (H/K)(K_{i\,\text{high pH}}/K_{i\,\text{low pH}})]\} \tag{70}$$

The point at which the curve levels out represents the pK of the inhibitor or enzyme group in the EI complex, and thus is shifted by the same factor as the log of the ratio of K_i values for the two ionization states.

When deprotonation of a group on the inhibitor or enzyme prevents binding, the pK_i profile drops at high pH and attains a slope of -1 above a point 1 pH unit above the pK:

$$\text{Apparent p}K_i = \text{p}K_i - \log(1 + K/\text{H}) \tag{71}$$

Again, the profile will level off at high pH if deprotonation does not totally prevent binding:

$$\text{app p}K_i = \text{p}K_{i\ \text{low pH}} - \log\{(1 + K/\text{H})/[1 + (K/\text{H})(K_{i\ \text{low pH}}/K_{i\ \text{high pH}})]\} \tag{72}$$

2. *Log(V/K) Profile for a Nonsticky Substrate*

The V/K profile is sensitive to the pK values of groups on the substrate or the enzyme form with which it combines that have a required protonation state for binding and/or catalysis. Since catalysis is usually more sensitive to correct protonation state than binding, one tends to see few partial effects in V/K profiles; that is, when they begin to drop they keep on decreasing. For nonsticky substrates the profiles have simple shapes, and the pK values seen are the correct ones. Thus, if the physiological substrate is known to be sticky, one should always look for a slower alternate substrate to use for pH studies.

When protonation of the substrate or a group on the enzyme prevents binding and/or catalysis, the V/K profile is given by

$$\log(V/K) = \log[(V/K)_{\text{high pH}}] - \log(1 + \text{H}/K) \tag{73}$$

where K is the acid dissociation constant of the group whose protonation prevents binding and/or catalysis. When deprotonation of a group kills activity,

$$\log(V/K) = \log[(V/K)_{\text{low pH}}] - \log(1 + K/\text{H}) \tag{74}$$

If activity falls at both low and high pH,

$$\log(V/K) = \log[(V/K)_{\text{pH ind}}] - \log(1 + \text{H}/K_1 + K_2/\text{H}) \tag{75}$$

where K_1 and K_2 are the acid dissociation constants of the two groups controlling activity. As long as pK_1 and pK_2 are at least 2 pH units apart, a fit to Eq. (75) will usually determine the pK values accurately, but when the pK values are closer than 2 pH units one can determine the average of the values accurately but not their individual values. This is because the pK separation depends on the sharpness of the transition from the asymptote at low pH with slope 1 to the asymptote at high pH with slope -1. If pK_1 = pK_2 in Eq. (75), the peak of the profile is lower than the crossover point of the symptotes by log 3 = 0.48, whereas if the pK values are 0.6 pH units apart, the separation is log 4 = 0.60. In practice, the pK values in Eq. (75) should have a minimum separation of 0.6 pH units if this profile results from independent ionization of two groups, since the pK values in Eq. (75) are apparent ones, and the second log term in this equation when written in terms of the true pK values is $(1 + K_{2t}/K_{1t} + H/K_{1t} + K_{2t}/H)$ where K_{1t} and K_{2t} are the true pK values of independently ionizing groups. If pK_{1t} = pK_{2t}, then pK_1 = pK_{1t} − 0.3 and pK_2 = pK_{2t} + 0.3. The difference between true and apparent pK values disappears when $K_1 > K_2$, or as long as the pK values are at least a pH unit apart. The equations for this case are presented fully in Ref. (*2*).

When a V/K profile is represented by Eq. (75) we cannot *a priori* tell whether the group with pK_1 has to be ionized and the group with pK_2 protonated, or vice versa, since the shape of the pH profile will be exactly the same. For example, fumarase (fumarate hydratase) has an imidazole (pK 7.1) that must be protonated and a carboxyl group (pK 5.85) that must be ionized for reaction of malate, whereas the imidazole must be neutral and the carboxyl protonated for reaction of fumarate (*78*). It is common for such reverse protonation to be required in one direction of an enzymic reaction.

When reverse protonation is involved, the proportion of enzyme and/or substrate in the correct form to react is reduced by $(1 + K_1/K_2)$, and, thus, if pK_1 and pK_2 are separated by several pH units and the pH-independent V/K value is high enough, one can rule out reverse protonation. For example, if the pK separation is 3 pH units, and the apparent pH-independent V/KE_t value is $10^7\ M^{-1}\ sec^{-1}$, reverse protonation would require the true pH-independent value of V/KE_t to be $10^{10}\ M^{-1}\ sec^{-1}$, which exceeds the diffusion-limiting rate ($\sim 10^8\ M^{-1}\ sec^{-1}$ for nucleotide-size reactants, $\sim 10^9\ M^{-1}\ sec^{-1}$ for smaller substrates).

Comparison of pK_i and log(V/K) profiles may enable one to identify catalytic groups unambiguously. Thus, any pK seen in the V/K profile, but not in a pK_i profile, corresponds to a group whose protonation is important for catalysis but not for binding. The reverse is not true; groups seen in both profiles may still be involved in catalysis as well as binding. Be sure that the inhibitor used for the pK_i profile and the substrate have the same number of ionizing groups, however. For example, the V/K profile for isocitrate with isocitrate dehydrogenase shows a decrease below the third pK of the substrate, since only the trianion is bound, but oxalylglycine, a competitive inhibitor, does not show this pK in its pK_i profile because it is a dicarboxylic acid and its pK values are much lower (*79*).

3. *Log(V) Profiles*

The pK values seen in log(V) profiles are those of groups involved in the rate-limiting unimolecular steps in the mechanism. In most cases these groups are involved in catalysis in some way, but the pK values are perturbed from where they were in free enzyme or substrate. It is also important to note which pK

values are *not* seen in the log(V) profile. For example, if a group in substrate or enzyme has a required protonation state for binding, the pK is seen in the log(V/K) profile but not in the V profile. Saturation with the substrate forces the enzyme–substrate complexes into the correct protonation states, and the pK values are displaced to very low or very high pH where they cannot be observed. This is a common occurrence; the V for reaction of NADPH, α-ketoglutarate, and ammonia catalyzed by glutamate dehydrogenase is pH independent, although the pK_i for α-ketoglutarate dissociation from E–NADPH–α-ketoglutarate decreases above a pK of 5, and then again to a slope of -2 above a pK of 8, and V/K for ammonia decreases a factor of 10 per pH unit below the pK of ammonia (*80*). Combination of α-ketoglutarate prevents loss of protons from the groups with pK values of 5 and 8 in E–NADPH, whereas only NH_3, and not NH_4^+, will combine with the E–NADPH–α-ketoglutarate complex.

The pK values seen in the log(V) profile may be perturbed from their position in free enzyme or substrate for two reasons. First, the combination of the substrate may perturb the pK either by specific hydrogen-bonding interactions or by solvent exclusion from the active site. For example, with fumarase (fumarate hydratase) the pK of the imidazole group (pK 7.1 in free enzyme) is seen at 4.9 in the V profile for fumarate and at 9.0 in the V profile for malate (*78*). These shifts are typical of hydrogen-bonding interactions and show the selectivity of binding to correctly protonated species. In this case the imidazole needs to be protonated to permit elimination of water from malate, and thus the hydrogen bond to the hydroxyl of malate perturbs the pK upward. The imidazole needs to be neutral to accept a proton from the specifically bound water to permit it to react with fumarate, and thus the pK is perturbed downward by the hydrogen bond to this water.

In contrast, the pK of the carboxyl group on fumarase is raised by either substrate (from 5.85 to 6.4 by malate, and to 7.0 by fumarate) (*78*). These effects are basically solvent perturbations resulting from dehydrating this carboxyl group, which cannot form hydrogen bonds to C-3 of the substrate. Note that the presence of fumarate makes the pK values of the two groups cross in order to promote the needed state of protonation for the reaction (protonated carboxyl, neutral imidazole).

The second reason pK values are perturbed in log(V) profiles is that there has been a shift in rate-limiting steps. For example, with malic enzyme the pK values of 6 and 8 seen in the V/K profile for malate are shifted outward in the log(V) profile by approximately 1.3 pH units because at neutral pH NADPH release from E–NADPH, rather than catalysis, is the rate-limiting step for V. Only when incorrect protonation has slowed catalysis by a factor of about 20 does it become equal to the rate of NADPH release, and thus the pK values are perturbed outward by log 20, or 1.3, pH units (*81*). The deuterium isotope effects on V/K of 1.5 and of 1.0 on V at neutral pH support this interpretation [the isotope effect on V does become equal to that on V/K when catalysis becomes rate limiting at very low or very high pH (*82*)].

4. *pH Profiles for Sticky Substrates*

Although the pH profiles for nonsticky substrates have simpler shapes and are easier to interpret, there is useful information in the pH profiles for a sticky substrate, and thus one must know what to expect, and how to interpret the results. The pK values may be perturbed in the V/K profile for a sticky substrate, and both the V/K and V profiles may show a "hollow," or at least a flatter shape, near the pK. Consider the following mechanism in which protonation of a group with a pK less than 7 prevents catalysis:

$$\begin{array}{ccccccc} EH & \underset{k_8}{\overset{k_7 A}{\rightleftharpoons}} & EAH & & & & \\ K_1 \updownarrow & & k_6 H \uparrow\downarrow k_5 & & & & \\ E & \underset{k_2}{\overset{k_1 A}{\rightleftharpoons}} & EA & \xrightarrow{k_3} & EP & \xrightarrow{k_9} & E \end{array} \tag{76}$$

As long as EA and EAH can be interconverted directly (i.e., k_5 and k_6 have finite values) and EAH forms, the low-pH asymptote with a slope of 1 intersects the high-pH plateau at [pK_1 $-$ log$(1 + k_3/k_2)$]. Thus, the apparent pK is displaced outward by the stickiness of the substrate. In a mechanism more complicated than Mechanism (76) the displacement is log$(1 + S_r')$, where S_r' is a stickiness ratio that is slightly different from that determined by the isotope partitioning method in Section VI,A above [the stickiness of the first product released also contributes to the ratio; see Ref. (*3*)], but in a simple mechanism such as Mechanism (76) S_r and S_r' are the same.

If EA and EAH are rapidly interconverted in Mechanism (76) (i.e., $k_5 \gg k_8$), the shape of the V/K profile is not altered from that predicted by Eq. (73), except for the displacement of the pK. If $(1 + k_8/k_5) > k_7/k_1$, however, which will occur when $k_5 < k_8$, the shape of the profile is altered. The curve begins to drop at pK_1 but then starts to level out before finally dropping with an asymptote that corresponds to the perturbed pK, so that a hollow is seen in the curve. The equation for the profile is given in Refs. (*2*) and (*3*).

If EAH does not form in Mechanism (76) (in which case the pK is seen in the V/K profile, but not in the V profile), or if EAH is only a dead-end complex and k_5 and k_6 are zero, the V/K profile has the simple shape predicted by Eq. (73), and the pK is not perturbed by the stickiness of the substrate and is seen in its correct position. Thus, the displacement of a pK by the stickiness of a substrate requires that incorrectly protonated E–substrate complexes form and that, without dissociation of substrate, they be interconverted at a reasonable rate to a correctly protonated complex able to undergo the catalytic reaction. This requirement should be kept in mind while reading the rest of this section below.

When the pK in Mechanism (76) is above 7, the interconversion of EA and EAH will involve reaction with hydroxide, rather than with protons, k_6 will be unimolecular, and $k_5[OH^-]$ will replace k_5. The pK now is displaced by the stickiness of the substrate only when EA and EAH are rapidly equilibrated, and the degree of displacement decreases as k_6 becomes equal to or less than k_2 [see Ref. (*3*) for a fuller discussion].

When V/K decreases at high pH and the pK is above 7, we have

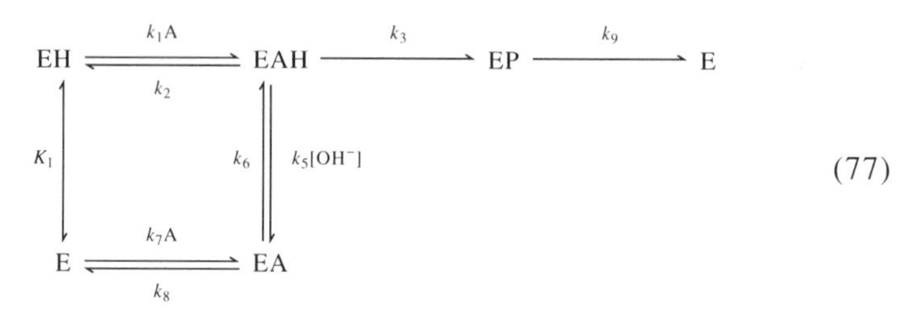

(77)

As long as EA forms, and EA and EAH can be directly interconverted, the apparent pK, as given by the intersection of the low-pH plateau and the high-pH asymptote with a slope of -1, is displaced to higher pH by $\log(1 + k_3/k_2)$. As long as $k_6 \gg k_8$, the shape is the one predicted by Eq. (74), but when $k_6 \leqq k_8$, the curve will have a hollow. The V/K profile for creatine phosphate with creatine kinase showed a prominent hollow at 12°C, a barely visible one at 25°C, and no hollow at 35°C, although in each case the apparent pK was displaced to higher pH (by 1.1, 0.7, and 0.4 pH units at 12°, 25°, and 35°C) (*40*). The value of k_8/k_6 was estimated to be 9.3 and 2.5 at 12° and 25°C, respectively, and thus it is clear that the stickiness of the proton is most temperature sensitive, with the stickiness of the substrate being less so. The actual V/K values had an even lower ΔH value (*40*). In contrast, with alanine dehydrogenase, where alanine is a sticky substrate and serine a slow, nonsticky one, protons on the acid–base catalytic groups were rapidly equilibrated, and no hollows were seen in the V/K profile for alanine, although the pK values were displaced outward by about 1 pH unit (*83*).

Mechanism (77) applies to mechanisms where pK_1 is above 7. If it is below 7 and equilibration of EA and EAH involves protons, rather than hydroxide, the stickiness of the substrate will displace the pK in the V/K profile only if the interconversion of EA and EAH is fast (*3*).

The $\log(V)$ profile can also show a more complicated shape with a hollow if both the substrate and the proton on the catalytic group are sticky. If the protonation equilibrium of the catalytic group is rapid in the E–substrate complex, however, no hollow will be seen (*2*, *3*).

5. pK_i and pK_m Profiles for Metal Ions

Because metal ion activators that combine directly with enzymes frequently have imidazole, carboxyl, sulfhydryl, or possibly other protonatable groups as ligands, their binding is prevented, or severely decreased, by protonation of such groups. The pK values of such ligands should show up in the pK_i (absence of substrate) or pK_m (presence of substrate) profiles for the metal ion, but not in V or V/K profiles for the substrates when the metal ion is saturating. For example, pyruvate kinase catalyzes the metal ion-dependent decarboxylation of oxaloacetate in the absence of nucleotides or K^+, and the enzyme shows equilibrium ordered kinetics, with the metal ion adding prior to oxaloacetate (*84*). The pK_i profiles for the various metal ions decrease below a pK of 7 and increase above a pK of 8.9 (*84, 85*). The pK of 7 is clearly a ligand for the metal ion [thought to be a glutamate on the basis of the X-ray structure (*86*)] and does not show up in V or V/K profiles for oxaloacetate.

The increase in the pK_i value for Mg^{2+}, Mn^{2+}, and Co^{2+} above pH 9 reflects tighter binding of the metal ion. The group responsible may be the lysine thought to be the acid–base catalyst for enolization during the physiological reaction, and binding of the metal ion will lower the pK of this group by the degree to which the metal ion binds more tightly when the lysine is deprotonated. The pK of this lysine is 8.2 in the presence of Mg^{2+}, approximately 7.3 with Co^{2+}, and probably has an intermediate value with Mn^{2+} (*85*). The increase in metal ion binding above pH 9 was originally thought to result from ionization of water bound to the metal ion, but the discovery that the pK in the pK_i profile did not differ for the various metal ions and that the pK of water bound to the E–Co^{2+} complex was 7.3 led to a correct assignment (*85*).

To date few pK_i profiles for metal ion activators have been determined. However, this is a powerful tool for identifying ligands for the metal ion, even if an X-ray structure is not available.

6. Identification of Groups with pK Values Seen in pH Profiles

It is often possible to deduce the role of catalytic groups whose pK values are seen in pH profiles, but, as noted earlier, identification of the group is harder. If an X-ray structure is available, one should be able to deduce which groups are responsible for the pK values, and in favorable cases the pK values of histidines on the enzyme can be determined by NMR. It is useful, however, to have some independent way to identify the groups when no structure is available or when one wants to confirm deductions based on structure.

Two kinetic methods have been used to identify catalytic groups from pH profiles. First, the temperature coefficients of the pK values of the various ionizing groups differ considerably. Carboxyl and phosphate groups have ΔH_{ion} values near zero, whereas histidine, cysteine, and tyrosine groups have values of approximately 6 kcal/mol and amino groups and water coordinated to metal ions have values of about 12 kcal/mol. To use this method one runs the pH profiles at various temperatures, using nonsticky substrates so that the change in stickiness with temperature does not perturb the result. The p*K* is then plotted versus the reciprocal of absolute temperature, and the slope is $\Delta H_{ion}/2.3R$. The p*K* values are higher at lower temperature, with ΔpK between 0° and 25°C being 0.4 or 0.8 pH units for ΔH_{ion} values of 6 and 12 kcal/mol.

The problem with this method is that if a conformation change in the protein accompanies ionization of the group (or the conformational equilibrium is perturbed by the ionization) the ΔH of this conformation change is wholly or partly added to ΔH_{ion}. This usually results in too high a value, but it could in theory give too low a one. This method is probably most useful when ΔH_{ion} is near zero, and the acid–base catalyst a carboxyl group, as in several kinases (*87, 88*).

The second method of identifying groups is solvent perturbation. In these studies one uses organic solvents such as dimethylformamide, dimethyl sulfoxide, dioxane, formamide, or various alcohols at as high a level as can be used without overly reducing reaction rates. The p*K* values of neutral acid groups (carboxyl, phosphate, sulfhydryl, tyrosine, metal-bound water), which ionize to form H^+ and an anion, are elevated by replacing water with the solvent, even when the dielectric constant is not much changed. The effect arises because water hydrates both H^+ and the new anion, stabilizing the charge separation.

In contrast, cationic acids (histidine, lysine, arginine) are protonated at low pH and ionize to H^+ and a species with one less positive charge. There is thus no charge separation during ionization, and the number of charges stays the same. As a result the p*K* is not appreciably perturbed by replacement of water by an organic solvent.

Although one can simply run pH profiles with and without solvent and determine changes in the p*K* values, this requires accurate pH measurement in the presence of the solvent. A technique which avoids this problem involves the use of two sets of buffers, one made of cationic acids and the other of neutral acids. One measures the pH values before addition of solvent, then plots profiles based on these pH values. The solvent raises the p*K* (and thus the pH in solution, since the actual ratio of buffer forms stays the same) of a neutral acid buffer, but not that of a cationic acid buffer.

Thus, when the group on the enzyme is a neutral acid, no net change in p*K* is seen after solvent addition when the buffer is a neutral acid, but the p*K* will be raised by solvent when a cationic acid buffer is used. When the group on the enzyme is a cationic acid, no change is seen after solvent addition when the buffer is a cationic one, but the p*K* will appear lower after solvent addition when a neutral acid buffer is used (this is really a frame shift in the pH scale). This method was first used to identify the histidines that are involved in ribonuclease (*89*).

This method is a sound one, and the only caution is that one must be sure that the group of interest is exposed to solvent under the conditions of the experiment. If the group is buried under a substrate, its p*K* cannot be perturbed by solvent. Thus, no perturbation in the p*K* of the catalytic aspartate on hexokinase is seen in the *V*/*K* profile for MgATP when glucose is truly saturating (*90*). In theory one should try to study this group in the free enzyme, but a large number of presumably carboxyl groups are seen in the *V*/*K* profile for glucose and obscure the p*K* of the catalytic aspartate. Protonation of these carboxyl groups is prevented by glucose binding (or more likely by the large conformation change which accompanies it), so that only the p*K* of the catalytic aspartate is seen in the *V*/*K* profile for MgATP (*87*).

B. Isotope Effects

Isotope effects are one of the most powerful tools available to the enzymologist, since they give information about the catalytic reaction itself. By properly chosen experiments one can determine kinetic mechanism, rate-limiting steps, chemical mechanism, and transition state structure; in short, isotope effects can do it all. It is thus not surprising that they are being actively used in our as well as other laboratories to investigate enzyme mechanisms.

Physical organic chemists have used isotope effects for years to give information on mechanisms and transition state structures, and the interested reader should consult standard books on the subject (*91, 92*). It was only in 1975, however, when Northrop discovered the way to combine deuterium and tritium isotope effects on *V*/*K* to obtain an estimate of the intrinsic isotope effect on the bond-breaking step, that isotope effects became a practical tool for the enzymologist (*93*). We give a quick sketch of the methods here, but the reader may wish to consult other recent reviews (*94–96*).

In describing isotope effects, we shall use the notation of Northrop (*97*), where leading superscripts identify the isotopic atom, with D, T, 13, 14, 15, 17, and 18 standing for deuterium, tritium, ^{13}C, ^{14}C, ^{15}N, ^{17}O, and ^{18}O. The isotope effect is the ratio of parameters for unlabeled and labeled species. Thus ^{D}V is a deuterium isotope effect on V (i.e., V_H/V_D), and $^{13}(V/K)$ is a ^{13}C isotope effect on V/K. The same notation is used for equilibrium isotope effects, with $^{15}K_{eq}$ being an ^{15}N isotope effect on K_{eq} (i.e., $K_{eq\ ^{14}N}/K_{eq\ ^{15}N}$).

Isotope effects are called primary when a bond is made or broken to the isotopic atom during the reaction and secondary when it is not. Secondary deuterium or tritium isotope effects are called α when the isotopic hydrogen is attached to a primary carbon undergoing bond cleavage and β or γ when the hydrogen is on a carbon once or twice removed from the primary carbon. Primary isotope effects are almost always normal, or greater than unity, whereas secondary isotope effects can be either normal or inverse (i.e., less than unity).

In fact, a secondary isotope effect is normal in one direction and inverse in the other if the transition state structure is intermediate between that in substrate and product.

Isotope effects result from changes in the stiffness of bonding of isotopic atoms. Kinetic isotope effects reflect differences between reactant and transition state, and equilibrium isotope effects come from differences between reactant and product. Equilibrium isotope effects are easily measured experimentally, and they can also be calculated from the force fields derived for small molecules from vibrational frequencies. They are commonly expressed as fractionation factors, which are equilibrium isotope effects for exchange of isotope with a reference molecule such as water or ammonia. Tables of fractionation factors are in Refs. (*96*) and (*98*) and further values are in Ref. (*99*).

The reason primary kinetic isotope effects tend to be large is that one vibrational mode becomes the reaction coordinate motion and has no restoring force in the transition state. The loss of this vibrational mode lowers the fractionation factor in the transition state and leads to discrimination against the heavy isotope.

Although isotopic substitution in one position normally causes effects that are nearly independent of isotopic substitution elsewhere, there is one exception to this rule. When the reaction coordinate motion involves the coupled motions only of hydrogen atoms, the first deuterium substitution in the system causes a larger isotope effect than if another deuterium were already present. Theory shows that this requires tunneling on the part of the hydrogens (*100, 101*), but it is likely that tunneling always occurs with hydrogen motions, since this coupled motion effect is seen in all nonenzymic and enzymic reactions where one would expect to see it (*102–107*). The motions can be bending ones as well as primary transfers. Thus glucose-6-phosphate dehydrogenase shows coupled motions of the primary hydrogen (which is transferred to $NADP^+$ as a hydride ion), the hydrogen on the hydroxyl group at C-1 [which is transferred to a carboxyl group on the enzyme (*108*)], and the α-secondary hydrogen at C-4 of $NADP^+$, which must bend 55° out of the plane of the nicotinamide ring during the reaction (*102*). This phenomenon was first discovered with formate dehydrogenase, where the secondary deuterium isotope effect at C-4 of the nucleotide was 1.23, even though the equilibrium isotope effect for the reduction of NAD^+ is 0.89 (*103*). The normal value shows that the bending motion of this hydrogen is part of the reaction coordinate motion, and the reduction in this value to 1.07 with deuterated formate shows that tunneling is involved in the coupled hydrogen motions.

1. *Measurement of Isotope Effects*

There are three ways to determine an isotope effect on an enzymic reaction, and it is important to understand the differences in these methods and their advantages and disadvantages.

a. *Direct Comparison of V or V/K Values.* The simplest method of measurement is to compare V or V/K values from reciprocal plots with labeled and unlabeled substrates. The ratio of slopes (labeled/unlabeled) is $^{D}(V/K)$, and the ratio of the vertical intercepts is ^{D}V. The advantage of this method is its simplicity and the fact that it is the only way to determine an isotope effect on V. The disadvantage is that it is not very sensitive and can be used only for isotope effects greater than 1.1, unless extreme care is used. Its use is thus largely limited to primary deuterium isotope effects or to larger secondary ones.

b. *Equilibrium Perturbation.* In the equilibrium perturbation method one incubates enzyme with a labeled substrate and unlabeled product (called the perturbants), together with any other reactants involved, at levels near equilibrium. If initial concentrations are correctly picked, the reaction will be perturbed to one side (toward the labeled perturbant, for a normal isotope effect), then will return to the starting point. The perturbation is driven by the faster initial reaction of the unlabeled perturbant than of the labeled one, but after isotopic mixing the system returns to both chemical and isotopic equilibrium. The method requires that there be a change during reaction in color or in some other easily followed parameter such as optical rotation or CD.

The size of the perturbation at the maximum point is

$$(A_{\max} - A_{o})/A'_{o} = \alpha^{-1/(\alpha-1)} - \alpha^{-\alpha/(\alpha-1)} \tag{78}$$

where α is the apparent isotope effect and $(A_{\max} - A_{o})$ is the perturbation size. A'_{o} is the reciprocal of the sum of reciprocals of the perturbant concentrations, plus a contribution for the concentration of any other reactants present at a level lower than that of the most dilute perturbant. See Ref. (*98*) for the exact equation, as well as a more detailed discussion of the method.

The apparent isotope effect α in Eq. (78) is related to the isotope effects in forward and reverse reactions by

$$^{D}(\mathrm{Eq.P.})_{r} = \alpha(1 + K/^{D}K_{eq})/(1 + K) \tag{79}$$

$$^{D}(\mathrm{Eq.P.})_{f} = {}^{D}(\mathrm{Eq.P.})_{r}\,^{D}K_{eq} \tag{80}$$

where the subscripts "f" and "r" stand for forward and reverse, and K is the ratio at equilibrium of the concentrations of *unlabeled* perturbants (i.e., P/A). This is not the same as the actual starting levels in the experiment, because of the equilibrium isotope effect, and K is thus calculated in practice from $^{D}K_{eq}$ and the concentrations of the perturbants used.

This method is more sensitive than the direct comparison one, and it can be used to measure isotope effects as low as 1.01. The disadvantage is that one must be at equilibrium, and it is tricky to adjust the starting concentrations so that the reaction returns to the starting point. It also does not measure isotope effects on

V. The ^{D}(Eq.P.) value may be the same as $^{D}(V/K)$, or may differ (see below). This method works very well for deuterium isotope effects and has been used to measure ^{18}O isotope effects on fumarase (fumarate hydratase) (*109*), and ^{13}C ones on malic enzyme (*110*).

c. *Internal Competition*. The internal competition method involves the use of labeled and unlabeled reactants at the same time in competition with each other, with the isotope effect being determined from changes in the ratio of the two species in product or residual substrate. The method works best for irreversible reactions. It is the only one available for tritium or ^{14}C isotope effects, since tritium and ^{14}C are used as trace labels, and is useful for heavy atom isotope effects, where the 1.1% ^{13}C, 0.36% ^{15}N, and 0.2% ^{18}O natural abundances are convenient trace labels. The method can be made quite sensitive, especially for heavy atom isotope effects where an isotope ratio mass spectrometer is used to determine the mass ratio. Isotope effects can be determined to 0.1% or better with the isotope ratio mass spectrometer, although the only molecules that are commonly used in such a machine are H_2, CO_2, and N_2. The disadvantage is that one only obtains a V/K isotope effect.

As noted above, one can determine the isotope effect from the mass ratio or specific activity in product (R_p) or residual substrate (R_s), relative to the value for initial substrate (R_o), which may be determined by 100% conversion to product, if it is easier to determine the mass ratio in product than in substrate. If x is the isotope being used, and f is the fractional conversion to product at the time of measurement of R_s or R_p, comparison of R_o and R_p gives

$$^{x}(V/K) = \log(1 - f)/\log(1 - fR_p/R_o) \tag{81}$$

Comparison of R_o and R_s gives

$$^{x}(V/K) = \log(1 - f)/\log[(1 - f)(R_s/R_o)] \tag{82}$$

whereas comparison of R_s and R_p gives

$$^{x}(V/K) = \log(1 - f)/\log[(1 - f)/(1 - f + fR_p/R_s)] \tag{83}$$

Equations (81)–(83) apply when a trace label is being used; if the labeled and unlabeled reactants are of comparable concentration, the equations are more complex (*96*). For an isotope effect greater than 2, most of the label stays behind in the substrate, and the use of Eq. (82) is impractical. Equation (83) has proved very useful for measurement of heavy atom isotope effects, where f is around 0.5, and one isolates both residual substrate and product for analysis. With tritium isotope effects, however, one uses Eq. (81) and as low an f value as practical.

The use of natural abundance ^{13}C or other heavy atom labels is limited by the ease with which the appropriate atoms are isolatable as CO_2 or N_2 for insertion in the isotope ratio mass spectrometer. The nitrogen of interest is easily converted to ammonia by the Kjeldahl method if there is only one nitrogen in the molecule, and ammonia is converted to N_2 by hypobromite. Carbon and oxygen labels, however, are more of a problem.

The solution is the remote label method introduced by O'Leary and Marlier (*111*) in which one uses changes in mass ratio in the remote labeled position to determine discrimination by the isotopic atom of interest. For example, to measure secondary ^{18}O isotope effects on the hydrolysis of glucose-6-P, the substrate was a mixture of 1% glucose-6-P containing ^{13}C at C-1 and ^{18}O in the nonbridge oxygens of the phosphate group and 99% glucose-6-P with ^{12}C (i.e., carbon depleted in ^{13}C) at C-1 (*112*). Mixing these species restored the natural abundance of ^{13}C at C-1 (important to minimize errors in determining mass ratios in the isotope ratio mass spectrometer, since contamination with atmospheric CO_2 otherwise causes large errors). Discrimination caused by ^{18}O changed the ratio of the two species in residual substrate, and this was determined by conversion to ribulose-5-P and CO_2 by glucose-6-P and 6-phosphogluconate dehydrogenases. Any ^{13}C isotope effect at C-1 was corrected for by running the reaction with unlabeled glucose-6-P containing only the natural abundance of ^{13}C at C-1.

Nitrogen can also be used as a remote label; the exocyclic amino group of adenine, which is readily inserted into adenosine by reaction of chloropurine riboside with ammonia and easily removed by adenosine deaminase, is being used in our laboratory for measurement of ^{18}O isotope effects in the γ-phosphate of ATP. In this case, 0.36% of the ^{15}N-, ^{18}O-labeled species is mixed with 99.6% of the ^{14}N-labeled material to restore the natural abundance of ^{15}N in the amino group of adenine. Depleted ^{12}C and ^{14}N are commercially available.

2. *Equations for Isotope Effects*

If only one step in an enzymic reaction is isotope sensitive, the observed isotope effects are

$$^{x}(V/K) \text{ or } ^{x}(\text{Eq.P.}) = (^{x}k + c_f + {}^{x}K_{eq}c_r)/(1 + c_f + c_r) \tag{84}$$

$$^{x}V = (^{x}k + c_{Vf} + {}^{x}K_{eq}c_r)/(1 + c_{Vf} + c_r) \tag{85}$$

where x is the isotope of interest, and thus ^{x}k is the intrinsic isotope effect in the forward direction on the isotope-sensitive step (the intrinsic isotope effect on this step in the reverse direction is $^{x}k/^{x}K_{eq}$).

The constants c_f and c_r in Eqs. (84) and (85) are "commitments" in the forward and reverse directions and represent partition ratios for the complexes that undergo the isotope-sensitive step in the forward and reverse directions, respectively. Thus, c_f is the ratio of the rate constant for the isotope-sensitive step to the net rate constant for release of a substrate back off of the enzyme. The substrate involved in this calculation is the variable one in a direct comparison study, the perturbant in an equilibrium perturbation study, or the labeled substrate in an internal competition experiment. These need not be the same substrate, and thus

c_f may differ when different methods are used to determine the same isotope effect. For example, when a dehydrogenase with an ordered mechanism (NAD^+ adding first) is studied with a deuterium or tritium label in the 4 position of the nicotinamide ring of the nucleotide, c_f is calculated for NAD^+ release in an internal competition (i.e., with tritium) or equilibrium perturbation (with deuterium) experiment, but for whichever substrate is the variable one in a direct comparison between unlabeled and deuterated substrates. Thus, the commitment would be calculated for the second substrate if it were the variable one, even though the label was in NAD^+.

The c_r value is calculated for the first product released in a direct comparison or internal competition study, but for the perturbant in an equilibrium perturbation experiment. For example, with isocitrate dehydrogenase when deuterated isocitrate was used, $^D(V/K)$ was near unity at neutral pH because the forward commitment for isocitrate is large, and certainly much larger than the reverse commitment for CO_2, the first product released. In an equilibrium perturbation experiment, however, the perturbant on the right-hand side of the reaction is NADPH, which has an even higher reverse commitment than the forward one for isocitrate, so that the observed isotope effect of 1.15 was nearly equal to $^DK_{eq}$ (*113*).

The constant c_{Vf} in Eq. (85) is not a commitment but represents a comparison between the rate constant for the isotope-sensitive step and rate constants for all other forward unimolecular steps. Specifically, it is the sum of the ratios of the catalytic rate constant (multiplied by a precatalytic factor) and each other net rate constant in the forward direction, treating the step prior to the isotope-sensitive one as being irreversible.

It is readily seen from Eq. (84) that if c_f is large the isotope effect becomes unity. No discrimination occurs because every collision produces reaction, and the substrate does not redissociate prior to reaction. Conversely, when c_r is very large the catalytic step comes to equilibrium, and we see the equilibrium isotope effect. From Eq. (85) we see that a large c_r value again brings the catalytic reaction to equilibrium, but a large c_{Vf} produces no isotope effect. Steps after release of the first product can still affect c_{Vf}, and in fact ^DV values are unity for a number of dehydrogenases because release of the second substrate is rate limiting.

In the following mechanism

$$\mathrm{E} \underset{k_2}{\overset{k_1\mathrm{A}}{\rightleftharpoons}} \mathrm{EA} \underset{k_4}{\overset{k_3}{\rightleftharpoons}} \mathrm{EA^*} \underset{k_6}{\overset{k_5}{\rightleftharpoons}} \mathrm{EPQ^*} \underset{k_8}{\overset{k_7}{\rightleftharpoons}} \mathrm{EPQ} \xrightarrow{k_9} \mathrm{EQ} \xrightarrow{k_{11}} \mathrm{E} \tag{86}$$

where k_5 and k_6 are for the isotope-sensitive step,

$$c_f = (k_5/k_4)(1 + k_3/k_2) \tag{87}$$

$$c_r = (k_6/k_7)(1 + k_8/k_9) \tag{88}$$

$$c_{Vf} = k_5[k_3/(k_3 + k_4)][1/k_3 + (1/k_7)(1 + k_8/k_9) + 1/k_9 + 1/k_{11}] \quad (89)$$

c_f and c_r have internal (in) and external (ex) parts, with the latter involving reactant release, and the former not. Thus, in Mechanism (86),

$$c_{f\text{-in}} = k_5/k_4 \quad (90)$$

$$c_{f\text{-ex}} = k_3k_5/(k_2k_4) \quad (91)$$

$$c_{r\text{-in}} = k_6/k_7 \quad (92)$$

$$c_{r\text{-ex}} = k_6k_8/(k_7k_9) \quad (93)$$

The external commitment can often be eliminated by altering pH, but usually the internal commitments are pH independent [alcohol dehydrogenases are an exception (*114*)].

Equation (84) can be converted to the comparable equation for the back reaction by dividing both sides by ${}^xK_{eq}$. c_f and c_r now become reverse and forward commitments in the back reaction, and ${}^xk/{}^xK_{eq}$ is the intrinsic isotope effect in the reverse reaction. However, this does not work with Eq. (85); the comparable equation for the back reaction has a similar form, with c_{Vr} and c_f as the constants, where c_{Vr} is given by an equation similar to Eq. (89).

3. *Determination of Intrinsic Isotope Effects*

It is obvious that in order to use Eq. (84) one must know the size of ${}^xk$ as well as the commitments (${}^xK_{eq}$ can be readily measured). It was the discovery by Northrop in 1975 that the deuterium and tritium isotope effects on V/K could be compared to estimate ${}^Dk$ that opened up the field of isotope effects on enzyme-catalyzed reactions (*93*):

$$[{}^D(V/K) - 1]/[{}^T(V/K) - 1] = [{}^Dk - 1]/[({}^Dk)^{1.44} - 1] \quad (94)$$

Equation (94) assumes that in Eq. (84) either $c_r = 0$ or ${}^DK_{eq} = 1.0$, and that ${}^Tk = ({}^Dk)^{1.44}$, the so-called Swain–Schaad relationship, which seems to hold for isotope effects of greatly different sizes (*115*). The disadvantage of this method is that the solution is not well conditioned if the commitments are so large that ${}^D(V/K)$ cannot be measured accurately. It has been used successfully to determine ${}^Dk$ values for alcohol dehydrogenases, where c_r does appear to be near zero (*114*).

When ${}^DK_{eq}$ is not unity and c_r is finite, Northrop's method does not give an exact answer, and one divides ${}^D(V/K)$ by ${}^DK_{eq}$ and ${}^T(V/K)$ by ${}^TK_{eq}$ [$=({}^DK_{eq})^{1.44}$] to give parameters in the back reaction and then repeats the calculation. The true value of ${}^Dk$ lies between the apparent value calculated in the forward direction and the apparent value in the back reaction times ${}^DK_{eq}$, whereas the true value in the back reaction lies between these limits divided by ${}^DK_{eq}$. This method when applied to malic enzyme gave ^{D}k values of 5–8 in the forward direction (*82*); the value now accepted is 5.7 (*116*).

A more precise method for estimating intrinsic isotope effects involves measurement of ^{13}C isotope effects with a deuterated and unlabeled substrate, as well as the deuterium isotope effect itself. The ^{13}C isotope effects are measured with the isotope ratio mass spectrometer. The deuterium and ^{13}C isotope effects are given by Eq. (84) with x = D or 13; however, when the ^{13}C isotope effect is measured with a deuterated substrate, c_f is reduced by ^{D}k, since k_5 in Eq. (87) is reduced by this amount. c_r is reduced by $^{D}k/^{D}K_{eq}$, the intrinsic isotope effect in the back reaction. By assuming various ratios of c_r/c_f between zero and infinity, one can solve the three equations for the observed isotope effects simultaneously to give values of ^{D}k, ^{13}k, c_f, and c_r (*117*). This method is a precise one because of the accuracy with which one can measure ^{13}C isotope effects with the isotope ratio mass spectrometer, and it gives very narrow limits on ^{13}k, somewhat wider ones on ^{D}k, and greater errors on c_f and c_r, although their sum is well determined. This method works best when the commitments are small, and if $^{D}K_{eq}$ is unity it gives an exact solution for ^{13}k, and if $^{13}K_{eq}$ is unity, an exact solution for ^{D}k. If a secondary deuterium isotope effect can be measured on the same step, this value plus the ^{13}C isotope effect with a secondary deuterated substrate (commitments reduced by the intrinsic secondary deuterium isotope effects) gives a total of five equations in five unknowns which will give an exact solution for all parameters. To date this has been done only with glucose-6-P and liver alcohol dehydrogenases (*102, 118*).

The above method requires that all isotope effects be on the same step. When this is not the case, ways can still be found to determine intrinsic isotope effects in favorable cases, and this has been done with malic enzyme by using intermediate partitioning (*116*).

4. *Determination of Kinetic Mechanism*

Isotope effects are an excellent way to distinguish ordered from random mechanisms and to determine the relative rates of release of substrates in a random mechanism. In the following mechanism,

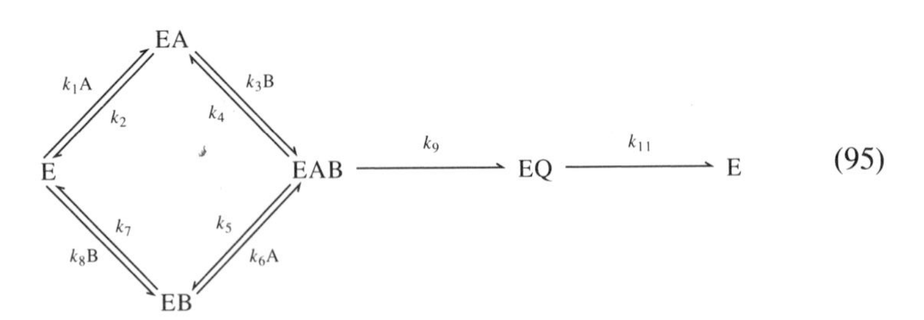

(95)

the V/K isotope effect when A is varied, or when A is the labeled substrate or perturbant in internal competition or equilibrium perturbation experiments, is

$$\text{apparent } {}^{x}(V/K)_{\text{A}} = ({}^{x}k_9 + c_{\text{f-ex}})/(1 + c_{\text{f-ex}}) \tag{96}$$

where

$$c_{\text{f-ex}} = k_9/[k_5 + k_4k_2/(k_2 + k_3B)] \tag{97}$$

We do not mean to imply that ${}^{x}k_9$ is an intrinsic isotope effect; it will be given by an equation similar to Eq. (84) but containing only the internal portion of the forward commitment ($c_{\text{f-in}}$). The level of B will affect $c_{\text{f-ex}}$ as long as k_4 is as big or bigger than k_5. If k_5 is zero (an ordered mechanism), $c_{\text{f-ex}}$ becomes infinite at very high B, and no isotope effect is seen on V/K_{a} itself. At very low B, on the other hand, $c_{\text{f-ex}} = k_9/k_4$. The level of B at which the isotope effect is half-suppressed is $K_{\text{ia}}K_{\text{b}}/K_{\text{a}}$. For a random mechanism, $c_{\text{f-ex}}$ varies from $k_9/(k_4 + k_5)$ at low B to k_9/k_5 at high B.

When B is the varied substrate, or the labeled one or perturbant, Eq. (96) holds for apparent ${}^{x}(V/K)_{\text{B}}$, and now

$$c_{\text{f-ex}} = k_9/[k_4 + k_5k_7/(k_7 + k_6\text{A})] \tag{98}$$

If $k_5 = 0$ (ordered mechanism), $c_{\text{f-ex}} = k_9/k_4$ at all times, and one sees the full ${}^{x}(V/K_{\text{b}})$ isotope effect regardless of A level. This is also the isotope effect when B is varied at low A (see above). In a random mechanism, $c_{\text{f-ex}}$ varies from $k_9/(k_4 + k_5)$ at low A to k_9/k_4 at very high A. Thus, in a rapid equilibrium random mechanism (k_4, $k_5 \gg k_9$), one sees ${}^{x}k_9$ as the isotope effect regardless of which substrate is varied or labeled. When there is some stickiness to one or both substrates, however, the above analysis will demonstrate it.

An understanding of these principles allows one to enhance isotope effects by proper choice of conditions. Thus, in equilibrium perturbation studies of liver alcohol dehydrogenase with the nucleotides as perturbants, the use of subsaturating levels of alcohol and ketone permitted the commitments to be those of the nonsticky alcohol and ketone, rather than those of the nucleotides (*104, 119*).

This approach is a very powerful one both for determining kinetic mechanism and for measuring the stickiness of substrates. References (*104*) and (*119*) give examples of its use, and Ref. (*120*) gives a description of some of the experimental problems caused by impurities in labeled or unlabeled substrates, and how to avoid them.

5. *Determining Rate-Limiting Steps*

In the previous section we showed how isotope effects can be used to determine stickiness. Another approach is to determine the change in isotope effects with pH. If incorrect protonation of the enzyme–substrate complex is possible, the apparent rate of the catalytic reaction slows down, but the rate of substrate release will not slow or will be accelerated. Thus, the external commitment will be abolished, although internal commitments will normally remain. Thus, the V/K isotope effect should rise as V/K drops and reach a value limited only by any internal forward commitment plus any reverse commitment that remains. In most cases the isotope effect on V will also increase as the catalytic step becomes more rate limiting, and the isotope effects on V and V/K should become the same as both parameters decrease (*113*).

With alanine dehydrogenase, $^{D}(V/K_{alanine})$ was 1.4 at the pH optimum and increased to 2.0 below pH 7, whereas $^{D}(V/K_{serine})$ was 2.0 at all pH values (serine is a slow, nonsticky substrate) (*83*). The internal commitments appear to be the same for both substrates and are sizeable, since the intrinsic isotope effect is likely to be 5 or 6. In cases like this, a rough estimate of the stickiness of the substrate can be obtained from the following equation:

$$S_r'' = [^{D}(V/K)_{low\ pH} - {}^{D}(V/K)_{pH\ opt}]/[^{D}(V/K)_{pH\ opt} - 1] \qquad (99)$$

If deprotonation causes a drop in V/K, the first term in the numerator is $^{D}(V/K)_{high\ pH}$. S_r'' in Eq. (99) is slightly different from S_r, the stickiness factor determined from the isotope partition method, or S_r', the value from displacement of apparent pK values in V/K profiles, but to a first approximation will have the same value. See Ref. (*3*) for a more complete discussion. A recent review gives other examples of determining rate-limiting steps by means of isotope effects (*94*).

6. *Determination of Chemical Mechanism*

Isotope effects have been used in a number of ways to help deduce chemical mechanisms. We mention only a few here; the reader should consult Refs. (*94–96*). One of the most useful techniques is the use of isotope effects to distinguish concerted from stepwise reactions by measuring the effect of a deuterium isotope effect on the size of another isotope effect (*117*). The method works especially well when it is a ^{13}C or other heavy atom isotope effect that is measured with the isotope ratio mass spectrometer.

As noted above in Section VII,B,3, when a deuterium and ^{13}C isotope effect are on the same step in the mechanism, deuteration reduces commitments and thus enhances the size of the observed ^{13}C isotope effect. For a stepwise mechanism, however, deuteration makes a step other than the ^{13}C-sensitive one more rate limiting, and thus decreases the size of the observed ^{13}C isotope effect. The ^{13}C isotope effect with deuterated and unlabeled substrates and the deuterium isotope itself are now no longer independent, but are related by the following equations.

In the direction where the deuterium-sensitive step comes first,

$$[^{13}(V/K)_H - 1]/[^{13}(V/K)_D - 1] = {}^{D}(V/K)/{}^{D}K_{eq} \qquad (100)$$

whereas in the direction where the ^{13}C-sensitive step comes first,

$$[^{13}(V/K)_H - {}^{13}K_{eq}]/[^{13}(V/K)_D - {}^{13}K_{eq}] = {}^D(V/K) \qquad (101)$$

One can thus distinguish the nature of the intermediate by seeing which equation is fitted by the data. This was first done with malic enzyme, using malate deuterated at C-2 and observing the ^{13}C isotope effect at C-4, and demonstrated that dehydrogenation occurs prior to decarboxylation (*117*). Isocitrate (*79*) and 6-phosphogluconate dehydrogenases (*121*) appear to have similar stepwise mechanisms, but prephenate dehydrogenase catalyzes a concerted oxidative decarboxylation because the energy released by aromatization is so great (*122*).

This method was also introduced by Knowles and co-workers for the case where one deuterium isotope effect is measured as a function of deuteration in another position (*123*). There is little difficulty if the mechanism is in fact stepwise, but there can be a real problem with concerted mechanisms because of the effects of coupled hydrogen motions (see Section VII,B above). This coupling gives lower isotope effects when another deuterium is present, and thus may lead one to conclude that the mechanism is stepwise.

Isotope effects helped greatly in sorting out the chemical mechanism of alcohol dehydrogenases (*114*). These have stepwise mechanisms in which proton removal from the alcohol following coordination to Zn^{2+} gives an alkoxide and a protonated base (His-51 for liver alcohol dehydrogenase). This base can lose a proton at high pH, thus committing the reaction to continue. Hydride transfer to oxidize the alkoxide to a ketone or aldehyde can occur with the base either protonated or ionized. Thus, at high pH the reaction has an infinitely high forward commitment and no isotope effect on V/K. In the reverse reaction at high pH hydride transfer comes to equilibrium, with protonation of the alkoxide being the slow step. The observed deuterium isotope effect is thus $^DK_{eq}$, which is 0.85 in this direction for a ketone. It is unusual for an intermediate on an enzyme to be able to be protonated or ionized, and this occurs in the present case because His-51 is exposed to solvent. For most enzymes the protonation states of reactants and catalytic groups are locked once the enzyme undergoes the conformation change that leads to catalysis.

Phenylalanine ammonia-lyase was shown to have a stepwise mechanism with a carbanion intermediate by measurement of ^{15}N isotope effects with deuterated and unlabeled dihydrophenylalanine, a slow alternate substrate (*99*). This was a particularly intriguing case because the study was undertaken because it was thought that the mechanism would *not* be a carbanion one.

Adenosine deaminase was studied by determining ^{15}N isotope effects in D_2O and H_2O using both adenosine and 8-oxoadenosine, a slow substrate (*75*). The reaction goes *faster* in D_2O, and yet the ^{15}N isotope effect was larger. It was concluded that a sulfhydryl group donated a proton to oxygen or nitrogen during the formation of the tetrahedral intermediate from which ammonia leaves, so that ${}^{D}K_{eq}$ for this step is strongly inverse (~0.4). The partition ratio of this intermediate was 1.4 with adenosine but 0.1 with 8-oxoadenosine, so that the formation of the intermediate was nearly at equilibrium with the slow substrate and C–N bond cleavage was the major rate limiting step. It was postulated that the sulfhydryl group donated a proton to N-1 during formation of the intermediate.

A number of enzymes have been postulated to have sulfhydryl groups as acid–base catalysts on the basis of the low fractionation factors of hydrogens on these groups (~0.5). These include proline racemase (*124*), enolase (*125*), adenosine deaminase (*75*), pyruvate carboxylase (*126*), and biotin carboxylase (*62*). Whether all of these assignments will turn out to be correct is not yet known, but the only structures with a fractionation factor of approximately 0.5 other than a sulfhydryl group are hydroxide (0.48, not a likely reactant in enzymic reactions) or a proton in a symmetrical hydrogen bond between two similar groups, one of which is ionized [e.g., R–O···H···O–R, where there is a single negative charge in this grouping (*127*)]. Whether the latter structure ever is involved in enzymic reactions as a catalytic group is not known.

7. *Determination of Transition State Structure*

Once intrinsic isotope effects are determined, one is in a position to deduce transition state structure, just as the physical organic chemist does for nonenzymic reactions. Unfortunately, in many cases workers have assumed, rather than proved, that commitments are zero and intrinsic isotope effects were being looked at. Transition state structures have been investigated in the formate and liver alcohol dehydrogenase reactions as the redox potential of the nucleotide substrate was changed (*103, 118*). Primary deuterium and ^{13}C, secondary deuterium, and for formate dehydrogenase ^{18}O isotope effects were determined. In both cases the transition states appeared to be late with NAD^+ and to become earlier as the redox potential of the nucleotide became more positive. So far the conclusions from such studies have been qualitative in nature, and there is room for much more work on these systems.

An effort is underway in our laboratory to determine transition state structures for phosphoryl transfer reactions by the use of secondary ^{18}O isotope effects measured by the remote label method. The hypothesis is that an associative mechanism would have single bonds from phosphorus to the nonbridge oxygens in the transition state, whereas a dissociative mechanism would have enhanced bond order. So far we have determined the secondary ^{18}O isotope effects on glucose-6-P hydrolysis at pH 4.5, 100°C (*112*), and by alkaline phosphatase (*128*). The chemical hydrolysis showed no isotope effect on P–O bond cleavage, which is consistent with a largely dissociative transition state without total conservation of bond order to phosphorus (i.e., a partial positive change exists on phosphorus).

With alkaline phosphatase the isotope effect was slightly inverse, and slightly more so at pH 6, where catalysis is not rate limiting (below the pK in the V/K profile), than at the pH optimum (*128*). These data are consistent with a dissociative mechanism for the enzymic reaction. Attempts are underway to determine similar isotope effects for kinases and for phosphodiester hydrolyses by ribonuclease and bisimidazoylcycloheptaamylose, a ribonuclease mimic (*129*). Preliminary results suggest that phosphate monoesters hydrolyze by a dissociative mechanism, while phosphodiesters show S_N2 reactions, and triesters have associative, but concerted mechanisms. The basic rule is thus that the charge on the non-bridge oxygens in the transition state is never more than minus one (*130*).

Actually, secondary isotope effects are more likely to give useful information on transition state structures than primary ones, except where coupled hydrogen motions distort the secondary isotope effects. Secondary isotope effects were quite useful in determining the carbanion mechanism for fumarase (*109*). The interested reader is referred to a recent review which gives a number of other examples (*96*).

VIII. Concluding Remarks

The reader who has read both Chapter 1 of Volume II and this chapter will have a broad introduction to the subject of steady-state kinetics. Although the tools of the kineticist have become more powerful and sophisticated, the number of people well versed in their use has not grown in parallel fashion. This is probably because textbooks still do not cover the subject adequately; moreover, the only useful information in print other than the primary literature is in chapters such as this one or other reviews. We hope this chapter will make it easier for enzymologists to learn these methods, which they now need more than ever to characterize the mechanisms of all of the altered enzymes that site-directed mutagenesis can provide.

References

1. Cleland, W. W. (1970). "The Enzymes," 3rd Ed., Vol. 2, p. 1.
2. Cleland, W. W. (1977). *Adv. Enzymol.* **45,** 273.
3. Cleland, W. W. (1986). *In* "Investigations of Rates and Mechanisms of Reaction" (C. Bernasconi, ed.), Vol. 6, p. 791. Wiley, New York.
4. Cleland, W. W. (1963). *BBA* **67,** 104.
5. Cleland, W. W. (1979). "Methods in Enzymology," Vol. 63, p. 103.
6. Cleland, W. W. (1967). *Adv. Enzymol.* **29,** 1.
7. King, E. L., and Altman, C. (1956). *J. Phys. Chem.* **60,** 1375.

8. Mahler, H. R., and Cordes, E. H. (1966). "Biological Chemistry." Harper, New York.
9. Cleland, W. W. (1975). *Biochemistry* **14,** 3220.
10. Cha, S. (1968). *JBC* **243,** 820.
11. Cleland, W. W. (1973). *JBC* **248,** 8353.
12. Janson, C. A., and Cleland, W. W. (1974). *JBC* **249,** 2562.
13. Viola, R. E., and Cleland, W. W. (1982). "Methods in Enzymology," Vol. 87, p. 353.
14. Rife, J. E., and Cleland, W. W. (1980). *Biochemistry* **19,** 2321.
15. Qureshi, N., Dugan, R. E., Cleland, W. W., and Porter, J. S. (1976). *Biochemistry* **15,** 4191.
16. Plowman, K. M., and Cleland, W. W. (1967). *JBC* **242,** 4239.
17. Tipton, P. A., and Cleland, W. W. (1988). *Biochemistry* **27,** 4317.
18. Cleland, W. W. (1982). "Methods in Enzymology," Vol. 87, p. 366.
19. Martensen, T. M., and Mansour, T. E. (1976). *JBC* **251,** 3664.
20. Raushel, F. M., and Cleland, W. W. (1973). *JBC* **248,** 8174.
21. Bar-Tana, J., and Cleland, W. W. (1974). *JBC* **249,** 1263.
22. Pecoraro, V. L., Hermes, J. D., and Cleland, W. W. (1984). *Biochemistry* **23,** 5262.
23. Cleland, W. W., and Mildvan, A. S. (1979). *Adv. Inorg. Biochem.* **1,** 163.
24. Cleland, W. W. (1982). "Methods in Enzymology," Vol. 87, p. 159.
25. Dunaway-Mariano, D., and Cleland, W. W. (1980). *Biochemistry* **19,** 1506.
26. Merritt, E. A., Sundaralingam, M., Cornelius, R. D., and Cleland, W. W. (1978). *Biochemistry* **17,** 3274.
27. Cornelius, R. D., and Cleland, W. W. (1978). *Biochemistry* **17,** 3279.
28. Speckhard, D. C., Pecoraro, V. L., Knight, W. B., and Cleland, W. W. (1986). *J. Am. Chem. Soc.* **108,** 4167; Correction: Speckhard, D. C., Pecoraro, V. L., Knight, W. B., and Cleland, W. W. (1988). *J. Am. Chem. Soc.* **110,** 2349.
29. Shorter, A. L., Haromy, T. P., Scalzo-Brush, T., Knight, W. B., Dunaway-Mariano, D., and Sundaralingam, M. (1987). *Biochemistry* **26,** 2060.
30. Lu, Z., Shorter, A. L., Lin, I., and Dunaway-Mariano, D. (1988). *Inorg. Chem.* **27,** 4135.
31. Eckstein, F., Romaniuk, P. J., and Connolly, B. A. (1982). "Methods in Enzymology," Vol. 87, p. 197.
32. Frey, P. A., Richard, J. P., Ho, H.-T., Brody, R. S., Sammons, R. D., and Sheu, K.-F. (1982). "Methods in Enzymology," Vol. 87, p. 213.
33. Northrop, D. B. (1969). *JBC* **244,** 5808.
34. Rendina, A. R., and Orme-Johnson, W. H. (1978). *Biochemistry* **17,** 5388.
35. McClure, W. R., Lardy, H. A., Wagner, M., and Cleland, W. W. (1971). *JBC* **246,** 3579.
36. Tsai, C. S., Burgett, M. W., and Reed, L. J. (1973). *JBC* **246,** 3579.
37. Katiyar, S. S., Cleland, W. W., and Porter, J. W. (1975). *JBC* **250,** 2709.
38. Rudolph, F. B., and Fromm, H. J. (1970). *Biochemistry* **9,** 4660.
39. Knight, W. B., and Cleland, W. W. (1989). *Biochemistry* **28,** 5728.
40. Cook, P. F., Kenyon, G. L., and Cleland, W. W. (1981). *Biochemistry* **20,** 1204.
41. Northrop, D. B., and Cleland, W. W. (1974). *JBC* **249,** 2928.
42. Grimshaw, C. E., and Cleland, W. W. (1981). *Biochemistry* **20,** 5650.
43. Danenberg, P. V., and Danenberg, K. D. (1978). *Biochemistry* **17,** 4018.
44. Fujioka, M. (1975). *JBC* **250,** 8986.
45. Danenberg, K. D., and Cleland, W. W. (1975). *Biochemistry* **14,** 28.
46. Fisher, L. M., Albery, W. J., and Knowles, J. R. (1986). *Biochemistry* **25,** 2529.
47. Silverman, D. N., and Vincent, S. H. (1983). *Crit. Rev. Biochem.* **14,** 207.
48. Anthony, R. S., and Spector, L. B. (1971). *JBC* **246,** 6129.
49. Cedar, H., and Schwarz, J. H. (1969). *JBC* **244,** 4122.
50. Hass, L. F., and Byrne, W. L. (1960). *J. Am. Chem. Soc.* **82,** 947.

51. Nandi, D. L. (1978). *JBC* **253,** 8872.
52. Morrison, J. F., and Cleland, W. W. (1966). *JBC* **241,** 673.
53. Fromm, H. J., Silverstein, E., and Boyer, P. D. (1964). *JBC* **239,** 3645.
54. Gulbinsky, J. S., and Cleland, W. W. (1968). *Biochemistry* **7,** 566.
55. Raushel, F. M., and Cleland, W. W. (1977). *Biochemistry* **16,** 2176.
56. Ainslie, G. R., Jr., and Cleland, W. W. (1972). *JBC* **247,** 946.
57. Uhr, M. L., Thompson, V. W., and Cleland, W. W. (1974). *JBC* **249,** 2920.
58. Hsuanyu, Y., and Wedler, F. C. (1987). *ABB* **259,** 316.
59. Hsuanyu, Y., and Wedler, F. C. (1988). *JBC* **263,** 4172.
60. Midelfort, C. G., and Rose, I. A. (1976). *JBC* **251,** 5881.
61. Raushel, F. M., and Villafranca, J. J. (1988). *Crit. Rev. Biochem.* **23,** 1.
62. Tipton, P. A., and Cleland, W. W. (1988). *Biochemistry* **27,** 4325.
63. Raushel, F. M., and Garrard, L. J. (1984). *Biochemistry* **23,** 1791.
64. Hester, L. S., and Raushel, F. M. (1987). *Biochemistry* **26,** 6465.
65. Hester, L. S., and Raushel, F. M. (1987). *JBC* **262,** 12092.
66. Britton, H. G., and Clarke, J. B. (1968). *BJ* **110,** 161.
67. Fisher, L. M., Albery, W. J., and Knowles, J. R. (1986). *Biochemistry* **25,** 2538.
68. Belasco, J. G., Bruice, T. W., Fisher, L. M., Albery, W. J., and Knowles, J. R. (1986). *Biochemistry* **25,** 2564.
69. Rose, I. A., O'Connell, E. L., Litwin, S., and Bar-Tana, J. (1974). *JBC* **249,** 5163.
70. Viola, R. E., Raushel, F. M., Rendina, A. R., and Cleland, W. W. (1982). *Biochemistry* **21,** 1295.
71. Brouwer, A. C., and Kirsch, J. F. (1982). *Biochemistry* **21,** 1302.
72. Hardy, L. W., and Kirsch, J. F. (1984). *Biochemistry* **23,** 1275.
73. Kurz, L. C., Weitkamp, E., and Frieden, C. (1987). *Biochemistry* **26,** 3027.
74. Kurz, L. C., and Frieden, C. (1983). *Biochemistry* **22,** 382.
75. Weiss, P. M., Cook, P. F., Hermes, J. D., and Cleland, W. W. (1987). *Biochemistry* **26,** 7378.
76. Grissom, C. B., and Cleland, W. W. (1988). *Biochemistry* **27,** 2927.
77. Cleland, W. W. (1982). "Methods in Enzymology," Vol. 87, p. 390.
78. Brandt, D. A., Barnett, L. B., and Alberty, R. A. (1963). *J. Am. Chem. Soc.* **85,** 2204.
79. Grissom, C. B., and Cleland, W. W. (1988). *Biochemistry* **27,** 2934.
80. Rife, J. E., and Cleland, W. W. (1980). *Biochemistry* **19,** 2328.
81. Schimerlik, M. I., and Cleland, W. W. (1977). *Biochemistry* **16,** 576.
82. Schimerlik, M. I., Grimshaw, C. E., and Cleland, W. W. (1977). *Biochemistry* **16,** 571.
83. Grimshaw, C. E., Cook, P. F., and Cleland, W. W. (1981). *Biochemistry* **20,** 5655.
84. Dougherty, T. M., and Cleland, W. W. (1985). *Biochemistry* **24,** 5870.
85. Kiick, D. M., and Cleland, W. W. (1989). *ABB* **270,** 647.
86. Muirhead, H., Clayden, D. A., Barford, D., Lorimer, C. G., Fothergill-Gilmore, L. A., Schiltz, E., and Schmitt, W. (1986). *EMBO J.* **5,** 475.
87. Viola, R. E., and Cleland, W. W. (1978). *Biochemistry* **17,** 4111.
88. Raushel, F. M., and Cleland, W. W. (1977). *Biochemistry* **16,** 2176.
89. Findley, D., Mathias, A. P., and Rabin, B. R. (1962). *BJ* **85,** 139.
90. Grace, S., and Dunaway-Mariano, D. (1983). *Biochemistry* **22,** 4238.
91. Melander, L., and Saunders, W. H., Jr. (1980). "Reaction Rates of Isotopic Molecules." Wiley, New York.
92. Collins, C. J., and Bowman, N. S. (eds.) (1970). "Isotope Effects in Chemical Reactions." Van Nostrand-Reinhold, New York.
93. Northrop, D. B. (1975). *Biochemistry* **14,** 2644.
94. Cleland, W. W. (1982). *Crit. Rev. Biochem.* **13,** 385.

95. Cleland, W. W. (1987). *Bioorg. Chem.* **15,** 283.
96. Cleland, W. W. (1987). *In* "Isotopes in Organic Chemistry" (E. Buncel and C. C. Lee, eds.), Vol. 7, p. 61. Elsevier, Amsterdam.
97. Northrop, D. B. (1977). *In* "Isotope Effects of Enzyme-Catalyzed Reactions" (W. W. Cleland, M. H. O'Leary, and D. B. Northrop, eds.), p. 122. Univ. Park Press, Baltimore, Maryland.
98. Cleland, W. W. (1980). "Methods in Enzymology," Vol. 64, p. 104.
99. Hermes, J. D., Weiss, P. M., and Cleland, W. W. (1985). *Biochemistry* **24,** 2959.
100. Huskey, W. P., and Schowen, R. L. (1983). *J. Am. Chem. Soc.* **105,** 5704.
101. Saunders, W. H., Jr. (1984). *J. Am. Chem. Soc.* **106,** 2223.
102. Hermes, J. D., and Cleland, W. W. (1984). *J. Am. Chem. Soc.* **106,** 7263.
103. Hermes, J. D., Morrical, S. W., O'Leary, M. H., and Cleland, W. W. (1984). *Biochemistry* **23,** 5479.
104. Cook, P. F., Oppenheimer, N. J., and Cleland, W. W. (1981). *Biochemistry* **20,** 1817.
105. Welsh, K. M., Creighton, D. J., and Klinman, J. P. (1980). *Biochemistry* **19,** 2005.
106. Limbach, H.-H., Hennig, J., Gerritzen, D., and Rumpel, H. (1982). *Faraday Discuss. Chem. Soc.* **74,** 229.
107. Ostovic, D., Roberts, R. M. G., and Kreevoy, M. M. (1983). *J. Am. Chem. Soc.* **105,** 7629.
108. Viola, R. E. (1984). *ABB* **228,** 415.
109. Blanchard, J. S., and Cleland, W. W. (1980). *Biochemistry* **19,** 4506.
110. Schimerlik, M. I., Rife, J. E., and Cleland, W. W. (1975). *Biochemistry* **14,** 5347.
111. O'Leary, M. H., and Marlier, J. F. (1979). *J. Am. Chem. Soc.* **101,** 3300.
112. Weiss, P. M., Knight, W. B., and Cleland, W. W. (1986). *J. Am. Chem. Soc.* **108,** 2761.
113. Cook, P. F., and Cleland, W. W. (1981). *Biochemistry* **20,** 1797.
114. Cook, P. F., and Cleland, W. W. (1981). *Biochemistry* **20,** 1805.
115. Swain, C. G., Stivers, E. C., Reuwer, J. F., Jr., and Schaad, L. (1957). *J. Am. Chem. Soc.* **80,** 5885.
116. Grissom, C. B., and Cleland, W. W. (1985). *Biochemistry* **24,** 944.
117. Hermes, J. D., Roeske, C. A., O'Leary, M. H., and Cleland, W. W. (1982). *Biochemistry* **21,** 5106.
118. Scharschmidt, M., Fisher, M. A., and Cleland, W. W. (1984). *Biochemistry* **23,** 5471.
119. Cook, P. F., and Cleland, W. W. (1981). *Biochemistry* **20,** 1790.
120. Grimshaw, C. E., and Cleland, W. W. (1980). *Biochemistry* **19,** 3153.
121. Rendina, A. R., Hermes, J. D., and Cleland, W. W. (1984). *Biochemistry* **23,** 6257.
122. Hermes, J. D., Tipton, P. A., Fisher, M. A., O'Leary, M. H., Morrison, J. F., and Cleland, W. W. (1984). *Biochemistry* **23,** 6263.
123. Belasco, J. G., Albery, W. J., and Knowles, J. R. (1983). *J. Am. Chem. Soc.* **105,** 2475.
124. Belasco, J. G., Bruice, R. W., Albery, W. J., and Knowles, J. R. (1986). *Biochemistry* **25,** 2558.
125. Weiss, P. M., Boerner, R. J., and Cleland, W. W. (1987). *J. Am. Chem. Soc.* **109,** 7201.
126. Attwood, P. V., Tipton, P. A., and Cleland, W. W. (1986). *Biochemistry* **25,** 8197.
127. Kreevoy, M. M., Liang, T.-M., and Chang, K.-C. (1977). *J. Am. Chem. Soc.* **99,** 5207.
128. Weiss, P. M., and Cleland, W. W. (1989). *J. Am. Chem. Soc.* **111,** 1928.
129. Breslow, R., Doherty, J. B., Guillot, G., and Lipsey, C. (1978). *J. Am. Chem. Soc.* **100,** 3227.
130. Cleland, W. W. (1990). *FASEB J.* **4,** 2899.

4

Analysis of Protein Function by Mutagenesis

KENNETH A. JOHNSON* • STEPHEN J. BENKOVIC[†]

**Department of Molecular and Cell Biology*
[†]*Department of Chemistry*
The Pennsylvania State University
University Park, Pennsylvania 16802

Introduction

The mysterious role of primary amino acid sequence, particularly specific residues, in dictating the structural characteristics, physical properties, and dynamic behavior of biological molecules has long held us in fascination. Recently, the combination of site-directed mutagenesis, high-resolution structural infor-

THE ENZYMES, Vol. XIX

mation, and in-depth kinetic analysis has allowed direct, quantifiable assessment of the contributions of individual amino acids toward enzymic specificity and catalysis. During the preceding decade, chemical modification was the primary method available to define the amino acids that are directly involved in catalysis. Site-specific mutagenesis has several obvious advantages over chemical modification, including absolute specificity, the ability to make very minor structural changes, and the complete conversion of one amino acid to another; however, it suffers at present from the limited repertoire of possible substitutions. Many errors and misassignments from chemical modification studies have been corrected by site-directed mutagenesis (*1–3*), and complementary use of the two methods is now beginning (*4*).

The goal of this chapter is to provide a review of the design and analysis of mutated proteins, particularly with regard to their importance in understanding fundamentals of enzymic catalysis. We have not tried to be exhaustive in our coverage of the pertinent literature, but rather have selected examples that fit our purpose, revealing our own prejudices in their choice. Other reviews offer additional information and differing viewpoints (*5–7*). This chapter is divided into three sections as outlined below. In the first two, we consider the contributions of individual amino acids toward enzyme structure and function as revealed by analysis of mutant proteins containing a small number of amino acid substitutions. In the final section, we extend these studies to consider the substitution of larger domains of protein structure to produce enzymes with altered specificity.

1. *Contributions of amino acids to protein structure and stability.* In Section I we consider amino acids that do not interact directly with the substrate but confer structural stability and examine the malleability of enzymes to allow single-site substitutions. There are variable degrees to which substitutions are allowed at various positions, reflecting the packing of side chains or their participation in the formation of ion pairs or hydrogen bonds which stabilize protein structure. In general, however, careful substitutions of individual amino acids can be made without propagation of long-range structural changes.

2. *Amino acids contributing to substrate binding and catalysis.* As discussed in Section II, mutagenesis and crystallography work together to identify those amino acids which participate in the formation of hydrophobic, ionic, or hydrogen bonds with the substrate and thereby effect a chemical reaction by transition state stabilization. A subset of these residues are amino acids which participate directly in acid–base or nucleophilic catalysis. Although mutagenesis cannot allow one to distinguish whether a given amino acid participates as a nucleophile, acts as an acid–base catalyst, or merely contributes binding energy to stabilize the transition state, quantitative analysis can provide an estimate of the thermodynamic contribution of each amino acid. Moreover, proper kinetic analysis can distinguish whether a given amino acid contributes primarily toward ground state binding or transition state stabilization. Specificity in substrate recognition can

be manifest in binding the ground state, the transition state, or both. Proper alignment of the reactants at the active site is achieved only by the precise constellation of key residues forming the active site surface. Misalignment has a more profound effect on catalysis than on substrate binding because of the more stringent requirements for transition state stabilization as compared to ground state binding. We also consider the involvement of specific residues in the regulation of enzyme function; site-directed mutagenesis has allowed definition of amino acids which transmit structural information from one site to another, ultimately serving to couple a change in conformation to alter substrate binding and catalytic activity at a distant site.

3. *Alteration of enzyme structure required to engineer new specificities*. In Section III, we draw heavily on an analysis of structural variations in protein families that has occurred during evolution and its implications for protein structure and function. This analysis provides a perspective for our conjectures about the future directions of mutagenesis experiments, guiding the way toward the engineering of new specificities. These studies of protein families serve to define the degree of structural change that is required to achieve new specificities as well as the variation in structure that is allowed among proteins with identical specificities. A few examples of attempts to engineer altered enzyme activities are described.

I. Structural Aspects of Mutant Enzymes

A. Structure Determination

In order to interpret the results of site-directed mutagenesis, it is of foremost importance to establish the structural consequences of single amino acid substitutions. X-Ray structure analyses have addressed the extent of secondary and tertiary structural change in the mutant enzymes. The crystal structures of several mutant enzymes have been determined, including subtilisin (*8*), T4 lysozyme (*9, 10*), dihydrofolate reductase (*11*), and tyrosyl-tRNA synthetase (tyrosine–tRNA ligase) (*12*). In each case, the change brought about by the mutation was accommodated locally such that changes in structure were not propagated throughout the molecule. For example, high-resolution X-ray structures of single-site mutants of *Escherichia coli* dihydrofolate reductase, involving substitutions of Ser or Asn for Asp-27, reveal a structural perturbation only at the site of replacement (*13*). Analysis of the structures and stability of 13 mutants of T4 lysozyme at residue Thr-157 have shown the enzyme to be remarkably tolerant toward amino acid substitutions (*10*). In high-resolution structures of T4 lysozyme, movements of up to 1 Å in the side-chain atoms at or close to the site of the substitution have been observed, with adjustments in the backbone of less than 0.5 Å (*14*). In several cases, it has been shown that water is bound to or displaced from the site to accommodate the change in volume of the substituted amino acid (*15, 16*). Such amino acid substitutions which have only a localized perturbation of structure have been referred to as nondisruptive.

The major limitation of crystallography is that the sensitivity of the method is insufficient for detection of conformational changes less than 0.5 Å (*13*). The use of two-dimensional (2D) NMR methods, particularly 2D NOESY spectra, extends the sensitivity of the analysis somewhat but not without the attendant difficulties of spectral assignment and interpretation (*17*). This approach, guided by knowledge of the X-ray structure (*18*), has been used to demonstrate that the introduction of Asp or Ser for the active site Glu-43 in staphylococcal nuclease produces a conformational change that extends into regions of the protein molecule up to 15–30 Å from the position of the substitution (*19, 20*). On steric and polar grounds, the Asp replacement is structurally conservative, suggesting that, at least in this protein, subtle conformational changes can be propagated over considerable distance. However, the magnitude of this perturbation is unknown. Although the uncertainties in solution of the structure from 2D NMR data limit the resolution to 0.5 Å, it is possible that observed changes in NMR spectra of a mutant can arise from considerably smaller movements. Such small shifts in structure can have significant effects on catalysis, as discussed in a later section.

The net conclusion of this analysis is that protein structures are rather malleable and that one can perform single, conservative amino acid substitutions without grossly affecting protein stability or folding in most instances. Nonetheless, quantitative interpretations of results need to be subject to a cautionary note reflecting the very real possibility that subtle changes in structure can alter the alignment of residues, leading to an overestimation of the thermodynamic contribution of any given amino acid.

B. Thermodynamic Stability

Thermodynamic stability is a global property of the enzyme structure, and contributions of individual amino acids toward the free energy of folding are additive and highly cooperative. Analysis of mutant proteins has defined the contributions of various amino acids toward the overall stability of the protein. Replacements which alter the formation of ion pairs, hydrogen bonds, van der Waals contacts, or hydrophobic interactions each tend to destabilize the folded protein by a "qualitatively comparable" amount (*14*). In one approach to this problem, site-specific substitution of amino acids has provided new approaches toward dissecting the kinetic mechanisms of protein folding (*21*).

New insights into protein folding have been obtained by analysis of the locations of temperature-sensitive mutations in T4 lysozyme (*10*). These particularly destabilizing mutations appear to have only one feature in common; they occur at sites of low mobility and low solvent accessibility in the folded state, as illustrated in Fig. 1. Any substitution of an amino acid at these locations perturbs the

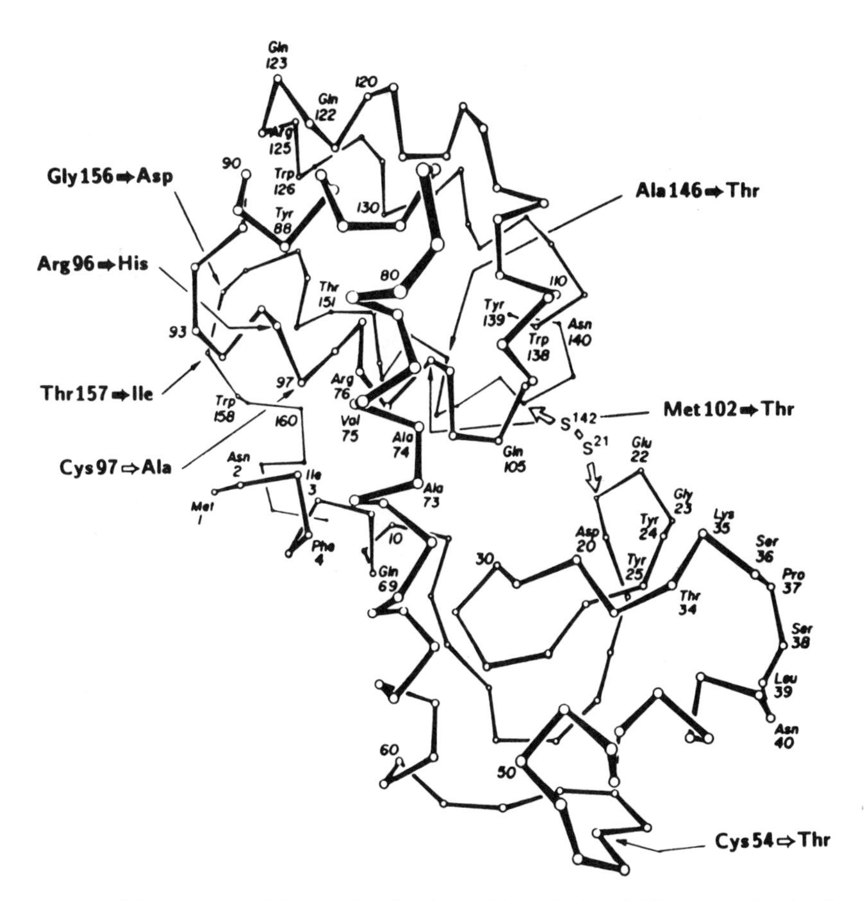

FIG. 1. Schematic view of the α-carbon backbone of bacteriophage T4 lysozyme, showing the locations of temperature-sensitive mutations, indicated by the heavy arrows (Gly-156 → Asp, Arg-96 → His, Thr-157 → Ile, Ala-146 → Thr, and Met-102 → Thr). In addition, the mutations that permit the formation of a disulfide bridge across the mouth of the active-site cleft are shown (Thr-21 → Cys—Thr-142 → Cys, Cys-97 → Ala, and Cys-54 → Thr). [Reprinted with permission from Refs. (*9*) and (*154*).

packing of amino acids in the folded state with significant global affects on protein structure which transcend the anticipated thermodynamic contribution of a single amino acid. None of these destabilizing mutations are located in the mobile, surface-exposed residues. These observations argue that the behavior of mutants is dominated by the energetics of the folded state, and the effects of amino acid substitutions on the unfolded state can be largely ignored. An alternate view has been proposed based on studies of solvent-induced denaturation (*22*); however, it is perhaps not surprising that solvent-induced denaturation should be dominated by alterations in the ability of the solvent to interact with the unfolded state.

Matthews has suggested that genetic screens for temperature-sensitive mutants may select for substitutions that are atypical with respect to the magnitude of their destabilizing effects (*14, 15*). For example, random mutagenesis led to selection of a Thr → Ile mutant that disrupted a hydrogen-bonding network involving the γ-hydroxyl of the threonine side chain (Fig. 2). Site-directed mutagenesis showed that of the 13 substitutions examined at this position, Ile provided the greatest destabilization (*15*). An interesting observation was also made in the case of the glycine substitution where the lack of a side chain allowed a water molecule to restore the hydrogen-bonding network by occupying the position of the Thr γ-hydroxyl group and thereby giving a stability closest to wild type.

C. Increasing Protein Stability

It is of obvious commercial importance to engineer proteins with increased stability, and several approaches have been taken, with mixed success. Because of the evidence that disulfide bonds are important in the conformational stability of proteins (*23*), there have been a number of examples of introducing such linkages into various proteins, including dihydrofolate reductase (*24*), T4 lysozyme (*25*), and subtilisin (*26*). Sites for substitution were selected using computer modeling so that the sulfurs from the two cysteines come within the appropriate bond distance (2.0 Å) and the completed disulfide bond is able to adopt a favorable geometry without acquiring unfavorable contacts with neighboring atoms or requiring movement of the main chain (*27, 28*).

The stability gained by the added disulfide bond has been lower than might be anticipated and somewhat variable. Disulfide linkages in wild-type proteins have special geometries which may be difficult to achieve in an engineered protein; the modest reduction in entropy in forming the disulfide may be offset by unfavorable distortion in the protein structure [see Ref. (*14*) and references cited

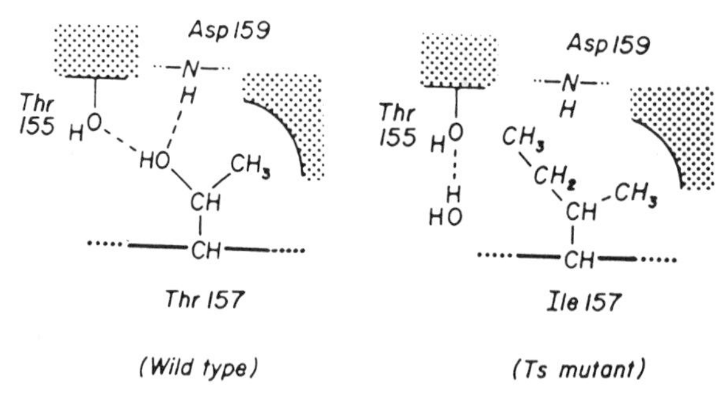

FIG. 2. Schematic illustration comparing the local environment of Thr-157 in wild-type lysozyme and Ile-157 in the temperature-sensitive mutant structure. Hydrogen bonds are indicated by dashed lines. [Reprinted with permission from Ref. (*10*).]

therein]. In one case the increased stability was only relative to the reduced cysteine precursor owing to disruption of secondary structural elements (*26*). In another study, the addition of a disulfide bond to dihydrofolate reductase led to increase stability toward denaturation by guanidine hydrochloride but decreased stability toward thermal denaturation (*11*), suggesting that, although it may slightly destabilize the folded state, this disulfide may have a greater effect on destabilizing the interaction with guanidine hydrochloride in the unfolded state.

It has also been possible to modulate enzyme function (*29*) by introduction of a disulfide bridge spanning the active site cleft of T4 lysozyme (Fig. 1). In order to avoid possible thiol–disulfide interchange with Cys-54 and Cys-97 in the native structure, these two residues were converted to threonine and alanine, respectively, with no loss in the activity or stability of the enzyme. The latter protein was then further modified by replacing Thr-21 and Thr-42 by cysteines that spontaneously oxidized to the desired disulfide. This oxidized enzyme form exhibited no detectable activity, although some activity (7% of wild type) was restored on reduction of the linkage. This represents a novel use of the disulfide bond to modulate catalytic activity.

Another approach toward stabilization is to reduce the entropy of the unfolded state by substitution to achieve amino acids with more restricted configurational freedom. Glycine contributes the greatest configurational freedom toward the unfolded state and therefore produces more entropy loss on folding, whereas proline is the most restricted and contributes the least toward the entropy of the unfolded state. Thus, careful substitution with reference to steric constraints in the native structure can decrease the entropy of unfolding and thereby increase the thermostability. In particular, substitutions of Gly $\rightarrow$ Ala and Ala $\rightarrow$ Pro are expected to enhance the stability by approximately 1 kcal/mol. These substitutions have increased the thermostability of several proteins including T4 lysozyme (*21*), the λ repressor (*30*), and neutral protease from *Bacillus stearothermophilus* (*31*).

Reduction in the area of exposed hydrophobic surfaces can also enhance thermodynamic stability. Chothia has estimated a proportionality constant of 24 cal/mol of hydrophobic free energy per square angstrom of solvent-exposed surface area (*32*). Substitutions at Ile-3 of T4 lysozyme enhance the stability by amounts that agree surprisingly well with this prediction (*33*). However, there is some debate over the choice of the proper hydrophobicity scale to quantitate the contributions of each hydrophobic residue, and it is perhaps an oversimplification to expect such a simple relationship to hold for all amino acids (*34*).

D. Irreversible Denaturation

The desire to improve overall stability, particularly for enzymes employed in bioreactors, has broadened the search for causes of the deleterious changes in protein structure induced by temperature changes, oxidation, and exposure to that organic solvents which result in the irreversible loss of activity. Inactivation, for example, has been traced to the deamidation of asparagine residues, the oxidation of methionines, the β-elimination of cysteine residues, and the hydrolysis of aspartic acid residues that accompany or follow the conformational changes associated with solvent or thermal-induced denaturation (*35*). Two model systems for attenuating these processes feature studies on triose-phosphate isomerase and subtilisin.

Solutions of dimeric yeast triose-phosphate isomerase undergo aggregation and loss of activity on heating owing to deamidation of three asparagine residues (*36*). The crystal structure of the enzyme reveals that three asparagine residues (Asn-14, -68, and -78) of each subunit are present at the interfacial van der Waals surface. Deamidation of such residues results in the formation of charged aspartic acid residues, thereby promoting dissociation to the catalytically inactive monomers (*36, 37*). It is noteworthy that the enzyme from the thermophile *Bacillus stearothermophilus* does not have asparagine at these three positions (*38*). Single and double mutant forms of the yeast enzyme containing either the Asn-78 → Thr or the Asn-14 → Thr and Asn-78 → Ile substitutions exhibited increased half-lives toward irreversible inactivation from 13 min to 17 and 25 min, respectively. As predicted, mutation of Asn-78 → Asp slightly destabilized the protein, giving a half-life of 11 min.

Subtilisin, because of its industrial applications, has been a particularly attractive candidate protein for increasing its overall stability. The enzyme is oxidatively inactivated by formation of methionine sulfoxide at Met-222 (*39*). Mutagenesis to introduce substitutions with all 19 amino acids produced two favorable mutants, Ala-222 and Ser-222, that were resistant to oxidation by peroxide and retained at least 60% of the activity of the wild-type enzyme (*40*). The choice of alanine or serine as optimal for the replacement could not have been made from the existing three-dimensional structure of the protein (*41*); predictions based on homologous exchanges of amino acids in related proteins (*42*) would suggest leucine or valine as the most appropriate substitutions for methionine. Thus, in our current learning phase, random mutagenesis has still provided some surprises.

Nevertheless, the need for an interplay between three-dimensional structure and site-specific mutagenesis to improve the odds for achieving the engineering objective is exemplified by the production of an engineered subtilisin that is catalytically active and stable in dimethylformamide (*43*). Six mutations were introduced into the molecule, and their resultant contributions toward improved stabilization were evaluated by X-ray structural analysis, with the results summarized in Fig. 3. Increased hydrophobic interactions and van der Waals contacts resulted from Met-50 → Phe, Gly-169 → Ala, and Gln-206 → Cys; improved hydrogen bonding stemmed from Tyr-47 → Lys and Asn-182 → Ser; and the change of Asn-76 → Asp improved the affinity of the Ca^{2+}-binding site. The

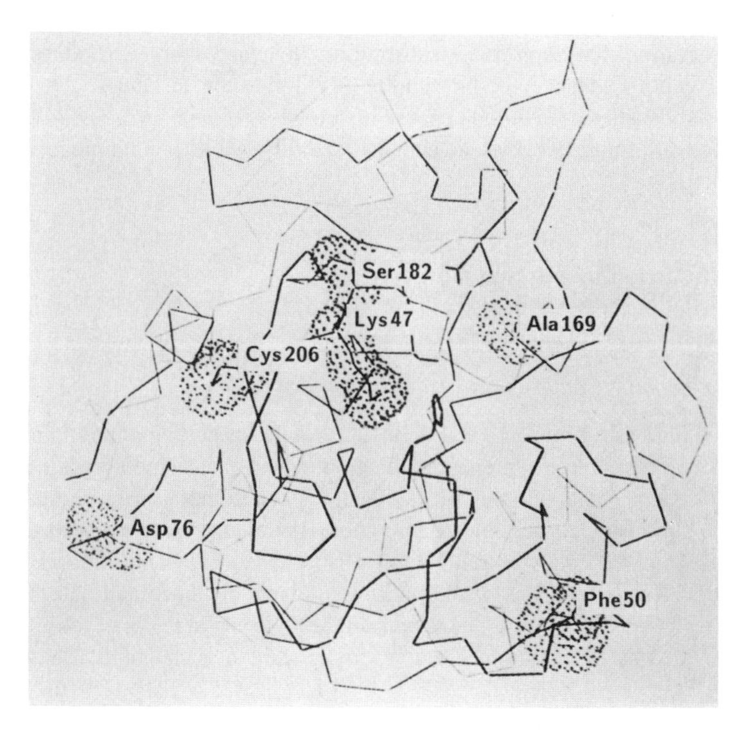

FIG. 3. Engineered subtilisin 8350 produced by Genex (*44*) is catalytically active and stable in dimethylformamide. The six positions of mutation and their improved stabilizing interactions determined by X-ray structural analysis are as follows: Met-50 → Phe and Gly-169 → Ala, hydrophobic; Asn-76 → Asp, Ca^{2+} binding; Gln-206 → Cys, van der Waals contact; and Tyr-47 → Lys and Asn-182 → Ser, hydrogen bonding. [Reprinted with permission from Ref. (*43*).]

choice of these residues was a combination of both rational design and random screening. In parallel work, a second weak Ca^{2+}-binding site was adjusted to have improved affinity by introducing negatively charged side chains in the vicinity of the bound Ca^{2+} by changing Pro-172 and Gly-132 to Asp residues (*44*). These changes singly increase the affinity of the Ca^{2+}-binding site by 2 to 3-fold and, when introduced together, some 6-fold. As a consequence of tighter binding, there was a marked improvement in the thermal stability of the protein at low Ca^{2+} levels.

E. FUTURE DIRECTIONS

Further avenues for enhancing protein stability include the elimination or utilization of any unsatisfied hydrogen-bonding groups within the protein and the addition of new favorable interactions in rigid parts of the structure (*10*). In addition, the potential for favorable interactions with α-helix dipoles has led to the observation that appropriate substitutions to place charged residues at the ends of helices enhance thermostability (*45*). In some instances, the gain in thermostability has been at the expense of some loss of enzyme activity; the real challenge for future work is to enhance stability while maintaining enzymic efficiency.

II. Substrate Binding and Catalysis

A. Active Site Catalytic Residues

1. *Nucleophilic Catalysis*

Serine proteases provide a well-defined example for participation of an amino acid as a nucleophile in an enzyme reaction. Site-directed mutagenesis has allowed further insights into the roles of each residue in the Asp-His-Ser catalytic triad. In subtilisin, all possible combinations involving replacement of each of the residues of the catalytic triad, Asp^{32}-His^{64}-Ser^{221} (Fig. 4), have been examined (*46*). Kinetic analysis of the multiple mutants demonstrated that the residues interact synergistically. The results are best viewed in terms of the acceleration provided by the addition of each catalytic residue as shown in Table I. The wild-type enzyme accelerates catalysis by a factor of 2×10^{10} over the rate of the uncatalyzed reaction. The enzyme lacking the catalytic triad (triple Ala mutant), accelerates the reaction by a factor of 2700, which probably reflects transition state stabilization owing to substrate binding interactions, including con-

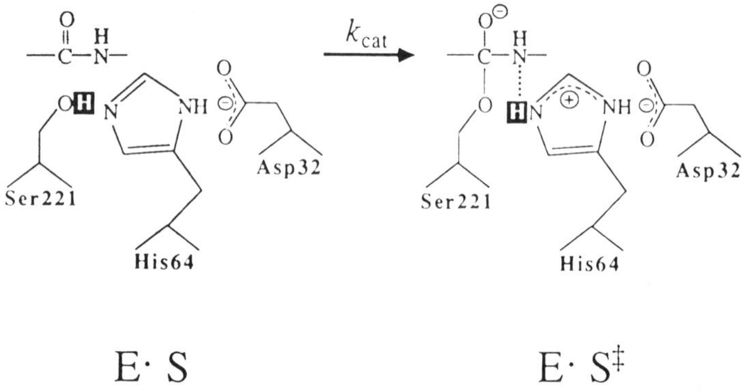

E·S $\xrightarrow{k_{cat}}$ E·S‡

FIG. 4. Schematic showing the rate-limiting acylation step in the hydrolysis of peptide bonds by subtilisin. In going from the Michaelis enzyme–substrate complex (E·S) to the transition state complex (E·S‡), the proton on Ser-221 (darkly shaded) is transferred to His-64, thus permitting nucleophilic attack on the scissile peptide bond. [Reprinted with permission from Ref. (*46*).]

TABLE I

SUBTILISIN CATALYTIC TRIAD[a]

Active site			k_{cat}/k_{uncat}	$k_{cat}/k_{\_\_\_\_}$
—	—	—	2.7×10^3	1
—	His	—	2.5×10^3	0.9
—	—	Asp	2.5×10^3	0.9
—	His	Asp	3.0×10^3	1.1
Ser	—	—	2.3×10^4	8.4
Ser	—	Asp	3.4×10^3	1.2
Ser	His	—	2.0×10^5	385
Ser	His	Asp	1.9×10^{10}	1.9×10^6

[a] The ratio k_{cat}/k_{uncat} is the value of k_{cat} for each enzyme divided by the rate of the uncatalyzed reaction. For the ratio $k_{cat}/k_{\_\_\_\_}$, the rate observed for each mutant is divided by that observed for the triple mutant.

tributions from stabilization of the developing negative charge on the carbonyl oxygen of the substrate by hydrogen bonding to Asn-155 and the backbone amide of residue 221. In the absence of Ser-221, the other members of the triad have no effect on acceleration of the catalytic rate. The addition of Ser-221 brings the rate acceleration up approximately 8-fold, and the rate becomes more markedly dependent on pH, suggesting that Ser-221 acts as a catalytic nucleophile. Very little of the full catalytic potential of the triad is realized until all three residues are in place. Thus, it is unreasonable to attempt to ascribe a certain rate acceleration to each of the residues; rather, together as a unit they contribute a factor of 2×10^6 toward the overall catalytic rate.

Detailed analysis of the role of the aspartic acid residue in the catalytic triad has been provided by studies on trypsin (*47*). As shown by crystallography, replacement of the aspartic acid residue with asparagine reverses the direction of hydrogen-bonding contacts, resulting in a 10^4-fold reduction in rate at pH 7. However, loss of the aspartate base from the charge relay system can be overcome at pH 10, leading to a rate that is within 10% of the wild type. Studies on a synthetic chemical model of the catalytic triad (*48*) and subsequent theoretical analysis (*49*) argue against proton transfer from His to Asp; rather, they support an electrostatic role in stabilization and alignment of the protonated form of His formed during catalysis (Fig. 4).

2. *Acid–Base Catalysts*

In several cases, site-directed mutagenesis has helped to identify residues directly involved in acid–base catalysis. Deletion of residues which participate in acid–base catalysis leads to a marked reduction in catalytic activity at neutral pH and an alteration in the pH dependence for catalysis similar to results described above for serine proteases. In studies on dihydrofolate reductase, the k_{cat} value of an Asp-27 → Asn mutant was reduced 300-fold relative to wild type at pH 7.0 but with a drastically altered pH profile; activity was restored at low pH as the substrate was protonated according to its known pK_a (*13, 24*). These data provide quantitative support for the role of Asp-27 in stabilizing the protonated form of the substrate at the active site.

The importance of optimal distance for proton transfer has been emphasized by work on triose-phosphate isomerase. An essential base, Glu-165, has been replaced by Asp, effectively increasing the bond distance for proton transfer by 1 Å (*50*). The rates of the enzyme-catalyzed enolization steps are reduced 1000-fold (*50*) relative to wild type. Although the mutant is impaired, its activity is still substantial considering that the wild-type enzyme accelerates the reaction 10^9-fold relative to acetate ion in solution. Attempts to select for second-site revertants which restore catalytic activity have met with only modest success (*51, 52*), but they begin to address the important questions pertaining to the evolution of the optimal geometry of the constellation of amino acids around the active site.

Further studies on triose-phosphate isomerase have focused on the role of His-95, which appears to act as an electrophile by stabilizing the developing negative charge on the carbonyl oxygen of the substrate during enolization. A mutant His-95 → Gln was constructed and shown to have catalytic activity 400-fold less than wild type (*53*). However, detailed mechanistic studies suggested that the mutant catalyzed the reaction by an alternate mechanism involving an enediolate intermediate rather than the stabilized enediol seen in the wild type. These studies point to the importance of complete mechanistic analysis of mutant enzymes. Moreover, it appears to be the rule rather than the exception that alteration of these residues which participate directly in catalysis is likely to lead to a new catalytic mechanism.

3. *Electrostatic Effects*

Effects of electrostatic interactions on substrate binding have been evaluated by amino acid substitution at two residues in the P1 binding cleft of subtilisin, Glu-156 and Gly-166. The two residues are involved in two modes of substrate binding dependent on the properties of the P1 substrate, thus contributing to the broad specificity of subtilisin. Measurements of k_{cat}/K_m ratios showed the anticipated increasing reaction toward glutamyl substrates and decreasing reaction toward lysyl P1 substrates with increasing positive charge of the binding pocket, although the effects were not symmetrical, with a bias toward negatively charged substrates. Overall, electrostatic interactions contributed 2 ± 0.6 kcal/mol to the binding of charged substrates (*54*). From the estimated distance between the ions of opposite charge, an apparent dielectric constant of 50–60 was calculated.

The importance of salt bridges at the NADPH-binding site of dihydrofolate reductase has been examined by a thorough analysis of two mutations, His-45

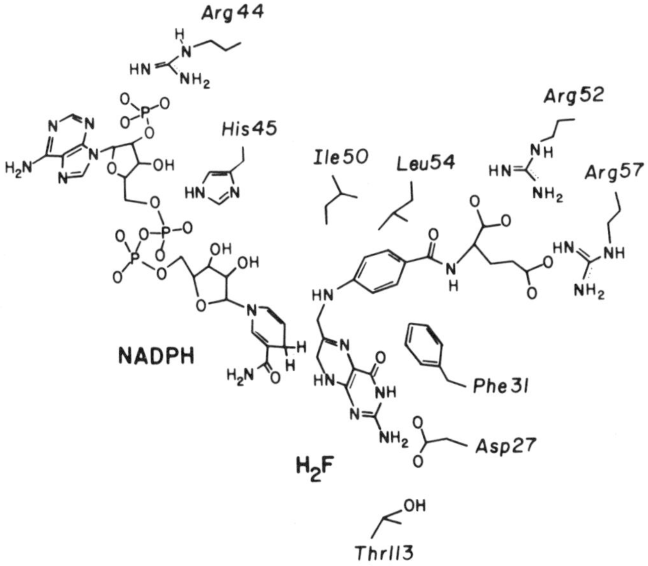

FIG. 5. Active-site structure of dihydrofolate reductase, showing selected amino acids involved in substrate binding. [Redrawn combining information from Refs. (*70*), (*136*), and (*155*).]

→ Gln and Arg-44 → Leu, which disrupt ionic contacts with the 2′-phosphate and the pyrophosphoryl moiety of NADPH (*136*) (see Fig. 5). The binding of NADPH was weakened by 1.1 and 1.5 kcal/mol and the transition state was destabilized by 0.6 and 1.8 kcal/mol for His-45 → Gln and Arg-44 → Leu, respectively. Both mutants transmitted their effects beyond the immediate environment to a position 25 Å away by elevating the pK_a of Asp-27 by 1.0 and 1.9 units for His-45 → Gln and Arg-44 → Leu, respectively. However, these changes presumably arise from structural rather than electrostatic effects; a dielectric constant of 5 would be required to account for the 2.5 kcal/mol effect over a distance of 25 Å.

The general effect of electrostatic charge on catalytic activity has been assessed by observations on the change in the pK_a of an active site His-64 of subtilisin occurring owing to alteration of charged residues on the enzyme surface (*56, 57*) as illustrated in Fig. 6. The pK_a of this histidine was perturbed by +0.08 to −1.00 pK_a units by the introduction or deletion of charged residues within 12–18 Å (Table II). The value of ΔpK_a was used to calculate the effective dielectric constant, D_{eff}, according to $D_{eff} = 244/\Delta_q r\ \Delta pK_a$ where r is the separation in angstroms and Δ_q the change in charge. The calculated ΔpK_a and D_{eff} values were determined from the Warwicker–Watson algorithm, in which the protein/solvent is divided into cubes, typically of dimension 1 Å, and a dielectric constant (D) is assigned to each cube which corresponds to the protein (D =

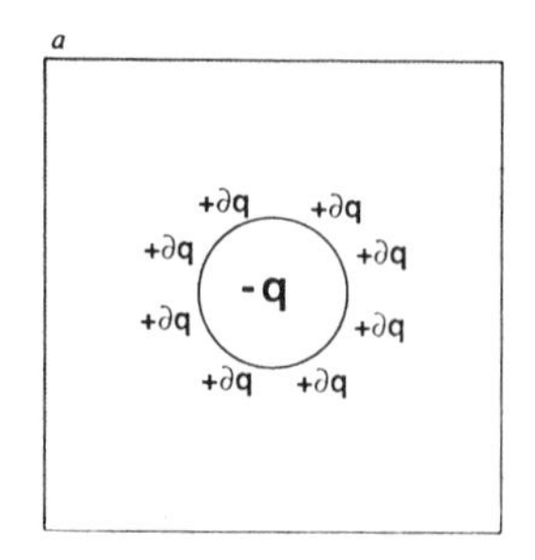

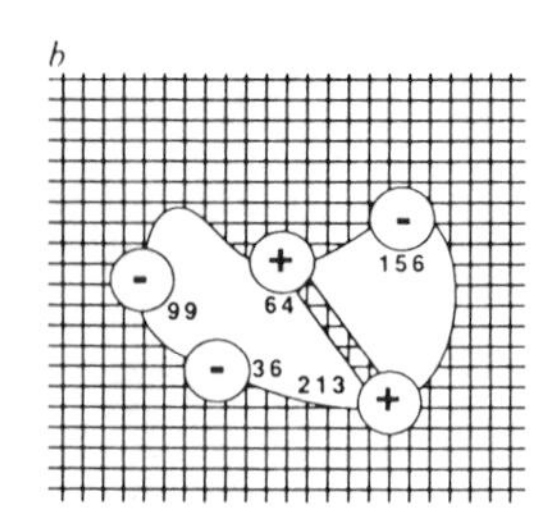

FIG. 6. (a) Physics of the dielectric effect. A charge q in a continuous medium of dielectric D induces a surface polarization, $\Sigma\ \delta q$, which partially neutralizes the original charge. The effective charge inside the dielectric, $-(1 - \Sigma\ \delta q)$ is equal to $-1/D$. (b) Calculation of electrostatic effects in mutant subtilisins. The Warwicker–Watson algorithm uses a grid and associates dielectrics and charges appropriately. The positions of the charged residues illustrate roughly whether the interaction is mainly through protein or solvent. [Reprinted with permission from Ref. (*57*).]

TABLE II

pK_a SHIFTS ARISING FROM MUTATIONS IN SUBTILISIN[a]

Mutant	Mean distance (Å)	ΔpK_a	D_{eff}^{obs}	ΔpK_a	D_{eff}^{calc}
Asp-99 → Ser	12.6	−0.40	48	−0.31	62
Glu-156 → Ser	14.4	−0.38	45	−0.35	48
Ser-99 → Lys	15.0	(−0.25)	65	−0.25	65
Ser-156 → Lys	16.5	(−0.25)	59	−0.30	49
Lys-213 → Thr	17.6	+0.08	173	+0.19	73
Asp-36 → Gln	15.1	−0.18	90	−0.16	101
Asp-99 → Lys	(13.8)	−0.64	55	−0.56	63
Gly-156 → Lys	(18.5)	−0.63	50	−0.65	48
Asp-99 → Ser, Glu-156 → Ser	(13.5)	−0.63	57	−0.66	55
Asp-99 → Lys, Glu-186 → Lys	(14.7)	−1.00	66	−1.21	55

[a]Numbers in parentheses denote experimental values calculated from two or more other values rather than directly determined. Distances for mutants were obtained assuming the side chain to be fully extended. Values for the pK_a shifts are the mean values on the two histidine imidazole nitrogens. Distances are the average distance from the side-chain nitrogens or oxygens to the two histidine imidazole nitrogens of His-64.

3.5), the solvent ($D = 80$), or an intermediate value for the boundary (Fig. 6). Each charge is spread to the eight corners of the cube, and a finite difference solution of Poisson's equation is then obtained (*58*). The calculated values are in general agreement with those determined experimentally. The greatest deviation occurred with the Lys-213 → Thr mutation, which is in a flexible region of the molecule so that the separation distance is less certain (Fig. 6b). Nonetheless, the agreement between the calculated and observed effects can be cited as further evidence for the retention of tertiary structure in the mutants.

4. *Difficulties in Reversal of Ion Dipoles*

Two attempts toward reversal of substrate charge specificity have largely failed. In studies on trypsin, Asp-189, which lies at the bottom of the substrate-binding pocket and confers specificity for positively charged amino acid substrates, was replaced by Lys (*59*). Confirming the role of Asp-189 in substrate binding, the Lys-189 mutant showed no activity toward arginyl or lysyl substrates; however, there was no compensatory shift in specificity toward acidic substrates. Rather, a low rate of reaction toward hydrophobic residues was the only activity detected. The results could be rationalized by computer modeling studies, which suggested that the positively charged amino group of Lys-189 was directed outside of the substrate-binding pocket by interaction with nearby main-chain carbonyls, thereby leaving a hydrophobic pocket.

In an attempt to reverse the substrate charge specificity for aspartate aminotransferase, an Arg-292 residue which forms a salt bridge with the carboxylate on the side chain of the substrate aspartate was modified to an Asp (*60*). Model building studies based on the crystal structure of the wild-type enzyme indicated that the Arg-292$^+$:Asp$^-$ ion pair could be replaced by an Asp-292$^+$:Arg$^-$ pair in the mutant (Fig. 7). The specificity of the Arg-292 → Asp mutant in terms of

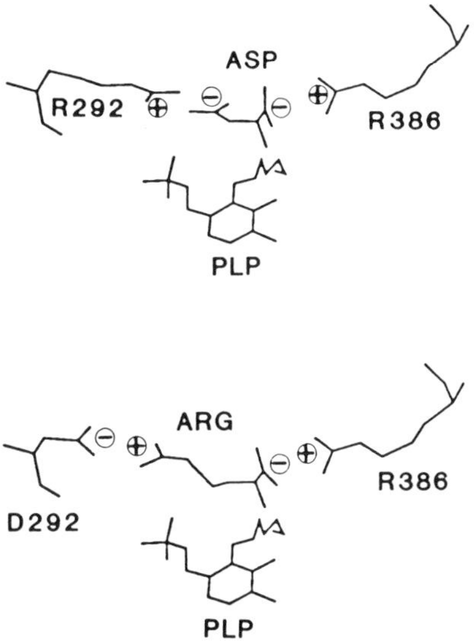

FIG. 7. (*Top*) View of the active-site structure of chicken heart mitochondrial aspartate aminotransferase with bound L-aspartate. (*Bottom*) Computer graphics model of the hypothetical active-site structure of *E. coli* Arg-292 → Asp (R292D) mutant aspartate aminotransferase with L-arginine bound. [Reprinted with permission from Ref. (*60*).]

the k_{cat}/K_m ratio for aspartic acid was reduced by 5 orders of magnitude (7 kcal/mol); however, the activity of the mutant toward arginine as a substrate was only 10-fold greater than wild type. Although one might conclude that the mutation has produced an inversion of the substrate charge specificity, a more accurate assessment of the results would be that the charge specificity was largely lost. The wild type prefers aspartate over arginine by a factor of 1.2×10^6, whereas the mutant prefers arginine over aspartate by only 6-fold.

The marginal specificity of the mutant aminotransferase toward cationic amino acids could be due to the formation of a salt bridge between Asp-292 and a positively charged amino acid near the active site analogous to that proposed for the trypsin mutant described above. Alternatively, it has been argued that such attempts at reversal of ion pair polarity are predestined to fail because the average electrostatic potential created by distant charged residues tends to favor a particular polarity at the active site (*61*). Accordingly, multiple additional mutations of distant charged residues would be required to complete the charge reversal. This postulate is reasonable and directly testable.

5. *Interfacial Active Sites*

Site-directed mutagenesis has been used to establish that the active site lies at the interface between subunits of certain oligomeric enzymes (*62–64*). The analysis relies on restoration of activity on forming a hybrid from proteins containing mutations at two positions. Studies of this type were first performed on aspartate transcarbamylase (aspartate carbamoyltransferase) (*62, 63*), where an active hybrid catalytic trimer was isolated from a mixture of two inactive mutants. The rationale for this analysis is shown in Fig. 8, illustrating work done on ribulose-bisphosphate carboxylase (*64*). Two mutant enzymes, each unable to carry out catalysis, were recombined to form hybrids. Based on random association of monomers to form the catalytic dimer as shown in Fig. 8, it is expected that 50% of the trimers should form one wild-type active site (B, C), such that the mixture of the hybrids exhibits 25% of the wild-type activity. This complementation demonstrates that the active site must be at the interface between the subunits.

Because these studies can provide definitive results without the aid of a crystal structure, intersubunit complementation is likely to become a standard method in prospecting for active sites in oligomeric enzymes.

B. Thermodynamics of Catalysis

The goal of a complete analysis is to construct a free energy profile for the enzyme-catalyzed reaction by measurement of the rate and equilibrium constants for each step in the pathway. Changes in the reaction profile with mutation can then serve to quantify the roles of each amino acid substitution toward each step

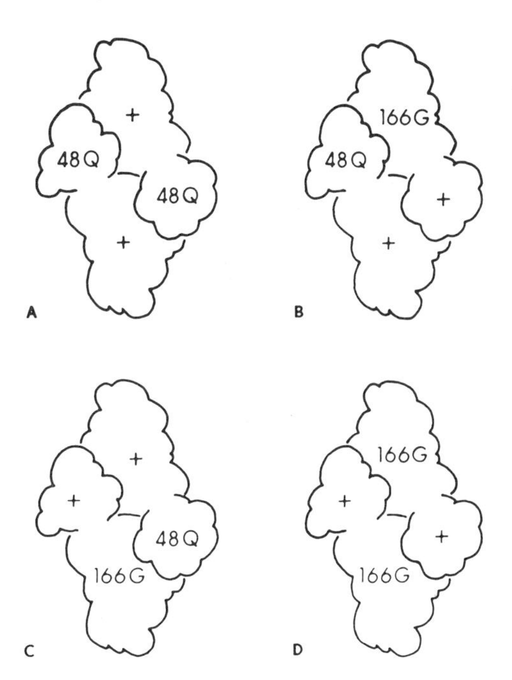

FIG. 8. Different homodimers (A and D) and heterodimers (B and C) that can form in a cell expressing both the Glu-48 → Gln (E48Q) and Lys-166 → Gly (K166G) alleles. If the active site of the enzyme is formed by an intersubunit contact of distinct domains, each of the heterodimers (B and C) will contain one functional active site composed of wild-type domains (+ adjacent to +). [Reprinted with permission from Ref. (*64*).]

in catalysis, namely, substrate ground state binding, transition state stabilization, alteration of the equilibrium constant for the chemical reaction, or product release. Such analysis has provided unambiguous definition of the roles of individual amino acids in catalysis for a number of enzymes. We discuss in depth the results obtained with two enzymes, tyrosyl-tRNA synthetase (tyrosine–tRNA ligase) and dihydrofolate reductase.

1. *Kinetic Analysis of Mutant Enzymes*

The initial kinetic characterization of a mutant enzyme consists of comparison to wild type with respect to the two fundamental steady-state kinetic constants: k_{cat}, the maximum rate at saturating substrate, and k_{cat}/K_m, the apparent second-order rate constant for substrate binding, often referred to as the specificity constant. The Michaelis constant, K_m, as the ratio of the maximum rate divided by the apparent substrate binding rate, should be considered as a derivative of the two fundamental enzymic constants.

It is unfortunate that k_{cat} and K_m have often been interpreted as the rate of catalysis and the substrate dissociation constant, respectively, because it is rare that this is actually the case. Often, k_{cat} is a function of rate-limiting substrate binding or product release steps. Studies on dihydrofolate reductase have pointed clearly to potential errors in assigning k_{cat} to the rate of the chemical reaction at the active site. Steady-state turnover by dihydrofolate reductase is limited by product release at a rate of 12 sec^{-1}, whereas the rate of the catalytic step at the active site is nearly 1000 sec^{-1}. Several mutations have been described which lead to an increase in k_{cat} and a decrease in the rate of the chemical reaction (*65, 66*) such that chemistry becomes rate limiting in steady-state turnover by the mutant enzymes. Comparison of k_{cat} values for mutant and wild-type enzymes is a particularly misleading indicator of the effect of the mutation in the absence of further detailed mechanistic studies.

2. *Utilization of Binding Energy in Catalysis*

Hydrogen-bonding interactions between the substrate and enzyme have been explored in depth in the catalytic mechanism of tyrosyl-tRNA synthetase (tyrosine–tRNA ligase). The enzyme-catalyzed reaction occurs in two steps:

$$\begin{aligned} \text{ATP} + \text{Tyr} &\rightleftharpoons \text{Tyr-AMP} + \text{PP}_i \\ \text{Tyr-AMP} + \text{tRNA} &\rightleftharpoons \text{Tyr-tRNA} + \text{AMP} \end{aligned} \tag{1}$$

The tyrosyl adenylate (Tyr~AMP) is formed and then held tightly at the active site until it reacts with tRNA in the second step. The equilibrium for the formation of Tyr~AMP is unfavorable in solution ($K_{eq} = 3.5 \times 10^{-7}$) but is near unity at the active site ($K_{eq} = 2.3$). Thus, the enzyme must contribute 9.3 kcal/mol toward stabilization of the products at the active site. This is accomplished by the increasing strength and number of hydrogen-bonding interactions between the reactants and active site amino acids as the reaction proceeds.

One must bear in mind that the first the half-reaction by itself is a rather unusual enzyme-catalyzed reaction, where the primary function of the enzyme is to pull the reactants uphill and hold on to them tightly. In fact, the first step of the tyrosyl-tRNA synthetase reaction represents a clear-cut case of the classic theoretical problem in enzymology where the enzyme binds the substrates and products so tightly that they fall in to a thermodynamic well and turnover is very slow. As we shall see, this has a direct bearing on the results obtained by mutagenesis. Moreover, because the enzyme catalyzes only a single turnover in the absence of tRNA, it has been relatively easy to measure the equilibrium constants for substrate binding and the rate of the catalytic step for the wild-type and mutant enzymes. This has then led to a complete thermodynamic description relating binding energy to catalysis for a large number of mutants.

Figure 9 shows the theoretical transition state built into the active site of tyrosyl-tRNA synthetase, inferred from the crystal structure of the E·Tyr~AMP complex. Inspection of the active site reveals that there are no residues in position to participate in acid–base catalysis, so the crystal structure provided no obvious clues as to the reaction mechanism. However, analysis of the free energy

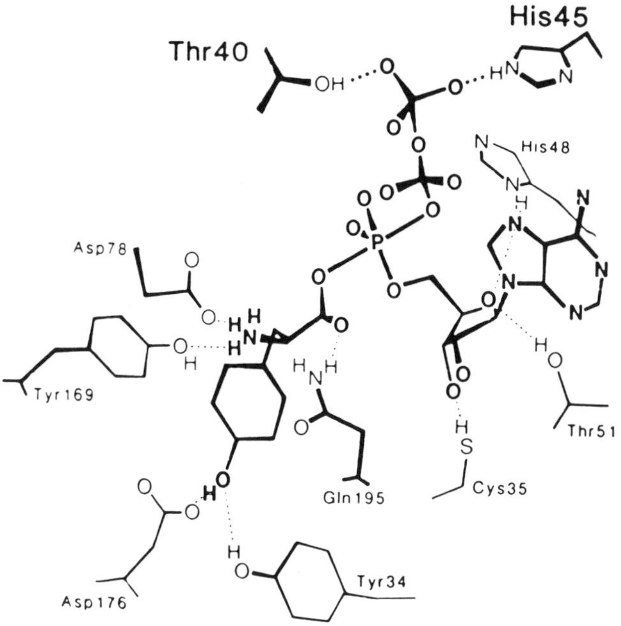

FIG. 9. Model building of the pentacoordinate transition state of the reaction into the crystallographic structure of tyrosyl-tRNA synthetase. [Reprinted with permission from Ref. (*156*).]

profile for the reaction catalyzed by wild-type and mutant enzymes has provided a quantitative definition of the thermodynamic contributions of each amino acid toward catalysis.

The binding of substrates to tyrosyl-tRNA synthetase is random but can most easily be studied by the sequence:

$$\mathrm{E} \overset{K_1[\mathrm{Tyr}]}{\rightleftharpoons} \mathrm{E{\cdot}Tyr} \overset{K_2[\mathrm{ATP}]}{\rightleftharpoons} \mathrm{E{\cdot}Tyr{\cdot}ATP} \underset{k_{-3}}{\overset{k_3}{\rightleftharpoons}} \mathrm{E{\cdot}Tyr\text{-}AMP{\cdot}PP_i} \underset{K_4[\mathrm{PP_i}]}{\rightleftharpoons} \mathrm{E{\cdot}Tyr\text{-}AMP} \quad (2)$$

Measuring the equilibrium constant for Tyr binding to the enzyme provided a direct measurement of the standard state free energy change:

$$\Delta G^\circ = -RT \log(K_1^{\mathrm{wt}}) \quad (3)$$

where K_1^{wt} is the Tyr binding constant for the wild-type enzyme. Comparison of mutant and wild-type enzymes allows calculation of the change in free energy of binding arising from the amino acid substitution:

$$\Delta\Delta G = -RT \log(K_1^{\mathrm{wt}}/K_1^{\mathrm{mt}}) \quad (4)$$

where K_1^{mt} is the binding constant for the mutant enzyme. It is important to note that although ΔG° depends on the choice of the standard state (1 *M* Tyr), $\Delta\Delta G$, does not.

Measurement of the ATP concentration dependence of the rate of Tyr-AMP production in a single turnover provided estimates of both K_2 and k_3, assuming that the binding of ATP in Eq. (2) is a rapid equilibrium. Contributions of single amino acid substitutions to the apparent free energy of binding ATP were then estimated from the effect of the mutation on the value of K_2, and the apparent free energy of activation of the reaction was calculated according to transition state theory:

$$\Delta G^{\circ\ddagger} = -RT\log(k_3^{\mathrm{wt}}) + RT\log(k_\mathrm{B}T/h) \tag{5}$$

Comparison to the mutant enzyme gives

$$\Delta\Delta G^{\ddagger} = RT\log(k_3^{\mathrm{wt}}/k_3^{\mathrm{mt}}) \tag{6}$$

providing an estimate of the apparent contribution of the mutated amino acid toward the transition state stabilization.

Consider, for example, the free energy profile comparing the wild type with the Cys-35 → Gly mutation shown in Fig. 10. This analysis reveals that the hydrogen bond between Cys-35 and the ribose of ATP develops along the reaction coordinate to the transition state and becomes stronger in binding the products; there was no difference between the mutant and wild-type enzymes in the free energy of binding Tyr or ATP in the ground state. This study clearly defined the role of Cys-35 in providing binding energy to stabilize the transition state and to drive the reaction forward.

In contrast, as shown in Fig. 11, substitution of Phe for Tyr-34 or Tyr-169 leads to a loss of substrate binding energy for Tyr with no further changes on the binding of ATP or formation of products. These data indicate that the hydrogen bonds contribute to the ground state binding energy of tyrosine at the active site, but do not contribute to the binding of ATP, the stabilization of the transition state, or the formation of products. In contrast, His-48 forms a hydrogen bond with ATP that is increased in strength in the transition state and again with the formation of the products.

Similar analysis of mutations of other amino acids at or near the active site has led to a more complete description of the changes in strength of hydrogen bonds occurring at each stage of the reaction. These results are summarized in Fig. 12, where hydrogen bonds are designated by dashed lines. The general trend is toward the formation of more hydrogen bonds as the reaction progresses, thus accounting for the thermodynamic driving force for the reaction.

One of the first questions to address is whether one can account for the stabilization of Tyr-AMP from the summation of the contributions of the individual amino acids toward the differential binding of Tyr-AMP relative to Tyr and ATP, $\Sigma\ \Delta\Delta G_{\mathrm{E\cdot T\cdot ATP}} - \Delta\Delta G_{\mathrm{E\cdot T\text{-}AMP\cdot PP_i}}$, where the sum is taken over all contributing amino acids. In an ideal case, this sum should equal the 9.3 kcal/mol stabilization of Tyr-AMP by the enzyme. In addition, it is expected that the total transi-

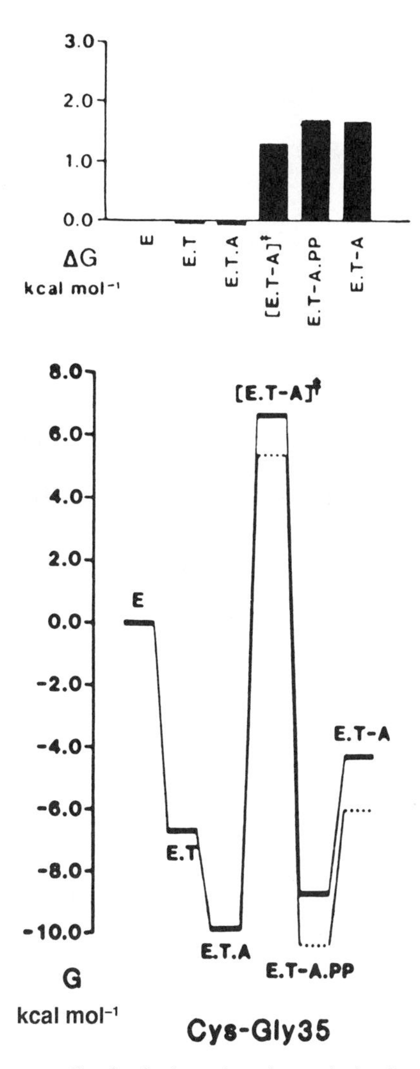

FIG. 10. Gibbs free energy profiles for the formation of tyrosyl adenylate and pyrophosphate, as defined in Eq. (2), by wild-type (energy levels in dashed lines) and mutant (energy levels in solid lines) tyrosyl-tRNA synthetases, using standard states of 1 *M* for tyrosine, ATP, and pyrophosphate. [Reprinted with permission from Ref. (*151*).]

tion state stabilization should equal that predicted from the acceleration of the catalytic reaction.

The total transition state stabilization, calculated from the sum of effects seen on all mutants, is approximately 25 kcal/mol, predicting an enhancement of the enzyme-catalyzed reaction of approximately 10^{18} relative to the uncatalyzed rate.

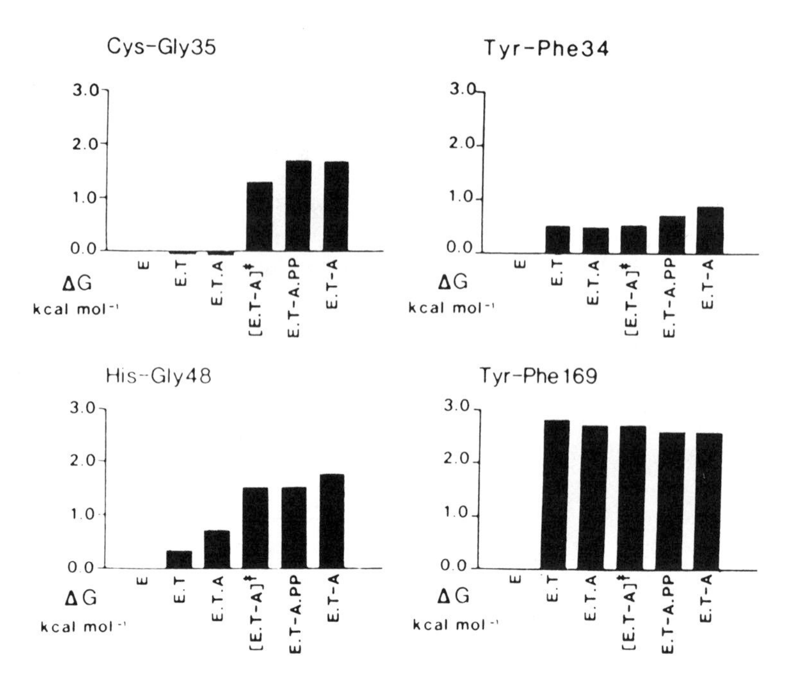

FIG. 11. Gibbs free energies of enzyme-bound complexes of mutant tyrosyl-tRNA synthetase relative to those of wild-type enzyme. [Reprinted with permission from Ref. (*151*).]

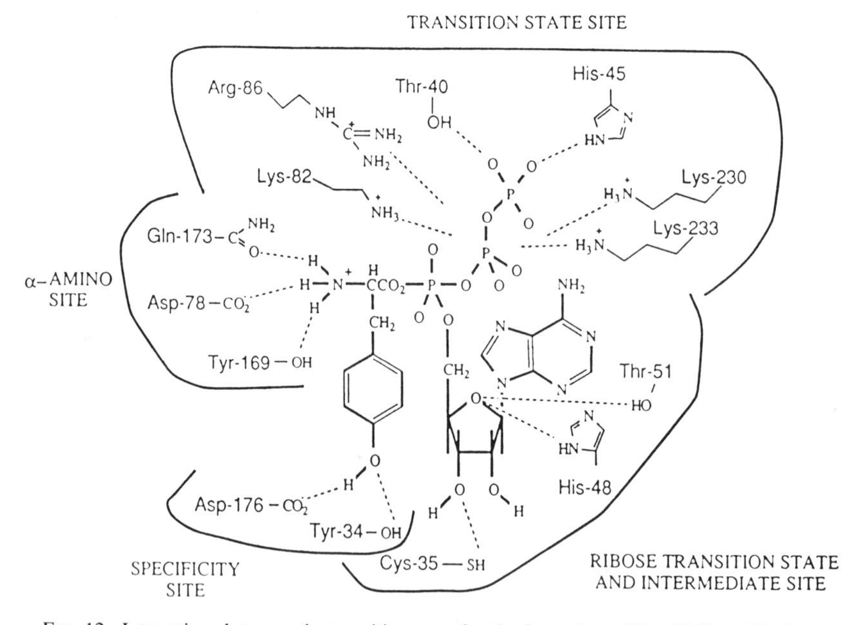

FIG. 12. Interactions between the transition state for the formation of Tyr-AMP and hydrogen-bonding (or charged) side chains that were deduced from site-directed mutagenesis of tyrosyl-tRNA synthetase. The overall site may be classified in terms of subsites as indicated and described in the text. [Reprinted with permission from Ref. (*157*).]

This free energy exceeds all reasonable estimates by at least 60%. The total stabilization of products relative to reactants at the active site is known with less certainty because of the difficulty of making measurements with some mutants, but the calculated sum is greater than 13 kcal/mol, exceeding the predicted value by 40%. Thus, even though the present totals do not include all amino acids, the measured free energies overestimate the contributions of each amino acid toward the net binding energy. Accordingly, it is important to stress that the change in free energy of binding observed with each amino acid substitution is an apparent free energy change. For example, in the present case, the changes occurring on substitution are not limited simply to removing a given hydrogen bond, but one must also consider the effects of local reorientation of residues and changes in hydration of the free enzyme (*67*). Orientation of groups at the active site can have profound effects on the rate of reaction which exceed the effects on binding thermodynamics. Therefore, the apparent $\Delta\Delta G^{\ddagger}$ value can greatly overestimate the changes in binding energy attributable to the interaction with a single amino acid. Recent molecular dynamics calculations have shown, in fact, that $\Delta\Delta G^{\ddagger}$ represents a small difference between several large opposing interaction terms (*68*).

3. *Linear Free Energy Relationships*

Linear free energy relationships have been noted which relate the rate of reaction to an equilibrium constant at the active site of an enzyme as the structure of the enzyme is varied by site-directed mutagenesis (*69, 70*). This analysis has been useful in that it demonstrates that measurements on individual mutants represent observations on a continuous free energy surface with implications for enzymic catalysis in general. Moreover, closer analysis of those points which deviate from such linear free energy relationships might provide deeper insights into understanding the unusual properties of single amino acid substitutions. It is of foremost importance to understand the basis for prediction of a linear relationship between the rate and the free energy of a reaction.

The classic analysis of transition state structure in physical organic chemistry is based on the premise that the rate constant for a given reaction should be a simple function of the equilibrium constant for the reaction when the structures of the reactants are systematically varied such that the following relationship holds:

$$k = AK^{\beta} \quad \Rightarrow \quad \log k = \log K + \text{constant} \tag{7}$$

where A and β are constants such that a plot of log k against log K should be a straight line with slope β. Accordingly, Brønsted or Hammett plots have been used to examine the dependence of a rate of reaction on the equilibrium constant as the structures of the reactants are varied, and thereby to infer the nature of the transition state relative to the structural variation explored.

In reality, because $K = k_+/k_-$, the relationship tests for the interdependence of k_+ and k_- according to

$$\log k_+ = \frac{\beta}{\beta - 1} \log k_- + \mathit{constant} \tag{8}$$

For the usual case, β falls in the range of 0–1. When β is 0, k_+ is a constant, and the variation in the equilibrium constant is attributable solely to changes in k_-, implying that the transition state for the reaction is close to the reactants in structure. Conversely, a β value of 1 implies that the transition state for the reaction more closely resembles the products.

What does this mean for enzymology when the structure of the catalyst is varied? Fersht has shown such a linear relationship for a carefully selected series of mutations in tyrosyl-tRNA synthetase (Fig. 13A). The correlation is quite good, with a β value of 0.8 relating the change in the rate of the catalytic step to the equilibrium constant at the active site of the enzyme. The first important conclusion is that the enzyme mutants appear to obey simple physical laws and that the attenuation of the substrate binding energy at the active site is directly proportional to a reduction in catalytic power. Second, as the mutations further weaken the hydrogen-bonding interactions which serve to stabilize the product, the reduction in the equilibrium constant for the reaction is largely attributable to a decrease in the rate of the forward reaction. According to the simple Brønsted-type analysis, this would imply that the transition state resembles the products more closely than the reactants.

The validity of the analysis leading to a linear free energy relationship for tyrosyl-tRNA synthetase has been criticized on the basis of the method of data analysis (*71*). Estell has pointed out that in plotting $\log(k_+)$ versus $\log(k_+/k_-)$, the presence of the same variable on both axes can lead to an apparently linear correlation from a randomly distributed sample set, if most of the variation in k_+/k_- is due to changes in k_+. Fersht has countered these arguments by pointing out that the assumption that k_+ dominates the variation in k_+/k_- is tantamount to assuming a β value close to 1 (*71*). Moreover, plots of k_+ versus k_- for tyrosyl-tRNA synthetase also demonstrate the linear free energy relationship, although there is clearly more scatter revealed in the data (Fig. 13B) and the correlation coefficient has dropped from 0.99 to 0.69. This interchange (*71*) has served to clarify the need for accurate statistical analysis of independent parameters and careful interpretation of results within the framework of chemistry and physics.

However, the most important criticism of the linear free energy relationships is that this simple and apparently satisfying relationship is followed for only a small number of the amino acids examined, namely, Cys-35, His-48, and Thr-51, each of which is hydrogen bonded to the ribose of ATP and shows an increase in bond strength during the reaction. Mutations at Thr-34 and Thr-169

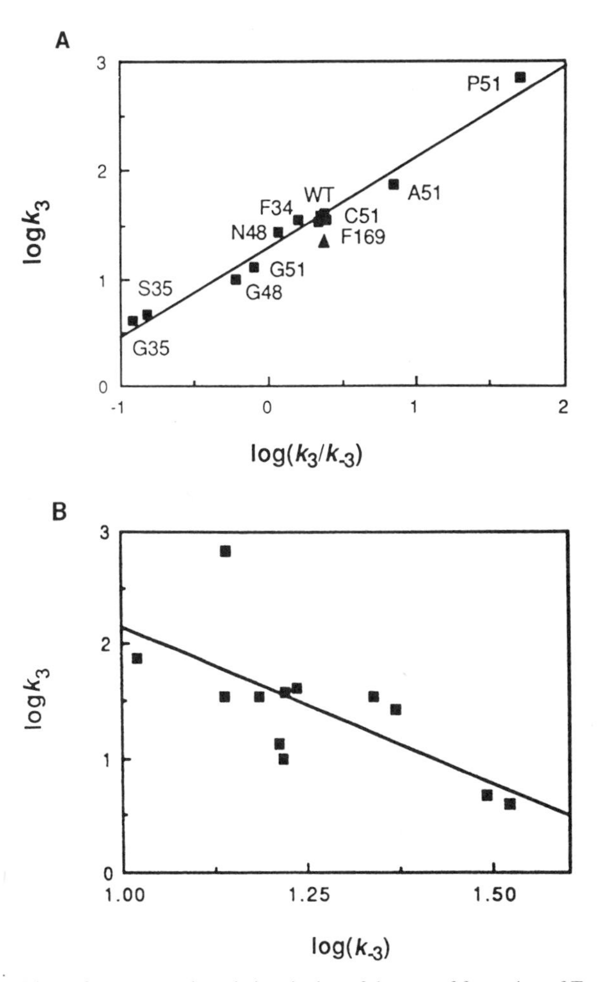

FIG. 13. (A) Linear free energy plot relating the log of the rate of formation of Tyr-AMP (k_3) to the log of the equilibrium constant (K_3) at the active site of tyrosyl-tRNA synthetase for several mutants. The slope is 0.83 with a correlation coefficient of 0.99. (B) The same data as shown in (A) is plotted to illustrate the correlation between log k_3 and log k_{-3}. The slope provides a β value of 0.8 with a correlation coefficient of 0.69. [Reprinted with permission from Refs. (*69*) and (*71*).]

are included in the free energy plots (*69*), but the variation from wild type in the rate and equilibrium is so small that they add little weight to the correlation. Moreover, closer analysis of the free energy profiles for substitutions of these amino acids (Phe-34 or Phe-169) reveals that Thr-34 and Thr-169 are principally involved in ground state binding to tyrosine, with insignificant changes during catalysis. Thus, one cannot justify their inclusion in the linear free energy relationship.

4. *Vertical Free Energy Relationships*

There is no *a priori* reason to suspect that changes in the rate of reaction must be accompanied by a change in the equilibrium constant for the catalytic step at the active site of the enzyme. Indeed, significant deviations from the linear free energy relationship have been seen for those residues that accelerate the reaction to the greatest extent because they are involved in transition state stabilization without altering the equilibrium constant for the reaction at the active site. Thr-40 and His-45 bind to the γ-phosphate of ATP, and mutations at these two positions lead to changes in the rate of reaction spanning nearly 6 orders of magnitude with no reported change in the equilibrium constant (*69*). Thus, these two amino acids stabilize the transition state for the reaction without altering the net equilibrium, and their true catalytic power is correlated with their failure to follow a linear free energy relationship! Other amino acids which bind to the pyrophosphate portion of ATP (Lys-230, Lys-233, Lys-82, and Arg-86) also appear to fall into this class; they contribute 3–6 kcal/mol of stabilization energy to the transition state with little effect on the equilibrium constant, although the effects on the equilibrium constant are not known with certainty in all cases (*72*).

Lys-82, Arg-86, Lys-230, and Lys-233 are on two exposed loops which have very high temperature factors (*73*). To reconcile the crystallography of the ground state structure with the kinetic data there must be a large rearrangement of these mobile loops following substrate binding in an induced fit mechanism. Based on this and other observations, it is quite likely that the rate-limiting step in the tyrosyl-tRNA synthetase reaction is a conformational change in the enzyme. Accordingly, the complete reaction sequence following substrate binding includes the following steps:

$$\mathrm{E{\cdot}Tyr{\cdot}ATP} \underset{k_{-3}}{\overset{k_3}{\rightleftharpoons}} \mathrm{E'{\cdot}Tyr{\cdot}ATP} \underset{k'_{-3}}{\overset{k'_3}{\rightleftharpoons}} \mathrm{E{\cdot}Tyr{\cdot}AMP{\cdot}PP_i} \underset{K_4[\mathrm{PPP_i}]}{\rightleftharpoons} \mathrm{E{\cdot}Tyr{\cdot}AMP} \qquad (9)$$

The rate measured as k_3 is then a change in the conformation of the E·Tyr·ATP complex, which is then followed by a much more rapid, and therefore unresolved, catalytic step, k'_3. The transition state stabilization and the linear free energies described by Fersht may be a function of a rate-limiting conformational change, not chemistry. Changes in hydrogen bond strength leading to a change in the rate of the reaction then reflect the effects of hydrogen bonds in driving the conformational change in the enzyme. This conformational change is then also responsible for stabilization of the products at the active site of the enzyme.

5. *Dihydrofolate Reductase*

Transient state and equilibrium measurements have led to a complete kinetic scheme for dihydrofolate reductase which accounts for the steady-state parameters and predicts the full time course kinetics under a variety of substrate concentrations and at various pH values. The pH-independent pathway is shown in

Scheme I. The preferred pathway is represented by the closed loop (heavier arrows), which bypasses free enzyme. Following the release of $NADP^+$, the release of H_4F from the E·H_4F complex is too slow to account for turnover, but it is enhanced 10-fold by the binding of NADPH at the neighboring site. Studies on dihydrofolate reductase have provided a complete analysis of the effects of point mutations on the rate and equilibrium constants for each step in the reaction sequence. The active site structure of dihydrofolate reductase is shown schematically in Fig. 5, illustrating some of the amino acids interacting with the substrates that have been mutated.

A linear free energy relationship has been observed between the rate of hydride transfer and the equilibrium constant for binding dihydrofolate for a series of mutations in the folate binding site (Phe-31 → Tyr, Phe-31 → Val, Thr-113 → Val, and the double mutant Phe-31 → Val, Leu-54 → Gly) (*70*). However, what appears at first glance as a simple linear free energy relationship is, in fact, somewhat of a puzzle. A linear decrease in rate with a change in free energy for binding the ground state implies that these mutations must have exactly a 2-fold greater effect on the free energy of binding the transition state (differential binding). This is a surprise indeed because there is no reason that there should be precisely a 2-fold difference in binding energy of the transition state versus the ground state. However, analysis of these same mutants on a Brønsted plot produces a vertical free energy plot; that is, changes in the rate of the catalytic step over three orders of magnitude do not lead to changes in the equilibrium constant for hydride transfer at the active site, so the plot of log k versus log K is nearly vertical. This analysis, more than any other, implicates the mutated residues in selective transition state stabilization.

It is obvious from these studies that linear free energy relationships are not global; in other words, they rationalize the behavior of only a select few amino acids at an active site, and, thus, they are not reliable indicators of the role of active site residues in catalysis. The more definitive analysis is based on measurement of changes in ground state binding and rates of individual enzymic steps defining the complete free energy profile for each mutation. Even then results must be interpreted with careful consideration for the nature of the change in structure.

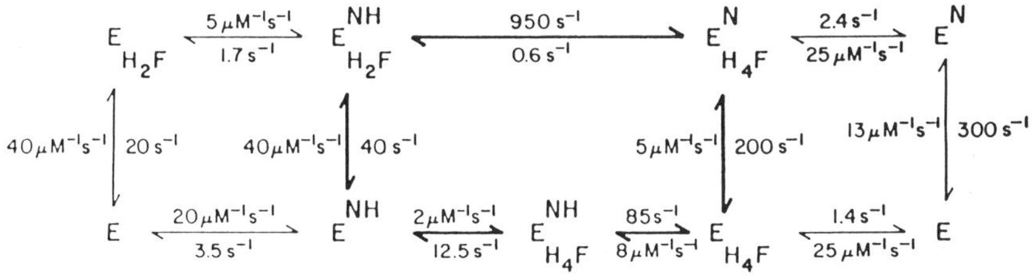

$E^{}_{H_2F}$ ⇄ (5 μM⁻¹s⁻¹ / 1.7 s⁻¹) $E^{NH}_{H_2F}$ ⇄ (950 s⁻¹ / 0.6 s⁻¹) $E^{N}_{H_4F}$ ⇄ (2.4 s⁻¹ / 25 μM⁻¹s⁻¹) E^{N}

Vertical steps: E_{H_2F} ⇄ E (40 μM⁻¹s⁻¹ / 20 s⁻¹); $E^{NH}_{H_2F}$ ⇄ E^{NH} (40 μM⁻¹s⁻¹ / 40 s⁻¹); $E^{N}_{H_4F}$ ⇄ E_{H_4F} (5 μM⁻¹s⁻¹ / 200 s⁻¹); E^{N} ⇄ E (13 μM⁻¹s⁻¹ / 300 s⁻¹)

E ⇄ (20 μM⁻¹s⁻¹ / 3.5 s⁻¹) E^{NH} ⇄ (2 μM⁻¹s⁻¹ / 12.5 s⁻¹) $E^{NH}_{H_4F}$ ⇄ (85 s⁻¹ / 8 μM⁻¹s⁻¹) E_{H_4F} ⇄ (1.4 s⁻¹ / 25 μM⁻¹s⁻¹) E

SCHEME I

The free energy profiles for dihydrofolate reductase wild type and the Thr-113 → Val mutant are compared in Fig. 14 (*65*). Thr-113 is a strictly conserved residue that forms a hydrogen bond to the active site Asp-27 and to the folate substrate. This single substitution leads to a 2.3 kcal/mol decrease in the binding affinity for folate ligands owing to both a decrease in the rate constant for binding and an increase in the rate of dissociation. In addition, the pK_a of Asp-27 increases from 6.5 to 7.05, and the pH-independent rate of hydride transfer is decreased 6-fold (1 kcal/mol), although the equilibrium constant for the chemical reaction at the active site is unchanged in the mutant. Owing to the multiplicity of effects, Thr-113 cannot be classified simply as contributing toward ground state or transition state binding. From this and related studies, it appears likely that strictly conserved active-site residues may exert multiple effects on key ground and transition states, and one may only speculate as to how and when they arose during evolution.

A series of mutations at residue Leu-54 provided a nearly horizontal free energy plot relating the rate of hydride transfer to the binding constant for dihydrofolate. This strictly conserved residue forms part of the hydrophobic wall which

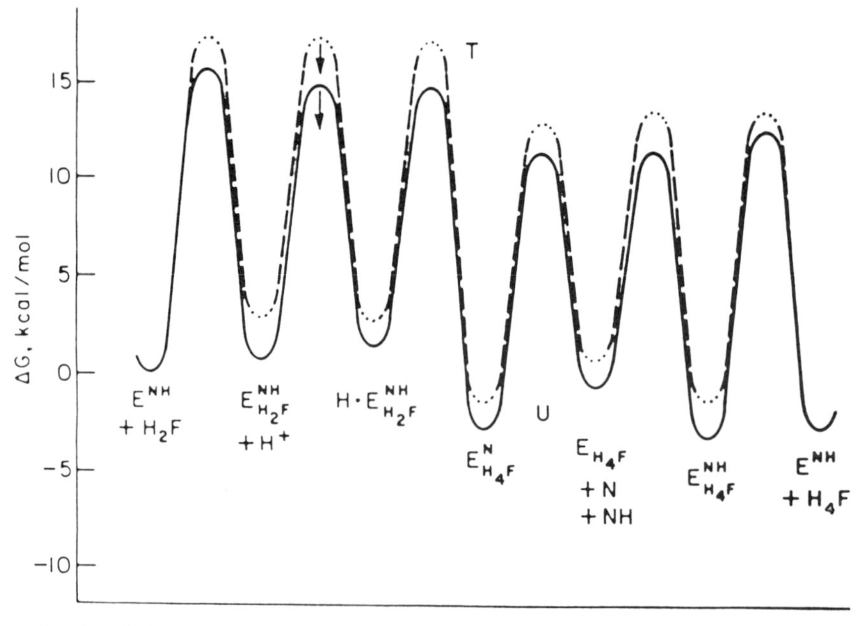

FIG. 14. Gibbs free energy coordinate diagrams for *E. coli* wild type (solid line) and for the Thr-113 → Val mutant (dashed line) of dihydrofolate reductase aligned at the substrate binary complex, E·NH, calculated for the following conditions: 1.0 m*M* NADPH, 1.5 m*M* $NADP^+$, 0.3 μM dihydrofolate, 13 μM tetrahydrofolate, and 0.1 *M* NaCl, pH 7.0, 25°C. [Reprinted with permission from Ref. (*65*).]

binds the *p*-aminobenzoyl side chain of dihydrofolate. Three mutants, Gly-54, Ile-54, and Asn-54, have been completely characterized (*74*). The binding of NADPH and $NADP^+$ is virtually unaffected, although the synergistic effect of NADPH on folate binding is lost in the Leu-54 → Ile mutant. The binding constant for dihydrofolate is weakened in each mutant, with effects varying between 10- and 1700-fold. Each mutant exhibited a 30-fold reduction in the rate of hydride transfer, independent of the K_d for dihydrofolate over the range spanned by the mutants, effectively uncoupling binding from catalysis. Thus, it is feasible that Leu-54 is strictly conserved because of its unique role in alignment of the dihydrofolate for reaction.

A quantitative contribution of alignment to catalysis has been addressed theoretically for dihydrofolate reductase, where calculations suggest an optimal carbon–carbon bond distance for hydride transfer of 2.6 Å. Extending this distance by 0.1 Å would increase the activation energy by 0.7 kcal/mol, whereas a 0.3-Å increase in bond distance would increase the activation energy by 5.0 kcal/mol (*70*). These translate to changes in rate of 3.3- and 4600-fold, respectively. Thus, changes of the order of less than 1 Å can account for relatively large changes in the rate of reaction caused by specific mutations.

6. *Key Residues in Catalysis*

There are certain residues that, when mutated, lead to enzymes largely or completely devoid of activity. In many instances, these residues are directly involved in acid–base or nucleophilic catalysis, and so the mutation merely confirms their crucial role. However, there are instances where mutation of a residue leads to loss of catalytic activity when there is no obvious role for that amino acid in the chemistry of the reaction. For example, mutations of Asp-176 in tyrosyl-tRNA synthetase have never led to an active enzyme. Because this residue binds the hydroxyl group of tyrosine at the active site, it serves to distinguish Phe from Tyr and is apparently a major determinant of the enzyme specificity. Substrate alignment may trigger a kind of cooperative process that organizes the active site for effective catalysis such that misalignment of a single residue can allow shifts in structure that propagate to weaken the effectiveness of the interaction of the transition state with other residues.

The substrate binding free energy can be decoupled from catalysis when the mutant residue leads to misalignment of the substrate away from the optimal configuration for reaction. A clear example of this is provided by the Arg-86 → Gln mutation in tyrosyl-tRNA synthetase. Gln binds the pyrophosphate in E·Tyr·ATP and E·Tyr-AMP·PP_i more tightly than Arg-86, but the rate of the reaction is decreased by nearly 4 orders of magnitude (*72*). Thus, the increased binding interaction between Gln-86 and pyrophosphate is nonproductive. The data argue that Arg-86 is a key interactive residue along the reaction coordinate and that proper alignment is crucial for effective translation of binding energy into catalytic efficiency. This example also illustrates the artifacts that can result when side chains containing functional polar groups are used as replacements in mutagenesis. A more quantitative assessment of the contribution of alignment to catalysis is discussed in the following section.

7. *Apparent Free Energy Contributions*

It is useful to take note of the magnitude of the apparent free energy changes in substrate binding or transition state stabilization observed for single amino acid substitutions. A summary of the observed changes for three of the enzymes previously discussed is given in Table III. The free energy contributed by a single hydrogen bond involving uncharged partners has been observed to fall in the range of 0.5–1.8 kcal/mol for nondisruptive amino acid substitutions in tyrosyl-tRNA synthetase. Greater changes are observed that can be rationalized via disruption of a hydrogen-bonding network, additional hydrophobic interactions, or steric effects, as noted. For example, the substitution Gln-173 → Ala in tyrosyl-tRNA synthetase disrupts a hydrogen-bonding network involving nearby amino acids, leading to a greater loss in binding affinity for the substrate.

Ionic bonds account for 1–1.5 kcal/mol in the two cases described for dihydrofolate reductase involving salt bridges to the phosphate and pyrophosphoryl moieties of NADPH. Ionic bonds involving transition state stabilization of the triphosphate moiety of ATP in tyrosyl-tRNA synthetase lead to considerably larger observed effects (*72*). These large effects have been attributed to the effects of bound water on hydrogen bonds involving charged residues (*67*). Alternatively, they may reflect the cooperativity of the process involving an induced fit type mechanism occurring in this portion of tyrosyl-tRNA synthetase, as described above.

Substitutions which remove hydrophobic interactions lead to a loss of 1–2 kcal/mol in the absence of further disruptive or steric effects. In substitutions of Leu-54 → Gly or Asn in dihydrofolate reductase, it is likely that the space caused by the substitution is filled by water in the mutant, leading to a greater loss of hydrophobic interaction energy.

It is important to note that any observed changes represent the small difference between opposing terms, including contributions from the bond in question, disruption of nearby amino acids, and the release of bound water (*68*). Nonetheless, these studies, taken together, indicate that hydrophobic, ionic, or hydrogen bonds each contribute 1–2 kcal/mol of interaction energy for each amino acid in the absence of dominating complications. Considerably larger changes are observed owing to a compounding of the effects near the site of the substitution, which can be rationalized in many cases. Certainly, this compounding of effects may be important for catalysis and specificity, and it seems reasonable to presume that enzymes have evolved to take advantage of such effects when the results are beneficial.

TABLE III

APPARENT FREE ENERGIES OF INTERACTION[a]

Substitution	$\Delta\Delta G$ (kcal/mol)	Comment	Enzyme	Ref.
Hydrogen bonds				
Uncharged				
His-48 → Gly	1.2	Ribose of ATP	TRS	*151*
Cys-35 → Gly	1.6	Ribose of ATP	TRS	*151*
Try-34 → Phe	0.7	Second H-bond donor	TRS	*151*
Trp-21 → Leu	3.5	Hydrophobic contribution?	DHFR	*135*
Gln-173 → Ala	4.5–5	Disrupts H-bond network	TRS	*152*
Thr-157 → Ala	1.4	Removes H bond, thermostability	Lysozyme	*10*
Thr-157 → His	2.1	Extra water in mutant	Lysozyme	*10*
Thr-157 → Ile	2.9	Additional steric effects	Lysozyme	*10*
Charged donor or acceptor				
Asp-38 → Ala	2–3	Disrupts H-bond network	TRS	*153*
Asp-78 → Ala	4–5	Disrupts H-bond network	TRS	*153*
Try-169 → Phe	2.6	NH_3^+ of tyrosine	TRS	*151*
Thr-113 → Val	2.9	Folate binding via Asp-27	DHFR	*65*
Hydrophobic bonds				
Phe-31 → Val	1.9	Aromatic interaction	DHFR	*66*
Phe-31 → Tyr	1.5	Steric effects	DHFR	*66*
Leu-54 → Ile	1.3	Phenyl ring of folate	DHFR	*74*
Leu-54 → Gly	4.4	Space filled by water?	DHFR	*74*
Leu-54 → Asn	3.5	Space filled by water?	DHFR	*74*
Ionic bonds				
Arg-44 → Leu	1.1	2′-Phosphate of NADPH	DHFR	*136*
His-45 → Gln	1.5	Pyrophosphoryl moiety of NADPH	DHFR	*136*
Lys-82 → Ala	3	Transition state stabilization	TRS	*72*
Lys-230 → Ala	3	Transition state stabilization	TRS	*72*
Lys-233 → Ala	4.5	Transition state stabilization	TRS	*72*

[a] Selected examples of apparent free energies of interaction attributable to single amino acid side chains are summarized for three enzymes, T4 lysozyme, dihydrofolate reductase (DHFR), and tyrosyl-tRNA synthetase (TRS) (see Figs. 1, 5, and 12, respectively, for structural information). In each case, the thermodynamic effect of the substitution was quantitated in terms of reduced binding affinity for the substrate or transition state, or reduced thermostability in the case of T4 lysozyme.

8. *Regulation of Enzyme Activity*

Aspartate transcarbamylase (aspartate carbamoyltransferase) is perhaps the best understood allosteric enzyme, and current work has taken advantage of the crystal structure to examine residues involved in the allosteric switch (*75–77*). The "240s loop" (residues 230–245) provides the key to the tertiary and quaternary structure change accounting for the concerted allosteric transition (Fig. 15). In the T state, the 240s loop is stabilized by interactions between Tyr-240 and

Asp-271 and by the intersubunit link between Glu-239 and both Lys-164 and Tyr-165; Arg-229 is bent out of the active site and interacts with Glu-272. In the R state, the 240s loop shifts to allow interactions between Glu-233 and both Lys-164 and Tyr-165 and between Glu-50 and both Arg-167 and Arg-234; the interaction between Arg-229 and Glu-272 is broken, and a new bond between Arg-229 and Glu-233 stabilizes the R state and helps to position Arg-229 to interact with the β-carboxylate of aspartate. Accordingly, the interaction of aspartate with Arg-229 breaks the Arg-229–Glu-272 interaction allowing movement of the 240s loop, which is transmitted to the neighboring subunit. This quaternary conformational change leads to loss of contacts that stabilize the T state and formation of new contacts to stabilize the R state.

Site-directed mutagenesis has allowed direct tests for the contributions of key

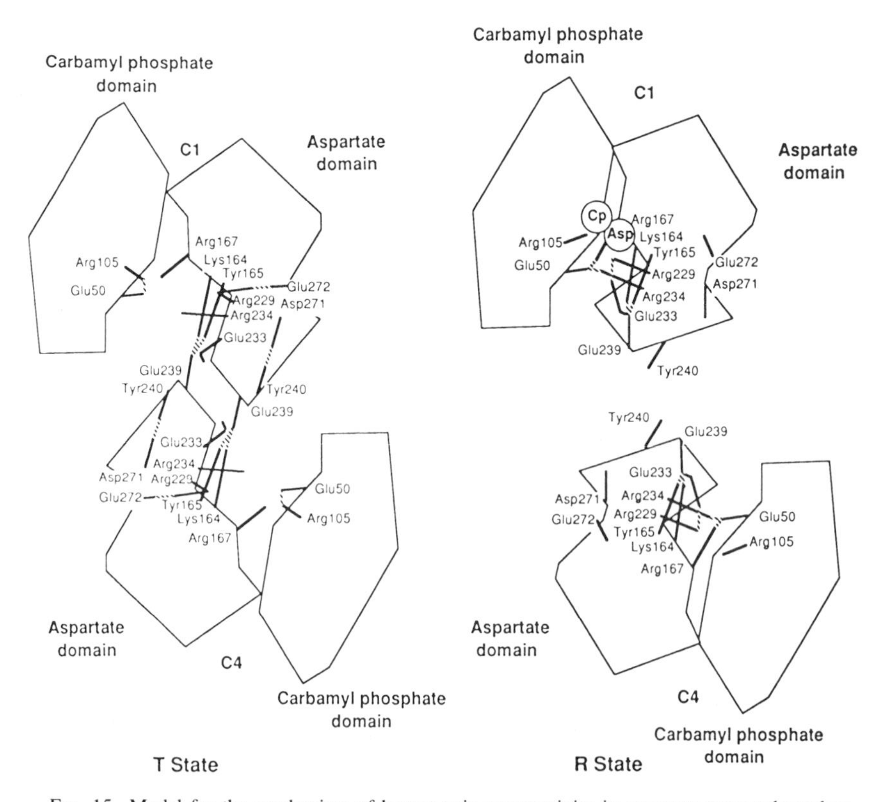

FIG. 15. Model for the mechanism of homotropic cooperativity in aspartate transcarbamylase. Shown schematically are the two extreme conformations in the T state and the R state. The binding of the substrates at one active site induces the domain closure in that catalytic chain and requires a quaternary conformational change which allows the 240s loops of the upper and lower catalytic chains to move to their final positions. The formation of the R state, in a concerted fashion, is further stabilized by a variety of new interactions as shown. [Reprinted with permission from Ref. (*1*).]

amino acids by characterization of mutants which led to stabilization of either the T or the R state (*75–77*). The mutation Glu-239 → Gln prevented interaction of Glu-239 with both Lys-164 and Tyr-165, thereby locking the enzyme in the high-activity R state and confirming the role of Glu-239 in stabilizing the T state. The Arg-229 → Ala mutant showed a 10,000-fold loss in maximal activity, but it retained substantial cooperativity, suggesting that Arg-229 is principally necessary for catalysis in binding aspartate. In contrast, the Glu-233 → Ser mutant exhibited only an 80-fold loss in maximal activity, but all substrate-induced cooperativity was lost. The bisubstrate analog PALA was able to activate the Glu-233 → Ser mutant, indicating that cooperative interactions still existed, although the free energy change for shifting from the T to the R state was too great to be overcome by the binding of aspartate. A Glu-272 → Ser mutant showed only slight reduction in maximal activity and reduced cooperativity, consistent with the role of Glu-272 in stabilizing the T state by positioning the side chain of Arg-229 to inhibit its interaction with the aspartate.

9. *Prospecting for Active Site Residues*

Until recently, knowledge of the tertiary structure has been prerequisite for a meaningful interpretation of data derived from mutagenesis experiments, primarily because of the nature of the questions posed. Nevertheless, there are situations where more intuitive tactics can lead to valuable insights, as illustrated by a collection of selected examples.

The human β_2-adrenergic receptor possesses an Asp-79 as part of a sequence domain that is conserved in various β-adrenergic, α-adrenergic (*78*), and muscarinic cholinergic receptors (*79, 80*) and all opsin proteins (*81*) sequenced to date. It has been proposed that the residue forms a counterion for the protonated Schiff base involved in retinaldehyde binding to opsin (*82*). The human β_2-adrenergic receptor has been stably expressed in a murine cell line lacking β-receptors and has shown the anticipated affinity for catecholamines, decreased affinity in the presence of guanine nucleotides, and isoproterenol stimulation of intracellular cyclic AMP levels. In contrast, cells bearing an Asp-79-substituted receptor had marked reductions (40 to 240-fold) in agonist binding, insensitivity to the presence of guanine nucleotides, and no coupling to intracellular cAMP levels (*83*). The abolition of the latter wild-type receptor properties suggests that the role of Asp-74 is not limited to agonist binding but extends to the coupling mechanism by which conformation changes arising from binding at the receptor are transmitted through this transmembrane segment to other proteins.

A second example is interleukin 1, which was first defined as a lymphocyte activity factor produced by macrophages and other cell types (*84, 85*). Interleukin activities are mediated by two distinct protease products (interleukin 1α and interleukin 1β) (*86, 87*). Plasmids containing interleukin 1β have been used to express both wild-type and mutant proteins in *E. coli* (*88*). The nucleotide sequence shows the presence of two cysteines located at amino acids 8 and 71.

Conversion of one or both to alanine does not result in a loss of the ability of the protein to inhibit the growth of melanoma cells. This finding, although undoubtedly somewhat disappointing to the investigators, nevertheless is readily interpretable; the conserved cysteines are not required for activity. In contrast, the implications of major activity loss in such prospecting mutagenesis experiments are by no means clear, as one must consider possible changes in protein conformation or stability.

A third example derives from the intense effort to understand viral infectivity. The major nucleocapsid protein of avian retroviruses, pp12, binds to virus RNA and is thought to be involved in packaging of the viral RNA (*89*). Site-directed mutagenesis was used to replace the lysine residues of pp12 that previous chemical modification studies had implicated as important in RNA binding (*90*). Single amino acid substitutions at Lys-36, Lys-37, and Lys-39 did not affect virus production; however, the double mutant Lys-36 → Ile, Lys-37 → Ile completely blocked viral replication. The isolated double-mutant pp12 protein has a low affinity for RNA; furthermore, a cell line expressing the mutant produces viral particles lacking RNA (*91*). The explanations for this lethality are numerous, as must be the case for studies of this type in the absence of tertiary structural data. Further experiments, however, led to an unexpected result. The phosphorylation of Ser-40 is necessary for RNA binding; thus, it was anticipated that its substitution by Ala would weaken or abolish RNA binding and, consequently, viral replication. However, substitution by Ala, Thr, Trp, Glu, or Gly did not alter RNA binding, suggesting that phosphorylation of serine removed a blocking effect rather than adding a specific salt–pair interaction critical for RNA binding (*92*).

The last example features the enzyme ribonucleotide reductase, which is thought to be composed of two subunits, B_1 and B_2, in a 1 : 1 complex, with four thiols at the active site (*93*). The conversion Cys-759 → Ser generated a mutant protein that could be tested for its ability to reduce cytosine diphosphate and to be coupled to an external reductant such as thioredoxin or dithiothreitol. The finding that the Ser-759 mutant retained the activity of the wild-type enzyme with dithiothreitol as the reductant unequivocally established that Cys-759 does not function in substrate reduction. A similar mutation at Cys-225 led to the unexpected self-cleavage of the B_1 subunit into fragments when the mutant protein was undergoing multiple turnovers, suggesting a drastic change in the nature of the reduction step.

The above examples should suffice to acquaint one with the main virtues of rapidly prospecting for influential binding-site residues. The importance of having both a direct physical as well as a biological assay to filter out artifacts cannot be overemphasized. The widespread use of this methodology encompasses the identification of the glycosaminoglycan attachment site in mouse invariant-chain proteoglycan core protein (*94*), amino acids important in the binding of human apolipoprotein E to its receptor (*95*), the implication of several amino acids involved in transport of ions across the colicin E1 ion channel (*96*), and the introduction of NMR-detectable amino acids in the linking domains of the pyruvate dehydrogenase multienzyme complex to monitor linking chain flexibility (*97*).

A question also arises as to whether certain key features of various catalytic mechanisms depend on factors that are not readily discerned by inspection of crystal structures. We have already noted the rearrangement of mobile loops in tyrosyl-tRNA synthetase. In another case, two cysteine proteases, actinidin and papain, were shown by crystallography to have similar active sites, each containing a reactive thiol and an imidazole (*98*). However, the two reacted differently toward a mechanism-based inhibitor which contained a disulfide capable of binding in the S_1–S_2 intersubsite regions of the active site of the enzyme, illustrated for papain in Fig. 16. The disulfide interchange reaction features a transition state that is pH sensitive owing to the $-S^-/IMH^+$ pair, the catalytically active

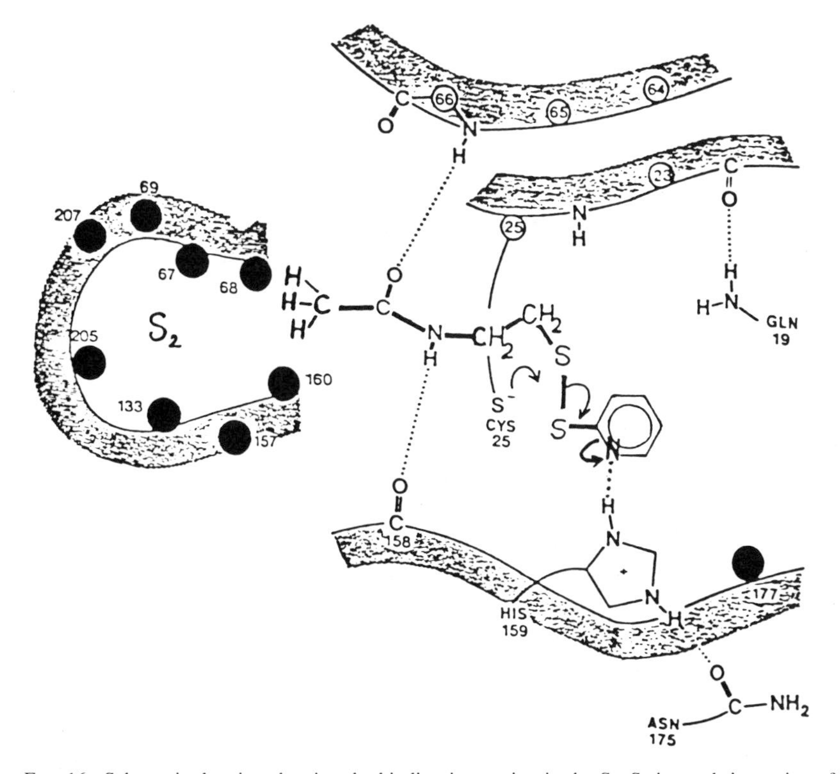

FIG. 16. Schematic drawing showing the binding interaction in the S_1–S_2 intersubsite region of papain postulated to provide for the transition state, involving nucleophilic attack by the thiolate anion of Cys-25 assisted by association of the pyridyl nitrogen atom of the leaving group with the imidazolium cation of His-159. [Reprinted with permission from Ref. (*99*).]

species. The reactions of this and other structurally related disulfide probes were carefully measured as a function of pH for both enzymes, revealing subtle differences in the pH-independent rate constant and controlling pK_a values. The differing response was attributed to a second ion pair His-81/Glu-52 that is remote from the active site in papain but absent in actinidin, where the analogous residue is an Asn (*99*). This also points to the potential importance of long-range electrostatic effects as discussed earlier.

A main challenge to practitioners of mutagenesis is to structurally manipulate an enzyme to introduce improved catalytic efficiencies or substrate affinity in the absence of three-dimensional structural information. Deletion of 414 amino acids from the carboxyl terminus of *E. coli* alanine-tRNA synthetase (alanine–tRNA ligase) gives a fragment, 461N, which catalyzes the adenylate synthesis reaction with the same efficiency as the native protein but has a substantially reduced k_{cat}/K_m ratio for tRNA in the aminoacylation reaction (*100, 101*). An Ala-409 → Val mutation partially compensates for the large deletion by having only a 5-fold lower value for k_{cat}/K_m (*100*), apparently through the addition of new compensatory interactions. Introduction of this same mutation into the full-length protein unaccountably improves the binding of tRNA relative to native protein (0.54 kcal/mol) to nearly the same extent as for the deletion (0.65 kcal/mol) (*102*). The effect on the k_{cat}/K_m ratio for the native protein is again 5-fold as noted in the deletion/mutant comparison. The generality of this method for engineering improved activities is at present unknown, but the increasing use of cassette mutagenesis methods should provide that information shortly (*103, 104*).

III. Engineering New Specificities

Naturally Occurring Protein Variants

We begin our analysis of the requirements and potential for engineering new enzymic specificities by considering the constructs derived by Nature in the course of evolution. We draw heavily from comparative studies of homologous proteins in order to discern the effect of amino acid changes on the kinetics, stability, and substrate specificity that constitute some of the primary determinants of biological behavior of that protein. Our focus is on classes of proteins deemed homologous on the basis of primary amino acid sequence adjustment, avoiding cases of borderline homology (*105, 106*). We sidestep the vexatious question in biological evolution as to whether changes in primary sequence within a protein family were governed by natural selection, implying a biological function, or were the result of random drift (*55, 107–110*).

The evolution of protein structure has been described in terms of movement across a surface of 20^n points corresponding to the 20 primary amino acids and *n*, the number of amino acids in the protein (*55*). Consequently, for a protein of only 100 amino acids, the surface is composed of some 10^{130} points. The high dimensionality of the surface results in a large number of paths between any two structures and ensures that at least some of the paths and intermediate structures between the two points will be allowed. Over the millennia, the rate of change has been influenced by environmental conditions or functional constraints. One can estimate an upper limit for this rate of 10^{-8}–10^{-9} changes/site/year by examining variations in DNA sequences such as pseudogenes or introns that do not themselves encode for catalytically active proteins (*111, 112*). This number is similar to the rate of divergence of albumins, a class of proteins generally thought not to be functionally constrained (*113*). Thus, presuming multicellular life to be present for over 100 million years, a structural divergence of 25% should be observed in the absence of restraints. Let us now examine how such changes are manifest in the primary sequence of several protein families.

1. *Effect of Changes in Primary Sequence on Secondary and Tertiary Structure*

Lesk and Chothia (*114*) have examined nine different globins in which sequence similarity ranges from 16–90%. Although the nine sequences vary in length between 136 and 154 amino acids, there are, taking into account insertions and deletions, a total of 171 possible residue positions. In only 116 of these positions do all nine globins contain amino acids. Crystal structures have been obtained for all the molecules. The structural patterns of two globins with only 16% sequence similarity, namely, human deoxyhemoglobin α strand and *Chironomus* erythrocruorin, are depicted in Fig. 17 (*115–117*). Despite the very different amino acid sequences, the two globins are remarkably similar in their secondary and tertiary structures.

Eight helices labeled A through H are common to all globins, with the exception that in some cases the small D helix is lacking (Fig. 17). The helices assemble in a common pattern, enclosing the heme group in a pocket of similar geometry, with the 116 amino acids being divided nearly equally between residues that form no intramolecular contacts and those that form helix-to-helix or helix-to-heme contacts. Only 5 positions of the 116 common ones have the same residues; two of these are involved in the binding of the heme to the protein and may be functionally constrained, and others lying on the surface apparently are not. Of the contact positions (59 amino acids), 50% are buried to form a hydrophobic core of nearly constant volume (± 15 Å^3) despite a mean variation in overall protein volume of 56 Å^3 (*118*). Thus, individual variations owing to variations in primary sequence cancel out in the critical core region.

Although the arrangements of the helices in the globins are topologically similar, changes in the volumes of the buried residues cause shifts in the relative

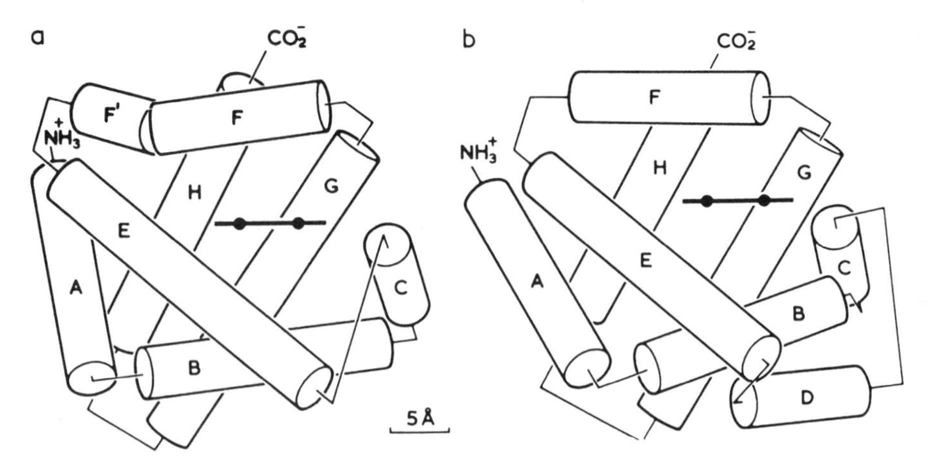

FIG. 17. Structural pattern of the globins. Cylinders represent the helical regions of the α strand of (a) human deoxyhemoglobin and (b) *Chironomus* erythrocruorin. [Reprinted with permission from Ref. (*114*).]

positions of homologous helices. In regions of contact, the values of interaxial distance and angle between helices vary by as much as 7 Å and up to 30°. In contrast, portions of the C, E, I, and G helices which form the sides and bottom of the heme pocket shift no more than 2–3 Å relative to the heme so that the changes in helix packing are not independent but coupled to maintain the structure of the heme pocket. These changes plus slight movements in the position of the heme lead to the maintenance of an identical binding site geometry across the globin family by coupling all shifts in the atomic loci. Chothia and Lesk have lately extended elements of their analysis to 226 globin sequences, finding only two residues that are absolutely conserved in all sequences; nevertheless, the proteins share the same basic three-dimensional structure.

In contrast to the case for globins, a similar analysis of the structures for various *c* cytochromes, namely, *c* (tuna), c_2 (*Rhodospirillum rubrum*), *c*-550 (*Paracoccus dentrificans*), and *c*-551 (*Pseudomonas aeruginosa*), reveals the presence of alternative structures for the heme pocket. These proteins contain 48 residues identifiable as homologous from superposition of the structures. The other 54 to 64 residues are in loops that vary greatly in sequence, length, and conformation or in helices that are found in only some of the structures (*119*). A section parallel to and including the heme plane for cytochromes *c* and *c*-551 (Fig. 18) illustrates the different means of forming the propionic acid side of the heme pocket. In cytochrome *c*, residues 39–79 form the pocket; in cytochrome *c*-551, residues 34–59 are visible with a differing packing pattern. Obviously these changes do not impede the electron transfer function of the cytochromes, which apparently have less spatial and geometric demands than the globins.

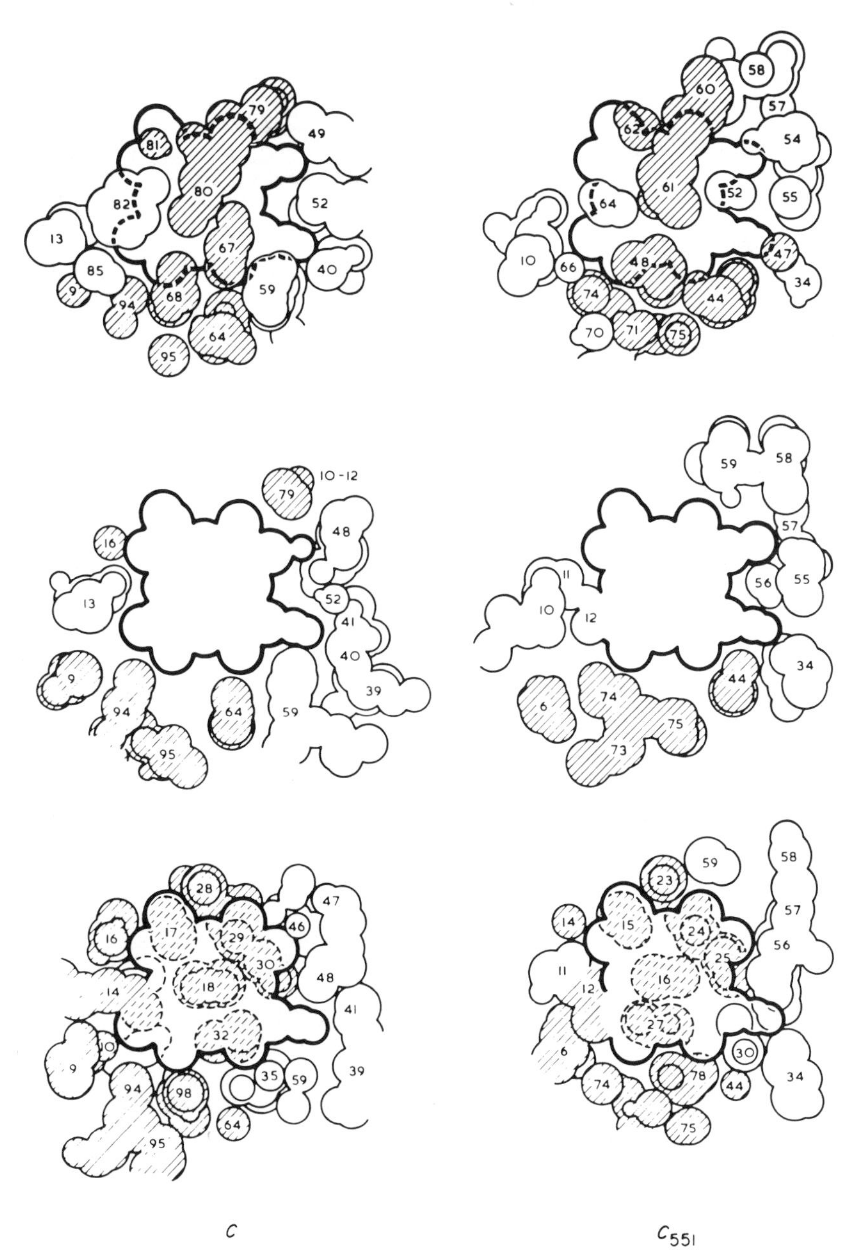

FIG. 18. Packing of residues about the heme group in *c* cytochromes. Serial sections, parallel to the heme plane, have been cut through van der Waals envelopes of atoms in cytochromes *c* and *c*-551. Three superposed sections 1 Å apart are shown: above (*top*), including (*middle*), and below (*bottom*) the heme plane. Shaded residues are part of the core structure. The shifts in the positions of the α helices, relative to the heme, can be seen by the different positions of homologous residues 64, 68, 94, and 98 in cytochrome *c* and 44, 48, 74, and 78 in cytochrome *c*-551. [Reprinted with permission from Ref. (*158*).]

2. *Rates of Change of Structural Elements during Evolution*

The implicit conclusion that tertiary structure changes more slowly than the primary amino acid sequences has been broadened by the comparison of core regions of 25 proteins derived from 8 different protein families. Core regions were defined in pairs of proteins by (i) superimposing the main-chain atoms comprising secondary structural elements such as α helices or β sheets and (ii) extending the superposition to include additional atoms at both ends (*120*). The extension was continued as long as the deviations in the positions of the residues in the last residue included were no greater than 3 Å. The collection of such regions defines a common fold in both proteins termed the core region. The extent of the structural divergence of two homologous proteins being compared was measured by superimposing their common cores and calculating the root mean square difference, Δ, in the position of their main-chain atoms. The results for selected examples are shown in Table IV.

The value of Δ, in turn, is exponentially related to the percentage of identical core residues so that a protein structure will conservatively provide a close general structural model for other proteins if the sequence similarity exceeds 50%. Consequently, there is an increasing use of this type of relationship to predict tertiary structure and secondary elements from primary sequences. Nevertheless, there are cases, such as the *Lactobacillus casei* and *E. coli* dihydrofolate reductases, where the primary sequence similarity is only 30% yet Δ is only 1.3 Å (*121*).

It is apparent from these and other cases that tertiary structure resists change despite numerous substitutions in primary amino acid sequence. In fact, there is increasing evidence that there are a small number of tertiary structural types; for

TABLE IV

CORES OF HOMOLOGOUS PROTEINS: SIZE, FIT, AND RESIDUE IDENTITY

Family	Protein pair	Number of residues	Residues in common core	Δ	% Identical residues
Globin	Human β-deoxyhemoglobin erythrocruorin	151:136	122	2.28	15
Cytochrome *c*	Tuna *c*: bacterial c_2	103:112	99	1.13	37
Serine proteases	Bovine γ-chymotrypsin: bovine trypsin	236:222	203	0.99	47
Immunoglobulin domains	Vλ2:Cλ	110:99	48	1.47	13
Lysozyme	Human:hen egg white	130:129	128	0.70	61
Thiolesterase	Papain:actinidin	212:218	206	0.77	49

example, the catenoid *n* β-barrel structure of triose-phosphate isomerase serves as the scaffolding for diverse reaction types (*122*). The bewildering number of substitutions must disguise residue changes that are neutral or conservative with respect to a specific position or coordinated across several positions. Methods are now being developed to locate coordinated changes in homologous sequences (*123, 124*). These require several sequences of proteins within a family and at least one crystallographic structure. The analysis depends on sequence alignment, followed by encoding of the amino acids at given positions in order to discern which combination of positions encodes identical patterns of amino acids. In a second step, the groupings are located in the crystallographic structure and the minimum distance between side chain atoms measured to identify spatially close pairs. The distance distributions of such pairs is further analyzed to exclude patterns that would arise randomly. The results at present are most significant when comparing amino acids whose side chains are less than 5 Å apart. Examples of complementary coordinated changes in the cysteine proteases are Val-32–Gly, Ala-162–Leu, Val-164–Leu, and Tyr-186–Phe. Less obvious is the pairing of substitutions Val-32–Gln and Ala-162–Arg unless compensated by other substitutions not detected by this method. It is necessary that such analyses be refined in order to utilize the enormous information latent in protein families for understanding the forces that stabilize secondary and tertiary structure so as to guide the engineering of proteins that retain structural stability but have altered specificities and, it is hoped, novel catalytic properties.

3. *Effect on Specificity and Dynamics*

There are many examples of a single natural variant which leads to altered binding properties. Sickle cell anemia results from a single amino acid change in hemoglobin causing the $\alpha_2\beta_2$ units to self-associate to form linear polymers (*125*). Pancreatic and seminal ribonucleases have 81% structural similarity, yet the former acts only on single-stranded nucleic acid whereas the latter is a dimer and acts on either single- or double-stranded substrates (*126–128*). The three alcohol dehydrogenase isozymes from yeast show 95% structural similarity and yet have different substrate specificities (*129*). Further changes in sequence similarity often result in changes in catalytic mechanisms as well as stereospecificity. The alcohol dehydrogenases from yeast and *Drosophila* show limited sequence similarity (25%) but transfer either the *pro-S* or *pro-R* hydrogen from the cofactor respectively. Furthermore, the enzyme from yeast utilizes a Zn^{2+} ion in catalysis, and the *Drosophila* enzyme does not (*130*). These data and others have led Benner and Ellington to propose an evolutionary clock defining the time required to achieve a desired change in a functional property (*55*). They suggest the following hierarchy of increasing evolutionary time required to alter a given property: binding $\cong$ turnover $<$ stereospecificity $\cong$ catalytic mechanism $<$ tertiary structure. It is reasonable to suppose that our efforts to achieve new enzyme activities by protein engineering will follow a similar pattern.

It would be quite useful to be able to sift through the mass of primary sequence data and identify residues involved in catalysis, defined earlier as those groups participating directly in the chemical step either as acid–base or nucleophilic catalysts (*131*). These residues might be identified by their appearing in localized regions of high sequence conservation that ensures their correct orientation at the active site. Alignment algorithms that examine a 4-Å sphere around known catalytic residues in a given enzyme family have shown a high degree of sphere homology (Table V) (*132, 133*). Apart from Glu-35 in lysozyme, His-119 in ribonuclease, and Arg-171 in lactate dehydrogenase, the values were over 0.8. These findings probably reflect to some extent the localized functional constraints on the protein structure. Other features of the catalytic residues that emerge from this analysis are not unexpected: histidine is 10 times more abundant at an active site than at large; the catalytic residues are not confined to a particular structural element; and the active site amino acids are generally less exposed than other polar amino acids.

The above analysis may be a corollary of a more extensive hypothesis, namely, that active-site surface for a given reaction is conserved. The dynamics of turnover for dihydrofolate reductase (*134, 135*) from *E. coli* and *L. casei* underscore the complexity of predictive active-site analysis. A close inspection of the active sites of the two enzymes showed that there was a low degree of

TABLE V

CONSERVATION OF SEQUENCE NEAR CATALYTIC RESIDUES[a]

Enzyme	Homology	Catalytic Residues	Sphere
Lysozyme	0.8	Glu-35	0.74
		Asp-52	0.94
Ribonuclease	0.6	His-12	0.89
		His-119	0.76
Lactate dehydrogenase	0.6	Arg-109, -171	0.86, 0.78
		His-195	0.84
Triose-phosphate isomerase	0.7	Glu-165	0.95
Trypsin	0.5	His-87	0.97
		Asp-102	0.94
		Ser-195	1.00
Superoxide dismutase	0.6	His-61	0.86
		Arg-141	0.81
Phospholipase A_2	0.4	His-48	0.85
Glyceraldehyde-phosphate dehydrogenase	0.7	Cys-149	0.95
		His-176	0.82

[a] Extent of sequence conservation at each position in the alignment was evaluated by assigning each amino acid a conservation value that decreases from 1.0 (the residue is invariant in all sequences) to 0.0 (the residue differs in all sequences) (*153*).

primary amino acid sequence similarity for the amino acids that form the active site pocket, similar to the degree of homology of the enzymes as a whole (*70*). Nonetheless, in each species, a single Asp functions as the sole acid–base residue, and the two active site regions could be superimposed by aligning the structures of the methotrexate complexes (Fig. 19). All residues within 7 Å of any point on the methotrexate surface were enjoined, forming an ensemble of 40 amino acids. Amino acids that either were identical or had only backbone interactions, with their side chains oriented away from the inhibitor/substrate binding

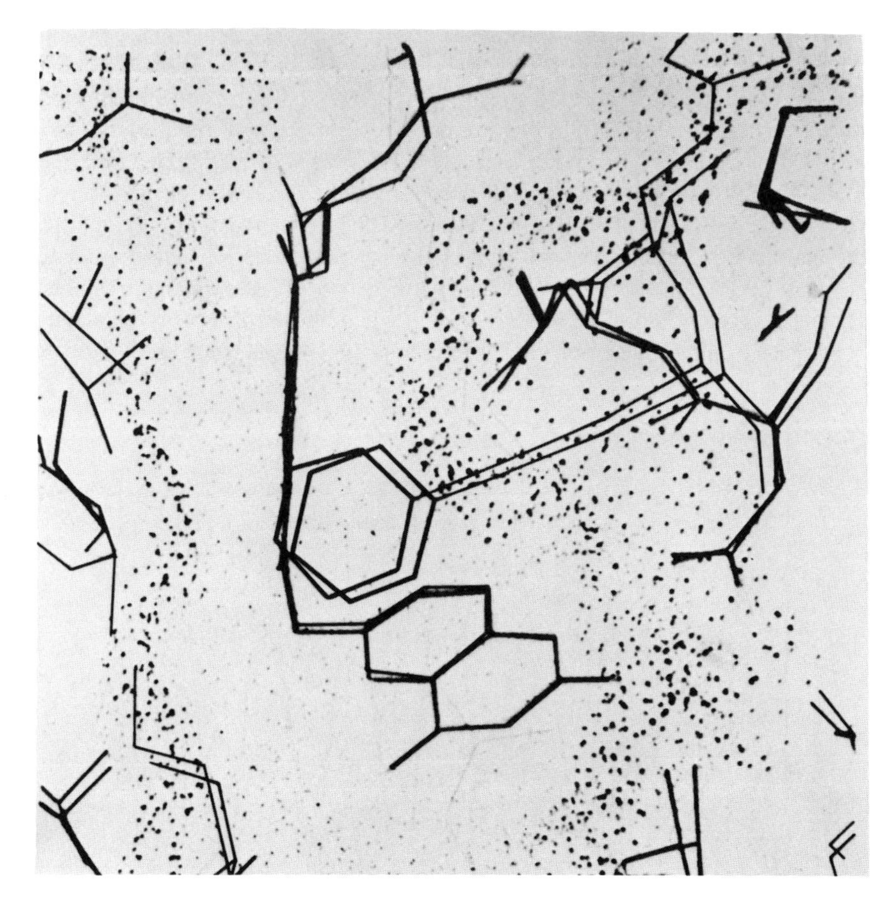

FIG. 19. Superpositioning of the active sites of the *E. coli* and *L. casei* dihydrofolate reductase with methotrexate as a center for alignment. All residues within 5 Å of the methotrexate are included. Key residues that are visible in the figure include Leu-28, Leu-27 (upper right); Asp-27, Asp-26 (middle right); Ala-7, Ala-6 (lower right); Thr-46, Thr-45 (lower left); Ile-50, Phe-49 (middle left); and Arg-52, Leu-51 (upper left). Phe-31 can be seen at center. See also Fig. 5. [Reprinted with permission from Refs. (*70*) and (*71*).]

site, were deemed homologous. By these standards, 18 amino acids (40%) in this ensemble differ in the two enzyme species. Similar analysis employing NADPH as the means of alignment has also been performed (*136*). It is apparent from the alignment of the two sites that, despite the loss in sequence similarity, the surfaces of the two active sites constructed from their van der Waals envelopes are remarkably similar. The important lesson from this analysis is that it is these surfaces which are conserved and not the amino acids.

A complete kinetic scheme has been established for the enzyme from both sources. The *L. casei* dihydrofolate reductase followed a reaction sequence identical to the *E. coli* enzyme (Scheme I); moreover, none of the rate constants varied by more than 40-fold! Figure 20 is a reaction coordinate diagram comparing the steady-state turnover pathway for *E. coli* and *L. casei* dihydrofolate reductase, drawn at an arbitrary saturating concentration (1 m*M*) of NADPH at pH 7. The two main differences are (i) *L. casei* dihydrofolate reductase binds NADPH more tightly in both binary (E·NH, −2 kcal/mol) and tertiary ($E{\cdot}NH{\cdot}H_2F$, −1.4 kcal/mol; $E{\cdot}NH{\cdot}H_4F$, −1.8 kcal/mol) complexes, and (ii) the internal equilibrium constant ($E{\cdot}NH{\cdot}H_2F \rightleftharpoons E{\cdot}N{\cdot}H_4F$) for hydride transfer is less favorable for the *L. casei* enzyme (1 kcal/mol). These changes, as noted later, are smaller than those observed for single amino acid substitutions at the active site of either enzyme. Thus, the overall kinetic sequence as well as the

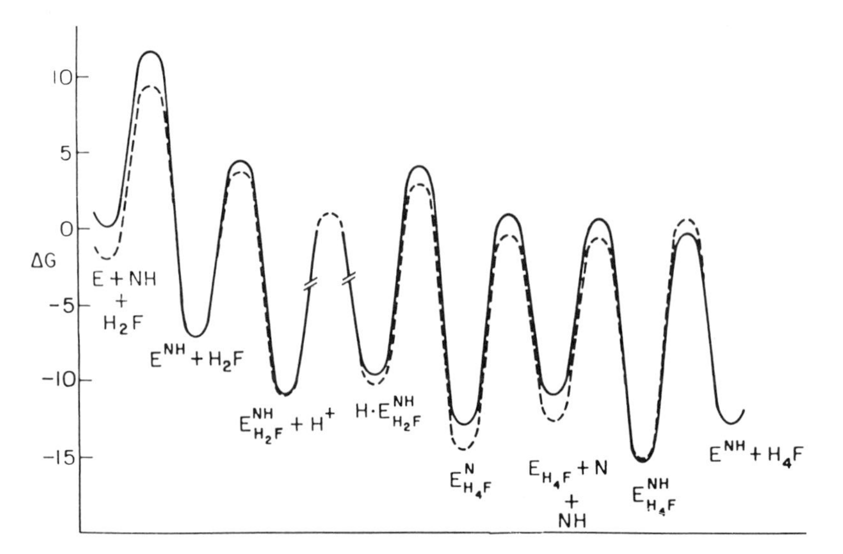

FIG. 20. Gibbs free energy coordinate diagrams for *E. coli* and *L. casei* dihydrofolate reductase aligned at the substrate ternary complex, $E{\cdot}NH{\cdot}H_4F$. Conditions are as in Fig. 14. Solid line, *L. casei*; dashed line, *E. coli*. [Reprinted with permission from Ref. (*70*).]

magnitude of the individual steps appear to reflect the common active-site surface, formed as an ensemble of compensatory and identical side chains. A similar comparative kinetic and structural analysis of triose-phosphate isomerase from chicken and yeast revealed nearly identical kinetic properties despite differences of 50% in their overall amino acid sequences (*137*). However, this perspective of the importance of the dimensions and surface of the active-site cavity does not per se rationalize the observation that the *L. casei* enzyme slowly catalyzes the reduction of the parent folic acid whereas the *E. coli* enzyme does not (*135*). The reduction of folate to tetrahydrofolate occurs with hydride transfer from the 4-*pro-R* position of NADPH to the *si* face of the pteridine in both cases (*138*).

4. *Engineering New Substrate Specificity*

The construction of enzymes with new substrate specificities is now a realistic goal, and some novel approaches have been presented. For example, removal of an active-site histidine by the His-64 → Ala mutation in subtilisin results in an enzyme with markedly reduced activity, but one which can be enhanced 400-fold with substrates containing histidine at the P1 site (*139*). Apparently, the substrate histidine assists catalysis by partially compensating for the role of the lost active-site His-64. In a similar study, mutation of Lys-258 to Ala in aspartate aminotransferase produces an enzyme whose activity can be restored by exogenous amines (*140*).

Holbrook and co-workers have successfully engineered lactate dehydrogenase to shift its substrate specificity from preference for the lactate ⇌ pyruvate reaction toward the malate ⇌ oxaloacetate reaction (*141*). The wild-type enzyme discriminates between the two keto acids, favoring pyruvate over oxaloacetate by a factor of 1000 in terms of k_{cat}/K_m. As shown in Fig. 21, this discrimination between a methyl group and a carboxymethyl group relies on contacts with Gln-102 and Thr-246. The double mutant Gln-102 → Arg, Thr-246 → Gly provides the required positive charge and space for binding the additional carboxylate in oxaloacetate. Accordingly, the mutant enzyme prefers oxaloacetate over pyruvate by a factor of 8400, leading to an overall shift in specificity of 8.4×10^6. This change in specificity can be understood in terms of three factors contributing to substrate recognition: (1) the maintenance of overall charge balance in the enzyme–substrate complex, (2) the congruence of substrate and active-site volumes, and (3) the electrostatic complementarity between the protein and ligand surfaces (*141*).

Subtilisins are serine proteases of broad specificity, although the order of preference for the P1 residue (the amino acid adjacent to the site of cleavage in the substrate) varies from one species to the next. Subtilisins from *Bacillus licheniformis* and *Bacillus amyloliquefaciens* differ in protein sequence by 31% and show a factor of 10–60 variation in substrate specificity. For 19 residues that are

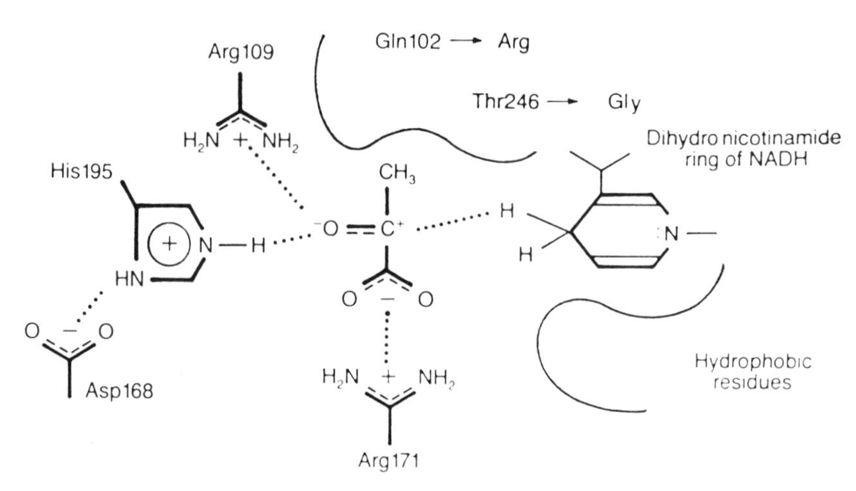

FIG. 21. Schematic summary of the mechanism of lactate dehydrogenase showing the active site with bound substrate and coenzyme in the proposed transition state, along with the residues in the protein which are important in this interaction. The transition state more closely resembles the NADH–pyruvate structure, but with a highly polarized, reactive carbonyl bond in the substrate. [Reprinted with permission from Ref. (*141*).]

within van der Waals contact of the model substrates in the crystal structures, the only differences are at residues 156 (Glu versus Ser) and 217 (Tyr versus Leu). A third residue at position 169 (Gly versus Ala) is within 7 Å. When the three residues of *B. licheniformis* were substituted into *B. amyloliquefacients* subtilisin, the substrate specificity of the triple mutant approached that of the *B. licheniformis*.

There are few surprises in these cases. In each study, the substitutions were firmly grounded on substitutions suggested by examination of a homologous protein. Nonetheless, these are clear examples of the potential for subtle modifications of substrate specificity.

5. *Hybrid Enzymes*

It is apparent that Nature has constructed enzymes with new specificities by stitching together domains with the desired ligand-binding properties (*142*). Following the lead given by Nature, efforts are now underway to construct novel enzymes by making a hybrid of two related proteins.

New hybrids of plasminogen activator have been constructed in attempts to increase the specificity toward fibrin polymers and thereby improve the therapeutic value for treatment of coronary artery thrombosis (*143, 144*). Plasminogen activators are serine proteases that act on plasminogen to release plasmin, also a serine protease with broad specificity, which then degrades the fibrin polymers. In addition to the protease domain, plasminogen activators contain a distinct fibrin-binding domain, all within a single polypeptide; they are thought to have evolved by an "exon shuffling" mechanism by recombination of adjacent exons encoding autonomous structural and functional domains (*142*). Attempts to improve specificity have been based on recombination of the autonomous functional units involving different proteases and fibrin-binding domains to make new hybrids and have had modest success (*143, 144*).

The recognition of distinct domains with separate functions has been instrumental in the design and construction of DNA-binding proteins with altered specificities. A kind of semihybrid was formed by replacing the amino acids on the outside surface of the DNA recognition α helix of the 434 repressor with the corresponding amino acids from the recognition helix of P22 repressor (*145*). This construction maintained the internal contacts of the residues of the α helix while altering the DNA-binding surface in a predefined manner. The binding specificity of the hybrid protein followed that of the P22 repressor. This approach has proved to have general applicability (*146*) and has allowed conversion of a transcriptional inhibitor into an activator (*90*).

Loop structures and β turns play a significant role in determining the conformation of enzyme active sites by linking together elements of secondary structure. The ability to use β turns as a cassette to join structural domains would be important for protein design. Modeling studies have indicated that the conformations of short segments can be predicted when the ends of the segment are constrained to join the flanking domains (*147*). The effect of substituting one β-turn geometry for another has been evaluated by constructing a hybrid where the five-residue type I β turn of concanavalin A replaced the type I′ β turn in staphylococcal nuclease (*10*). This represented a most favorable case because structural studies had indicated that the β strands leading away from the β turns could be coaligned in the two structures. The hybrid nuclease showed "full" enzymic activity, but its stability was reduced. The decreased stability could be understood in terms of the partial exposure to solvent of two hydrophobic residues adjacent to the turn. Importantly, the crystal structure of the hybrid protein showed that the backbone hydrogen-bonding contacts in the turn were conserved.

The active-site cavities of the murine, avian, *E. coli,* and *L. casei* dihydrofolate reductases show similar surface contours, yet the bacterial enzymes only slowly reduce folate and are inhibited to a greater extent by trimethoprim (*148*). One notable structural difference is the greater length of an internal loop element in the avian and murine dihydrofolate reductases in a sequence spanning residues 51 to 59 (in *E. coli*) and connecting the C α helix and β sheet that flank the aromatic side chain of the ligands. The influence of this loop on the enzymic activity has been assessed by insertion of the following extended segment to replace amino acids 51–59 in *E. coli* dihydrofolate reductase (*149*):

Pro Glu Lys Asn Arg Pro Leu Lys Asp Arg Ile Asn
Gly^{51} — — — Arg Pro Leu Pro Gly Arg Lys Asn^{59}

The semihybrid *E. coli* enzyme showed properties remarkably similar to the avian enzyme: (i) the K_d value for folate and dihydrofolate decreased from 2000 and 220 μM to 165 and 25 μM, respectively, compared to a K_d for avian dihydrofolate reductase of 67 and 14 μM for folate and dihydrofolate, respectively; (ii) the ratio of the rates of hydride transfer and tetrahydrofolate release at pH 6.0 and 9.0 were maintained; and (iii) the maximal activity of the enzyme was retained at 9 versus 6–12 sec^{-1}. However, the ability to reduce folate was only marginally increased by 50%. It is striking that the introduction of additional residues that might be expected to lead to weaker substrate binding instead strengthen the affinity. Moreover, the tighter binding occurred without loss of catalytic efficiency, thus requiring similar decreases in the transition state energy for the critical rate-limiting steps.

IV. Conclusions

Site-specific mutagenesis has proved to be a powerful tool; it has become the principal means of dissecting the active site of an enzyme to identify and quantitate the role of specific amino acids in ligand binding, specificity, and catalysis. Some of the more noteworthy triumphs are the unambiguous identification of active-site residues, the evaluation of hydrophobic and hydrogen bond strengths, the quantitation of electrostatic contributions toward catalysis, the definition of active sites at subunit interfaces, the evaluation and improvement of enzyme stability, and dissection of the subtle interplay between substrate binding and transition state stabilization.

In more and more cases, mutagenesis has extended the crystallographic view of the active site by revealing the importance of movement of remote residues into the catalytic process, concomitantly verifying the transition state theory that underpins our rationale for enzymic catalysis (*150*). The successes and many unpublished failures have underscored the delicate yet powerful precision of active-site geometry, and they have suggested that efficient catalysis is a matter of tenths of an angstrom in the juxtapositioning of side-chain residues and substrates. Moreover, this precise alignment is achieved in balance with a retention of considerable chain flexibility so that ligands can come and go at reasonable rates.

The examination of protein families reveals that these active-site structures can be constructed from a variety of primary amino acid sequences but fewer secondary and tertiary structural elements. What these combinations are we are only beginning to understand. Initial, halting steps have been taken to progress from disassembly to assembly of proteins with new specificities and improved activities or stability. Given the already high efficiency of enzymes toward their natural substrates, progress should come more easily in engineering new specificities. To date, the kinetic properties of hybrid proteins have not been remarkable, except for the fact that the proteins function at all. Clearly, this is the area with the most challenge and the most promise.

REFERENCES

1. Kantrowitz, E. R., and Lipscomb, W. N. (1988). *Science* **241,** 659.
2. Miles, E. W., Kawasaki, H., Ahmed, S. A., Morita, H., Morita, H., and Nagata, S. (1989). *JBC* **264,** 6280.
3. Profy, A. T., and Schimmel, P. (1988). *Prog. Nucleic Acid Res. Mol. Biol.* **35,** 1.
4. Kato, H., Tanaka, T., Nishioka, T., Kimura, A., and Oda, J. (1988). *JBC* **263,** 11646.
5. Gerlt, J. A. (1987). *Chem. Rev.* **87,** 1079.
6. Shaw, W. V. (1987). *BJ* **246,** 1.
7. Knowles, J. R. (1987). *Science* **236,** 1252.
8. Katz, B., and Kossiakoff, A. (1986). *JBC* **261,** 15480.
9. Alber, T., and Matthews, B. W. (1987). "Methods in Enzymology," Vol. 154, pp. 511–533.
10. Alber, T., Sun, D. P., Nye, J. A., Muchmore, D. C., and Matthews, B. W. (1987). *Biochemistry* **26,** 3754.
11. Villafranca, J. E., Howell, E. E., Oatley, S. J., Xuong, N., and Kraut, J. (1987). *Biochemistry* **26,** 2182.
12. Brown, K. A., Brick, P., and Blow, D. M. (1987). *Nature (London)* **326,** 416.
13. Howell, E. E., Villafranca, J. E., Warren, M. S., Oatley, S. J., and Kraut, J. (1986). *Science* **231,** 1123.
14. Matthews, B. W. (1987). *Biochemistry* **26,** 6885.
15. Grutter, M. G., Gray, T. M., Weaver, L. H., Wilson, T. A., and Matthews, B. W. (1987). *JMB* **197,** 315.
16. Gray, T. M., and Matthews, B. W. (1987). *JBC* **262,** 16858.
17. Hibler, D. W., Stolowich, N. J., Reynolds, M. A., Gerlt, J. A., Wilde, J. A., and Bolton, P. H. (1987). *Biochemistry* **26,** 6278.
18. Colton, F. A., Hazer, E. E., and Legg, M. J. (1979). *PNAS* **76,** 2551.
19. Wilde, J. A., Bolton, P. H., DellAcqua, M., Hibler, D. W., Pourmotabbed, T., and Gerlt, J. A. (1988). *Biochemistry* **27,** 4127.
20. Hibler, D. W., Stolowich, N. J., Reynolds, M. A., Gerlt, J. A., Wilde, J. A., and Bolton, P. H. (1987). *Biochemistry* **26,** 6278.
21. Matthews, B. W., Nicholson, H., and Becktel, W. J. (1987). *PNAS* **84,** 6663.
22. Shortle, D., and Meeker, A. K. (1986). *Proteins* **1,** 81.
23. Anfinsen, C. B. (1973). *Science* **181,** 223.
24. Villafranca, J. E., Howell, E. E., Voet, D. H., Strobel, M. S., Ogden, R. C., Abelson, J. N., and Kraut, J. (1983). *Science* **222,** 782.
25. Perry, L. J., and Wetzel, R. (1984). *Science* **226,** 555.
26. Wells, J. A., and Powers, D. B. (1986). *JBC* **261,** 6564.
27. Richardson, J. S. (1981). *Adv. Protein Chem.* **34,** 167.

28. Thornton, J. M. (1981). *JMB* **151,** 261.
29. Matsumura, M., and Matthews, B. W. (1989). *Science* **243,** 792.
30. Hecht, M. H., Sturtevant, J. M., and Sauer, R. T. (1986). *Proteins: Struct. Funct. Genet.* **1,** 43.
31. Imanak, T., Shibasaki, M., and Takagi, M. (1986). *Nature (London)* **324,** 695.
32. Chothia, C. (1974). *Nature (London)* **248,** 338.
33. Matsumura, M., Becktel, W. J., and Matthews, B. W. (1988). *Nature (London)* **334,** 406.
34. Rose, G. D. (1987). *Proteins* **2,** 79.
35. Alber, T., Bell, J. A., Sun, D. P., Nicholson, H., Wozniak, J. A., Cook, S., and Matthews, B. W. (1988). *Science* **239,** 631.
36. Ahern, T. J., Casal, J. I., Petsko, G. A., and Klibanov, A. M. (1987). *PNAS* **84,** 675.
37. Waley, S. G. (1973). *Biochemistry* **135,** 165.
38. Lu, H. S., Yuan, M. P., and Gracy, R. W. (1984). *JBC* **259,** 11958.
39. Stauffer, C. E., and Etson, D. (1969). *Biol. Chem.* **244,** 5333.
40. Estell, D. A., Graycar, T. P., and Wells, J. A. (1985). *JBC* **260,** 6518.
41. Wright, C. S., Alder, P. A., and Kraut, J. (1969). *Nature (London)* **221,** 235.
42. Dayhoff, M. O., Schwartz, R. M., and Orcutt, B. C. (1978). *Natl. Biomed. Res. Found.* **5,** 345.
43. Wong, C. H. (1989). *Science* **244,** 1145.
44. Pantoliano, M. W., Whitlow, M., Wood, J. F., Rollence, M. L., Finzel, B. C., Gilliland, G. L., Poulos, T. L., and Bryan, P. N. (1989). *Biochemistry* **27,** 8311.
45. Nicholson, H., Becktel, W. J., and Matthews, B. W. (1988). *Nature (London)* **336,** 651.
46. Carter, P., and Wells, J. A. (1988). *Nature (London)* **332,** 564.
47. Sprang-Standing, T., Fletterick, R. J., Stroud, R. M., Finer-Moore, Xuong, N. H., Hamlin, R., and Rutter, W. J. (1987). *Science* **237,** 905.
48. Rogers, G. A., and Bruice, T. C. (1974). *J. Am. Chem. Soc.* **96,** 2473.
49. Warshel, A., Naray-Szabo, G., Sussman, F., and Hwang, J.-K. (1989). *Biochemistry* **28,** 3629.
50. Raines, R. T., Sutton, E. L., Straus, D. R., Gilbert, W., and Knowles, J. (1986). *Biochemistry* **25,** 7142.
51. Hermes, J. D., Blacklow, S. C., Gallo, K. A., Bauer, A. J., and Knowles, J. R. (1987). *In* "Protein Structure, Folding, and Design," 2nd Ed., pp. 257–264. Alan R. Liss, New York.
52. Hermes, J. D., Blacklow, S. C., and Knowles, J. R. (1987). *Cold Spring Harbor Symp. Quant. Biol.* **52,** 597.
53. Nickbarg, E. B., Davenport, R. C., Petsko, G. A., and Knowles, J. R. (1988). *Biochemistry* **27,** 5948.
54. Wells, J. A., Powers, D. B., Bott, R. R., Graycar, T. P., and Estell, D. A. (1987). *PNAS* **84,** 1219.
55. Benner, S., and Ellington, A. D. (1988). *Crit. Rev. Biochem.* **23,** 369.
56. Russell, A. J., Thomas, P. G., and Fersht, A. R. (1987). *JMB* **193,** 803.
57. Sternberg, M. J. E., Hayes, F. R. F., Russell, A. J., Thomas, P. G., and Fersht, A. R. (1987). *Nature (London)* **330,** 86.
58. Warwicker, J., and Watson, H. R. (1982). *JMB* **174,** 527.
59. Graf, L., Craik, C. S., Patthy, A., Roczniak, S., Fletterick, F. J., and Rutter, W. J. (1987). *Biochemistry* **26,** 2616.
60. Cronin, C. N., and Kirsch, J. F. (1988). *Biochemistry* **27,** 4572.
61. Hwang, J. K., and Warshel, A. (1988). *Nature (London)* **334,** 270.
62. Yang, Y. R., and Schachman, H. K. (1987). *Anal. Biochem.* **163,** 188.
63. Wente, S. R., and Schachman, H. K. (1987). *PNAS* **84,** 31.
64. Larimer, F. W., Lee, E. H., Mural, R. J., Soper, T. S., and Hartman, F. C. (1987). *JBC* **262,** 15327.

65. Fierke, C. A., and Benkovic, S. J. (1989). *Biochemistry* **28,** 478.
66. Chen, J.-T., Taira, K., Tu, C.-P. D., and Benkovic, S. J. (1987). *Biochemistry* **26,** 4093.
67. Fersht, A. R. (1988). *Biochemistry* **27,** 1577.
68. Gao, J., Kuczera, K., Tidor, B., and Karplus, M. (1989). *Science* **244,** 1069.
69. Fersht, A. R., Leatherbarrow, R. J., and Wells, T. N. C. (1987). *Biochemistry* **26,** 6030.
70. Benkovic, S. J., Fierke, C. A., and Naylor, A. M. (1988). *Science* **239,** 1105.
71. Estell, D. A. (1987). *Protein Eng.* **1,** 441.
72. Fersht, A. R., Knill-Jones, J. W., Bedouelle, H., and Winter, G. (1988). *Biochemistry* **27,** 1581.
73. Brick, P., and Blow, D. M. (1987). *JMB* **194,** 287.
74. Murphy, D. J., and Benkovic, S. J. (1989). *Biochemistry* **28,** 3025.
75. Ladjimi, M. M., and Kantrowitz, E. R. (1988). *Biochemistry* **27,** 276.
76. Middleton, S. A., Stebbins, J. W., and Kantrowitz, E. R. (1989). *Biochemistry* **28,** 1617.
77. Gouaux, J. E., and Lipscomb, W. N. (1989). *Biochemistry* **28,** 1798.
78. Kobilka, B. K., Dixon, R. A., Friella, T., Dohlman, H., Bolanowski, M. A., Sigal, I. S., Yang-Feng, T. L., and Francke, V. (1987). *PNAS* **84,** 46.
79. Kubo, T., Fukuda, K., Mokam, A., Maeda, A., Takahashi, H., Mishina, M., Haga, T., and Haga, K. (1986). *Nature (London)* **323,** 411.
80. Peraluta, E. G., Winslow, J. W., Peterson, G. L., Smith, D. H., Ashkenazi, A., Ramachandran, S., Schimerlik, M. I., and Capon, D. S. (1987). *Science* **236,** 600.
81. Nathans, J., and Hogness, D. S. (1984). *PNAS* **81,** 4851.
82. Appleburg, M. L., and Hargrave, P. A. (1986). *Vision Res.* **26,** 1881.
83. Chung, F. Z., Wang, C. D., Potter, P. O., Venta, J. C., and Fraser, C. M. (1988). *JBC* **263,** 4052.
84. Mizel, S. B., and France, J. J. (1979). *Cell. Immunol.* **48,** 433.
85. Oppenheim, J. J., Kovacs, E. J., Matsushima, K., and Duram, S. K. (1986). *Immunol. Today* **3,** 45.
86. Auron, P. E., Webb, A. C., Rosenwasser, L. J., Mucci, S. F., Rich, A., Wolff, S. M., and Dinanello, C. A. (1984). *PNAS* **81,** 7907.
87. March, C. J., Mosley, B., Larsen, A., Cerreth, D. P., Brandt, G., Prica, V., Gillis, S., and Henney, C. S. (1985). *Nature (London)* **315,** 641.
88. Kamogashira, T., Masui, Y., Ohmoto, Y., Hirato, T., Nagamura, K., Mizuno, K., Hong, Y. M., and Kikumoto, Y. (1988). *BBRC* **150,** 1106.
89. Meric, C., and Spahr, P. F. (1986). *J. Virol.* **60,** 453.
90. Ma, J., and Ptashne, M. (1988). *Cell (Cambridge, Mass.)* **55,** 443.
91. Fu, X., Kata, R. J., Skalke, A. H., and Leis, J. (1988). *JBC* **263,** 2140.
92. Fu, X., Tuazon, P. T., Trangh, J. A., and Leis, J. (1988). *JBC* **263,** 2134.
93. Mao, S. S., Johnston, M. I., Bollinger, J. M., and Stubbe, J. (1989). *PNAS* **86,** 1485.
94. Miller, J., Hatch, J. A., Simonis, S., and Cullen, S. E. (1988). *PNAS* **85,** 1359.
95. Lalazan, A., Weisgraber, K. H., Rall, S. O., Giladi, H., Innerarity, T. L., Levanon, A. Z., Boyles, J. K., and Amit, B. (1988). *JBC* **263,** 3542.
96. Shriver, J. W., Cohen, F. S., Merill, A. R., and Cramer, W. A. (1988). *Biochemistry* **27,** 8421.
97. Texta, F. L., Radford, S. E., Laue, E. D., Perham, R. N., Miles, J. S., and Guest, J. R. (1988). *Biochemistry* **27,** 289.
98. Brocklehurst, K., Kowlessur, D., O'Driscoll, M., Patel, G., Quenby, S., Salih, E., Templeton, W., and Thomas, E. (1987). *BJ* **244,** 173.
99. Brocklehurst, K., Brocklehurst, S. H., Kowlessur, D., O'Driscoll, M., Patel, G., Salih, E., Templeton, W., and Thomas, E. (1988). *BJ* **256,** 543.
100. Ho, C., Jasin, M., and Schimmel, P. (1985). *Science* **229,** 389.

101. Regan, L., Bowie, L., and Schimmel, P. (1987). *Science* **235,** 1651.
102. Regan, L., Buxbaum, L., Hill, K., and Schimmel, P. (1988). *JBC* **263,** 18598.
103. Reidhaar-Olson, J. F., and Sauer, R. T. (1988). *Science* **241,** 53.
104. Schultz, S. C., and Richards, J. H. (1986). *PNAS* **83,** 1588.
105. Doolittle, R. F. (1981). *Science* **214,** 149.
106. Feng, D. F., Johnson, M. S., and Doolittle, R. F. (1985). *J. Mol. Evol.* **21,** 112.
107. Wu, T. T., Fitch, W. M., and Margoliash, E. (1974). *Annu. Rev. Biochem.* **43,** 539.
108. Kimura, M. (1968). *Nature (London)* **217,** 624.
109. Jukes, T. H., and Holmquist, R. (1972). *Science* **177,** 530.
110. Benner, S., and Ellington, A. D. (1988). *Biochemistry* **27,** 369.
111. Vanin, E. F. (1984). *BBA* **782,** 231.
112. Miyata, T., and Hayashida, H. (1981). *PNAS* **78,** 5739.
113. Gitlin, D., and Gitlin, J. D. (1975). *In* "The Plasma Proteins" (F. W. Putnam, ed.), Vol 2. Academic Press, New York.
114. Lesk, A. M., and Chothia, C. (1980). *JMB* **136,** 225.
115. Fermi, G. (1976). *JMB* **97,** 237.
116. Huber, R., Epp, O., and Frumanek, H. (1970). *JMB* **52,** 349.
117. Huber, R., Epp, O., Steigemann, W., and Formanek, H. (1971). *EJB* **19,** 42.
118. Lim, V. I., and Ptitsyn, O. B. (1970). *Mol. Biol.* **4,** 372.
119. Dickerson, R. E. (1980). *UCLA Forum Med. Sci.* **21,** 173.
120. Chothia, C., and Lesk, A. M. (1986). *EMBO J.* **5,** 823.
121. Bolin, J. T., Filmer, D. J., Matthews, D. A., Hamlin, R. C., and Kraut, J. (1982). *JBC* **257,** 13650.
122. Blum, Z., Lidin, S., and Andersson, S. (1988). *Angew. Chem. Int. Ed. Engl.* **27,** 953.
123. Altschuh, D., Lesk, A. M., Bloomer, A. C., and Klug, A. (1987). *JMB* **193,** 693.
124. Altschuh, D., Vernet, T., Berti, P., Maras, D., and Nagai, K. (1988). *Protein Eng.* **2,** 193.
125. Perutz, M. F. (1984). *Adv. Protein Chem.* **36,** 213.
126. Beintena, J. J., Fitch, W. M., and Carsana, A. (1986). *Mol. Biol. Evol.* **3,** 262.
127. Piccoli, R., and D'Alessio, G. (1984). *JBC* **259,** 693.
128. Reddy, E. S. P., Sitaram, N., Bhargava, P. M., and Scheit, K. H. (1979). *JMB* **135,** 525.
129. Joernvall, H. (1977). *EJB* **72,** 443.
130. Joernvall, H., Persson, M., and Jeffrey, J. (1981). *PNAS* **78,** 4226.
131. Karpeiskii, M. Y. (1976). *Mol. Biol.* **10,** 973.
132. Zvelebil, M. J. J. M., and Sternberg, M. J. E. (1988). *Protein Eng.* **2,** 127.
133. Barton, G. J., and Sternberg, M. J. E. (1987). *JMB* **198,** 327.
134. Fierke, C. A., Johnson, K. A., and Benkovic, S. J. (1987). *Biochemistry* **26,** 4085.
135. Andrews, J., Fierke, C. A., Birdsall, B., Ostler, G., Fenney, J., Roberts, G. C. K., and Benkovic, S. J. (1989). *Biochemistry* **28,** 5743.
136. Benkovic, S. J., Adams, J. A., Fierke, C. A., and Naylor, A. M. (1989). *Pteridines* **1,** 37.
137. Nickbarg, E. B., and Knowles, J. R. (1988). *Biochemistry* **27,** 5939.
138. Charlton, P. A., Young, D. W., Birdsall, B., Feeney, J., and Roberts, G. C. K. (1985). *J. Chem. Soc.*, p. 1349.
139. Carter, P., and Wells, J. A. (1987). *Science* **237,** 394.
140. Toney, M. D., and Kirsch, J. F. (1989). *Science* **243,** 1485.
141. Clarke, A. R., Atkinson, T., and Holbrook, J. J. (1989). *TIBS* **14,** 145.
142. Gilbert, W. (1985). *Science* **228,** 823.
143. Haber, E., Quertermous, T., Matsueda, G. R., and Runge, M. S. (1989). *Science* **243,** 51.
144. de Vries, C., Veerman, H., Blasi, F., and Pannekoek, H. (1988). *Biochemistry* **27,** 2565.
145. Wharton, R. P., and Ptashne, M. (1985). *Nature (London)* **316,** 601.
146. Wharton, R. P., Brown, E. L., and Ptashne, M. (1984). *Cell (Cambridge, Mass.)* **38,** 361.

147. Chothia, C., Lesk, A. M., Levitt, M., Amit, A. G., Mariuzza, R. A. Phillips, S. E. V., and Poljak, R. J. (1986). *Science* **233,** 755.
148. Blakley, R. L. (1984). *In* "Folates and Pterins" (R. L. Blakley and S. J. Benkovic, eds.), Vol. 1, Wiley, New York.
149. Bethell, R., and Benkovic, S. J. Unpublished results.
150. Jencks, W. P. (1980). *In* "Chemical Recognition in Biology." (F. Chapeville and A. L. Haenni, eds.), pp. 3–25. Springer-Verlag, New York.
151. Wells, T. N. C., and Fersht, A. R. (1986). *Biochemistry* **25,** 1881.
152. Lowe, D. M., Winter, G., and Fersht, A. R. (1987). *Biochemistry* **26,** 6038.
153. Zvelebil, M. J., Barton, G. J., Taylor, W. R., and Sternberg, M. J. E. (1987). *JMB* **195,** 957.
154. Matsumura, M., and Matthews, B. W. (1989). *Science* **243,** 792.
155. Taira, K., Fierke, C. A., Chen, J.-T., Johnson, K. A., and Benkovic, S. J. (1987). *TIBS* **12,** 275.
156. Leatherbarrow, R. J., Fersht, A. R., and Winter, G. (1985). *PNAS* **82,** 7840.
157. Fersht, A. R. (1987). *Biochemistry* **26,** 8031.
158. Chothia, C., and Lesk, A. M. (1985). *JMB* **182,** 151.

5

Mechanism-Based (Suicide) Enzyme Inactivation

MARK A. ATOR* • PAUL R. ORTIZ DE MONTELLANO[†]

**Department of Medicinal Chemistry*
Smith Kline & French Laboratories
King of Prussia, Pennsylvania 19406

[†]Department of Pharmaceutical Chemistry
School of Pharmacy
University of California, San Francisco
San Francisco, California 94143

THE ENZYMES, Vol. XIX

I. Introduction

A long-standing goal of enzymologists has been the design of specific enzyme inhibitors and inactivators for use as mechanistic probes and as therapeutic agents. Exploitation of the high degree of discrimination exhibited by enzymes in the binding of substrates traditionally has been the primary consideration in the development of such molecules. This approach is exemplified by reversible inhibitors and by active site-directed inhibitors or affinity labels. Reversible inhibitors are substrate analogs that block the active site by competing with substrate for binding to the enzyme, whereas active site-directed inhibitors typically utilize the binding of a substrate analog to position a reactive electrophilic functional group proximal to an active site nucleophile. Covalent modification of the enzyme then results in its inactivation. The specificity of these inhibitors is often limited, however, by the substantial latitude in binding of small molecules to a variety of enzymes and by the promiscuous reaction in solution of the activated functional group of the active site-directed inhibitor with macromolecules other than the target enzyme.

A greater measure of selectivity can be attained by combining the specificity of substrate binding with the chemical specificity implicit in enzymic catalysis, leading to the relatively recent development of mechanism-based enzyme inactivators. A mechanism-based enzyme inactivator is a molecule that binds to the target enzyme as a substrate and is initially processed by the normal catalytic mechanism of the enzyme. At some point in the process, however, a latent functional group of the substrate analog is unmasked, generating a chemically reactive intermediate. Covalent modification of an active site amino acid residue or the prosthetic group of the enzyme by this species leads to inactivation of the enzyme. The prospects for unique modification of the target enzyme are therefore high because the substrate analog is activated exclusively at its active site. In contrast, an active site-directed inhibitor is free to react indiscriminately with any enzyme or small molecule in solution prior to reaching its target. The action of mechanism-based enzyme inactivators has been described in the variety of appelations assigned to these compounds, including suicide substrates, k_{cat} inhibitors, "Trojan horse" inhibitors, and enzyme-activated irreversible inhibitors.

There is vast potential clinical utility in the application of mechanism-based enzyme inactivators to areas of medical interest. In addition to the specificity promised by this approach, the irreversible inactivation of the target provides advantages of potency and duration of action over reversible inhibitors. Reversible inhibitors require a steady-state level of the drug to be maintained in the face of drug metabolism and elimination, whereas a single dose of a mechanism-based inactivator might be sufficient to diminish rapidly the level of the target enzyme until new protein synthesis occurs. The increased concentration of substrate which necessarily results after a metabolic blockade can ultimately destroy the efficacy of a reversible competitive inhibitor, whereas an irreversibly inactivated enzyme is immune to such effects. Although the principles of mechanism-based enzyme inactivation have not yet guided the design of any drugs on the market, several clinically useful compounds have been realized in retrospect to utilize this mechanism of action. For example, tranylcypromine exerts its antidepressant action by inactivation of monoamine oxidase [amine oxidase (flavin-containing)] (Silverman, 1985), and 5-fluorouracil is an antitumor agent by virtue of its conversion to 5-fluoro-2′-deoxyuridine monophosphate (5-fluoro-dUMP), which subsequently inactivates thymidylate synthase (Santi *et al.*, 1974; Danenberg *et al.*, 1974).

Mechanism-based enzyme inactivators are also powerful tools in the determination of enzyme mechanisms. Because some understanding of the enzyme mechanism is required for the design of an inactivator, the success of the compound provides support for the mechanistic hypothesis. Analysis of the intermediates and products of an inactivation reaction can be extremely useful in illuminating the normal mechanism of enzymic catalysis. For example, the covalent modification of an enzyme by a mechanism-based inactivator facilitates isolation of active site peptides and identification of catalytically relevant amino acids.

Since the initial description of a mechanism-based enzyme inactivator in the late 1960s, the field has been reviewed extensively from a variety of perspectives (Abeles and Maycock, 1976; Walsh, 1982, 1984; Rando, 1984; Silverman and Hoffman, 1984; Palfreyman *et al.*, 1987; Silverman, 1988). Rather than to provide a comprehensive summary of the entire literature in this area, the goal of this chapter is to describe the classic approaches to inactivation of a variety of enzymes, drawing illustrations from the best understood examples. This review begins with a discussion of the criteria which define a mechanism-based enzyme inactivator and the strategies for design of such compounds.

A. Criteria for Mechanism-Based Enzyme Inactivation

In order for a compound to be defined as a mechanism-based enzyme inactivator, its reaction with the target enzyme must meet a simple set of kinetic and mechanistic criteria. The kinetics of the inactivation reaction should conform to those expected for the minimal scheme shown in Eq. (1).

$$\mathrm{E + I} \underset{k_{-1}}{\overset{k_1}{\rightleftharpoons}} \mathrm{E{\cdot}I} \xrightarrow{k_2} \mathrm{E{\cdot}X} \xrightarrow{k_3} \mathrm{E + P} \qquad \mathrm{E{\cdot}X} \xrightarrow{k_4} \mathrm{E{-}X} \tag{1}$$

In direct analogy to the Michaelis–Menten mechanism for reaction of enzyme with a substrate, the inactivator, I, binds to the enzyme to produce an E·I complex with a dissociation constant K_I. A first-order chemical reaction then produces the chemically reactive intermediate with a rate constant k_2. The activated species may either dissociate from the active site with a rate constant k_3 to yield product, P, or covalently modify the enzyme (k_4). The inactivation reaction should therefore be a time-dependent, pseudo-first-order process which displays saturation kinetics. This is verified by measuring the apparent rate constant for the loss of activity at several fixed concentrations of inactivator (Fig. 1A). The rate constant for inactivation at infinite [I], k_{inact} (a function of k_2, k_3, and k_4), and the K_I can be extracted from a double reciprocal plot of $1/k_{obs}$ versus $1/[I]$ (Fig. 1B) (Kitz and Wilson, 1962; Jung and Metcalf, 1975). A positive vertical

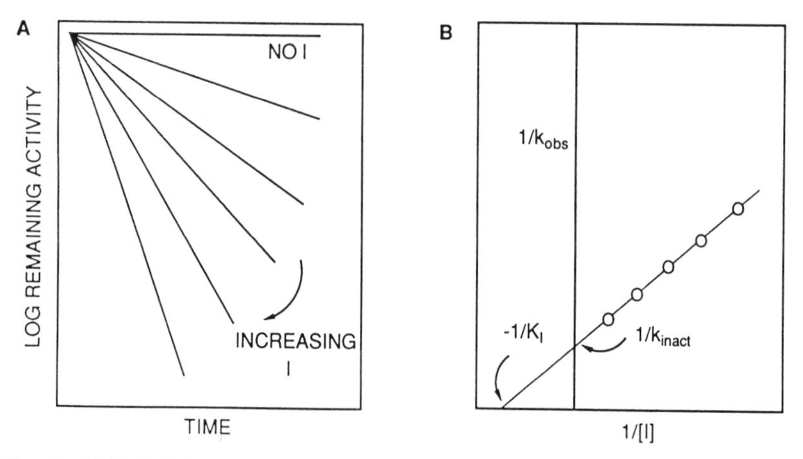

Fig. 1. (a) Typical time-dependent inactivation kinetics for a mechanism-based enzyme inactivator. (b) Replot of inactivation kinetics.

intercept confirms saturation kinetics. More sophisticated treatments of the steady-state kinetics of mechanism-based enzyme inactivation have been presented by Waley (1980, 1985), and Tatsunami *et al.* (1981). Tsou (1988) has described a method of determination of kinetic constants of inactivation reactions by analysis of progress curves obtained in the presence of the inactivator.

The second criterion for mechanism-based enzyme inactivation concerns the stoichiometry of enzyme modification. If mechanism-based inactivation of an enzyme is the result of specific covalent modification of an essential active site amino acid residue, then one radiolabeled inactivator molecule should be incorporated per active site. The stoichiometry of specific radiolabeling should also correlate with the extent of inactivation. In multimeric enzymes which display negative cooperativity, it is possible to observe complete inactivation following substoichiometric modification of the enzyme (Johnston *et al.*, 1979).

In some inactivation reactions, the activated species diffuses from the active site at a rate comparable to or greater than the rate of its reaction with the enzyme. This process, which is contrary to the rigorous definition of a mechanism-based enzyme inactivator, allows nonspecific modification of the enzyme, leading to a stoichiometry of enzyme modification exceeding one. In the absence of a radiolabeled substrate analog, release of the activated intermediate also can be diagnosed by the presence of a lag in the onset of inactivation or by protection against inactivation by small nucleophiles such as 2-mercaptoethanol which can compete for trapping of an electrophilic species. Such nonspecific reactions make analysis of the mechanistic relevance of the labeling difficult and decrease the *in vivo* potential of a compound, since the released inactivating species could also modify other cellular components. The critical parameter in determining the specificity of a mechanism-based enzyme inactivator is therefore the partition ratio, a term coined by Walsh which indicates the number of times product is released from the active site per inactivation event, or the ratio k_3/k_4, as defined in Eq. (1). The ideal partition ratio of zero is actually attained only rarely. A finite partition ratio can aid in elucidation of the mechanism of an inactivation reaction, however, since the identity of the product contains information regarding the catalytic processing of the inactivator and the nature of the inactivating species.

The crucial distinction between mechanism-based inactivators and other classes of enzyme inhibitors is the requirement for the catalytic action of the target enzyme to unmask the reactive functionality of the inactivator. Because the kinetics of inactivation by mechanism-based and affinity label approaches are identical, other methods must be used to demonstrate the necessity of enzyme action. The cofactors and additional substrates compulsory for turnover of normal substrates should be essential for the inactivation reaction, and the stereochemical constraints required of substrates should also be reflected in inactivator structures. Protection against inactivation in the presence of substrate demonstrates that the inactivator must bind to the active site of the enzyme, providing further evidence for a catalytic role for the enzyme. The catalytic action of the enzyme can also be verified by more sophisticated methods, such as observation of an isotope effect on inactivation reactions which require breakage of a bond during processing of the inactivator.

Irreversible covalent inactivation has previously been suggested as a prerequisite for consideration of a compound as a mechanism-based enzyme inactivator (Abeles and Maycock, 1976; Walsh, 1977). Silverman and Hoffman (1984) have argued, however, that rearrangement of the enzyme–inactivator adduct to release the inhibitor and restore activity is independent of the inactivation reaction and should not reflect on its classification as a mechanism-based process. Inactivation which is apparently irreversible is not a property unique to mechanism-based inactivation. Transition state or reaction intermediate analogs, which are stable mimics of unstable reaction intermediates, often bind with such slow dissociation rates that they are effectivley irreversibly bound under normal experimental time scales. Reaction intermediate analogs have been recently reviewed (Morrison and Walsh, 1987; Schloss, 1988) and are not covered here.

B. Strategies for Design of Mechanism-Based Inactivators

The successful design of a mechanism-based inactivator depends on the ability of the enzymologist to disguise within the framework of a substrate analog a latent functional group which can be unveiled by the catalytic action of the enzyme. It is also important that the activated chemical moiety thus produced be sufficiently reactive to modify the active site rapidly, in order to minimize the partition ratio. Although the chemical possibilities for the design of mechanism-based inactivators are virtually limitless, a number of strategies have met with repeated success. These approaches typically involve generation of electrophilic alkylating or acylating agents that capitalize on the nucleophilic nature of amino acids, although nucleophilic species have been designed to modify electrophilic cofactors.

The most common covalent bond-forming reaction is the Michael addition of an enzyme nucleophile to an activated double or triple bond. The enzyme-catalyzed production of a Michael acceptor has been achieved by a variety of mechanisms encompassing several classes of enzymes. Isomerization of an acetylene which is β to a carbonyl can occur by deprotonation at the α-carbon and reprotonation at the γ-carbon, yielding an electrophilic conjugated allene [Eq. (2a)]. Alternately, a deprotonation–reprotonation sequence can be used to bring a double bond into conjugation with an activating group [Eq. (2b)]. Activation of a multiple bond can be achieved by deprotonation to generate a carbonyl equivalent, as in the deprotonation of the α-carbon of an amino acid analog in an aldimine adduct with pyridoxal phosphate [Eq. (2c)]. Elimination reactions also provide a facile mechanism for production of a Michael acceptor [Eq. (2d)].

(2a)

(2b)

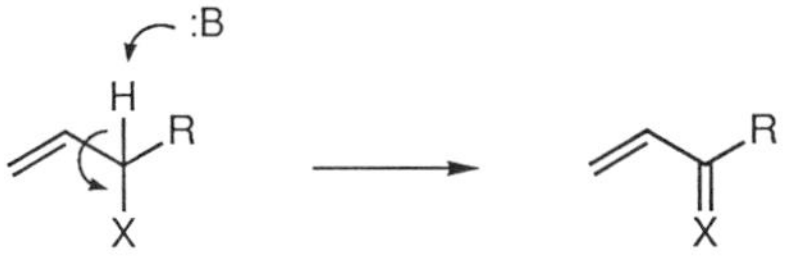

(2c)

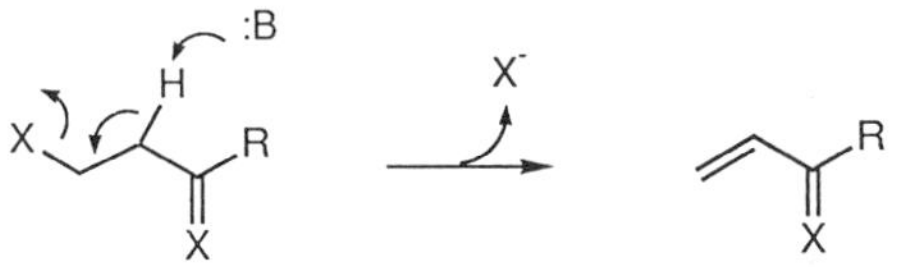

(2d)

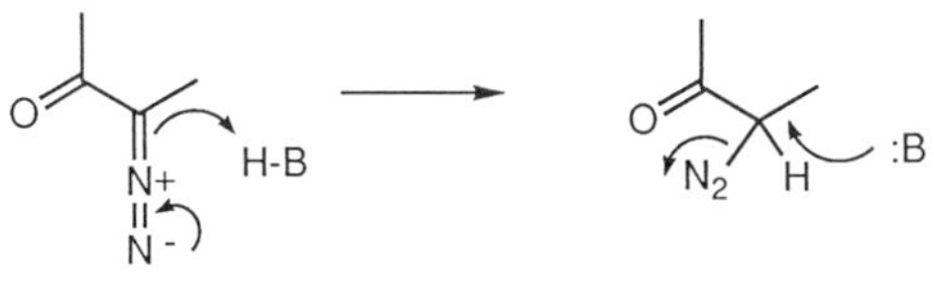

(2e)

Covalent enzyme modification can also be achieved by S_N2 substitution reactions. Enzymic protonation of a diazoketone yields a diazonium salt which then serves as a leaving group in a substitution reaction with an enzyme nucleophile [Eq. (2e)]. An indistinguishable mechanistic possibility in similar reactions is the liberation of nitrogen gas from the diazonium salt to produce a carbocation which then reacts with the nucleophile in an S_N1 process.

Inactivation of hydrolytic enzymes such as proteases and β-lactamases has been achieved through acyl-enzyme intermediates which deviate from the normal mechanism of hydrolysis. Generation of an acyl enzyme with electronic properties that render it resistant to hydrolysis results in effectively irreversible inactivation. A second approach involves production of a highly reactive moiety such as a quinone methide as a consequence of acyl-enzyme formation. The reactive group is tethered in the active site by the acyl enzyme, promoting specific reaction with the target.

II. Dehydrases and Isomerases

The study of mechanism-based enzyme inactivation was inaugurated with the investigation by Bloch and co-workers of the inactivation of β-hydroxydecanoyl thioester dehydrase (3-hydroxydecanoyl-[acyl-carrier-protein] dehydratase) by 3-decynoyl-*N*-acetylcysteamine, **1** (Fig. 2) (Helmkamp *et al.*, 1968; Bloch, 1986). The dehydrase, which is the first enzyme in bacterial unsaturated fatty acid biosynthesis, catalyzes two reactions: dehydration of 3-hydroxydecanoyl thioesters to (*E*)-2-decenoyl thioesters and isomerization of (*E*)-2-decenoyl and (*Z*)-3-decenoyl thioesters. When the enzyme is presented with **1,** the acetylenic analog of its deconjugated enoyl substrate, it isomerizes the unsaturation, resulting in formation of a conjungated allene (Fig. 2). Alkylation of an active site histidine residue by the electrophilic allene leads to enzyme inactivation (Helmkamp and Bloch, 1969). The isotope effect on the rate of dehydrase inactivation by $[2\text{-}^2H_2]$**1** ($k_H/k_D = 2.66$) is comparable to the isotope effect on dehydration of 2-^2H_2-labeled substrates, confirming the catalytic involvement of the enzyme in generation of the allene (Endo *et al.*, 1970). The nature of the enzyme–inactivator adduct has been elucidated in an elegant NMR study by Schwab and colleagues (Schwab *et al.*, 1986), who demonstrated that the thioester dienolate adduct is initially protonated at C-2 to yield an (*E*)-3-(N^{im}-histidinyl)-3-decenoyl thioester which then slowly isomerizes to the conjugated 2-decenoyl isomer (Fig. 2).

An analogous approach has been used to inactivate the Δ^5-3-ketosteroid isomerase (steroid Δ-isomerase) from *Pseudomonas testosteroni*, an enzyme which

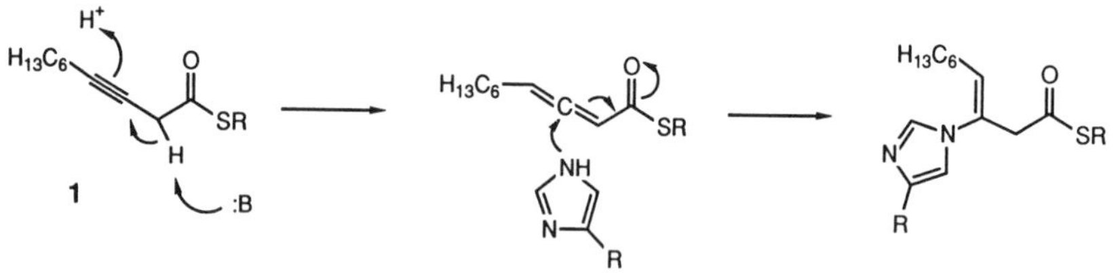

FIG. 2. Mechanism proposed for inactivation of β-hydroxydecanoyl thioester dehydrase by 3-decynoyl-*N*-acetycysteamine (**1**).

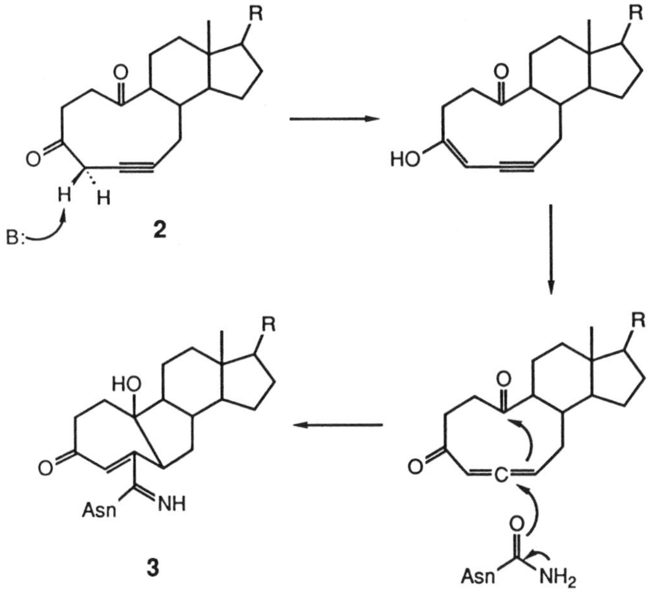

FIG. 3. Mechanism proposed for inactivation of Δ^5-3ketosteroid isomerase by 3-keto-5, 10-secosteroids (e.g., **2**).

catalyzes the intramolecular transfer of the 4β- and 6β-hydrogens of substrates via the intermediacy of a $\Delta^{3,5}$-dienol. Acetylenic 3-keto-5,10-secosteroids such as **2** (Fig. 3) are rapidly converted by the enzyme to the conjugated allene, which dissociates from the enzyme and then subsequently rebinds and inactivates the isomerase (Batzold and Robinson, 1975; Covey and Robinson, 1976). In spite of a rather daunting partition ratio of 6×10^5, **2** has been demonstrated to stoichiometrically label the enzyme, with the modification localized to an asparagine residue (Penning *et al.*, 1981; Penning and Talalay, 1981). Based on the spectroscopic properties and chemical stability of the covalently modified isomerase, the transannular enol imidate adduct **3** (Fig. 3) has been suggested as the structure of the derivatized enzyme.

III. Pyridoxal Phosphate-Dependent Enzymes

The pyridoxal phosphate-dependent enzymes have been a major focal point in the development of mechanism-based inactivators. Pyridoxal phosphate (PLP) is utilized in resonance stabilization of carbanions at the α- and β-carbons of amino acids in a variety of reactions which lead to chemical transformations at the α-, β-, and γ-carbons of the substrate (Walsh, 1979). These carbanion equivalents have been conscripted to take part in elimination, isomerization, and deprotonation–reprotonation sequences, resulting in the elaboration of chemically activated species. Examples of the most significant approaches to inactivation of PLP-dependent enzymes are discussed within the framework of the reactions normally catalyzed by each type of enzyme.

A. Inactivation Following α-Carbanion Formation: Decarboxylases, Transaminases, and Racemases

Specific decarboxylases are known for a majority of the amino acids, and several are prime targets for inactivation by virtue of their substantial medicinal importance. These include aromatic-amino-acid decarboxylase, which is responsible for the production of dopamine (DOPA); orithine decarboxylase, which supplies the polyamine putrescine; and glutamate decarboxylase, which converts glutamate to the inhibitory neurotransmitter γ-aminobutyric acid (GABA). The accepted mechanism of these enzymes involves decarboxylation of the amino acid to yield a resonance-stabilized carbanion at the α-carbon of the substrate. The intermediate is then protonated with retention of configuration to yield product (Walsh, 1979, p. 800).

The α-halomethyl analogs of amino acids have proved to be potent inactivators of their respective decarboxylases, as exemplified by the reaction of α-fluoromethyl-DOPA (**4,** Fig. 4) with aromatic-amino-acid decarboxylase (Maycock

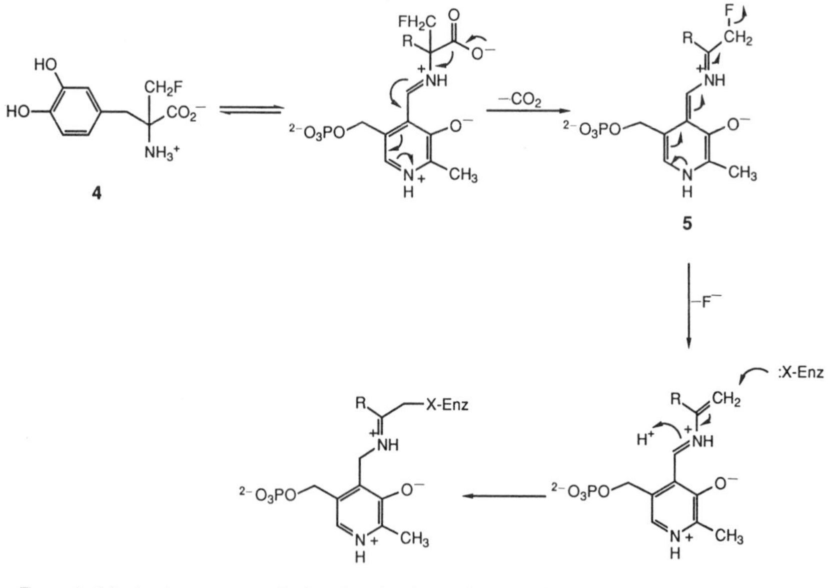

Fig. 4. Mechanism proposed for inactivation of aromatic-amino-acid decarboxylase by α-fluoromethyl-DOPA (**4**).

et al., 1980). Inactivation is observed only with the (*S*) isomer of **4**, which corresponds to the absolute configuration of the natural substrates, and is accompanied by stoichiometric incorporation of label from [*G*-^{3}H]**4** but no radiolabeling by the 1-^{14}C analog. The reaction proceeds with the loss of 1.3 equivalents of fluoride and 1 equivalent of CO_2 from C-1 of **4**, indicating that the partition ratio is close to zero. A mechanism consistent with these results proposes that elimination of fluoride from the quinoid intermediate **5** generates a Michael acceptor which covalently modifies the enzyme (Fig. 4).

Difluoromethyl analogs of amino acids are also highly effective inactivators of PLP-dependent decarboxylases. Among the best studied is α-difluoromethylornithine (DFMO), which appears to inhibit ornithine decarboxylase by a mechanism analogous to that depicted in Fig. 4 (Metcalf *et al.*, 1978; Pegg *et al.*, 1987). DFMO is highly effective against African trypanosomiasis and is undergoing clinical evaluation (Sjoerdsma and Schechter, 1984).

The catalytic action of a decarboxylase can also be used to bring an isolated and relatively unreactive olefin into conjugation, creating an electrophilic center. The mammalian glutamate decarboxylase is inactivated by 2-methyl-3,4-didehydroglutamate (**6**), whereas the bacterial enzyme is not affected (Chrystal *et al.*, 1979). A likely mechanism is shown in Fig. 5. The α-methyl group of **6** is proposed to add specificity by preventing reaction of the substrate analog with transaminases (aminotransferases) which utilize glutamate as a substrate.

Analogs of products as well as substrates have been successfully utilized as mechanism-based inactivators, exploiting the principle of microscopic reversibility. Bacterial glutamate decarboxylase is stereospecifically inactivated by the 4-aminobutyric acid analog, (*R*)-4-aminohex-5-ynoic acid (**7**, Fig. 6) (Jung *et al.*, 1978). The inactivation is rationalized by reversal of the protonation step in the decarboxylation of glutamate, leading to formation of a propargylic anion which could be protonated to produce either a conjugated allene (path A, Fig. 6) or a conjugated acetylene (path B). The observed stereospecificity of the reaction is consistent with the known stereochemistry of protonation of the decarboxylation reaction. Paradoxically, the mammalian glutamate decarboxylase is inhibited stereospecifically by the (*S*) isomer of the acetylenic compound, even though the stereochemistry of the decarboxylation reaction of the bacterial and mammalian enzymes is identical (Bouclier *et al.*, 1979). The mammalian enzyme is proposed to retain a vestigial transaminase activity inherited from an

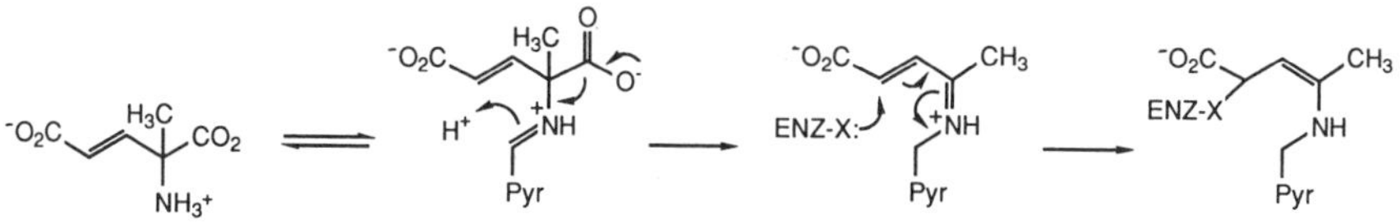

FIG. 5. Mechanism proposed for inactivation of glutamate decarboxylase by 2-methyl-3,4-didehydroglutamate (**6**).

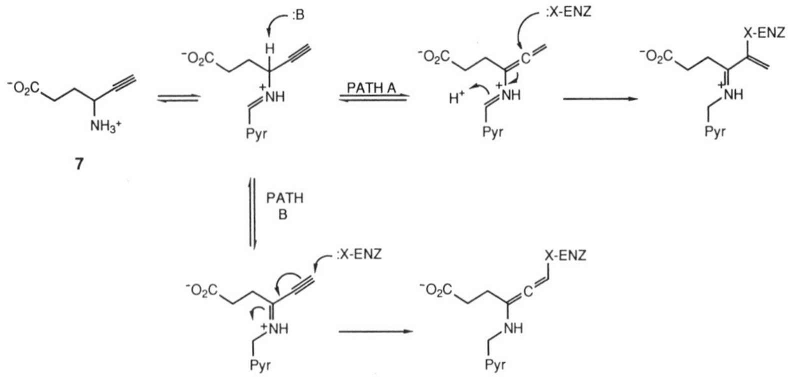

FIG. 6. Mechanism proposed for inactivation of glutamate decarboxylase by 4-aminohex-5-ynoic acid (**7**).

evolutionary ancestor which allows it to abstract the C-4 hydrogen of **7,** leading to inactivation.

Catalysis by PLP-dependent transaminases differs from the decarboxylases in that the first step of the reaction involves abstraction of the α-proton of the substrate rather than decarboxylation (Walsh, 1979, p. 781). The resultant carbanionic intermediates are directly analogous, however, allowing similar strategies to be employed for the inactivation of the two classes of enzymes. For example, inactivation of GABA transaminase (4-aminobutyrate aminotransferase) by 4-aminohex-5-ynoic acid is postulated to occur by the same mechanism as the reaction with glutamate decarboxylase (Jung and Metcalf, 1975). In contrast, the corresponding vinyl derivative, 4-aminohex-5-enoic acid (vinyl-GABA) is a potent inactivator of GABA transaminase but has no effect on glutamate decarboxylase (Lippert *et al.*, 1977). Vinyl-GABA is currently in clinical trials as a treatment for some types of epilepsy (Schechter, 1986).

A unique example of mechanism-based inactivation is the reaction of GABA transaminase with gabaculine (**8,** Fig. 7), a natural product isolated from *Streptomyces toyocaensis* (Rando, 1977). Removal of the hydrogen from C-5, as evidenced by an isotope effect on the inactivation reaction with [4,5-$^2H_2$]-gabaculine, produces an intermediate which could reasonably be expected to act as a Michael acceptor. Instead, loss of a hydrogen from C-6 yields a nonhydrolyzable aromatized adduct which has been isolated following denaturation of the modified transaminase and shown to be identical to standard material. The same adduct is obtained with double-bond isomers of gabaculine (Metcalf and Jung, 1979). The inactivation therefore proceeds by modification of the cofactor rather than the protein. A survey of a variety of transaminases reveals a correlation between sensitivity to gabaculine and the ability of the enzyme to catalyze exchange of the β-protons of its substrate, suggesting that aromatization of the analog is an enzyme-catalyzed event (Soper and Manning, 1982).

Confirmation of the molecular structure of the enzyme–inactivator adduct has been obtained for few modified PLP-dependent enzymes. In the case of the reaction of aspartate transaminase (aspartate aminotransferase) with L-serine *O*-sulfate, the surprising result thus obtained by Metzler and co-workers has forced reevaluation of the mechanism of similar inactivators (Ueno *et al.*, 1982). Conventional wisdom argued that the reaction should involve elimination of sulfate from the inactivator followed by addition of an enzyme nucleophile to the resulting double bond (Fig. 8). When subjected to high pH, however, the inactivated enzyme releases a yellow PLP adduct which has been identified as the aldol product of the cofactor and C-3 of pyruvate (**9,** Fig. 9) as previously prepared by

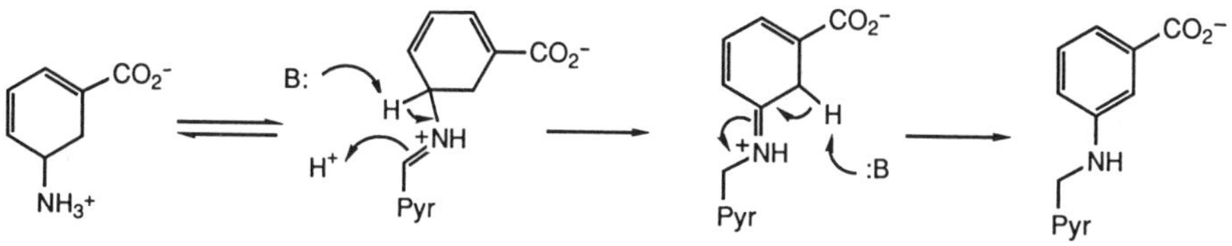

FIG. 7. Mechanism proposed for inactivation of GABA transaminase by gabaculine (**8**).

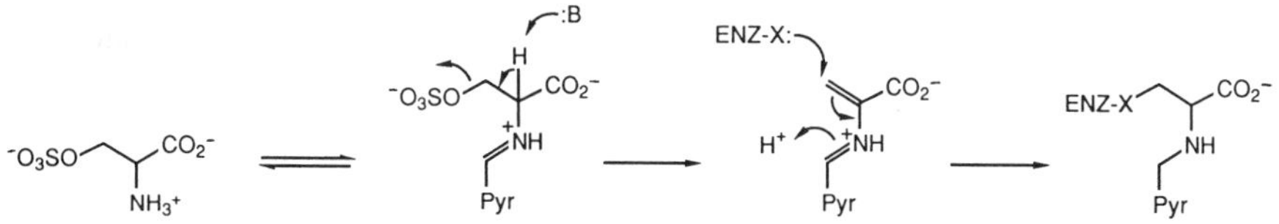

FIG. 8. Original mechanism proposed for inactivation of aspartate transaminase by serine *O*-sulfate.

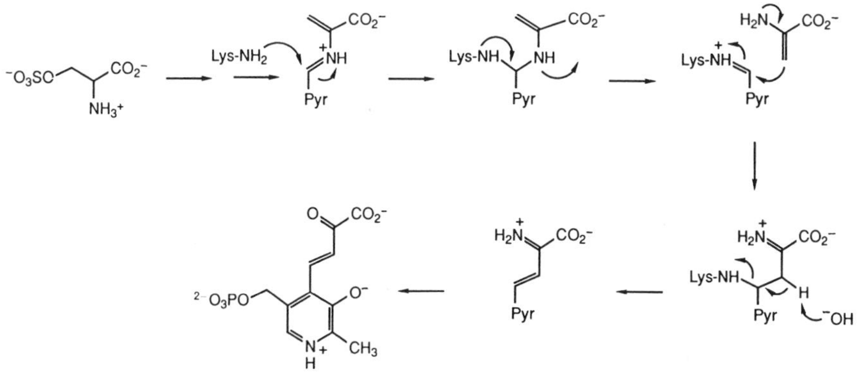

FIG. 9. Revised mechanism proposed for inactivation of aspartate transaminase by serine *O*-sulfate.

Schnackerz *et al.* (1979). The same adduct was isolated from glutamate decarboxylase following inactivation with serine *O*-sulfate (Likos *et al.*, 1982). The mechanism proposed to account for this observation involves elimination of sulfate from the inactivator followed by transimination with the active site lysine of the enzyme (Fig. 9). The released aminoacrylate then rotates in the active site and acts as a nucleophile, attacking the electrophilic imine. Under basic conditions the enzyme lysine is eliminated, generating the free cofactor adduct. This reaction is therefore unusual in that the inactivator is ultimately the nucleophilic partner in the inactivation reaction, in contrast to all of the reactions described to this point, in which the enzyme provides the nucleophile.

The factors that govern whether an inactivator of PLP-dependent enzymes operates by a Michael addition or an enamine mechanism have not yet been discerned, largely because the actual mechanism of inactivation has been firmly established in very few cases. Based on examination of the reaction of GABA transaminase with a series of fluorinated GABA analogs, Silverman and George (1988a) have presented a hypothesis for predicting the mechanism of transaminase inactivation. The reactions of GABA transaminase with 4-amino-5-fluoropentanoic acid (Silverman and Levy, 1981; Silverman and Invergo, 1986) and its unsaturated analog, 4-amino-5-fluoropent-2-enoic acid (Silverman *et al.*, 1986; Silverman and George, 1988b), have been shown to involve enzyme-catalyzed elimination of fluoride followed by modification of the cofactor by an enamine mechanism. In contrast, inactivation by 4-amino-2-fluorobut-2-enoic acid appears to proceed by azallylic isomerization of the initial Schiff base with subsequent addition of the enzyme to the resulting Michael acceptor (Silverman and George, 1988a). Elimination of fluoride also occurs in this example, but it is proposed to occur from the enzyme–inactivator adduct rather than as a step in activation of the analog. Silverman proposes as a general hypothesis that compounds which have a suitably positioned leaving group will undergo transaminase-mediated elimination followed by alkylation of PLP via the enamine mechanism. In compounds which do not contain a leaving group positioned for elimination following proton abstraction, azallylic isomerization will occur. If this isomerization unmasks an electrophile, trapping of an appropriately poised nucleophile may result; otherwise, transamination and product release occur. Although this theory is consistent with the currently verified mechanisms of transaminase inactivation, further experimentation is clearly necessary.

The PLP-dependent racemization of amino acids also occurs with the abstraction of the α-proton of the substrate, allowing similar inactivation approaches to be pursued. Walsh and co-workers have examined the reaction of alanine racemases from several sources with β-substituted alanines (Roise *et al.*, 1984; Badet *et al.*, 1984; Wang and Walsh, 1978, 1981). The partition ratio was found to be invariant with the stereochemistry of the α-carbon and the nature of the leaving group, indicating that a common intermediate was formed which did not recall either of these parameters. Analysis of the inactivated enzyme revealed that the aminoacrylate mechanism was operative, suggesting that aminoacrylate was the common intermediate. In contrast, alanine racemase is inactivated by β,β,β-trifluoroalanine to yield an inactivated enzyme species which is spectrally distinct from those obtained with monohaloalanines (Faraci and Walsh, 1989). Inactivation of the *B. stearothermophilus* racemase by β,β,β-trifluoro[1-^{14}C]-alanine occurs with elimination of 2 mol of fluoride per mole of enzyme and radiolabeling of the enzyme with a stoichiometry of 1. On denaturation of the derivatized enzyme, $^{14}CO_2$ and an additional 1 equivalent of fluoride are released. Reduction of the enzyme with [^{3}H]$NaBH_4$ and analysis of peptides obtained by tryptic digestion reveal that lysine-38, the residue which forms a Schiff base with the cofactor in the resting state of the racemase, bears the labeled group. The mechanism proposed for the inactivation is analogous to that depicted in Fig. 8. Elimination of fluoride yields a highly electrophilic β-difluoro-α,β-unsaturated imine, which alkylates the amino group of lysine with subsequent elimination of a second equivalent of fluoride. Faraci and Walsh hypothesize that the electrophilicity of the β-carbon, as determined by its substituents, controls whether inactivation of alanine racemase by β-substituted alanines occurs by the Michael addition or enamine mechanisms.

B. Inactivation Following Abstraction of the β-Hydrogen

Several PLP-dependent enzymes catalyze elimination and replacement reactions at the γ-carbon of substrates, an unusual process which provides novel routes for mechanism-based inactivation. An example of this class of enzymes is cystathionine γ-synthase [*O*-succinylhomoserine (thiol)-lyase], which converts *O*-succinyl-L-homoserine and L-cysteine to cystathionine and succinate as part of the bacterial methionine biosynthetic pathway (Walsh, 1979, p. 823). Formation of a PLP-stabilized α-carbanion intermediate activates the β-hydrogen for abstraction, yielding β-carbanion equivalents and allowing elimination of the γ-substituent. The resulting β,γ-unsaturated intermediate serves as an electrophilic acceptor for the replacement nucleophile. Suitable manipulation of the β-carbanion intermediate allows strategies for the design of inactivators which do not affect enzymes which abstract only the α-hydrogen.

The inactivation of cystathionine γ-synthase by the antibiotic natural product propargylglycine (**10,** Fig. 10) utilizes the β-carbanion intermediate to trigger a propargylic rearrangement, resulting in a conjugated *p*-quinoid PLP-allene (Johnston *et al.*, 1979). The enzyme is efficiently alkylated with a partition ratio of 4 to yield an adduct of undetermined structure. It is interesting to note that the enzyme is not inactivated by the conjugated olefinic intermediate which is on the normal catalytic pathway (the adduct of PLP and allylglycine), indicating that the allene is a more reactive electrophile or, perhaps, that addition of an enzyme

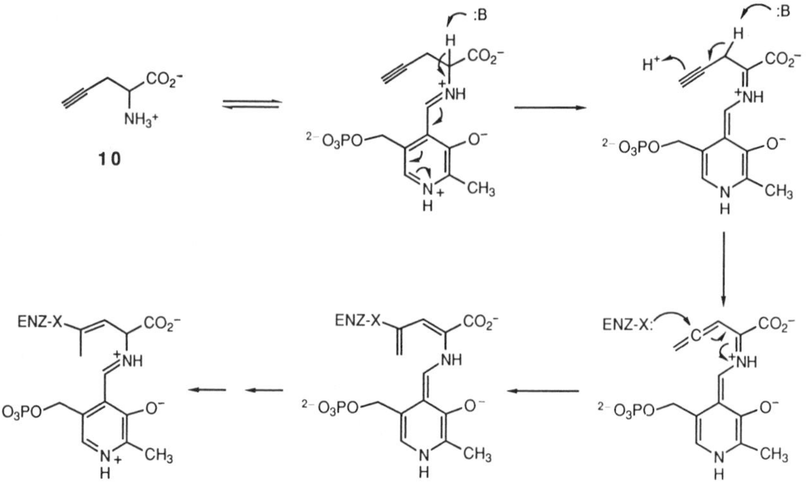

FIG. 10. Mechanism proposed for inactivation of cystathionine γ-synthase by propargylglycine (**10**).

nucleophile is too slow to compete with addition of cysteine. The inactivation of the γ-elimination enzymes γ-cystathionase (cystathionine γ-lyase) (Washtien and Abeles, 1977) and methionine γ-lyase (Johnston *et al.*, 1979) by propargylglycine is proposed to proceed by a similar mechanism.

A novel strategy for mechanism-based inactivation involves enzyme-catalyzed production of an intermediate which is not itself an electrophile, but which undergoes spontaneous rearrangement to reveal an electrophilic species (Johnston *et al.*, 1980). Cystathionine γ-synthase and methionine γ-lyase are proposed to abstract the α-hydrogen of 2-amino-4-chloro-5-(p-nitrophenylsulfinyl)pentanoic acid (**11,** Fig. 11) in a normal manner, activating the compound for labilization of the β-hydrogen and elimination of chloride. The resultant allyl sulfoxide undergoes a facile 2,3-sigmatropic rearrangement to yield a highly electrophilic allyl sulfenate. Inactivation results from transfer of the p-nitrophenylthio moiety to the enzyme, most likely producing a mixed disulfide adduct with a cysteine residue. The identity of the adduct is supported by the observations that the carbon chain of the compound is not bound to the inactive enzyme and that thiols reactivate the enzyme with stoichiometric release of p-nitrophenylthiol.

Several enzymes which perform reactions at the α-carbon of amino acids are also known to catalyze an apparently unrelated exchange of the β-hydrogens of their substrates. This capability renders glutamic–pyruvic transaminase (alanine aminotransferase) susceptible to inhibition by propargylglycine (Marcotte and Walsh, 1975), presumably by the mechanism described above. Alanine transaminase is inactivated by β-cyano-L-alanine in an analogous manner, although

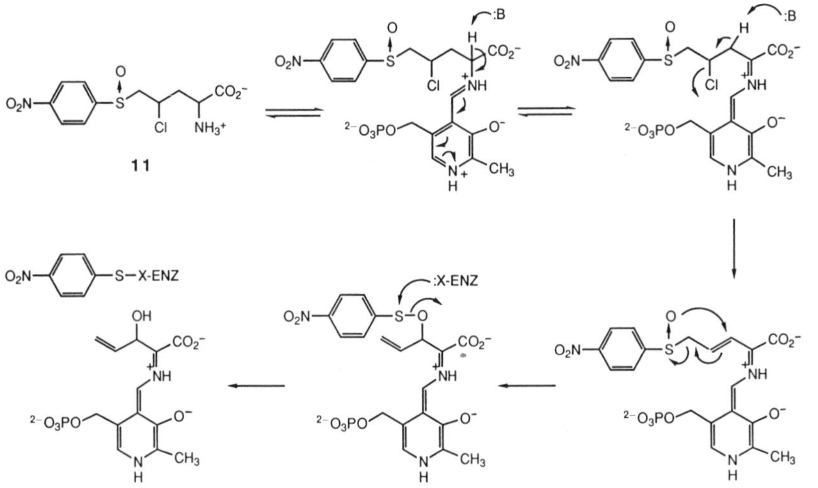

FIG. 11. Mechanism proposed for inactivation of cystathionine γ-synthase and methionine γ-lyase by 2-amino-4-chloro-5-(*p*-nitrophenylsulfinyl)pentanoic acid (**11**).

the adduct formed between the enzyme and the resultant ketenimine is labile, resulting in recovery of activity on dilution of the inactivated enzyme (Alston *et al.*, 1980).

IV. Flavin-Dependent Enzymes

A variety of enzymes use tightly bound flavin in the form of flavin mononucleotide (FMN) or flavin adenine dinucleotide (FAD) as an electron acceptor in the catalysis of oxidation reactions (Walsh, 1980; Bruice, 1980). Flavin is a highly versatile cofactor owing to its ability to accept and donate electrons in pairs or singly, leading to stable semiquinone intermediates, and by virtue of its ability to react with molecular oxygen, permitting formation of reduced oxygen species or the hydroxylation of substrates. The diverse mechanistic possibilities for flavin-dependent enzymes have made the design and study of their inactivators a particularly intriguing branch of the field.

A. OXIDASES AND DEHYDROGENASES OF α-AMINO AND α-HYDROXY ACIDS

Flavin-dependent oxidases and dehydrogenases mediate the net two-electron oxidation of their respective substrates with the formation of a reduced flavin intermediate. In a subsequent oxidative half-reaction, oxidases transfer two electrons to molecular oxygen, whereas dehydrogenases utilize one-electron recipients such as cytochromes or two-electron oxidants such as dithiols to reoxidize the cofactor. Tactics for mechanism-based inactivation of these enzymes typically involve generation of reactive intermediates in the reductive half-reaction, leading to modification of the protein or the flavin coenzyme.

A number of oxidases and dehydrogenases are believed to initiate catalysis by abstracting a hydrogen adjacent to an activating group, leading to carbanion intermediates which can be conscripted for inactivation reactions similar to those used for isomerases and PLP-dependent enzymes. For example, the acyl-CoA dehydrogenases are proposed to abstract the α-hydrogen of substrates as the first step in α,β-desaturation of fatty acids. This hypothesis is supported by the observation that the 3-acetylenic derivatives of fatty acyl-CoAs inactivate a number of fatty acyl-CoA dehydrogenases (Frerman *et al.*, 1980; Gomes *et al.*, 1981; Fendrich and Abeles, 1982) by a mechanism suggested to be identical to the propargylic rearrangement proposed for β-hydroxydecanoyl thioester dehydrase (Fig. 2). The rate of inactivation of the general acyl-CoA dehydrogenase by 2,3-octadienyl-CoA is independent of pH, whereas the rate with 3-butynoyl-CoA is found to decrease with increasing pH, suggesting that protonation of the allenic carbanion intermediate is required (Frerman *et al.*, 1980). Detailed chemical characterization of the adduct between butyryl-CoA dehydrogenase and (3-pentynoyl)pantetheine indicated that inactivation was the result of addition of a glutamate residue to C-3 of the rearranged substrate analog, with no evidence of flavin modification (Fendrich and Abeles, 1982). Similar results were obtained on incubation of (3-chloro-3-butenoyl)pantetheine with butyryl-CoA dehydrogenase, consistent with generation of an analogous allenic intermediate by β-elimination of chloride (Fig. 12) (Fendrich and Abeles, 1982).

Abstraction of the α-hydrogen is also implicated as the trigger for inactivation of the general acyl-CoA dehydrogenase by 3,4-pentadienoyl-CoA (Wenz *et al.*, 1985). Enzyme inactivation and reduction of the flavin are achieved with a stoichiometric quantity of inactivator, although addition of substrate causes the rapid regain of catalytic activity. Activity can also be regained over a longer time course with the release of 2,4-pentadienyl-CoA, although 20% of the enzyme appears to be irreversibly inactivated, apparently due to protein modification. The released 2,4-pentadienyl-CoA has no effect on activity, indicating that irreversible modification must precede product release. The mechanism offered to

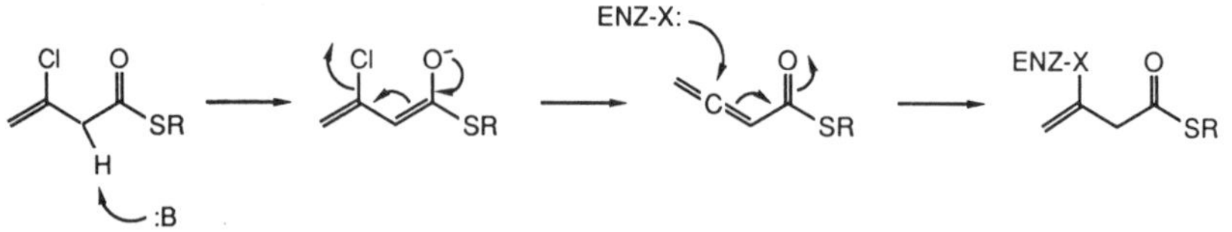

FIG. 12. Mechanism proposed for inactivation of butyryl-CoA dehydrogenase by (3-chloro-3-butenoyl)pantetheine.

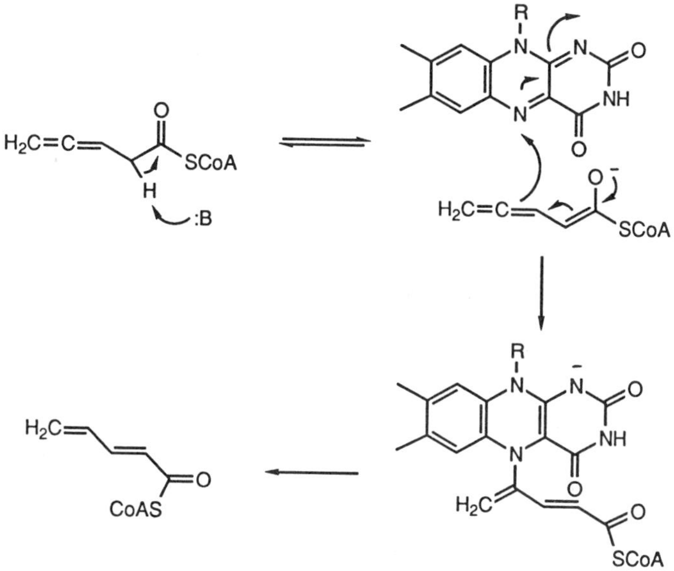

FIG. 13. Mechanism proposed for inactivation of general acyl-CoA dehydrogenase by 3,4-pentadienoyl-CoA.

account for these observations (Fig. 13) proposes that the delocalized carbanion resulting from abstraction of the α-hydrogen reversibly adds to the N-5 position of the cofactor. The mechanism of irreversible modification is unknown.

An incompletely understood but nevertheless intriguing inactivator of the general acyl-CoA dehydrogenase is methylenecyclopropylacetyl-CoA (**12,** Fig. 14). Hypoglycin A is a toxic amino acid which is found in unripe ackee fruit and causes severe hypoglycemia following ingestion, apparently as a result of its metabolism to **12.** The dehydrogenase is inactivated by the (R) isomer of **12** with a partition ratio of 4 (Baldwin and Parker, 1987). One equivalent of CoA adenine is incorporated per flavin, and the cofactor chromophore is altered (Wenz *et al.*, 1981). Denaturation of the modified enzyme under a variety of conditions allows isolation of a number of flavin products in variable yields, including 20% unmodified FAD and a C-4a–N-5 dihydroflavin derivative. The mechanism proposed for this process (Fig. 14) involves rearrangement of an α-carbanion to unmask a nucleophilic carbanion which modifies the electrophilic flavin.

A contrasting mode of flavoprotein reactivity with an acetylenic inactivator occurs in the reaction of 2-hydroxy-3-butynoate (**13,** Fig. 15) with a number of α-hydroxy acid oxidizing enzymes. This process is exemplified by the inactivation of L-lactate oxidase from *Mycobacterium smegmatis,* an enzyme which catalyzes the oxidative decarboxylation of lactate to yield acetate, carbon dioxide, and water (Walsh, 1979, p. 408). Incubation of **13** with lactate oxidase leads to inactivation of the enzyme with a partition ratio that varies from 110 in the

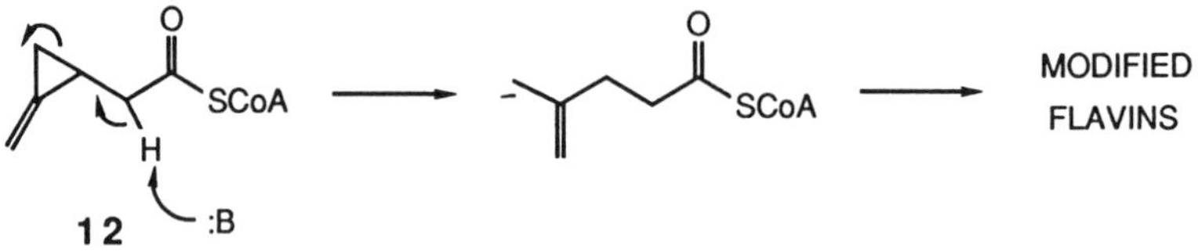

FIG. 14. Mechanism proposed for inactivation of general acyl-CoA dehydrogenase by methylene-cyclopropylacetyl-CoA (**12**).

presence of oxygen at 20% of saturation to 0 under anaerobic conditions, indicating the partitioning of a common intermediate (Ghisla *et al.*, 1976). No radioactivity from [2-^{3}H]**13** is found in the inactivated enzyme, consistent with the abstraction of that hydrogen during the reaction, whereas ^{3}H from [4-^{3}H]**13** is found in the adduct with a stoichiometry of one (Walsh *et al.*, 1972). Separation of the cofactor from the enzyme indicates that the FMN is modified with no labeling of the apoenzyme. Inactivation is also observed with 2-keto-3-butynoate and photochemically reduced lactate oxidase, the products expected by analogy with the first half-reaction of the normal catalytic process. The absorption spectrum of the inactivated enzyme is consistent with the presence of a 4a,5-disubstituted flavin, the structure of which was determined by spectroscopic and chemical methods to be compound **14** (Fig. 15) (Schonbrunn *et al.*, 1976). The observation that addition of the flavin to the inactivator occurred at C-2 and C-4 rather than C-3 eliminates the standard nucleophilic addition to an allenic intermediate as a mechanistic possibility. The mechanism proposed in Fig. 15 suggests that a nucleophilic allenic carbanion is formed which attacks the electrophilic C-4a position of the oxidized flavin in a reversal of the usual polarity, followed by cyclization of the adduct. An analogous process involving radical intermediates could be written, although there is no evidence for their existence in flavoprotein oxidase reactions. A second possible mechanism (Fig. 16) involves oxidation of **13** to 2-keto-3-butynoate, a Michael acceptor which could undergo nucleophilic attack by the reduced flavin, leading to the flavin adduct **14.** The fact that 2-hydroxy-3-butenoate is oxidized by lactate oxidase without enzyme inactivation argues against the mechanism in Fig. 16, since the resulting 2-keto-3-butenoate would also be a powerful electrophile but would not have the nucleophilic character necessary for the process described in Fig. 15.

The enzyme D-lactate dehydrogenase from *Megasphaera elsdenii* catalyzes the oxidation of D-lactate to pyruvate, with an electron-transferring flavoprotein serving as the ultimate oxidant. Its reaction is similar to the first step of the lactate oxidase reaction, but the two enzymes use enantiomeric substrates, leading to the proposal that the two enzymes utilize similar mechanisms but bind their substrates in opposite orientations (Ghisla *et al.*, 1976). Incubation of D-lactate dehydrogenase with D-**13** leads to enzyme inactivation with a partition ratio of 5 (Olson *et al.*, 1979). A novel pink chromophore formed concomitantly

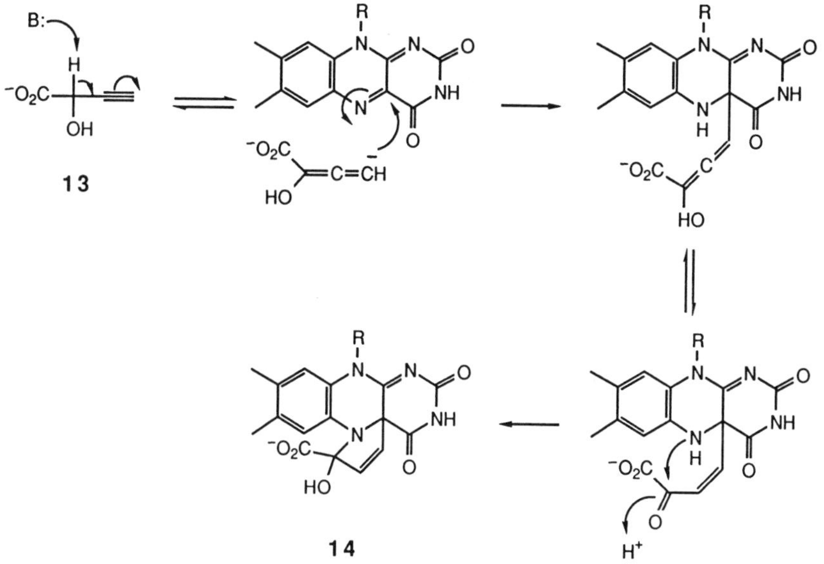

FIG. 15. Mechanism proposed for inactivation of L-lactate oxidase by 2-hydroxy-3-butynoate (**13**).

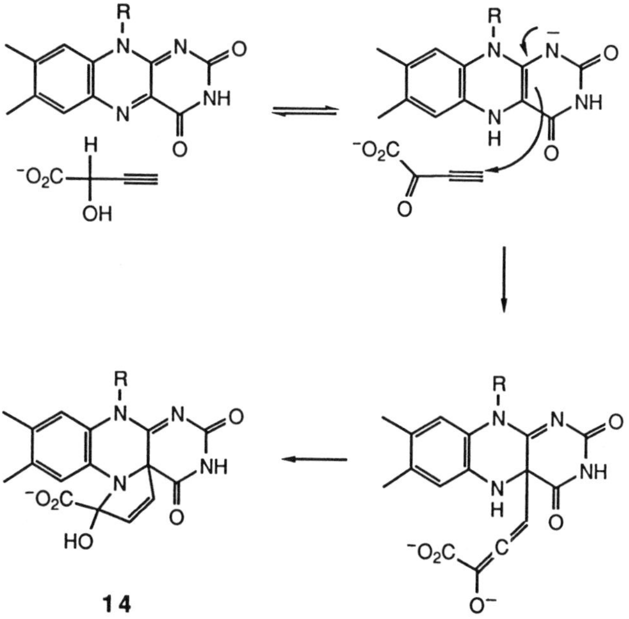

FIG. 16. Alternate mechanism proposed for inactivation of L-lactate oxidase by 2-hydroxy-3-butynoate.

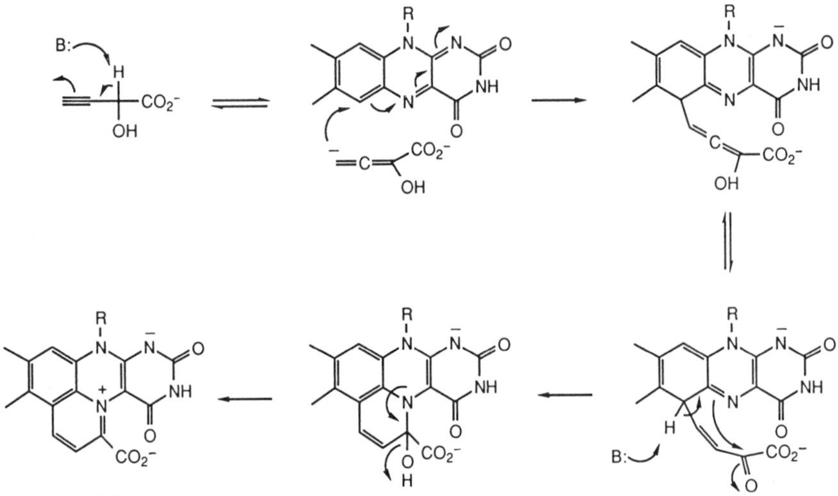

FIG. 17. Mechanism proposed for inactivation of D-lactate dehydrogenase by 2-hydroxy-3-butynoate.

with inactivation was shown to be an unusual 5,6-disubstituted flavin, **15** (Fig. 17) (Ghisla *et al.*, 1979). A likely mechanism for its inactivation was proposed to be addition of an allenic carbanion to C-6 of the oxidized flavin, as shown in Fig. 17. As in the case of lactate oxidase, a mechanism involving nucleophilic attack of reduced flavin on 2-keto-3-butynoate can be proposed, although it was disfavored on chemical grounds. The nature of the modified flavin therefore supports the hypothesis that lactate oxidase and D-lactate dehydrogenase are mechanistically similar but bind substrates in diastereotopic fashion.

B. Monoamine Oxidase

Monoamine oxidase [amine oxidase (flavin-containing)] catalyzes the oxidative catabolism of biogenic amines to the corresponding aldehydes, and it has proved to be a significant pharmacological target for the development of antidepressant agents. Retrospective analysis of the interaction of many of these compounds with the enzyme has indicated that they are, in fact, mechanism-based inactivators. Unlike the enzymes described in the previous section, monoamine oxidase oxidizes substrates that are not activated for an initial proton abstraction, necessitating a different mode of action. Based on analogy with electrochemical precedents and the distinctive reactivity of the enzyme with substrate analogs, a radical mechanism has been proposed for monoamine oxidase (Silverman and Hoffman, 1980), exemplifying the power of mechanism-based inactivators as probes of enzyme mechanism. In this mechanism, initial nitrogen radical cation formation by one-electron reduction of the flavin is proposed to render the α-hydrogen acidic. A subsequent one-electron transfer yields reduced flavin and the imine, which spontaneously hydrolyzes to the aldehyde.

A series of propargylamine analogs, including pargyline, clorgyline, and deprenyl (Fig. 18), are mechanism-based inactivators of monoamine oxidase. They have been particularly valuable pharmacological tools, as their study revealed the presence of isozymes of monoamine oxidase: monoamine oxidase A, which preferentially oxidizes serotonin and norepinephrine, is inactivated by clorgyline, and is relatively insensitive to deprenyl, whereas monoamine oxidase B, which oxidizes phenylethylamine and benzylamine, is selectively inactivated by deprenyl (Fowler *et al.*, 1982). The inactivation of bovine kidney monoamine oxidase by [7-^{14}C]pargyline results in reduction of the flavin and incorporation of 1 equivalent of radiolabel (Chuang *et al.*, 1974). A proteolytic fragment of the modified enzyme was isolated which contained the covalently bound flavin (bound through a thioether linkage between C-8a and a cysteine residue) and the radioactivity, leading to the suggestion that the flavin was modified, although its structure was not determined. In an analogous reaction, Maycock *et al.* (1976) demonstrated that the inactivation of monoamine oxidase by 3-dimethylamino-1-propyne resulted in formation of an N-5 flavin adduct. The mechanism proposed for this reaction (Fig. 19) is probably also applicable to inactivation of monoamine oxidase by pargyline.

Monoamine oxidase is also susceptible to inactivation by olefins such as allylamine. An isotope effect of 2.35 on inactivation by [1-$^{2}H_2$]allylamine and formation of a reduced flavin spectrum are consistent with monoamine oxidase-catalyzed oxidation of the compound (Rando and Eigner, 1977). Because the

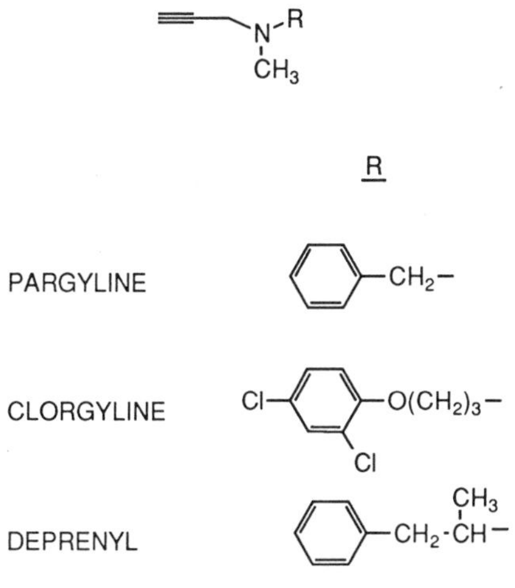

N-R
CH3

R

PARGYLINE CH2–

CLORGYLINE Cl– –O(CH2)3– Cl

DEPRENYL CH3 –CH2·CH–

FIG. 18. Propargylamine analogs that are inactivators of monoamine oxidase.

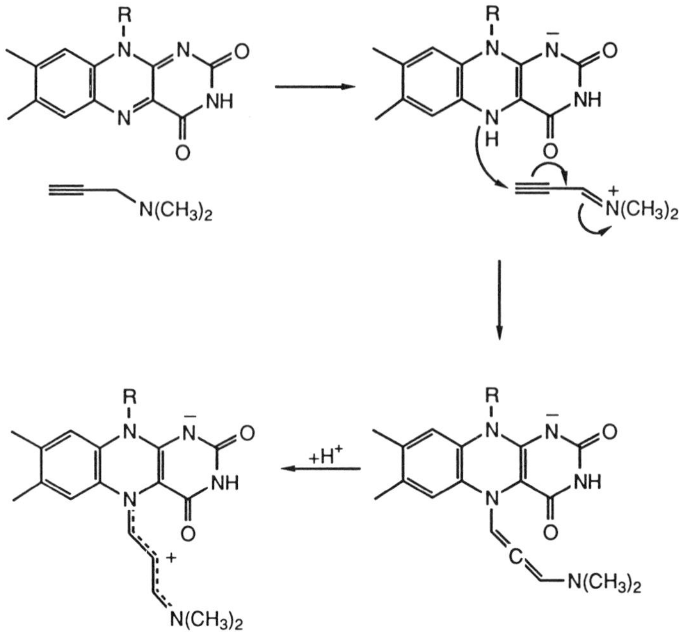

FIG. 19. Mechanism proposed for inactivation of monoamine oxidase by 3-dimethylamino-1-propyne.

reduced flavin spectrum was stable, inactivation was proposed to be due to alkylation of the reduced cofactor, in analogy to the reaction of 3-dimethylamino-1-propyne. Reexamination of the inactivation process by Silverman and co-workers (1985) revealed that the flavin is rapidly reoxidized on denaturation of the enzyme, indicating that the modification must reside on the enzyme. Treatment of the modified enzyme with benzylamine leads to recovery of oxidase activity and isolation of products consistent with addition of an amino acid residue to C-3 of propanal. In the mechanism suggested to account for the inactivation process (Fig. 20), the flavin semiquinone radical is in equilibrium with an amino acid radical, allowing modification of the enzyme to occur by radical recombination.

A great deal of attention has been focused on the interaction of monoamine oxidase with cyclopropylamine analogs, primarily through the efforts of Silverman and co-workers, which has provided much of the basis for the proposed radical mechanism for the enzyme. The clinically utilized antidepressant *trans*-2-phenylcyclopropylamine (2-PCPA, tranylcypromine) was recognized as a mechanism-based inactivator nearly 20 years after it was introduced to the market (Paech *et al.*, 1980). The mechanism initially proposed for the inactivation involved oxidation of 2-PCPA to 2-phenylcyclopropanone imine, which was suggested to alkylate a cysteine sulfhydryl. Silverman (1983) reported that acid denaturation of the inactive enzyme and treatment with 2,4-dinitrophenylhydrazine (2,4-DNP) yielded the 2,4-DNP derivative of cinnamaldehyde, not of 2-phenylcyclopropanone, indicating that cyclopropyl ring opening occurred during the modification process. The revised mechanism (Fig. 21) suggested oxidation of 2-PCPA to the amine radical cation, followed by subsequent cleavage of the cyclopropyl ring to yield a benzyl radical which alkylates an amino acid residue. Ring cleavage of cyclopropylamine radicals has been demonstrated to occur at rates too fast to measure in chemical model systems, providing a sound foundation for the mechanistic proposal (Maeda and Ingold, 1980).

Secondary amines such as *N*-cyclopropylbenzylamine (*N*-CBA) are also inactivators of monoamine oxidase. Inactivation by *N*-[*cyclopropyl*-^{3}H]N-CBA re-

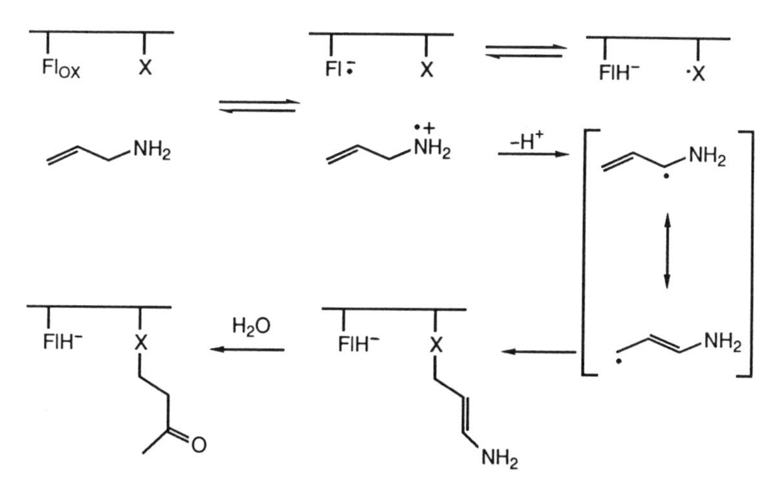

FIG. 20. Mechanism proposed for inactivation of monoamine oxidase by allylamine.

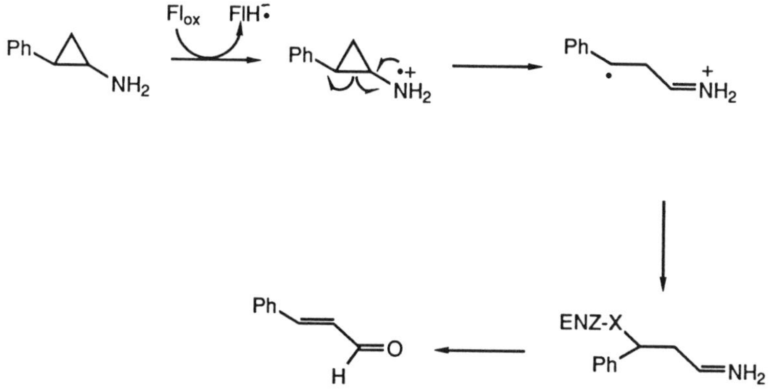

FIG. 21. Mechanism proposed for inactivation of monoamine oxidase by 2-phenylcyclopropylamine.

sults in formation of [^{3}H]acrolein and incorporation of 3 equivalents of label into the enzyme, indicating nonspecific labeling (Vazquez and Silverman, 1985). The reduced flavin of the inactivated enzyme is oxidized on denaturation of the enzyme, demonstrating that the flavin is not alkylated. Reactivation of the dead enzyme with benzylamine leads to release of a single equivalent of acrolein. A mechanism for inactivation was proposed which is analogous to that suggested for inactivation of monoamine oxidase by 2-PCPA (Fig. 22, pathway A). Although the nature of the adduct formed between *N*-CBA and monoamine oxidase is similar to the formed from allylamine, different amino acids must be modified, because the chemical stabilities of the two adducts differ.

Cyclopropylamine is also found as a product of the reaction of *N*-CBA with monoamine oxidase, demonstrating that the hypothesized secondary amine radical cation intermediate partitions equally between oxidation of the benzyl and cyclopropyl groups (Fig. 22, pathway B), a surprising result given the extremely rapid rearrangement of the cyclopropyl aminyl radical (Silverman, 1984). Methylation of the benzyl carbon to yield *N*-cyclopropyl-α-methylbenzylamine decreases the overall rate of product formation and alters the partitioning of the amine radical cation, with the rate of benzyl oxidation decreased to about 2% of the rate of cyclopropyl oxidation. The shift in partitioning of the intermediate is proposed to be due to an increase in the pK_a of the benzyl α-proton resulting from methylation of the carbon. The α,α-dimethyl compound is also a time-dependent inactivator, although it is proposed to modify the flavin as well as the enzyme.

Addition of a methyl group to C-1 of *N*-CBA to produce *N*-(1-methylcyclopropyl)benzylamine is a small structural change, but a significantly different outcome is observed in its reaction with monoamine oxidase (Silverman and Yamasaki, 1984). Radiolabeling experiments indicate that approximately 1.3 equivalents of the cyclopropyl and methyl groups are incorporated per equivalent of inactive enzyme, whereas the benzyl group is released as benzylamine and benzaldehyde. Analysis of the inactive enzyme is consistent with modification of the flavin at an unknown position by a methyl ketone-containing moiety, leading to proposal of the mechanism depicted in Fig. 23. Alternatively, the flavin semiquinone radical could attack the amine radical cation directly, leading to the same adduct. A rationale for the selectivity of various cyclopropylamine-containing inactivators for modification of the flavin or the protein has been put forth (Silverman and Zieske, 1985).

A great deal of interest has recently been expressed in the interaction of monoamine oxidase with 1-methyl-4-phenyl-1,2,3,6-tetrahydropyridine (MPTP), an agent which has been shown to cause symptoms of Parkinson's disease in humans. Treatment of animals with deprenyl or pargyline protects against the neurotoxic effects of MPTP, demonstrating that isozyme B is involved in its activation. MPTP is converted by monoamine oxidase to 1-methyl-4-phenyl-2,3-dihydropyridinium ion (MPDP$^+$), which is subsequently oxidized to yield 1-

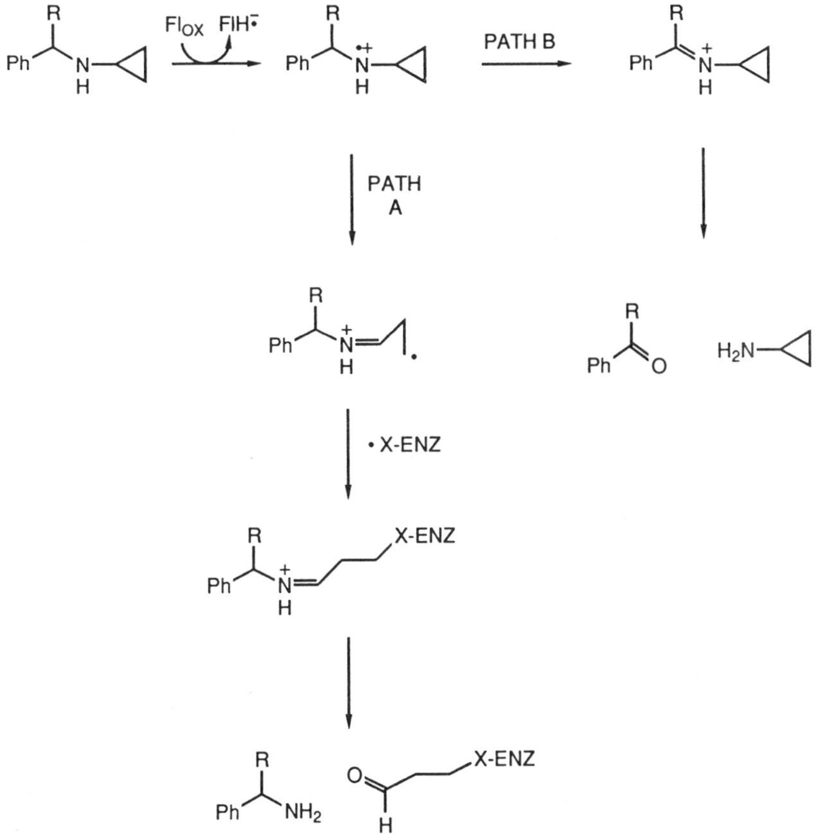

FIG. 22. Mechanism proposed for inactivation of monoamine oxidase by *N*-cyclopropylbenzylamine.

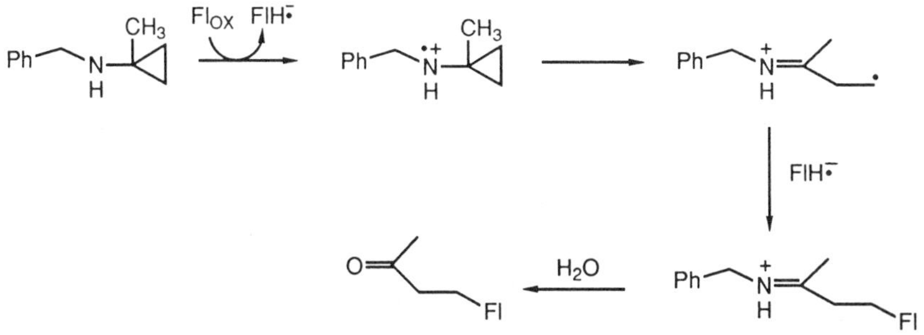

FIG. 23. Mechanism proposed for inactivation of monoamine oxidase by *N*-(1-methylcyclopropyl)benzylamine.

methyl-4-phenylpyridinium ion (MPP^+) by monoamine oxidase or by nonenzymic processes (Singer *et al.*, 1986). Both MPTP and $MPDP^+$ have been shown to inactivate the enzyme at comparable rates (Singer *et al.*, 1986). Inactivation with [*methyl*-^{3}H]MPTP leads to incorporation of 5 equivalents of radiolabel per equivalent of oxidase, all of which is bound to the protein. No mechanism has been proposed, although the electrophilic nature of $MPDP^+$ is likely to be involved. Buckman and Eiduson (1985), however, claim that the inactivation of monoamine oxidase B by MPTP occurs in a photoinduced process through the intermediacy of an unidentified metabolite which is characterized by an absorbance at 345 nm. Further studies clearly will be necessary to clarify the reaction of monoamine oxidase with MPTP.

C. Cyclohexanone Oxygenase

Cyclohexanone oxygenase (cyclohexanone monooxygenase) is an intriguing flavoprotein monooxygenase which converts a variety of cyclic ketones to lactones in a biological Baeyer–Villiger reaction. The enzyme is irreversibly inactivated by the 2-thia derivatives of C_5 to C_7 cyclic ketones as well as by ethylene monothiocarbonate (Latham and Walsh, 1987). Incubation of the oxygenase with 2-[^{35}S]thiacyclopentanone in the presence of oxygen and NADPH results in net incorporation of 0.9 equivalents of radioactivity, which is localized on the protein rather than the cofactor. A DTNB-tritatable thiol appears in the native enzyme at a rate identical of that of the inactivation process, whereas thiol titration of the ethylene monothiocarbonate-inactivated, denatured enzyme indicates a loss of two sulfhydryl groups. The mechanism proposed for inactivation by ethylene monothiocarbonate is shown in Fig. 24. Baeyer–Villiger oxidation of the

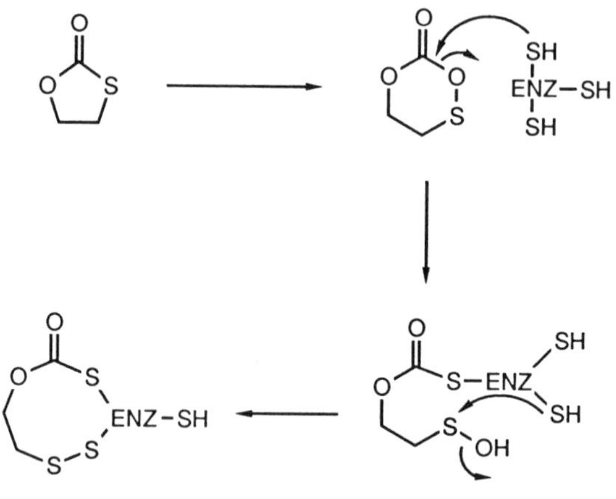

FIG. 24. Mechanism proposed for inactivation of cyclohexanone oxygenase by ethylene monothiocarbonate.

inactivator (or oxidation to an acyl sulfoxide) is followed by acylation of the enzyme. Subsequently, the attack of a second enzyme thiol of the sulfenic acid moiety cross-links the enzyme. The ability to form a cyclic adduct is apparently crucial, since *S*-phenylthioacetate is oxidized without enzyme inactivation.

V. Hemoprotein Enzymes

Hemoprotein enzymes primarily catalyze peroxidation, monooxygenation, and dioxygenation reactions. The reaction catalyzed by a given hemoprotein is determined by (1) the identity of the fixed axial iron ligand, (2) the ability of the hemoprotein to interact with physiological electron donors, (3) the degree of exposure of the heme group, and (4) the nature of the amino acid residues that interact with the heme and its ligands (Ortiz de Montellano, 1987). Mechanism-based inactivating agents have been developed for the monooxygenases and peroxidases, but the development of inactivators of hemoprotein dioxygenases is still in a rudimentary state and is not discussed here. The cytochrome *P*-450 enzymes and peroxidases activate oxygen by different mechanisms but are thought to express related oxidative species in which a ferryl complex [Fe^{IV}=O] is paired with a porphyrin or protein radical. Cytochrome *P*-450 enzymes insert an oxygen atom into their substrates, however, whereas the peroxidases abstract a single electron from theirs. Nevertheless, it is now thought that many of the two-electron oxidations catalyzed by cytochrome *P*-450 are achieved by two closely spaced one-electron steps (Ortiz de Montellano, 1986). Peroxidases are therefore usually inactivated by free radical products, whereas cytochrome *P*-450 monooxygenases can be inactivated either by free radical intermediates or by two-electron oxidized, electrophilic products. In general, free radicals tend to react with the prosthetic heme group and charged or neutral electrophilic products with the protein, although radicals may also react with the protein.

A. Peroxidases

Hemoprotein peroxidases cleave the dioxygen bond of hydrogen peroxide heterolytically to give a ferryl oxygen complex and a porphyrin radical cation or protein radical (Marnett *et al.*, 1986; Hewson and Hager, 1979; Dunford and Stillman, 1976). The ferryl oxygen is stabilized by hydrogen bonding and polar active site residues that hinder interaction of the ferryl oxygen complex with substrates. Recent work suggests, in fact, that substrates interact with, and deliver electrons to, the heme periphery rather than the ferryl oxygen (Ortiz de Montellano, 1987; Ator and Ortiz de Montellano, 1987). This rationalizes the fact that many of the mechanism-based inactivators of these enzymes add to the meso carbons of the prosthetic heme group rather than to the iron or nitrogen atoms usually involved in the reactions of cytochrome *P*-450. Horseradish peroxidase thus oxidizes cyclopropanol hydrate (Wiseman *et al.*, 1982), nitromethane (Porter and Bright, 1983), phenylhydrazine (Ator and Ortiz de Montellano, 1987), a variety of alkylhydrazines (Ator *et al.*, 1987), and sodium azide (Ortiz de Montellano *et al.*, 1988) to free radicals that react with the heme group to give the corresponding δ-meso-substituted heme adducts.

These reactions are thought to involve oxidation of the agents, in the case of the hydrazines after initial oxidation to the diazenes, to free radicals that react with the heme group (Fig. 25). Explicit evidence exists for the isoporphyrin intermediate (bottom right structure, Fig. 25) in the case of cyclopropanol hydrate (Wiseman *et al.*, 1982) and the alkylhydrazines (Ator *et al.*, 1987, 1989).

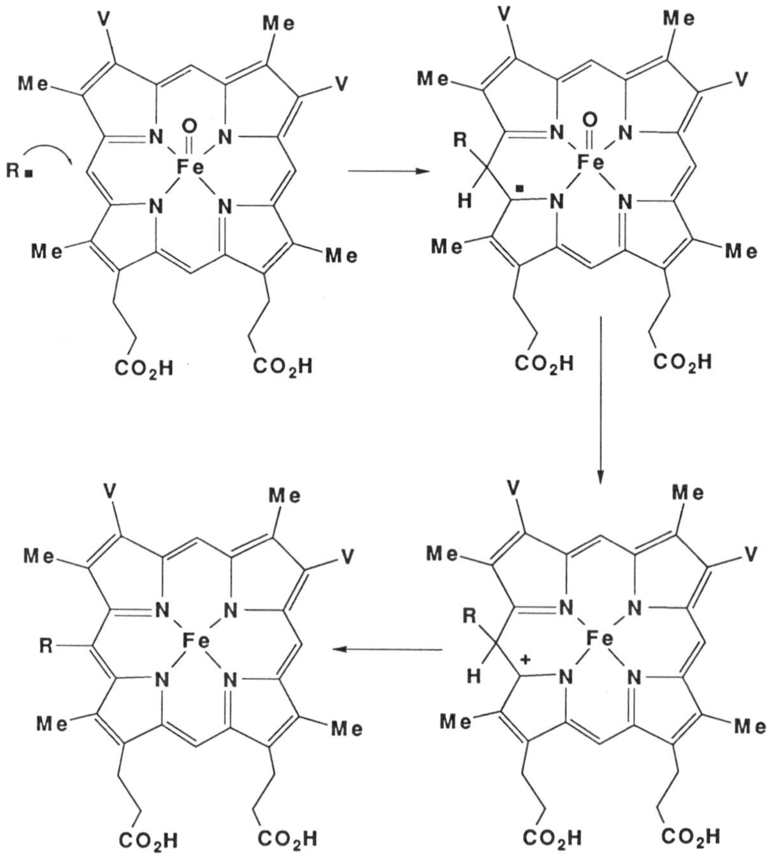

R = CH_2CH_2CHO, CH_2NO_2, Ph, Alkyl, N_3

FIG. 25. Mechanism proposed for inactivation of peroxidases by meso-alkylation of the prosthetic heme.

Enzyme inactivation by the alkyl hydrazines correlates well with formation of the isoporphyrin intermediate, but only in the case of methylhydrazine is catalytic activity recovered as the isoporphyrin decays to the meso-alkylated heme. The loss of activity is therefore caused by steric effects of the meso-substituent on the structure of the active site or, more probably, by interference with the approach of substrates to the heme edge. Analogous inactivation of the fungal peroxidase from *Coprinus macrorhizus* by sodium azide suggests that the inactivation mechanism may be relatively general (DePillis and Ortiz de Montellano, 1989). The fungal peroxidase, however, is not inactivated by alkyl hydrazines, a finding which suggests that the meso edge may be masked to different extents in different peroxidases and may make them more or less susceptible to inactivation by meso derivatization.

Horseradish peroxidase can be inactivated by protein as well as heme modifications. The reaction of horseradish peroxidase with phenylhydrazine is instructive in this regard (Ator and Ortiz de Montellano, 1987). Enzyme inactivation correlates well with covalent binding of 2 equivalents of radiolabeled phenylhydrazine but is associated with conversion of only a small fraction of the prosthetic group to the meso-phenyl adduct. Covalent binding of a metabolite of phenylhydrazine to the protein is therefore primarily responsible for enzyme inactivation, although the identity of the reactive species and the site at which it reacts are not known. The factors that determine whether the heme group (e.g., methylhydrazine, sodium azide) or the protein (e.g., phenylhydrazine) is the primary site of the inactivation reaction also remain elusive.

Thyroid peroxidase and lactoperoxidase, but not horseradish peroxidase (Petry and Eling, 1987), are inactivated by heterocyclic sulfur compounds such as propylthiouracil and 1-methylbenzimidazoline-2-thione (Fig. 26) (Neary *et al.*, 1984; Ohtaki *et al.*, 1985; Doerge, 1986). The inactivation of thyroid peroxidase

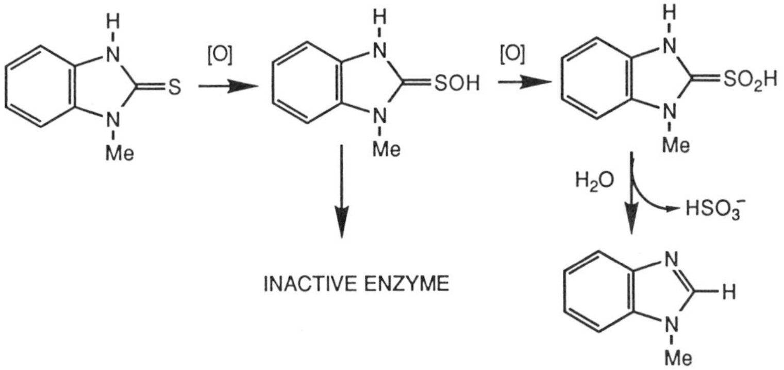

FIG. 26. Mechanism proposed for inactivation of thyroid peroxidase and lactoperoxidase by 1-methylbenzimidazoline-2-thione.

by these agents is of clinical utility in the treatment of hyperthyroidism (Gilman and Murad, 1975). Inactivation of lactoperoxidase by ^{14}C- and ^{35}S-labeled 1-methylbenzimidazoline-2-thione (Doerge, 1988a,b), and of thyroid peroxidase by ^{14}C-labeled 1-methylbenzimidazoline-2-thione or ^{35}S-labeled propylthiouracil (Engler *et al.*, 1982), correlates well with covalent binding of approximately 1 equivalent of each radiolabel. The chromophores of the two peroxidases are altered by these inactivation reactions, but no clear evidence is available on whether the inactivating agent is bound to the prosthetic group or the protein. The oxidation of 1-methylbenzimidazoline-2-thione associated with inactivation of lactoperoxidase yields 1-methylbenzimidazole and bisulfite as products (Doerge, 1988a,b). Doerge has proposed that lactoperoxidase-catalyzed oxygenation of the thiocarbamate sulfur atom produces a reactive species that reacts with the enzyme or is eventually further oxidized to yield sulfite (Fig. 26). It has been proposed that the sulfur is oxygenated via a radical cation intermediate because the relationships between the binding constants for substrates and inhibitors and the peak oxidation potentials of the inactivating agents are similar (Doerge *et al.*, 1987). The details of the inactivation mechanism remain to be clearly established, however. It is not known why thyroid peroxidase and lactoperoxidase, but not horseradish peroxidase, are susceptible to inactivation by heterocyclic sulfur compounds whereas all three enzymes are inactivated by phenylhydrazine or other hydrazines (Allison *et al.*, 1973; Hidaka *et al.*, 1970).

Chloroperoxidase is an unusual peroxidase in that it has a thiolate rather than imidazole iron ligand and catalyzes monooxygenation as well as halogenation and peroxidation reactions (Hewson and Hager, 1979; Ortiz de Montellano *et al.*, 1987; Colonna *et al.*, 1988). The enzyme self-inactivates very rapidly in the presence of H_2O_2 but absence of substrates, a process which makes the identification of mechanism-based inactivating agents difficult. It has been shown, nevertheless, that 1-aminobenzotriazole is a mechanism-based inactivator of the enzyme (Ortiz de Montellano *et al.*, 1984). This inactivation, which is probably mediated by the same mechanism as inactivation of cytochrome *P*-450 (see below), is unlikely to be relevant to classic peroxidases because it requires direct interaction of the inactivating agent with the ferryl oxygen.

B. Cytochrome-*P*-450 Monooxygenases

Cytochrome-*P*-450 monooxygenases reduce molecular oxygen to a molecule of water and what is thought to be a ferryl complex analogous to that of the peroxidases (Ortiz de Montellano, 1986; Guengerich and Macdonald, 1984). The activated oxygen species, however, has only been characterized by indirect methods and by analogy to model systems, so that its precise nature remains ambiguous. Nevertheless, the crystal structure of cytochrome $P\text{-}450_{cam}$ (Poulos *et al.*, 1987) and a wealth of evidence on the microsomal and mitochondrial enzymes conclusively establish that the substrate is bound close to the activated oxygen and reacts directly with it (Ortiz de Montellano, 1986). Furthermore, the evidence provided by studies of the reaction products, including evidence from studies with mechanism-based inhibitors, suggests that the reactions catalyzed by cytochrome *P*-450 can often be resolved into two one-electron steps (Ortiz de Montellano, 1986; Guengerich and Macdonald, 1984).

The first mechanism-based inactivator of cytochrome *P*-450 to be identified as such was 2-isopropyl-4-pentenamide, a potent porphyrinogenic agent used to produce in rodents a biochemical state comparable to that of the hepatic porphyrias in humans (De Matteis, 1978; Ortiz de Montellano *et al.*, 1978). Exploration of the structural requirements for mechanism-based inactivation of rat liver cytochrome *P*-450 by this agent has shown that the destructive potential is intrinsic to the terminal olefin group, so that olefins as simple as 1-octene or even ethylene are effective mechanism-based inactivating agents (Ortiz de Montellano and Mico, 1980). Terminal acetylenes, including acetylene itself, also convey the ability to inactivate cytochrome *P*-450 (Ortiz de Montellano and Kunze, 1980a). The inactivation of cytochrome *P*-450 by terminal olefins and acetylenes involves addition of the activated oxygen atom to the internal carbon and a nitrogen of the porphyrin to the terminal carbon of the π bond. This yields an *N*-(2-hydroxyalkyl) heme adduct in the case of olefins and an *N*-(2-oxoalkyl) adduct in the case of acetylenes (Fig. 27) (Ortiz de Montellano, 1986; Ortiz de Monte-

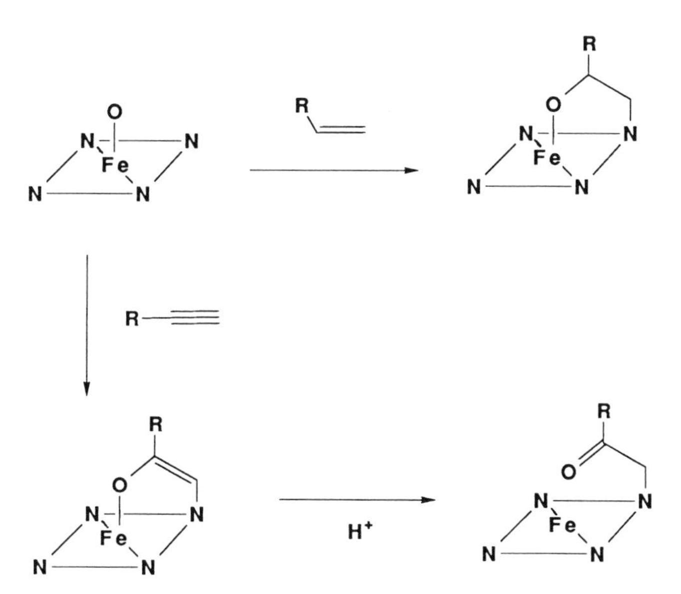

FIG. 27. Structure of alkylated heme species resulting from oxidation of olefins and acetylenes by cytochrome *P*-450.

llano and Reich, 1986). Our understanding of the mechanisms of these N-alkylation reactions is entangled in the uncertainties concerning the mechanisms by which π bonds are oxidized. The two strongest candidates are addition of the ferryl oxygen to the π bond to give a metallooxetane that subsequently rearranges to either the epoxide metabolite or the *N*-alkyl heme adduct. The alternative is addition of the ferryl oxygen to the olefin, possibly via initial electron abstraction, to give a carbon radical intermediate that closes to the epoxide or reacts with the heme nitrogen. These mechanisms are not mutually exclusive because the carbon radical and metallooxetane intermediates are, in principle, interconvertible (Fig. 28) (Ortiz de Montellano, 1986). The N-alkylation mechanism for terminal acetylenes is somewhat less ambiguous. The absence of substituent effects in alkylation of the heme by substituted phenylacetylenes and the observation of large primary kinetic isotope effects for migration of the acetylenic hydrogen to the adjacent carbon in the formation of ketene metabolites rule out radical cation intermediates and argue against metallooxetene intermediates (Komives and Ortiz de Montellano, 1987). Heme alkylation, in any case, has only been observed with terminal olefins and acetylenes and always involves addition of the oxygen to the internal carbon. Acetylenes appear to have lower partition coefficients than comparable olefins (Ortiz de Montellano, 1988) and have been used to construct isozyme-selective inactivating agents, some of which have potential clinical utility as inactivators of aromatase or other pharmacologically desirable targets (e.g., Metcalf *et al.*, 1981; CaJacob and Ortiz de Montellano, 1986).

The oxidation of alkyl and aryl hydrazines or 4-alkyl-3,5-bis(carbethoxy)-2,6-dimethyl-1,4-dihydropyridines by cytochrome *P*-450 results in addition of the

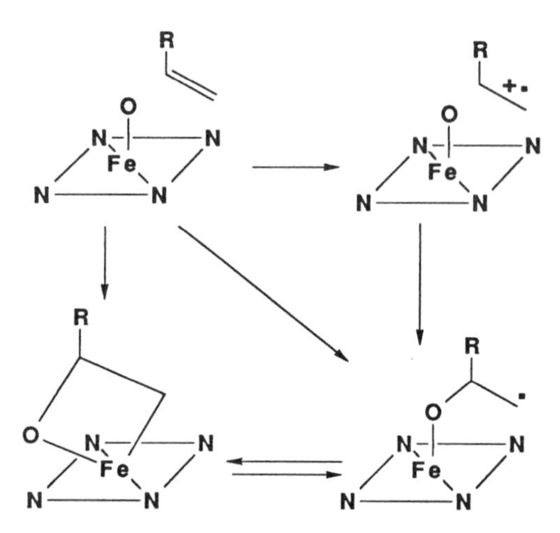

FIG. 28. Intermediates in the alkylation of the heme of cytochrome *P*-450 by olefins.

alkyl or aryl moieties to one of the heme nitrogen atoms (Fig. 29). *N*-Ethyl- and *N*-phenylethylprotoporphyrin IX have thus been isolated from incubations of cytochrome *P*-450 with the 4-ethyldihydropyridine and phenylethylhydrazine, respectively (Ortiz de Montellano *et al.*, 1981; Augusto *et al.*, 1982; Ortiz de Montellano *et al.*, 1983). Spin-trapping studies explicitly established that cytochrome *P*-450 catalyzes oxidative release of the ethyl and phenylethyl radicals, respectively, from these two substrates. Formation of free radicals from the alkylhydrazines involves oxidation to the alkyldiazenes followed by a one-electron oxidation and elimination of nitrogen gas. The formation of alkyl radicals from the 4-alkyl-1,4-dihydropyridines is explained by electron removal from the heterocycle to give a radical cation that aromatizes by extruding the 4-alkyl group as a free radical (Lee *et al.*, 1988). Structure–activity studies show that primary alkyl radicals are required for heme alkylation, although secondary radicals inactivate the enzyme by unidentified mechanisms that possibly involve reaction with the protein (Augusto *et al.*, 1982; McCluskey *et al.*, 1986). Primary radicals may also, in some instances, cause protein alkylation reactions (Correia *et al.*, 1987). The size of the alkyl group is not a critical determinant of inactivating activity but may alter the isozyme specificity of the reaction. The finding that the oxidation of 2,2-dialkyl-1,2-dihydroquinolines also results in heme N-alkylation shows that similar reactions can occur with partially saturated heterocycles other than the 4-alkyl-1,4-dihydropyridines (Lukton *et al.*, 1988).

The details of heme N-alkylation by alkyl radicals are unclear because no intermediates have been detected in the reaction. However, the fact that reaction

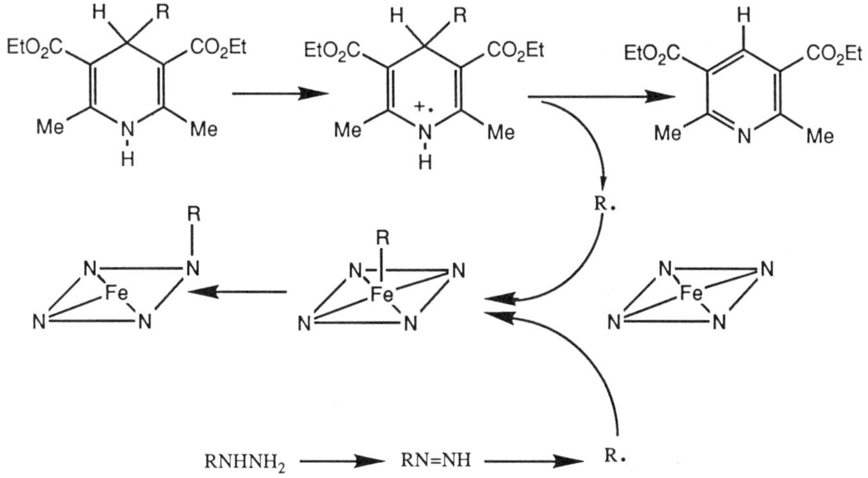

$RNHNH_2 \longrightarrow RN{=}NH \longrightarrow R\cdot$

FIG. 29. Mechanism proposed for inactivation of cytochrome *P*-450 by alkyl hydrazines and 4-alkyl-1,4-dihydropyridines.

of cytochrome *P*-450 with phenylhydrazine yields a complex with an absorption band at 480 nm that rearranges under oxidative conditions to *N*-phenyl heme (Jonen *et al.*, 1982; B. Swanson and P. R. Ortiz de Montellano, unpublished) suggests that heme alkylation by alkyl radicals may involve transient σ-bonded alkyl–iron complexes (Fig. 29). Several pieces of evidence indicate that the intermediate which absorbs at long wavelengths is a σ-bonded phenyl–iron complex: (1) the reaction of myoglobin with phenylhydrazine has been shown by several methods, including X-ray crystallography, to yield such a complex (Ringe *et al.*, 1984), (2) a chemical model exists for the phenyl–iron complex and for the proposed iron–nitrogen phenyl shift (Ortiz de Montellano *et al.*, 1982), and (3) the prosthetic group has been extracted from phenylhydrazine-inactivated cytochrome *P*-450 and shown recently to be the iron–phenyl heme complex (B. Swanson and P. R. Ortiz de Montellano, unpublished).

The oxidative formation of other species with free radical character affords alternative avenues to alkylation of the prosthetic heme group of cytochrome *P*-450. Two such species are benzyne and cyclobutadiene, in which the geometry of the molecule makes one of the π bonds highly reactive. Oxidation of 1-aminobenzotriazole, which is known to be chemically oxidized to benzyne, results in inactivation of the enzyme and formation of a heme adduct in which an ortho-disubstituted benzene ring bridges two of the heme nitrogen atoms (Fig. 30) (Ortiz de Montellano and Mathews, 1981a; Ortiz de Montellano *et al.*, 1984). As already noted, chloroperoxidase, which appears to have a cytochrome *P*-450-like active site, is also inactivated by 1-aminobenzotriazole. By analogy with the phenylhydrazine reaction, it is likely that benzyne adds initially to the iron and one of the porphyrin nitrogens and then undergoes an iron–nitrogen shift to give the second carbon–nitrogen bond (Fig. 30). Substituted 1-aminobenzotriazoles retain the ability to inactivate cytochrome *P*-450 (Ortiz de

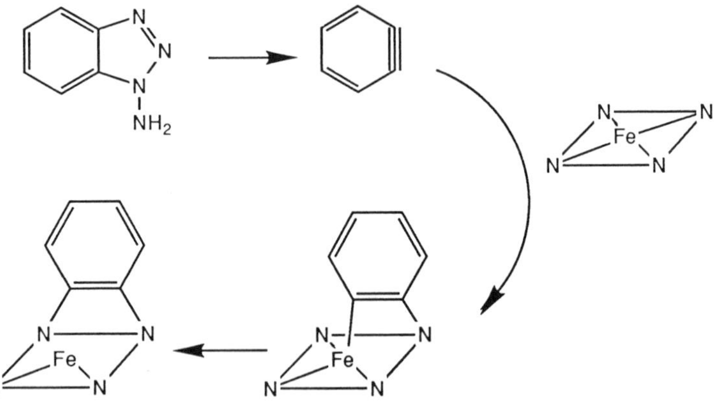

FIG. 30. Mechanism proposed for inactivation of cytochrome *P*-450 by 1-aminobenzotriazole.

Montellano *et al.*, 1984). The cytochrome *P*-450-catalyzed oxidation of 2,3-bis(carbethoxy)-2,3-diazabicyclo[2.1.0]hex-5-ene, a known cyclobutadiene precursor, results in enzyme inactivation and eventual isolation of *N*-(2-cyclobutenyl)protoporphyrin IX (Stearns and Ortiz de Montellano, 1985). The formation of this adduct is readily rationalized by oxidative release of cyclobutadiene (or its radical cation) which binds to a nitrogen of the heme and then abstracts a hydrogen from a protein residue (Fig. 31).

Carbenes constitute a further class of neutral reactive species that provide a route to alkylation of the prosthetic heme group. The oxidation of sydnones by cytochrome *P*-450, which yields an alkyl diazonium product, results in inactivation of the enzyme (Ortiz de Montellano and Grab, 1986; Grab *et al.*, 1988). This loss of activity is associated with the formation of *N*-alkyl heme adducts best rationalized by addition of a catalytically generated carbene or carbene equivalent to the heme group (Fig. 32). Model studies have shown that diazo compounds do react with reduced iron porphyrins to give N-alkylated products (Komives *et al.*, 1988). The details of the enzymic N-alkylation reactions remain to be defined, but evidence exists for two distinct reaction pathways. The first is

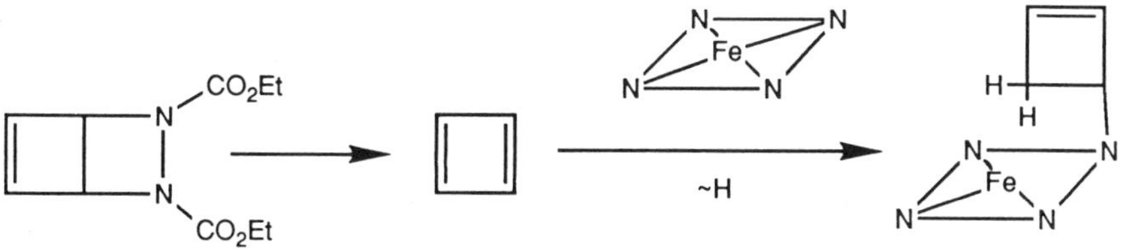

FIG. 31. Mechanism proposed for inactivation of cytochrome *P*-450 by 2,3-bis(carbethoxy)-2,3-diazabicyclo[2.1.0]hex-5-ene.

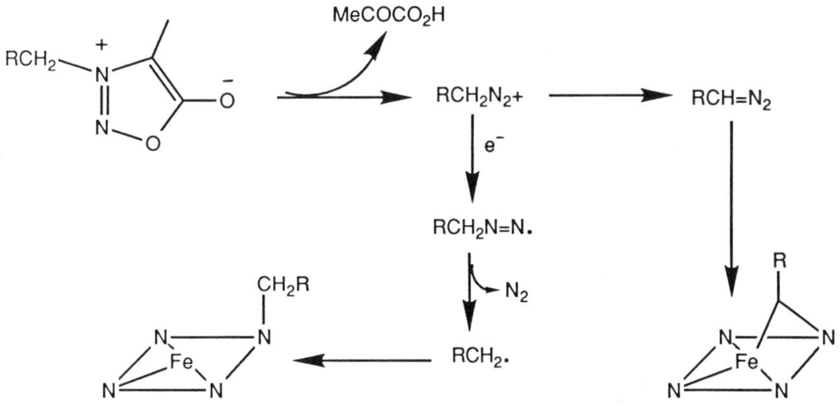

FIG. 32. Mechanism proposed for inactivation of cytochrome *P*-450 by sydnones.

one-electron reduction of the initial diazonium species to give the corresponding diazenyl radical. The alkyl free radical obtained by elimination of nitrogen from this intermediate reacts with the heme group to give the *N*-alkyl adduct (Grab *et al.*, 1988). The second pathway involves reaction of the diazoalkane, generated by deprotonation of the initial alkyl diazonium metabolite, with the iron atom of the prosthetic heme group. This probably yields an intermediate with the carbene carbon bridging the iron and one of the heme nitrogen atoms, although the products actually isolated from the reaction result from secondary reactions of this proposed bridged species. A carbene–iron complex may precede the bridged intermediate, but there is no actual evidence for such a complex (Ortiz de Montellano and Grab, 1986; Grab *et al.*, 1988). Little is known at this time about the structure–activity relationships for the inactivation of cytochrome *P*-450 by sydnones or diazo compounds.

The cytochrome *P*-450 inactivation reactions discussed so far have focused on alkylation of the heme group, but mechanism-based inactivating agents are also known that acylate or alkylate the protein framework (for reviews, see Ortiz de Montellano and Reich, 1986; Ortiz de Montellano, 1988). Broad substrate specificity, ability of the oxidizing agent to interact with multiple positions in the substrate, and extremely powerful oxidizing species make the cytochrome *P*-450-catalyzed formation of reactive metabolites inevitable. Only in a few cases, however, is there convincing evidence that cytochrome *P*-450 enzymes are alkylated by reactive species produced within their own active sites rather than by freely diffusible reactive species. The most thoroughly investigated of these agents is chloramphenicol, which owes its inactivating activity to the presence of a $COCHCl_2$ function (Reilly and Ivey, 1979; Halpert and Neal, 1980; Halpert, 1982). The acyl chloride produced by hydroxylation and elimination of HCl acylates one of the lysine residues or adds water to give the free acid metabolite (Fig. 33) (Halpert, 1981). The sequence position of the acylated lysine is not known, but it has been shown that its acylation interferes with the transfer of electrons from NADPH–cytochrome-*P*-450 reductase to the iron rather than with substrate binding or substrate oxidation (Halpert *et al.*, 1983, 1985). The acyl chloride intermediate thus appears to shift out of the substrate binding site before it reacts with the lysine residue. Structure–activity studies show that the $COCHCl_2$ function conveys the ability to inactivate cytochrome *P*-450 when inserted into different structures, the isozyme that is inactivated being determined by the carrier structure (Halpert *et al.*, 1986). Steroids with a 17-$COCHCl_2$ function, for example, are selective inactivators of the isozyme that catalyzes 21-hydroxylation of progesterone (Halpert *et al.*, 1988).

Addition of the ferryl oxygen to the internal carbon of a terminal acetylene results, as already discussed, in prosthetic heme alkylation. Addition of the oxygen to the terminal carbon, however, is associated with intramolecular 1,2-shift of the terminal hydrogen to give a ketene (Ortiz de Montellano, 1986). The

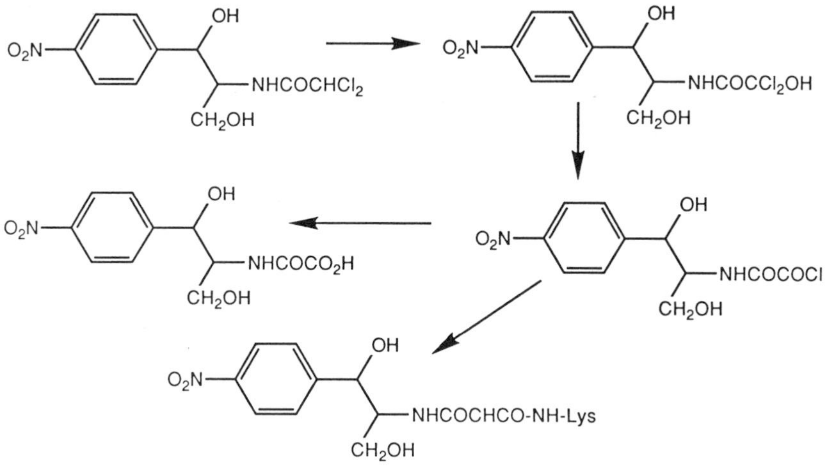

FIG. 33. Mechanism proposed for inactivation of cytochrome *P*-450 by chloramphenicol.

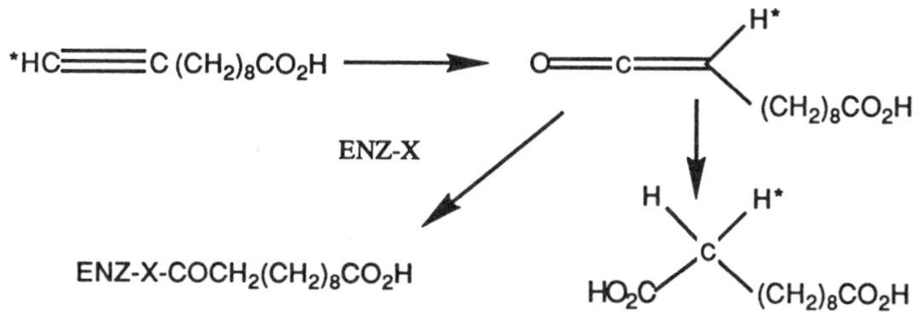

FIG. 34. Mechanism proposed for inactivation of cytochrome *P*-450 by terminal acetylenes.

ketene is normally hydrolyzed to the acid metabolite but may also inactivate the enzyme, presumably by acylating a protein residue (Fig. 34). Evidence for a protein modification reaction in the oxidation of acetylenes was first provided by the finding that the inactivation of microsomal enzymes by ethynyl-substituted polycyclic aromatic hydrocarbons substantially exceeds loss of the prosthetic heme group (Gan *et al.*, 1984). Subsequent studies with the lauric acid ω-hydroxylase purified from rat liver have demonstrated that inactivation of this enzyme by 10-undecynoic acid is due to a reaction with the protein rather than the heme group (CaJacob *et al.*, 1988). The inactivation appears to reflect acylation of the protein because labeled agent is bound to the protein with a very low partition ratio, but there is no direct information on the site or nature of the protein reaction.

The cytochrome *P*-450-catalyzed oxidation of sulfhydryl, thiocarbonyl, or thiophosphate groups produces reactive species that, in some instances, inactivate the enzyme in a true mechanism-based process. The oxidation of parathion, a thiophosphate, is the best characterized of these reactions (Halpert *et al.*, 1980; Halpert and Neal, 1981). Catalytic oxidation of parathion results in inactivation of the enzyme, loss of the heme group, and covalent binding of ^{35}S-labeled compound to the enzyme. The mechanism proposed for this reaction is oxygenation of the sulfur followed by its extrusion as a highly reactive sulfur atom (Fig. 35), but there is little evidence on the actual mechanism of the inactivation reaction. Replacement of the 19-methyl by a sulfhydryl group in a sterol has been used to construct a potent mechanism-based inactivating agent for aromatase, but the mechanism of the inactivating reaction is not known (Bednarski *et al.*, 1985). It is possible that the sulfhydryl group is oxidized to a sulfenic acid (SOH) that reacts with a sulfhydryl group or other nucleophile in the protein active site.

A variety of alternative routes, most of which are poorly understood but appear to differ in concept from those already discussed, have been reported for the mechanism-based inactivation of cytochrome-*P*-450 monooxygenases. The oxidation of allenes results in inactivation of the enzyme and loss of the heme group but not in the formation of detectable N-alkylated heme adducts (Ortiz de Montellano and Kunze, 1980b). Allenes have been used in the design of potentially useful inactivators of aromatase even though the nature of the allene-dependent inactivation reaction remains unclear (Metcalf *et al.*, 1981). The oxidation of 1,2,3-benzothiadiazoles is another example of a reaction that results in enzyme inactivation and prosthetic heme loss without the detectable formation of *N*-alkyl or *N*-aryl heme adducts (Ortiz de Montellano and Mathews, 1981b). Benzyl cyclopropylamines are oxidized, presumably to the corresponding cyclopropyl ring-opened radicals that alkylate and inactivate the enzyme, but the details of the inactivation reaction remain obscure (Hanzlik and Tullman, 1982; Macdonald *et al.*, 1982). The psoralens inactivate cytochrome *P*-450 by a mechanism that involves covalent binding of the agent to the protein, but the

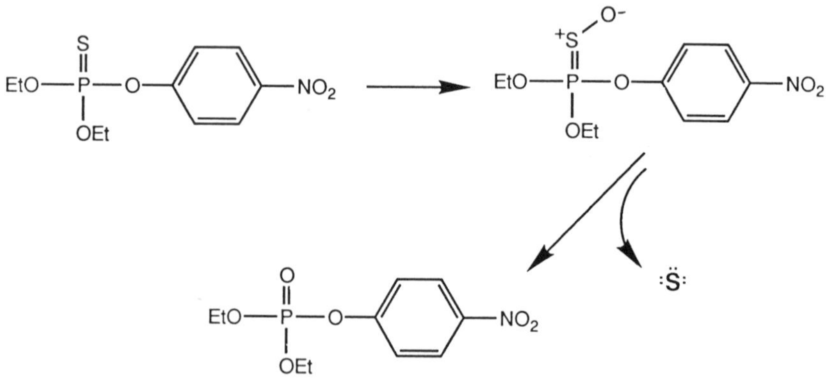

FIG. 35. Mechanism proposed for inactivation of cytochrome *P*-450 by parathion.

identity of the reactive species and the mechanism by which it is bound are not known (Fouin-Fortunet *et al.*, 1986). Placement of a trimethylsilyl group β to the position hydroxylated by the cytochrome *P*-450 enzyme that cleaves the cholesterol side chain has been shown to result in mechanism-based inactivation of the enzyme (Nagahisa *et al.*, 1984). This reaction may involve extrusion of the trimethylsilyl radical during the hydroxylation reaction, but the actual mechanism of the inactivation reaction is unknown. As these examples illustrate, the ability of cytochrome *P*-450 to oxidize most organic functionalities provides broad scope for the development of mechanism-based inactivating agents and leaves little doubt that new routes will be found for the inactivation of these enzymes.

VI. Nonheme Metalloenzymes

In analogy with their heme-containing counterparts, the nonheme metalloproteins appear to catalyze oxidation reactions by radical processes. In contrast to hemoproteins, however, the structural and mechanistic details of the nonheme metalloproteins, including the nature of their metal centers, are largely unknown. The study of mechanism-based inactivators has proved to be an important source of otherwise elusive mechanistic information on this class of enzymes.

A. Dopamine β-Hydroxylase

Dopamine β-hydroxylase (dopamine β-monooxygenase) is a monooxygenase which catalyzes the stereospecific benzylic hydroxylation of dopamine to generate norepinephrine. The tetrameric enzyme contains two atoms of Cu(II) per subunit which are reduced to Cu(I) by ascorbate and interact with dioxygen to produce an activated oxygen species of unknown and highly controversial structure. The tolerance of the enzyme for substitution on the phenyl ring and alterations to the side chain of substrates has allowed a variety of approaches to its mechanism-based inactivation (reviewed by Fitzpatrick and Villafranca, 1987). An unsaturated analog of the substrate tyramine, *p*-hydroxybenzylcyanide (**16,** Fig. 36), was the first reported inactivator of dopamine β-hydroxylase. The oxygen- and ascorbate-dependent turnover of **16** results in production of *p*-hydroxymandelonitrile (**17,** Fig. 36) and inactivation of the enzyme with a partition ratio of 8000 (Baldoni and Villafranca, 1980). The kinetics of inactivation by a selection of phenyl-substituted analogs and the conditions under which the modified enzyme could be reactivated were examined in detail (Columbo *et al.*, 1984a,b). The stoichiometry of labeling of the completely inactivated hydroxylase by [*ring*-^{3}H]**16** reaches a maximal value of 0.34 mol ^{3}H/mol of tetramer under 100% oxygen, indicating that inactivation must occur by more than one

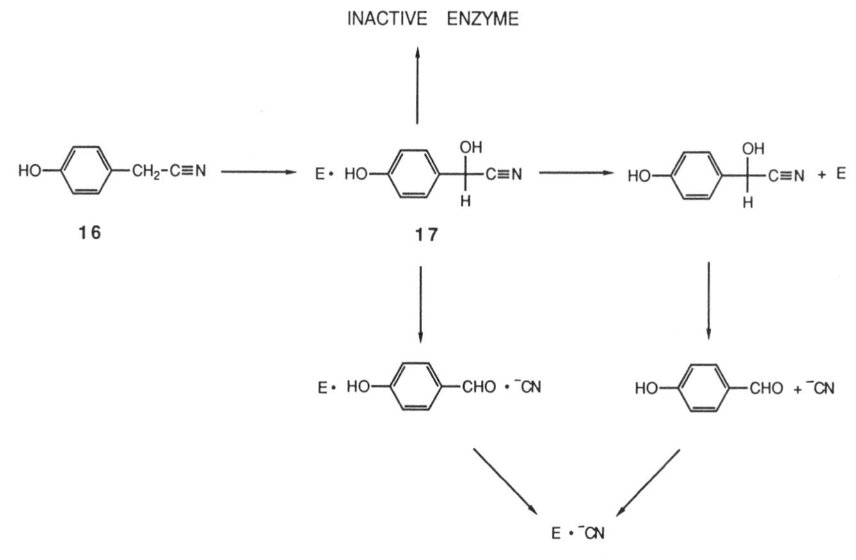

FIG. 36. Mechanism proposed for inactivation of dopamine β-hydroxylase by *p*-hydroxybenzyl-cyanide (**16**).

process (Colombo *et al.*, 1984c). The mechanism suggested to account for these observations (Fig. 36) proposes that covalent modification by **17** irreversibly inactivates the enzyme, while the decomposition of **17** (both at the active site and in solution) liberates cyanide, which reversibly complexes to the copper center of the hydroxylase. This hypothesis is supported by the binding of 2 mol of [^{14}C]CN/mol hydroxylase tetramer and by EPR studies which are consistent with the formation of Cu(II)–cyanide complexes (Colombo *et al.*, 1984b,c).

Olefinic analogs of phenethylamines also serve as mechanism-based inactivators of dopamine β-hydroxylase. The 2-X-3-(*p*-hydroxyphenyl)-1-propenes (X = H, Cl, Br) are hydroxylated to their 3-hydroxy analogs, with inactivation occurring every 40 to 100 turnovers (Rajashekhar *et al.*, 1984). A more detailed study of the reaction of dopamine β-hydroxylase with 2-bromo-3-(*p*-hydroxyphenyl)-1-propene led to the proposal (Fig. 37, path A) that elimination of HBr from the hydroxylated inactivator produces an α,β-unsaturated ketone which serves to alkylate the enzyme (Colombo *et al.*, 1984d). The lack of incorporation of radioactivity from [^{3}H]$NaBH_4$ into the modified enzyme casts doubt on the presence of a carbonyl in the adduct, however, and the identical partition ratios observed when X is H or Br argues against the necessity of halide elimination. An alternate mechanistic possibility emerged from the study of a series of ring-substituted 3-phenylpropenes (Fitzpatrick *et al.*, 1985). Deuteration of the benzylic position of 3-(*p*-hydroxyphenyl)propene results in an isotope effect of 2.0 on $k_{inact}/K(O_2)$ but no change in the partition ratio. This result indicates that the C–H bond is already broken in the intermediate which partitions between turn-

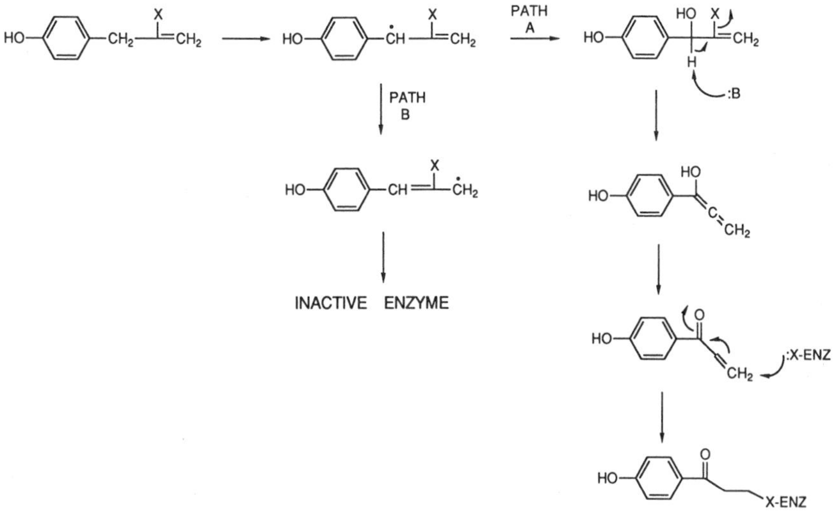

FIG. 37. Mechanism proposed for inactivation of dopamine β-hydroxylase by 2-substituted 3-(p-hydroxyphenyl)-1-propenes.

over and inactivation, and it suggests that C–H bond cleavage and hydroxylation are not concerted processes during catalysis. A linear free-energy plot of $V/K(O_2)$ versus σ^+ gave a good correlation with a r value of -1.2, but the partition ratio was relatively invariant with phenyl substitution. These results are consistent with radical abstraction of a benzylic hydrogen atom by an electrophilic species, leading to proposal of a radical mechanism for inactivation (Fig. 37, path B). This conclusion is further supported by an analysis of the partition ratios for inactivation of dopamine β-hydroxylase by 3-phenylpropynes and benzylcyclopropanes (Fitzpatrick and Villafranca, 1985).

Inactivation of dopamine β-hydroxylase is also reported to result from reaction with a variety of aryl allylamine analogs, typified by 1-phenyl-1-(aminomethyl)-ethene (Padgette *et al.*, 1985). The enzyme is covalently modified with a stoichiometry of approximately 5 per tetramer. The olefin is oxidized to the corresponding epoxide, although inactivation is demonstrated not to result from modification of the enzyme by the released product. Based on analogy with the mechanism of inactivation of cytochrome P-450 by olefins, dopamine β-hydroxylase is proposed to abstract one electron from the double bond to yield a radical cation, followed by recombination with the resulting $[Cu(III)=O]^+$ species to yield a carbocationic intermediate. This intermediate then partitions between cyclization to produce epoxide and alkylation and inactivation of the enzyme. Substitution of a thiazole or oxazole ring for the phenyl group of allylamine analogs gave significantly reduced K_I values (Bargar *et al.*, 1986).

Substitution of the β position of phenethylamine analogs also leads to inactivation of dopamine β-hydroxylase, apparently by radical routes. Dopamine β-hydroxylase converts β-chlorophenethylamine to α-aminoacetophenone and is inactivated with a partition ratio of 12,000 (Klinman and Krueger, 1982). The corresponding β-hydroxy compound is oxidized to the same product at comparable rate but is not an inactivator. Reaction of the enzyme with a α-aminoacetophenone leads to inactivation only in the absence of ascorbate, indicating that the Cu(II) form of the enzyme is affected. The mechanism of inactivation by β-chlorophenethylamine is proposed to involve dehydrohalogenation of a chlorohydrin intermediate at the active site to generate α-aminoacetophenone (Fig. 38) (Mangold and Klinman, 1984). The enol tautomer of the hydroxylase-bound product, which bears a resemblance to ascorbate, is suggested to reduce the enzyme with generation of a radical cation that acts as the inactivating species. The ability of α-aminoacetophenone to reduce the enzyme is supported by the observation that it can replace ascorbate as a reductant in the hydroxylation of tyramine. The hydroxylated product of β-hydroxyphenethylamine turnover, the *gem*-diol of α-aminoacetophenone, was demonstrated by isotopic labeling experiments to dissociate from the enzyme prior to dehydration, accounting for its inability to inactivate the hydroxylase.

This hypothesis was extended by a study of the inactivation of dopamine β-hydroxylase by phenylacetaldehyde (Bossard and Klinman, 1986). As predicted, phenylacetaldehyde inactivates the enzyme with concomitant production of β-hydroxyphenylacetaldehyde, which in turn can function as an inactivator or as a

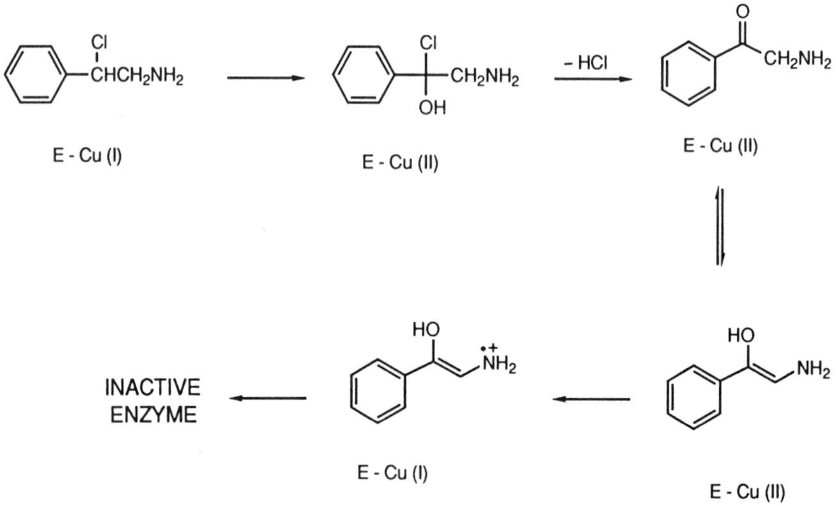

FIG. 38. Mechanism proposed for inactivation of dopamine β-hydroxylase by β-chlorophenethylamine.

reductant in substrate oxidation. A mechanism analogous to that shown in Fig. 38 is consistent with these observations. Phenylacetamides are also substrates and inactivators of the enzyme, although the hydroxylated products are incapable of enolization, a critical step in the mechanism depicted in Fig. 38. An alternate mechanism is proposed in which hydrogen atom abstraction from phenylacetamides leads directly to radical cation formation and enzyme inactivation without an intervening hydroxylation step.

In contrast to other inactivators of dopamine β-hydroxylase, *p*-cresol is unusual in that it does not contain a latent electrophile (Goodhart *et al.*, 1987). Oxidation of the benzylic carbon occurs, as evidenced by production of 4-hydroxybenzyl alcohol as well as the corresponding aldehyde. Inactivation with radiolabeled *p*-cresol leads to substoichiometric modification of the enzyme, with the majority of the radiolabel distributed between four tryptic peptides (DeWolf *et al.*, 1988). Analysis of two of the peptides indicates that they are of identical sequence, each containing modified tyrosine residues which differ in the structure of the *p*-cresol–amino acid adduct. A second pathway for inactivation is proposed which requires radical-mediated oxidation of the enzyme without incorporation of the inactivator.

B. Lipoxygenases

Lipoxygenases catalyze the conversion of *cis,cis*-1,4-diene units of polyunsaturated fatty acids such as arachidonic acid to *cis,trans*-pentadienyl hydroperoxides (Veldink and Vliegenthart, 1984). Mechanistic studies of the iron-containing soybean 15-lipoxygenase are consistent with a process in which a bisallylic hydrogen is abstracted by an enzyme base with concomitant reduction of the active site Fe(III) to Fe(II). The resultant delocalized pentadienyl radical reacts regio- and stereospecifically with oxygen to produce a hydroperoxyl radical which is subsequently reduced to the hydroperoxide with reoxidation of the iron. The mammalian 5-lipoxygenase catalyzes the first committed step in leukotriene biosynthesis, making approaches to its inactivation of substantial pharmaceutical interest.

Acetylenic analogs of arachidonic acid are inactivators of both 5- and 15-lipoxygenase. The 15-lipoxygenase is inactivated by 14,15-dehydroarachidonic acid (14,15-DHA), whereas the 5,6-, 8,9-, and 11,12-isomers of DHA have no effect (Corey and Park, 1982). The partition ratio of 260 is reduced only slightly to 160 in the presence of $NaBH_4$, indicating that loss of activity is not due to diffusible hydroperoxide products. Lipoxygenase is labeled with a stoichiometry of 0.87 following inactivation with [2-^{3}H]14,15-DHA, suggesting covalent modification of the enzyme. Inactivation is proposed to result from reaction of a vinyloxy radical or hydroxyl radical with the enzyme (Fig. 39).

Inactivation of 5-lipoxygenase from RBL-1 cells by 5,6-DHA is suggested to

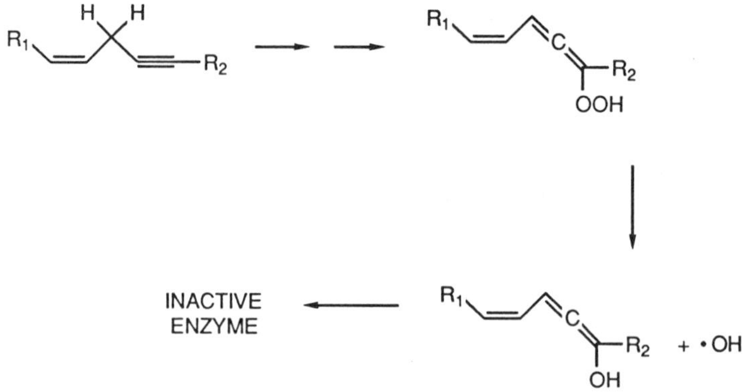

FIG. 39. Mechanism proposed for inactivation of 15-lipoxygenase by 14,15-dehydroarachidonic acid.

occur by an analogous mechanism (Corey and Munrow, 1982). The observation of an isotope effect of approximately 6 on the rate of inactivation by $[7R\text{-}^2H]$5,6-DHA provides additional support for this proposal (Corey *et al.*, 1984). Analogs in which the carboxylate of 5,6-DHA was replaced by a variety of polar phosphorus and sulfur sustituents were also inactivators of 5-lipoxygenase. An alternate mechanism for inactivation of 5-lipoxygenase by 5,6-DHA was proposed which involves abstraction of a proton from C-6 to yield an allenic intermediate which alkylates an active site residue (Sok *et al.*, 1982).

A similar compound, 5,8,11,14-eicosatetraynoic acid, is also an inactivator of 15-lipoxygenase (Kuhn *et al.*, 1984). In contrast to the results obtained with 14,15-DHA, no radiolabeling of the enzyme by the [*methyl*-^{14}C]methyl ester of the tetradehydro analog was observed. Instead, amino acid analysis of the inactive enzyme revealed the presence of 1 mol of methionine sulfoxide, suggesting that loss of activity resulted from oxidation of the active site rather than covalent modification.

Hexanal phenylhydrazone also serves as an inactivator of soybean lipoxygenase 1 (L-1) in a process which demonstrates kinetics more complex than those of standard mechanism-based inactivators (Galey *et al.*, 1988). Aerobic incubation of hexanal phenylhydrazone with L-1 leads to enzyme inactivation and conversion of the compound to its corresponding α-azo hydroperoxide, which is also an inactivator. Four equivalents of the α-azo hydroperoxide are sufficient to inactivate the enzyme completely, whereas the amount of the parent phenylhydrazone required to fully inactivate the enzyme increases from 13 to a maximum of 70 as the ratio of hexanal phenylhydrazone to L-1 increases. Since the partition ratio is normally independent of inhibitor and enzyme concentrations, a more complex mechanism is apparent. The addition to reaction mixtures of glutathione peroxidase, which reduces the α-azo hydroperoxide metabolite to the corresponding alcohol, suppresses about 80% of the inactivation. The α-azo hydroperoxide metabolite thus dissociates and rebinds to the enzyme before it succeeds in inactivating the enzyme. The increase in the partition ratio as the hexanal phenylhydrazone concentration increases is thus due to protection by the phenylhydrazone against binding of the hydroperoxide metabolite. If the 20% of the inactivation not suppressed by glutathione peroxidase is also mediated by the α-azo hydroperoxide, it must occur without release of the intermediate to the solution. Inactivation by [*phenyl*-^{14}C]phenylhydrazone leads to insignificant radiolabeling of the enzyme. The completely inactive enzyme contains 1.8 mol of methionine sulfoxide per mole of enzyme, supporting the mechanism for the inactivation proposed in Fig. 40.

A series of sulfur-substituted arachidonic acid analogs have been examined as lipoxygenase inactivators. The 7-thia- and 10-thiaarachidonic acid compounds are substrates for 15-lipoxygenase, whereas the enzyme is inactivated in an oxygen-dependent manner by 13-thiaarachidonate (Corey *et al.*, 1986). Inactivation is suggested to occur by the mechanism in Fig. 41. The corresponding sulfoxide is a competitive inhibitor but not an inactivator. The decreased electron density

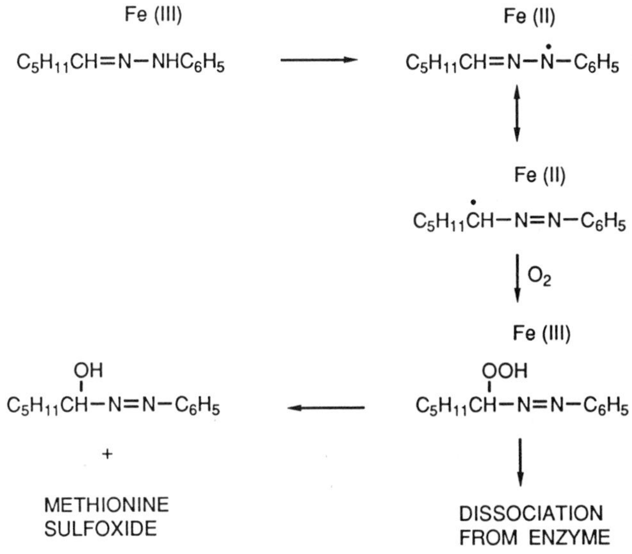

FIG. 40. Mechanism proposed for inactivation of 15-lipoxygenase by hexanal phenylhydrazone.

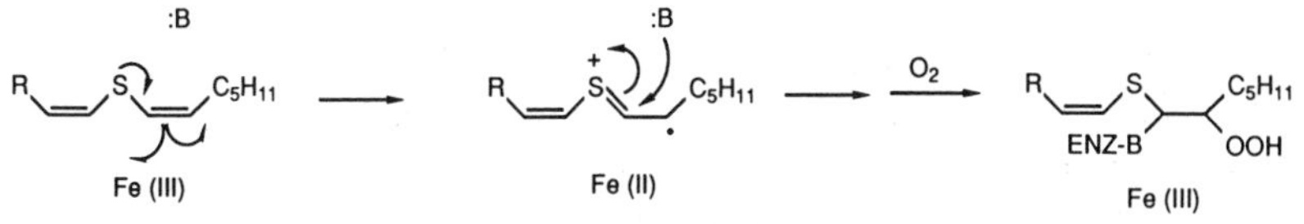

FIG. 41. Mechanism proposed for inactivation of 15-lipoxygenase by 13-thiaarachidonic acid.

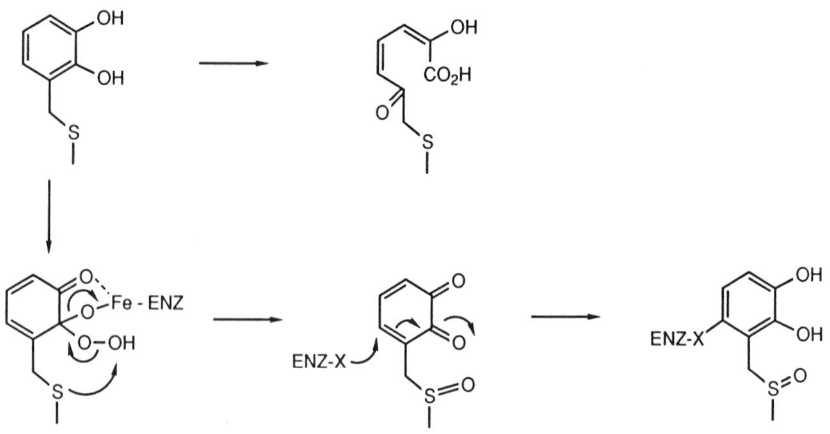

FIG. 42. Mechanism proposed for inactivation of catechol 2,3-dioxygenase by 3-[(methylthio)methyl]catechol.

of the sulfoxide sulfur is proposed as a rationale for its lack of activity. Analogous observations have been made with 5-lipoxygenase and 7-thiaarachidonate (Corey *et al.*, 1985).

C. CATECHOL 2,3-DIOXYGENASE

The nonheme iron enzyme catechol 2,3-dioxygenase catalyzes the cleavage of aromatic rings, apparently through the intermediacy of peroxidic species (Que, 1983). The dioxygenase is inactivated by 3-[(methylthio)methyl]catechol (Fig. 42) concomitantly with formation of the ring-opened product, 2-hydroxy-6-oxo-7-(methylthio)-2,4-hedptadienoic acid, with a partition ratio of 22,000 (Pascal and Huang, 1987). No inactivation is observed with the isosteric analog 3-propylcatechol, demonstrating the necessity of the thioether in that process. The enzyme is covalently modified with a stoichiometry of 1.6 by tritiated inactivator. The mechanism offered for the reaction proposes that sulfur oxidation by a peroxidic intermediate leads to formation of an electrophilic *o*-quinone, which modifies and inactivates the enzyme.

VII. Quinoproteins

A number of enzymes which catalyze oxidation reactions, including mammalian lysyl and plasma amine oxidases and bacterial alcohol dehydrogenases, have been determined to utilize pyrroloquinoline quinone (PQQ, methoxatin) as a cofactor (Duine *et al.*, 1987). Substrates of the amine oxidases appear to be activated for α-proton abstraction by formation of a Schiff base with PQQ, followed by hydrolysis of the imine and reoxidation of the cofactor with the release of ammonia (Williamson and Kagan, 1987). Most inactivators of the amine oxidases were studied prior to the identification of the cofactor and involve standard methods for generation of electrophilic intermediates. For example, lysyl oxidase is inactivated by β-aminopropionitrile without detectable formation of aldehyde products (Tang *et al.*, 1983). Equivalent amounts of radioactivity from 1,2-^{14}C- and 3-^{14}C-labeled compounds are bound to the inactive enzyme, demonstrating that the cyano group is not eliminated during the reaction. The stoichiometry of the modification varies, however, from only 0.07 to 0.1 equivalents per equivalent of enzyme, apparently owing to the presence of a substantial concentration of nonfunctional active sites. The mechanism suggested for this process involves alkylation of the enzyme by a ketenimine intermediate (Fig. 43). A similar hypothesis has been advanced for the inactivation of plasma amine oxidase by aminoacetonitrile (Maycock *et al.*, 1975).

The mechanism of PQQ-dependent alcohol dehydrogenases remains unknown, complicating the interpretation of results obtained with mechanism-based inactivators. Cyclopropanol has been examined as an inactivator of alcohol dehydrogenases from several bacterial sources. Complete inactivation of enzyme in the oxidized PQQ form is obtained with 1 equivalent of cyclopropanol per active site, leading to quantitative modification of the cofactor (Dijkstra *et al.*,

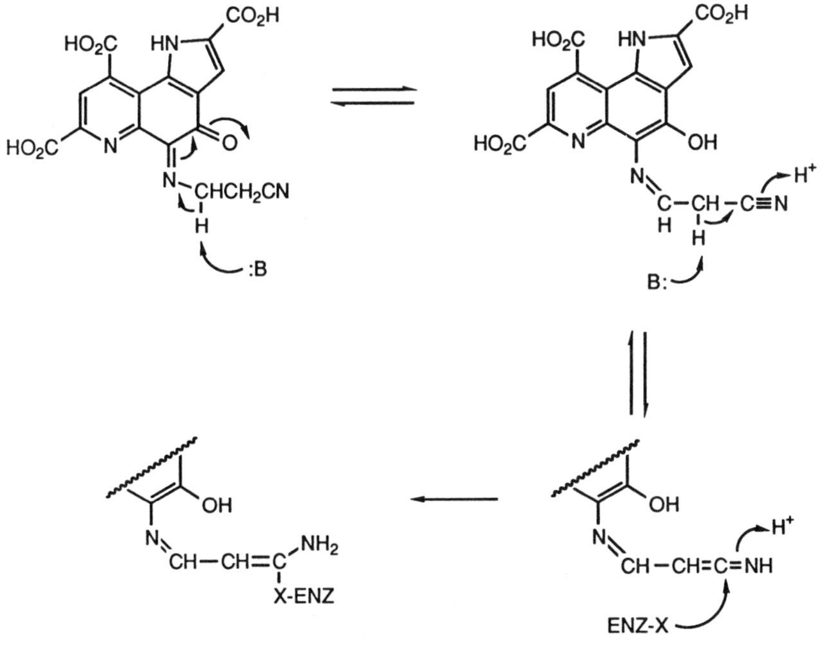

FIG. 43. Mechanism proposed for inactivation of lysyl oxidase by β-aminopropionitrile.

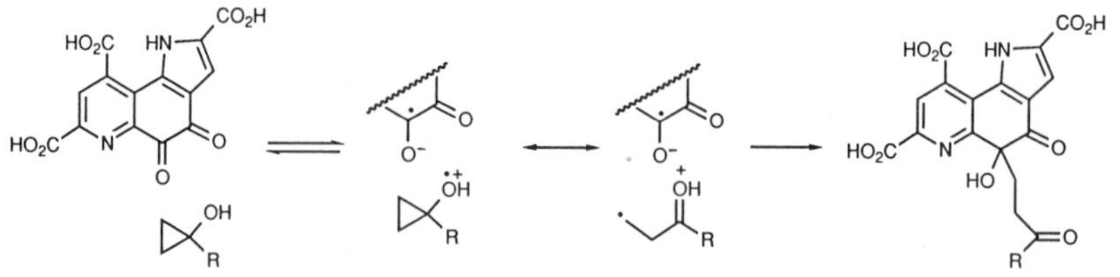

FIG. 44. Mechanism proposed for inactivation of PQQ-dependent alcohol dehydrogenase by cyclopropanol.

1984). Inactivation of the monomeric and dimeric dehydrogenases from *Pseudomonas* BB_1 is also observed with cyclopropanone hydrate and the corresponding ethyl hemiketal. Chromatography of the resolved cofactors yields distinct adducts, demonstrating that cyclopropanone is not an intermediate in the process. A mechanism involving radical ring cleavage of the cyclopropyl group was presented (Fig. 44). Abeles and co-workers initially hypothesized that the semiquinone form of PQQ was the active species, based on correlation of a stoichiometry of dehydrogenase labeling by [1-^{3}H]cyclopropanol of 0.14 and the presence in the resting enzyme of an ESR signal which integrated to 13% of the total enzyme concentration (Mincey *et al.*, 1981). A subsequent publication (Parkes and Abeles, 1984) concurred with the conclusion that oxidized cofactor was required for inactivation. The modified cofactor contains radioactivity from [1-^{14}C]- and [1-^{3}H]cyclopropanol which is released by treatment at pH 11, regenerating native PQQ. A structure for the modified cofactor identical to that shown in Fig. 44 was proposed, although no conclusions regarding radical versus ionic processes were drawn.

VIII. Hydrolytic Enzymes

A. PROTEASES

Although proteases historically have been the subject of intense investigation, leading to a relatively thorough understanding of their mode of action, their mechanism-based inactivation was pursued relatively rarely prior to the 1980s. Interest in the development of inactivators of proteases has increased as their potential utility as therapeutic agents has been more fully appreciated. The serine proteases, the largest and best understood class of proteolytic enzymes, have particularly been targeted for the design of inactivators. The inactivating molecules are typically heterocyclic compounds which acylate the active site serine of the enzyme, unmasking a reactive group which may modify other nucleophiles in the active site. Alternatively, electronic or geometric substituent effects stabilize the acyl enzyme to hydrolysis, rendering the modification essentially irreversible.

Isatoic anhydride (Fig. 45) is an example of a compound which inactivates serine proteases by formation of a stable acyl enzyme. Chymotrypsin is stoichiometrically modified by isatoic anhydride with the incorporation of one anthraniloyl group, as indicated by the absorbance spectrum of the derivatized enzyme (Moorman and Abeles, 1982). The anthraniloyl moiety is proposed to result from acylation of the active site serine followed by decarboxylation of the resulting carbamate (Fig. 45). The inactivation is irreversible by virtue of the electron-donating effect of the amino substituent of the benzoyl group, which is known to decrease dramatically the rate of deacylation of benzoylchymotrypsin (Caplow and Jencks, 1962). Substitution of the isatoic anhydride nucleus results in increased selectivity for trypsinlike proteases over chymotrypsin (Gelb and Abeles, 1986).

Derivatives of a simpler analog of isatoic anhydride, the 1,3-oxazine-2,6-diones, prepared in an effort to attain specificity for certain proteases, were assumed to operate by a mechanism analogous to that of the parent compound (Moorman and Abeles, 1982). A careful examination of the inactivation of chymotrypsin by 5-butyl-3*H*-1,3-oxazine-2,6-dione (**18,** R = *n*-butyl, Fig. 46) indicates that the mechanism is distinct from that of isatoic anhydride (Weidmann and Abeles, 1984). Incubation of chymotrypsin with [2-^{14}C]**18** yields inactive enzyme which contains 0.95 equivalents of ^{14}C per equivalent of enzyme as the

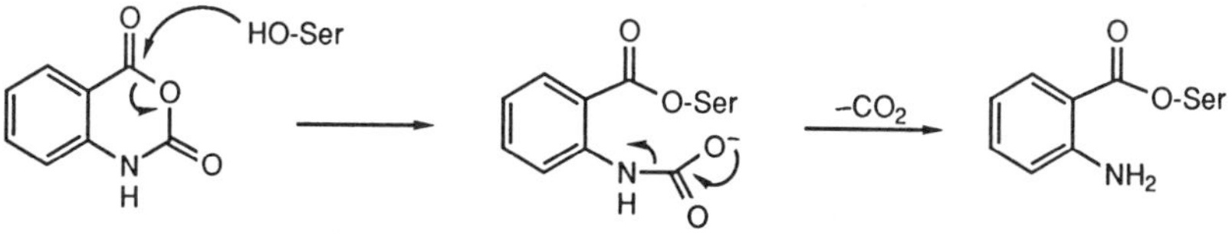

FIG. 45. Mechanism proposed for inactivation of chymotrypsin by isatoic anhydride.

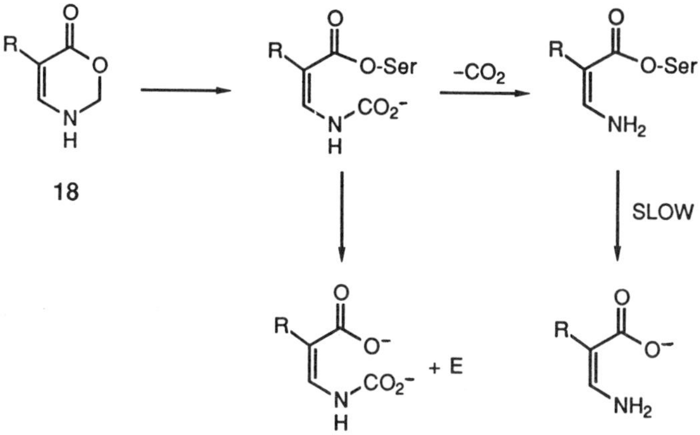

FIG. 46. Mechanism proposed for inactivation of chymotrypsin by 1,3-oxazine-2,6-diones (**18**).

carbamate carbon, demonstrating that decarboxylation is not rapid. The modified enzyme regains activity in a biphasic process, while ^{14}C is released from the enzyme in a monophasic process at a rate comparable to the fast phase of recovery of activity. After 5 hr, the enzyme contains 0.1 equivalent of radioactivity but has regained only 45% of its activity, indicating the existence of two enzyme–inactivator adducts. The revised mechanism (Fig. 46) suggests that the acyl enzyme can partition between hydrolysis (leading to recovery of activity) and release of CO_2 to give a more stable acyl enzyme with a partition ratio of 1.5 to 1.9.

A different strategy for serine protease inactivation is exemplified by the reaction of chymotrypsin with 3-benzyl-6-chloro-2-pyrone (**19,** Fig. 47) (Westkaemper and Abeles, 1983). Time-dependent inactivation is biphasic, apparently owing to heterogeneity of the enzyme, and reversible with a $t_{1/2}$ of 23 hr. No inactivation ensues with analogs which lack the chlorine or benzyl substituents. The pyrone chromophore is absent from the modified enzyme, ruling out 1,6-conjugate addition of a nucleophile to C-6. Acylation of the active site serine was suggested to unmask an acyl chloride, leading to covalent modification of an additional residue, such as histidine-57 (Fig. 47).

The subsequent studies of Gelb and Abeles (1984) concerning the structure of the chymotrypsin–pyrone adduct and the process of enzyme reactivation elegantly demonstrated that this mechanism was incorrect. ^{13}C NMR studies of chymotrypsin inactivated with ^{13}C-labeled **19** conclusively establish that the adduct contains a free carboxylate group which arises from C-6 of the pyrone, ruling out trapping of a nucleophile by the assumed acyl chloride intermediate. Analysis of the 2-benzyl-3,4-dehydroglutarate obtained from reactivation of the enzyme in $H_2{}^{18}O$ confirms that only one carboxylate is esterified in the adduct. The double bond in the adduct is between C-2 and C-3, as demonstrated by ^{13}C NMR, in contrast to the 3,4-dehydro product released following reactivation.

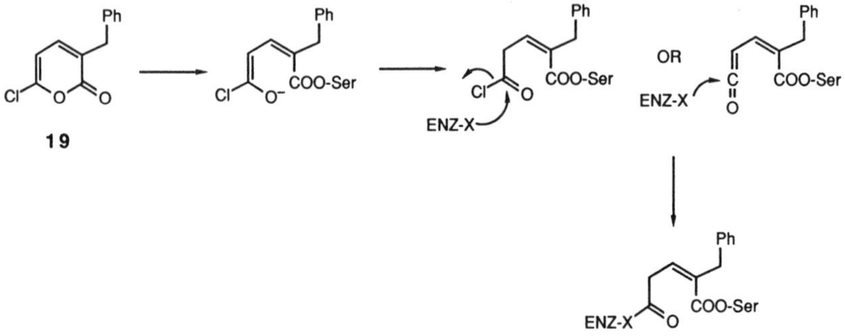

FIG. 47. Mechanism originally proposed for inactivation of chymotrypsin by 3-benzyl-6-chloro-2-pyrone (**19**).

The requirement for isomerization of the double bond in the reactivation reaction is supported by the demonstration of an isotope effect of 2.5 to 3.3 on the rate of recovery of activity following inactivation with [5-^{2}H]**19** and the observation that amines act as general bases rather than nucleophiles in their catalysis of the reactivation process. The revised mechanism for inactivation and reactivation is presented in Fig. 48. The terminal carboxylate is essential for the extremely slow deacylation rate, since the analogous adduct with a methyl group in place of the acid deacylates with a $t_{1/2}$ of 4.7 min. The X-ray crystal structure of chymotrypsin inactivated with **19** reveals that the terminal carboxylate of the adduct forms an ion pair with histidine-57, apparently suppressing hydrolysis of the acyl enzyme (Ringe *et al.*, 1986). Isomerization of the double bond of the adduct is proposed to disrupt the salt bridge, permitting rapid hydrolysis of the serine ester and reactivation of the enzyme.

In contrast, 5-benzyl-6-chloro-2-pyrone is not a substrate for chymotrypsin, but inactivates the enzyme with the formation of a shoulder in the absorbance spectrum at 320 nm (Westkaemper and Abeles, 1983). The chromophore, which is similar to that of the pyrone ring, appears with kinetics corresponding to the rate of inactivation and is lost on reactivation of the enzyme. The crystal structure and ^{13}C NMR studies of the modified enzyme demonstrate that the benzyl group is bound in the specificity pocket of the active site and that serine-195 is covalently attached to C-6 of the intact pyrone (Ringe *et al.*, 1985) (Fig. 49). Inactivation by the 5-benzyl analog is therefore an example of affinity labeling rather than mechanism-based inactivation and emphasizes the importance of the side chain of protease inactivators in proper orientation of the compound in the active site.

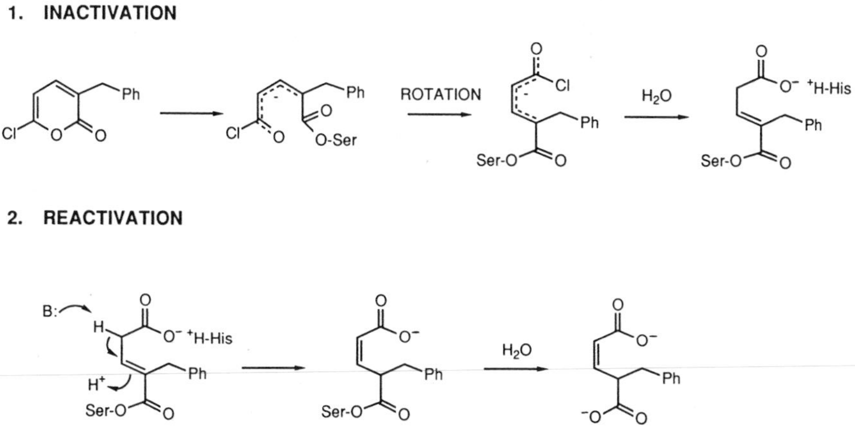

1. INACTIVATION

2. REACTIVATION

FIG. 48. Revised mechanism proposed for inactivation of chymotrypsin by 3-benzyl-6-chloro-2-pyrone and subsequent reactivation by amines.

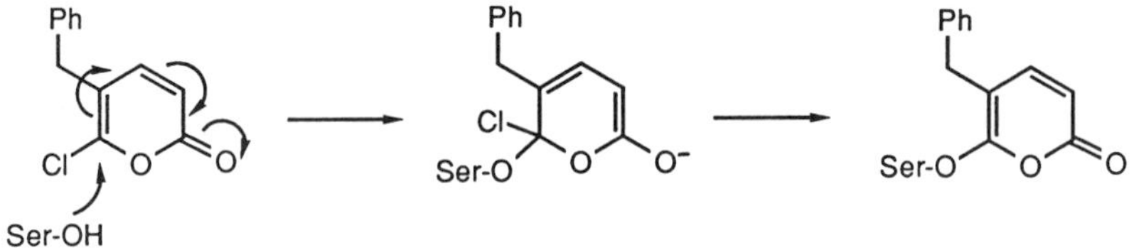

FIG. 49. Mechanism proposed for inactivation of chymotrypsin by 5-benzyl-6-chloro-2-pyrone.

The related 3-chloro- and 3,4-dichloroisocoumarins are general serine protease inactivators which are somewhat selective for elastases. They are believed to act via a mechanism similar to that originally proposed for 3-benzyl-6-chloro-2-pyrone (Harper *et al.*, 1983, 1985). The isocoumarin ring is opened concomitantly with inactivation, unveiling an acyl halide which is suggested to modify histidine-57 based on the stoichiometry of proton release during inactivation and reactivation, although the results cannot be distinguished from hydrolysis of the acyl halide and formation of a salt bridge as described above.

A second class of isocoumarin analogs, exemplified by 3-alkoxy-7-amino-4-chloroisocoumarins (Fig. 50), potently inactivates chymotrypsin and elastase (Harper and Powers, 1985). Acylation of the active site serine elaborates a 4-aminobenzyl chloride, which decomposes via elimination of HCl to yield a quinone imine methide. Alkylation of the enzyme, presumably at the histidine of the catalytic triad, by the quinone imine methide results in irreversible inactivation. Readily reversible inactivation occurs with the 7-nitro analog, emphasizing the essential intermediacy of the quinone methide imine in the irreversible inactivation process and arguing against covalent modification by nucleophilic displacement of chloride. The corresponding 7-guanidido derivatives demonstrate specificity for trypsin and a variety of enzymes in the blood coagulation cascade and lead to a decrease the rate of *in vitro* coagulation, although they are too hydrolytically labile to be of therapeutic use (Kam *et al.*, 1987, 1988). Inactivation of chymotrypsin by 3,4-dihydro-3,4-dibromo-6-bromomethylcoumarin is also suggested to proceed via a quinone methide intermediate which arises from elimination of the 6-bromo substituent (Bechet *et al.*, 1973, 1977a,b).

Ynenol lactones are also proposed to inactivate serine proteases irreversibly by alkylation of the active site histidine. Acylation of elastase by ynenol lactones produces an electrophilic allenone intermediate (Fig. 51) which covalently modifies and inactivates the enzyme with a partition ratio of 1.7 (Copp *et al.*, 1987). Direct addition of the allenone carboxylic acid is without effect, demonstrating that the inactivator must be tethered in the active site to allow reaction with the enzyme. Substitution α to the lactone carbonyl is required for loss of activity, whereas the rate of inactivation is decreased by substitution at the acetylene terminus, suggesting that allene formation is slowed or that nucleophilic attack on the allene is hindered.

Haloenol lactones were proposed as mechanism-based inactivators of pro-

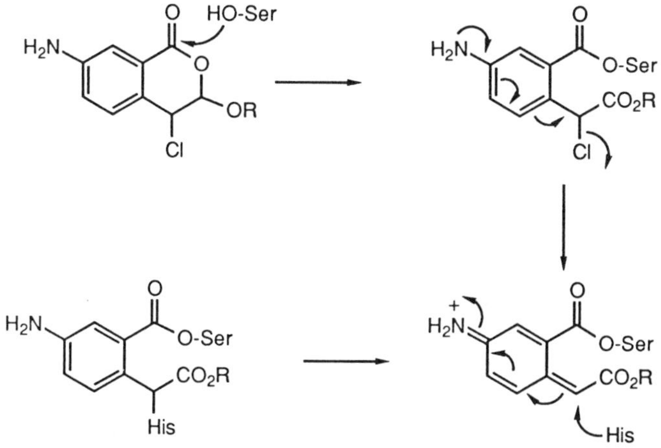

FIG. 50. Mechanism proposed for inactivation of chymotrypsin and elastase by 3-alkoxy-7-amino-4-chloroisocoumarins.

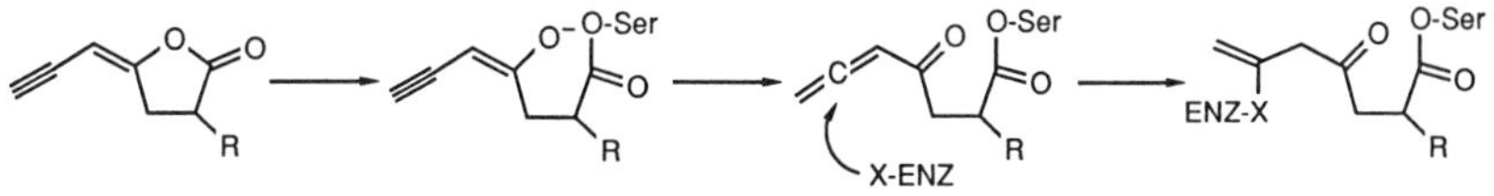

FIG. 51. Mechanism proposed for inactivation of elastase by ynenol lactones.

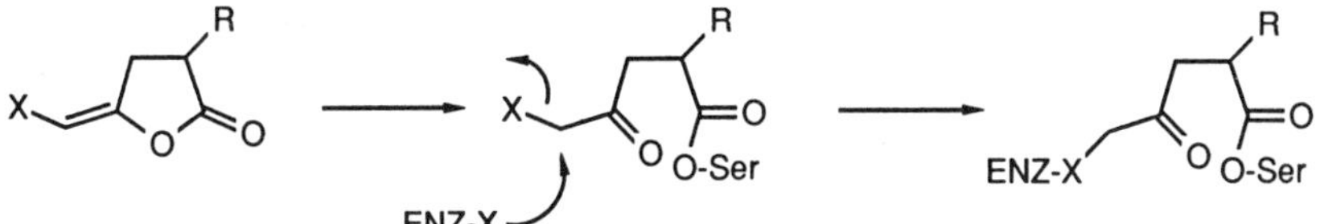

FIG. 52. Mechanism proposed for inactivation of chymotrypsin by haloenol lactones (**20**).

teases by Rando (1974) and have been synthesized and studied by Katzenellenbogen and co-workers (Daniels *et al.*, 1983; Daniels and Katzenellenbogen, 1986). Compounds such as **20** (Fig. 52) irreversibly and stoichiometrically inactivate chymotrypsin by the intermediacy of an α-haloketone, which is suggested to undergo S_N2 displacement by the active site histidine (Fig. 52). The effects of ring size, halogen, and hydrophobic substituents on the kinetic parameters of the inactivation reaction have been systematically examined.

B. β-LACTAMASES

The resistance of microbial pathogens to β-lactam antibiotics typically results from the expression of a β-lactamase by the organism. These enzymes hydrolyze the β-lactam moiety of the penicillin via an acyl-enzyme intermediate to yield the corresponding antibiotically inert penicilloic acid. Inactivators of β-lactamases therefore provide an opportunity to combat drug resistance in an approach exemplified by the marketing of Augmentin and Timentin (Beecham), which contain the β-lactamase inactivator sodium clavulanate in combination with the antibiotics amoxicillin and ticarcillin, respectively.

Although a large variety of β-lactamase inactivators have been developed, they fall into relatively few mechanistic classes. Inactivation of the TEM-2 β-lactamase from *E. coli* by penicillanic acid sulfone (**21,** Fig. 53), which has been studied in detail by Knowles and colleagues, provides a paradigm for the mechanism of other compounds including clavulanic acid and halopenams (Knowles, 1983). Inactivation of the enzyme occurs with a partition ratio of 7000, and activity is recovered when the enzyme is diluted into an assay mixture (Brenner and Knowles, 1981). The rates of both the hydrolysis and inactivation reactions are unexpectedly *increased* 3-fold with the 6,6-dideuterio analog of **21.** The mechanism proposed to account for this observation suggests that the initial acyl-enzyme intermediate **22** can partition between three pathways (Fig. 53). Deacylation followed by hydrolysis of the imine (path A) generates the ultimate products of the hydrolysis reaction, malonsemialdehyde and penicillanic acid

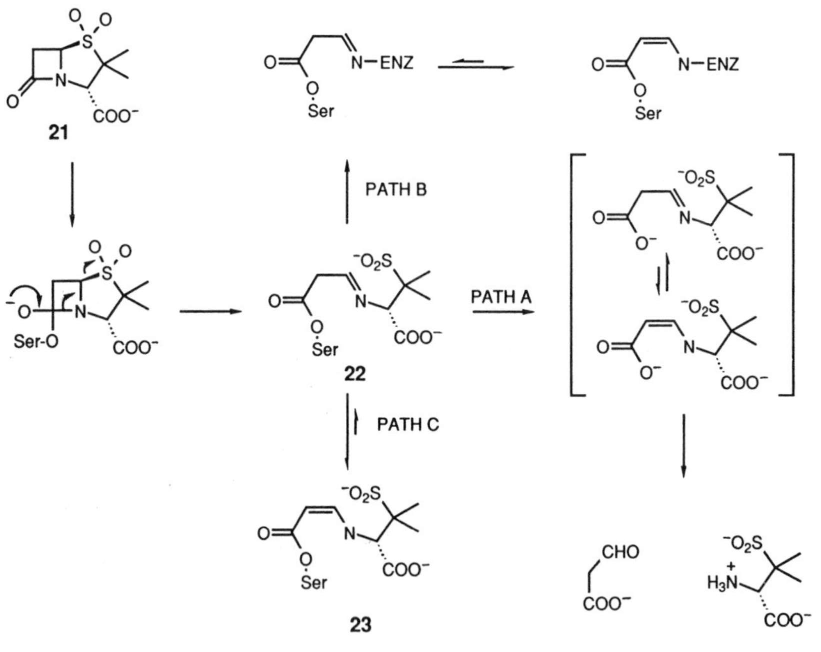

FIG. 53. Mechanism proposed for inactivation of β-lactamase by penicillanic acid sulfone (**21**).

sulfone. Attack of a nucleophile (suggested to be a lysine) on the imine leads to a covalently modified enzyme species which is believed to isomerize to the enamine, accounting for the observed increase in absorbance at 280 nm (path B, Fig. 53). Tautomerization of **22** to the β-aminoacrylate ester by abstraction of the 6β-hydrogen (path C) yields the transiently inhibited species **23,** which serves as a "waiting room" by virtue of the increased hydrolytic stability derived from conjugation of the enamine. Deuteriation of the 6-position decreases the amount of **23** formed, increasing the flux through the deacylation and inactivation pathways. This structure for the transiently inhibited enzyme is supported by the effect of pH on the partitioning of **22** (Kemal and Knowles, 1981). Inactivation with methyl-tritiated **21** does not lead to radiolabeling of the β-lactamase, consistent with modification of the enzyme by carbons 5–7 (Brenner and Knowles, 1984). The mechanism in Fig. 53 is further supported by a model study in which **21** was treated with methanol and diethylamine, yielding methyl β-(diethylamino)acrylate, which is directly analogous to the proposed structure for the inactivated enzyme (Brenner and Knowles, 1984).

As in the case of proteases, β-lactamases can be inactivated simply by formation of a stable acyl enzyme without an additional covalent modification. Stopped-flow spectroscopic studies demonstrate that a selection of cephalosporins form acyl-enzyme intermediates with the PC1 β-lactamase of *S. aureus* (Faraci and Pratt, 1985). This intermediate partitions between rapid deacylation and elimination of the 3′ leaving group to yield a 3-methylenedihydrothiazine (**24,** Fig. 54), which is sufficiently stable ($t_{1/2}$ 10 min at 30°C) to allow its isolation and spectral characterization. Transient inactivation therefore results from the slow hydrolysis of **24.** The identity of the 3′ leaving group determines the rate of elimination but has no effect on the deacylation rate and therefore controls the partition ratio, which reaches a lower limit of zero with good leaving groups such as acetate. Model studies and the spectral characteristics of the inactive enzyme show no evidence for addition of nucleophiles to the 3′-exomethylene group. The basis for the inert nature of **24** is therefore not clear, but it is hypothesized to derive from the differential ability of the two acyl enzymes to hydrogen bond (Faraci and Pratt, 1985). The absence of an N–H hydrogen bond donor in **24** might prevent the enzyme from undergoing a catalytically essential conformational change, or the presence of a hydrogen bond acceptor might drive the

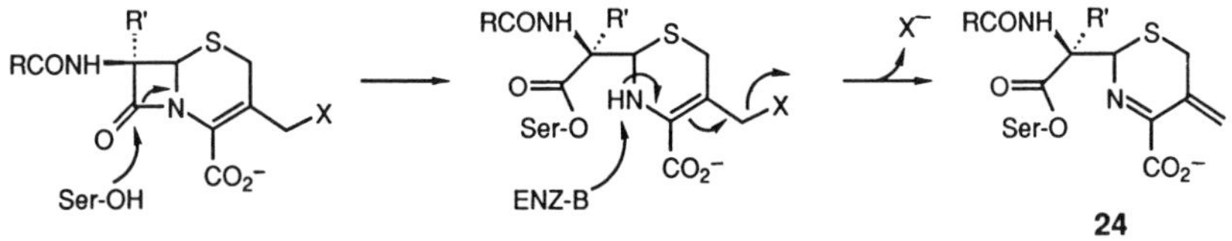

FIG. 54. Mechanism proposed for inactivation of β-lactamase by cephalosporins.

enzyme into an unproductive conformation. The inactivation of β-lactamases by olivanic acid derivatives also has been shown to involve rearrangement of the initial acyl enzyme intermediate, leading to increased hydrolytic stability (Charnas and Knowles, 1981; Easton and Knowles, 1982).

C. GLYCOSIDASES

Glycosidases which catalyze hydrolysis reactions with overall retention of configuration are believed to operate by protonation of the leaving group and stabilization of the resulting oxocarbonium ion by formation of a glycosyl glutamate intermediate (Sinnott, 1987). Hydrolysis of the glycosyl enzyme subsequently generates product. Mechanism-based inactivation of these enzymes has been achieved by use of the protonation step to generate a reactive intermediate and by the synthesis of analogs which yield stable glycosyl enzymes.

The *E. coli* β-galactosidase is inactivated by β-D-galactopyranosylmethyl(*p*-nitrophenyl)triazene (**25,** Fig. 55) with a partition ratio of four (Sinnott and Smith, 1976, 1978). Inactivation by [*methyl*-^{14}C]**25** leads to incorporation of 0.9 mol of radioactivity per mole of enzyme, which was determined to reside in an adduct with methionine-500 (Fowler *et al.*, 1978). The mechanism displayed in

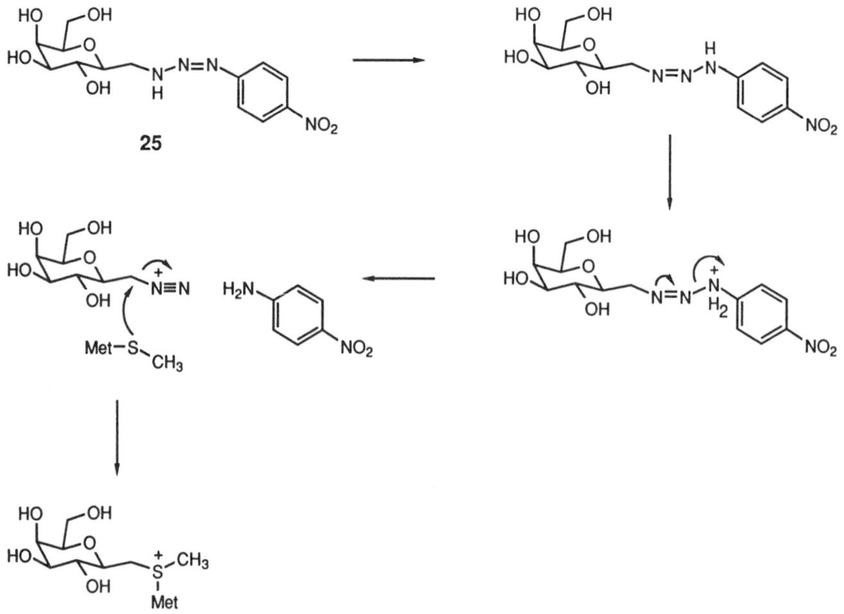

FIG. 55. Mechanism proposed for inactivation of β-galactosidase by β-D-galactopyranosylmethyl(*p*-nitrophenyl)triazene (**25**).

Fig. 55 (Sinnott and Smith, 1976) is consistent with the accepted route for chemical decomposition of triazenes, although substituent effects on inactivation of the β-galactosidase by aryl-substituted triazenes suggest that protonation of the aniline leaving group may not be required (Sinnott *et al.*, 1982). The ethyl derivative of **25** was synthesized as a probe to determine whether activation of the compound by the enzyme is required for inactivation: if the reaction is not enzyme catalyzed, then the inactivation rate should either be independent of the alkyl chain length or should increase, since the decomposition rate of the ethyl analog is 5 to 10 times greater than that of the methyl analog (Sinnott *et al.*, 1982). The actual rate of inactivation by the ethyl analog is significantly decreased, consistent with an enzyme-catalyzed reaction.

The 2-deoxy-2-fluoroglucosides have been described as inactivators of glycosidases (Withers *et al.*, 1987). The presence of the 2-fluoro substituent is proposed to destabilize the oxocarbonium ion transition states involved in glycosylation and deglycosylation of the active site glutamate residue. The substrate analog can be activated for the glycosylation step through the presence of a good leaving group such as 2,4-dinitrophenol, allowing formation of a stable 2-fluorodeoxyglucosyl enzyme intermediate. This approach leads to inactivation of the *Alcaligenes faecalis* β-glucosidase which is reversed only in the presence of substrates. Several 2-deoxy-2-fluoro-D-glycosyl fluorides have also been prepared and found to be inactivators, the efficacy of the compounds corresponding to the substrate specificity of the β-glucosidase (Withers *et al.*, 1988).

IX. Enzymes of Nucleotide Modification

A. Ribonucleotide Reductase

The *E. coli* ribonucleoside-diphosphate reductase catalyzes the conversion of nucleoside diphosphates to the corresponding 2′-deoxynucleotides. The hallmark of this intriguing enzyme is its unusual cofactor, which consists of a binuclear iron center and an organic free radical resulting from the oxidation of Tyr-122 of the B2 subunit (Sjoberg and Graslund, 1983; Larsson and Sjoberg, 1986). Although the presence of the tyrosyl radical is essential for enzyme activity, and the chemical mechanism proposed for ribonucleotide reduction invokes radical intermediates, its participation in the reaction has not been established (Ashley and Stubbe, 1985). However, in an example of the considerable utility of mechanism-based enzyme inactivators as probes of enzyme mechanism, analysis of the inactivation of the reductase by the 2′-azido-2′-deoxynucleoside diphosphates clearly demonstrates the involvement of the tyrosyl radical in a reductase-catalyzed reaction.

Inactivation of the reductase by 2′-azido-2′-deoxyuridine 5′-diphosphate (**26,** Fig. 56) is accompanied by loss of the tyrosyl radical (Thelander *et al.*, 1976)

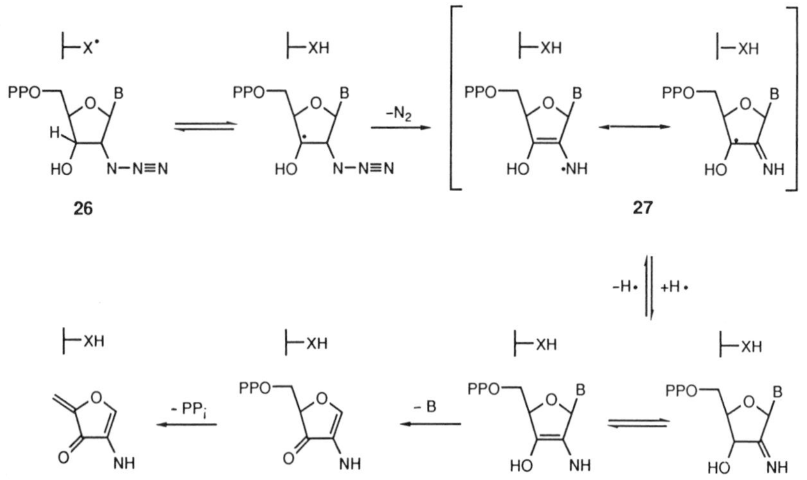

FIG. 56. Mechanism proposed for inactivation of ribonucleoside-diphosphate reductase by 2′-azido-2′-deoxyuridine 5′-diphosphate (**26**).

and formation of a new EPR signal (Sjoberg *et al.*, 1983). The hyperfine coupling pattern of the new signal is altered when inactivation is performed with [*azido*-^{15}N]**26,** indicating that the radical is localized on the nucleotide (Ator *et al.*, 1984). Complete inactivation is observed with 1 equivalent of **26,** and 0.2 equivalents of ^{3}H is released to the solvent from [3′-^{3}H]**26,** consistent with cleavage of the 3′-carbon–hydrogen bond with an istope effect of 5 (Salowe *et al.*, 1987). Deuteriation of the 3′-hydrogen of **26** decreases the rates of inactivation and loss of the tyrosyl radical, providing evidence for coupling of loss of the tyrosyl radical and abstraction of the 3′-hydrogen of the substrate analog. Stoichiometric amounts of uracil, inorganic pyrophosphate, and dinitrogen are produced in the reaction, and 1 mol of ^{3}H per mole of enzyme is incorporated into the B1 subunit when the inactivation is performed with [5′-^{3}H]**26.** These results were interpreted in terms of the mechanism in Fig. 56 (Salowe *et al.*, 1987). Abstraction of the 3′-hydrogen as a hydrogen atom yields a nucleotide radical which decomposes with the expulsion of dinitrogen to yield the EPR-active delocalized radical **27.** Elimination of uracil and pyrophosphate generates an unsaturated ketone which alkylates and inactivates the enzyme.

B. THYMIDYLATE SYNTHASE

Thymidylate synthase catalyzes the conversion of 2′-deoxyuridine 5′-phosphate to thymidine 5′-phosphate, and it is the sole source of this essential component of DNA. The reaction proceeds by transfer of a one-carbon fragment from the cofactor, 5,10-methylene tetrahydrofolate (CH_2H_4folate), of the enzyme to the C-5 position of the nucleotide, followed by reduction of the methylene group to the methyl oxidation state with conversion of the cofactor to dihydrofolate (Walsh, 1979, p. 838). Thymidylate synthase is inactivated by 5-fluoro-dUMP, which is the active metabolite of the anticancer drug 5-fluorouracil. The detailed studies of Heidelberger, Santi, Danenberg, and co-workers have probably subjected this reaction to closer scrutiny than any other mechanism-based inactivation, and they have been rewarded with a wealth of information regarding the catalytic mechanism of the enzyme.

Inactivation of the *Lactobacillus casei* thymidylate synthase by 5-fluoro-dUMP occurs with the formation of an isolable complex which contains equimolar amounts of 5-fluoro-[^{3}H]dUMP and [^{14}C]CH_2H_4folate (Danenberg *et al.*, 1974). When excess inactivator and cofactor are removed, intact 5-fluoro-dUMP dissociates from the enzyme and synthase activity is recovered with a $t_{1/2}$ of approximately 14 hr at 23°C (Santi *et al.*, 1974). The inactivation reaction requires the presence of the cofactor and reaches a maximum stoichiometry of 2 equivalents of 5-fluoro-dUMP bound per synthase dimer (Santi *et al.*, 1974). Radiolabel from 5-fluoro-dUMP remains bound to the enzyme through a variety of denaturation procedures, indicating that the complex is covalently bound to the enzyme. Inactivation is accompanied by loss of nucleotide absorbance at 269 nm, suggesting that the covalent attachment results from addition of a nucleophile to the 5,6-double bond (Danenberg *et al.*, 1974; Santi *et al.*, 1974). This hypothesis is supported by the observation of a large secondary isotope effect ($k_H/k_T = 1.23$) on the dissociation of 5-fluoro-[6-^{3}H]dUMP from the complex, consistent with a rate-limiting change in hybridization of C-6 from sp^3 to sp^2 (Santi *et al.*, 1974). The cofactor is also a component of the covalent complex, since proteolytic digestion of the complex produces a peptide which contains radiolabel from 5-fluoro-dUMP and CH_2H_4folate (Danenberg *et al.*, 1974).

The nucleophile which adds to C-6 of 5-fluoro-dUMP has been conclusively identified as a cysteine residue in elegant experiments by Danenberg and Heidelberger (1976). They treated thymidylate synthase that had been inactivated with 5-fluoro-[6-^{3}H]dUMP and [^{14}C]CH_2H_4folate with Raney nickel, a reagent which specifically reduces C–S bonds without affecting peptide bonds (Danenberg and Heidelberger, 1976). Both radioisotopes were liberated from the protein with identical kinetics, yielding an adduct whose behavior on thin-layer chromatography was intermediate between the nucleotide and the cofactor. Amino acid analysis of the reduced protein revealed an increase in the number of alanine residues at the expense of cysteine. When the analogous experiment was performed following inactivation by 5-fluoro-[6-^{3}H]dUMP of synthase isolated from *L. casei* grown on [^{35}S]cysteine, ^{3}H and ^{35}S were released at the same rate. Subsequent sequence analysis of peptides derived from the inactive protein demonstrated that Cys-198 was the labeled residue (Bellisario *et al.*, 1976, 1979).

The nature of the linkage of the cofactor to the inactive ternary complex has also been explored in detail. Early studies showed that folate analogs did not substitute for CH_2H_4folate in the formation of covalent complexes, suggesting that the methylene group of the cofactor was critical to the process (Danenberg *et al.*, 1974). In more recent ^{19}F NMR examinations of the complex, a ^{19}F–^{13}C coupling has been observed when $^{13}CD_2H_4$folate was used in its formation, demonstrating that the methylene group is covalently bound to C-5 of 5-fluoro-dUMP (Byrd *et al.*, 1978). Chemical degradation procedures which selectively react with N-5- and N-10-substituted folates indicate that the folate is N-5 substituted (Pellino and Danenberg, 1985).

A mechanism consistent with this substantial body of evidence is shown in Fig. 57. Addition of cysteine to C-6 followed by attack of the C-5 enolate on the methylene group of CH_2H_4folate occurs in analogy with the mechanism suggested for the normal methylation reaction. With 5-fluoro-dUMP, however, the synthase appears to become frozen in this covalent ternary complex, resulting in loss of activity.

C. *S*-Adenosylhomocysteine Hydrolase

The product of *S*-adenosylmethionine-dependent methylation reactions, *S*-adenosylhomocysteine (AdoHcy), is catabolized by AdoHcy hydrolase (adenosylhomocysteinase) to yield adenosine and homocysteine. The enzyme is believed to function via oxidation of the 3′ position of the adenosine moiety by a tightly bound molecule of NAD^+ to generate the 3′-ketonucleoside and NADH (Palmer and Abeles, 1979). Abstraction of the 4′ hydrogen allows elimination of homocysteine, followed by addition of water to the resulting α,β-unsaturated ketone to produce adenosine. AdoHcy is a potent product inhibitor of many methyltransferases, and its accumulation resulting from inhibition of AdoHcy hydrolase has a variety of profound effects. Among the most interesting is antiviral activity owing to interruption of methylation of the 5′-terminal cap of viral mRNAs (deClercq, 1985), providing pharmaceutical relevance for selective AdoHcy hydrolase inhibitors.

Hershfield (1979) reported that 2′-deoxyadenosine (2′-dAdo) and 9-β-D-

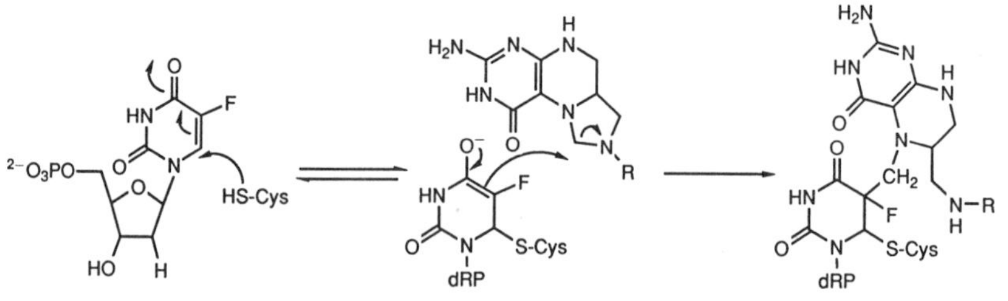

FIG. 57. Mechanism proposed for inactivation of thymidylate synthase by 5-fluoro-2′-deoxyuridine 5′-monophosphate.

arabinofuranosyladenine (Ara-A) inactivate AdoHcy hydrolase and that radioactivity from [^{3}H]2′-dAdo cochromatographs with the inactive enzyme on a gel filtration column. Detailed analysis of the reaction by Abeles and co-workers (Abeles *et al.*, 1980, 1982) revealed that stoichiometric amounts of [8-^{3}H]2′-dAdo bind to the hydrolase and that 80% of the radioactivity released on denaturation of the enzyme comigrates with adenine on HPLC. Inactivation is accompanied by the reduction of 2 of the 4 equivalents of bound NAD^+ to NADH and by the release to the solvent of 0.4 equivalents of ^{3}H per active site from [2′(*R*)-^{3}H]2′-dAdo, leading to the proposal that the enzyme contains two pairs of non-identical subunits. Reduction of the coenzyme by $NaBH_4$ suppresses the labilization of ^{3}H from [2′(*R*)-^{3}H]2′-dAdo but does not decrease the amount of adenine observed. A mechanism was proposed which hypothesizes that formation of a 3′-ketone intermediate allows transelimination of the 2′(*R*)-hydrogen and adenine, which causes inactivation by binding tightly to the enzyme (Fig. 58) (Abeles *et al.*, 1982). An additional process which involves enzyme-catalyzed hydrolysis of the glycosidic bond was proposed to account for the formation of adenine by the NADH form of the hydrolase.

AdoHcy hydrolase is inactivated by a variety of other nucleosides, including the cyclopentenyl analog of adenosine, neplanocin A (Borchardt *et al.*, 1984). Complete inactivation is observed with 0.5 equivalents of neplanocin A per active site, yielding 0.5 equivalents each of NADH and adenine (Wolfson *et al.*, 1986). The mechanism of the reaction was proposed to parallel that suggested for 2′-dAdo, although the elimination of adenine would require a cis-elimination, in contrast to the trans-elimination possible with a 2′-deoxy-3′-ketonucleoside. Subsequent studies concurred that NADH was formed in the reaction but demonstrated that the inactivation is reversible by addition of NAD^+, suggesting that inactivation is simply due to reduction of the coenzyme (Matuszewska and Borchardt, 1987).

D. DNA Polymerase

The *E. coli* DNA polymerase I is inactivated by adenosine 2′,3′-riboepoxide 5′-triphosphate (epoxyATP) in a reaction which requires Mg^{2+} and a comple-

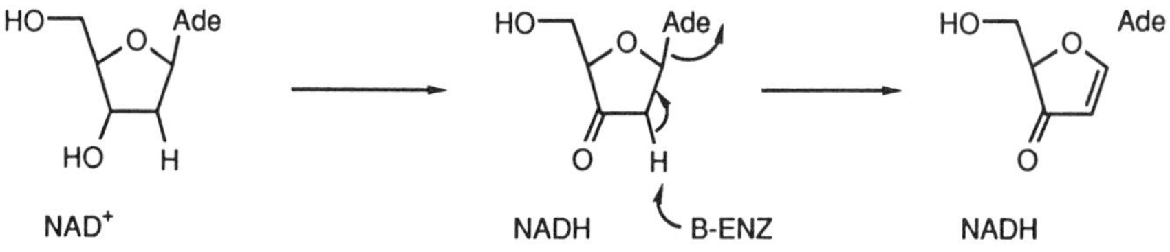

FIG. 58. Mechanism proposed for inactivation of *S*-adenosylhomocysteine hydrolase by 2′-deoxyadenosine.

mentary template (Abboud *et al.*, 1978). Inactivation is not the result of consumption of the template by chain termination or alkylation of the DNA by the reactive epoxide. Following a reaction which contained adenine-tritiated epoxyATP, the enzyme, inactivator, and template comigrate by gel filtration with 1 equivalent of nucleotide incorporated per equivalent of enzyme. These results were interpreted to imply that epoxyATP is a substrate for the polymerization reaction and is incorporated into the DNA. Subsequent alkylation of an active site residue by the epoxide is proposed to inactivate the enzyme by cross-linking it to the DNA.

Incorporation of a nucleotide analog into the DNA polymer by the herpes simplex virus DNA polymerase also leads to its demise (Furman *et al.*, 1984). Acyclovir triphosphate irreversibly inactivates the polymerase in a reaction which is also template dependent and is not the result of reversible binding of an acyclovir-terminated template. Since acyclovir does not contain any masked reactive groups, inactivation must result from noncovalent interactions. It was suggested that the enzyme might be "frozen" into a nonproductive conformation while attempting to excise the analog with its 3′,5′-exonuclease activity or while searching for a 3′-hydroxyl group for the next round of elongation.

X. Summary

Mechanism-based enzyme inactivation is now a mature field of endeavor in which basic chemical principles are applied to the construction of enzyme inhibitors. The design of mechanism-based inactivating agents is greatly facilitated by detailed knowledge of the catalytic mechanism of the target enzyme, but the chemical logic that links the catalytic mechanism to inactivation also makes such agents powerful probes of the catalytic process. A few general approaches, all of which involve introduction of a latent reactive functionality into a carrier structure that is recognized and processed by the enzyme, have emerged for the design of mechanism-based inactivating agents. The most common of these approaches is the insertion of a function that is catalytically converted to a Michael acceptor, but the full breadth of organic reactive species, including cations, anions, free radicals, and neutral species such as benzyne and cyclobutadiene, have been pressed into service in the design of mechanism-based inactivating agents. Novel and ingenious approaches will undoubtedly continue to be introduced as chemists strive to widen the scope of mechanism-based inactivation and to optimize the inactivation process by decreasing the partition ratio and increasing the rate of turnover of the inactivating agent. The continuing efforts in this area promise to lead not only to exciting new enzyme chemistry but also to the development of commercially useful agents.

References

Abboud, M. M., Sim, W. J., Loeb, L. A., and Mildvan, A. S. (1978). *JBC* **253,** 3415.
Abeles, R. H., and Maycock, A. L. (1976). *Acc. Chem. Res.* **9,** 313.
Abeles, R. H., Tashjian, A. H., Jr., and Fish, S. (1980). *BBRC* **95,** 612.
Abeles, R. H., Fish, S., and Lapinskas, B. (1982). *Biochemistry* **21,** 5557.
Allison, W. S., Swain, L. C., Tracy, S. M., and Benitez, L. V. (1973). *ABB* **155,** 400.
Alston, T. A., Porter, D. J. T., Mela, L., and Bright, H. J. (1980). *BBRC* **92,** 299.
Ashley, G., and Stubbe, J. (1985). *Pharmacol. Ther.* **30,** 301.
Ator, M. A., and Ortiz de Montellano, P. R. (1987). *JBC* **262,** 1542.
Ator, M., Salowe, S. P., Stubbe, J., Emptage, M., and Robins, M. J. (1984). *J. Am. Chem. Soc.* **106,** 1886.
Ator, M. A., David, S. K., and Ortiz de Montellano, P. R. (1987). *JBC* **262,** 14954.
Ator, M. A., David, S. K., and Ortiz de Montellano, P. R. (1989). *JBC* **264,** 9250.
Augusto, O., Beilan, H. S., and Ortiz de Montellano, P. R. (1982). *JBC* **257,** 11288.
Badet, B., Roise, D., and Walsh, C. T. (1984). *Biochemistry* **23,** 5188.
Baldoni, J. M., and Villafranca, J. J. (1980). *JBC* **255,** 8987.
Baldwin, J. E., and Parker, D. W. (1987). *J. Org. Chem.* **52,** 1475.
Bargar, T. M., Broersma, R. J., Creemer, L. C., McCarthy, J. R., Hornsberger, J.-M., Palfreyman, M. G., Wagner, J., and Jung, M. J. (1986). *J. Med. Chem.* **29,** 315.
Batzold, F. H., and Robinson, C. H. (1975). *J. Am. Chem. Soc.* **97,** 2576.
Bechet, J.-J., Dupaix, A., Yon, J., Wakselman, M., Robert, J.-C., and Vilkas, M. (1973). *EJB* **35,** 327.
Bechet, J.-J., Dupaix, A., and Blagoeva, I. (1977a). *Biochimie* **59,** 231.
Bechet, J.-J., Dupaix, A., Roucous, C., and Bonamy, A.-M. (1977b). *Biochemie* **59,** 241.
Bednarski, P. J., Porubek, D. J., and Nelson, S. D. (1985). *J. Med. Chem.* **28,** 775.
Bellisario, R. L., Maley, G. F., Galivan, J. H., and Maley, F. (1976). *PNAS* **73,** 1848.
Bellisario, R. L., Maley, G. F., Guarino, D. U., and Maley, F. (1979). *JBC* **254,** 1296.
Bloch, K. (1986). *J. Protein Chem.* **5,** 69.
Borchardt, R. T., Keller, B. T., and Patel-Thombre, U. (1984). *JBC* **259,** 4353.
Bossard, M. J., and Klinman, J. P. (1986). *JBC* **261,** 16421.
Bouclier, M., Jung, M. J., and Lippert, B. (1979). *EJB* **98,** 363.
Brenner, D. G., and Knowles, J. R. (1981). *Biochemistry* **20,** 3680.
Brenner, D. G., and Knowles, J. R. (1984). *Biochemistry* **23,** 5833.
Bruice, T. C. (1980). *Acc. Chem. Res.* **13,** 256.
Buckman, T. D., and Eiduson, S. (1985). *JBC* **260,** 11899.
Byrd, R. A., Dawson, W. H., Ellis, P. D., and Dunlap, R. B. (1978). *J. Am. Chem. Soc.* **100,** 7478.
CaJacob, C. A., and Ortiz de Montellano, P. R. (1986). *Biochemistry* **25,** 4705.
CaJacob, C. A., Chan, W. K., Shephard, E., and Ortiz de Montellano, P. R. (1988). *JBC* **263,** 18640.
Caplow, M., and Jencks, W. P. (1962). *Biochemistry* **1,** 883.
Charnas, R. L., and Knowles, J. R. (1981). *Biochemistry* **20,** 2732.
Chrystal, E., Bey, P., and Rando, R. R. (1979). *J. Neurochem.* **32,** 1501.
Chuang, H. Y. K., Patek, D. R., and Hellerman, L. (1974). *JBC* **249,** 2381.
Colombo, G., Rajashekhar, B., Giedroc, D. P., and Villafranca, J. J. (1984a). *JBC* **259,** 1593.
Colombo, G., Rajashekhar, B., Giedroc, D. P., and Villafranca, J. J. (1984b). *Biochemistry* **23,** 3590.
Colombo, G., Giedroc, D. P., Rajashekhar, B., and Villafranca, J. J. (1984). *JBC* **259,** 1601.
Colombo, G., Rajashekhar, B., Ash, D. E., and Villafranca, J. J. (1984d). *JBC* **259,** 1607.

Colonna, S., Gaggero, N., Manfredi, A., Casella, L., and Gullotti, M. (1988). *J. Chem. Soc., Chem. Commun.*, p. 1451.
Copp, L. J., Krantz, A., and Spencer, R. W. (1987). *Biochemistry* **26,** 169.
Corey, E. J., and Munroe, J. E. (1982). *J. Am. Chem. Soc.* **104,** 1752.
Corey, E. J., and Park, H. (1982). *J. Am. Chem. Soc.* **104,** 1750.
Corey, E. J., Lansbury, P. T., Jr., Cashman, J. R., and Kantner, S. S. (1984). *J. Am. Chem. Soc.* **106,** 1501.
Corey, E. J., Cashman, J. R., Eckrich, T. M., and Corey, D. R. (1985). *J. Am. Chem. Soc.* **107,** 713.
Corey, E. J., d'Alarcao, M., and Matsuda, A. P. T. (1986). *Tetrahedron Lett.* **27,** 3585.
Correia, M. A., Decker, C., Sugiyama, K., Caldera, P., Bornheim, L., Wrighton, S. A., Rettie, A. E., and Trager, W. F. (1987). *ABB* **258,** 436.
Covey, D. F., and Robinson, C. H. (1976). *J. Am. Chem. Soc.* **98,** 5038.
Danenberg, P. V., and Heidelberger, C. (1976). *Biochemistry* **15,** 1331.
Danenberg, P. V., Langenbach, R. J., and Heidelberger, C. (1974). *Biochemistry* **13,** 926.
Daniels, S. B., and Katzenellenbogen, J. A. (1986). *Biochemistry* **25,** 1436.
Daniels, S. B., Cooney, E., Sofia, M. J., Chakravarty, P. K., and Katzenellenbogen, J. A. (1983). *JBC* **258,** 15046.
deClercq, E. (1985). *Antimicrob. Agents Chemother.* **28,** 84.
De Matteis, F. (1978). *In* "Handbook of Experimental Pharmacology" (F. De Matteis and W. N. Aldridge, eds.), Vol. 44, pp. 95–127. Springer-Verlag, New York.
DePillis, G. D., and Ortiz de Montellano, P. R. (1989). *Biochemistry* **28,** 7947.
DeWolf, W. E., Jr., Carr, S. A., Varrichio, A., Goodhart, P. J., Mentzer, M. A., Roberts, G. D., Southan, C., Dolle, R. E., and Kruse, L. I. (1988). *Biochemistry* **27,** 9093.
Dijkstra, M., Frank, J., Jongejan, J. A., and Duine, J. A. (1984). *EJB* **140,** 369.
Doerge, D. R. (1986). *Biochemistry* **25,** 4724.
Doerge, D. R. (1988a). *Biochemistry* **27,** 3697.
Doerge, D. R. (1988b). *Xenobiotica* **18,** 1291.
Doerge, D. R., Pitz, G. L., and Root, D. P. (1987). *Biochem. Pharmacol.* **36,** 974.
Duine, J. A., Frank, J., and Jongejan, J. A. (1987). *Adv. Enzymol.* **59,** 169.
Dunford, H. B., and Stillman, J. S. (1976). *Coord. Chem. Rev.* **19,** 187.
Easton, C. J., and Knowles, J. R. (1982). *Biochemistry* **21,** 2857.
Endo, K., Helmkamp, G. M., Jr., and Bloch, K. (1970). *JBC* **245,** 4293.
Engler, H., Taurog, A., and Nakashima, T. (1982). *Biochem. Pharmacol.* **31,** 3801.
Faraci, W. S., and Pratt, R. F. (1985). *Biochemistry* **24,** 903.
Faraci, W. S., and Walsh, C. T. (1989). *Biochemistry* **28,** 431.
Fendrich, G., and Abeles, R. H. (1982). *Biochemistry* **21,** 6685.
Fitzpatrick, P. F., and Villafranca, J. J. (1985). *J. Am. Chem. Soc.* **107,** 5022.
Fitzpatrick, P. F., and Villafranca, J. J. (1987). *ABB* **257,** 231.
Fitzpatrick, P. F., Flory, D. R., Jr., and Villafranca, J. J. (1985). *Biochemistry* **24,** 2108.
Fouin-Fortunet, H., Tinel, M., Descatoire, V., Letteron, P., Larrey, D., Geneve, J., and Pessayre, D. (1986). *J. Pharmacol. Exp. Ther.* **236,** 237.
Fowler, A. V., Zabin, I., Sinnott, M. L., and Smith, P. J. (1978). *JBC* **253,** 5283.
Fowler, C. J., Mantle, T. J., and Tipton, K. F. (1982). *Biochem. Pharmacol.* **31,** 3555.
Frerman, F. E., Miziorko, H. M., and Beckmann, J. D. (1980). *JBC* **255,** 11192.
Furman, P. A., St. Clair, M. H., and Spector, T. (1984). *JBC* **259,** 9575.
Galey, J.-B., Bombard, S., Chopard, C., Girerd, J.-J., Lederer, F., Thang, D.-C., Nam, N.-H., Mansuy, D., and Chottard, J.-C. (1988). *Biochemistry* **27,** 1058.
Gan, L.-S. L., Acebo, A. L., and Alworth, W. L. (1984). *Biochemistry* **23,** 3827.
Gelb, M. H., and Abeles, R. H. (1984). *Biochemistry* **23,** 6596.

Gelb, M. H., and Abeles, R. H. (1986). *J. Med. Chem.* **29,** 585.
Ghisla, S., Ogata, H., Massey, V., Schonbrunn, A., Abeles, R. H., and Walsh, C. T. (1976). *Biochemistry* **15,** 1791.
Ghisla, S., Olson, S. T., Massey, V., and Lhoste, J.-M. (1979). *Biochemistry* **18,** 4733.
Gilman, A. G., and Murad, F. (1975). *In* "Pharmacological Basis of Therapeutics" (L. S. Goodman and A. Gilman, eds.), pp. 1410–1415. Macmillan, New York.
Gomes, B., Fendrich, G., and Abeles, R. H. (1981). *Biochemistry* **20,** 1481.
Goodhart, P. J., DeWolf, W. E., Jr., and Kruse, L. I. (1987). *Biochemistry* **26,** 2576.
Grab, L. A., Swanson, B. A., and Ortiz de Montellano, P. R. (1988). *Biochemistry* **27,** 4805.
Guengerich, F. P., and Macdonald, T. L. (1984). *Acc. Chem. Res.* **17,** 9.
Halpert, J. (1981). *Biochem. Pharmacol.* **30,** 875.
Halpert, J. (1982). *Mol. Pharmacol.* **21,** 166.
Halpert, J., and Neal, R. A. (1980). *Mol. Pharmacol.* **17,** 427.
Halpert, J., and Neal, R. A. (1981). *Drug Metab. Rev.* **12,** 239.
Halpert, J., Hammond, D., and Neal, R. A. (1980). *JBC* **255,** 1080.
Halpert, J., Naslund, B., and Betner, I. (1983). *Mol. Pharmacol.* **23,** 445.
Halpert, J. R., Miller, N. E., and Gorsky, L. D. (1985). *JBC* **260,** 8397.
Halpert, J. R., Balfour, C., Miller, N. E., and Kaminsky, L. S. (1986). *Mol. Pharmacol.* **30,** 19.
Halpert, J., Jaw, J.-Y., Balfour, C., Mash, E. A., and Johnson, E. F. (1988). *ABB* **264,** 462.
Hanzlik, R. P., and Tullman, R. H. (1982). *J. Am. Chem. Soc.* **104,** 2048.
Harper, J. W., and Powers, J. C. (1984). *Biochemistry* **24,** 7200.
Harper, J. W., Hemmi, K., and Powers, J. C. (1983). *J. Am. Chem. Soc.* **105,** 6518.
Harper, J. W., Hemmi, K., and Powers, J. C. (1985). *Biochemistry* **24,** 1831.
Helmkamp, G. M., Jr., and Bloch, K. (1969). *JBC* **244,** 6014.
Helmkamp, G. M., Jr., Rando, R. R., Brock, D. J. H., and Bloch, K. (1968). *JBC* **243,** 3229.
Hershfield, M. S. (1979). *JBC* **254,** 22.
Hewson, W. D., and Hager, L. P. (1979). *In* "The Porphyins" (D. Dolphin, ed.), Vol. 7, pp. 295–332. Academic Press, New York.
Hidaka, H., Udenfriend, S., Nagasaka, A., and DeGroot, L. J. (1970). *BBRC* **40,** 103.
Johnston, M., Jankowski, D., Marcotte, P., Tanaka, H. Esaki, N., Soda, K., and Walsh, C. (1979). *Biochemistry* **18,** 4690.
Johnston, M., Raines, R., Walsh, C., and Firestone, R. A. (1980). *J. Am. Chem. Soc.* **102,** 4241.
Jonen, H. G., Werringloer, J., Prough, R. A., and Estabrook, R. W. (1982). *JBC* **257,** 4404.
Jung, M. J., and Metcalf, B. W. (1975). *BBRC* **67,** 301.
Jung, M. J., Metcalf, B. W., Lippert, B., and Casara, P. (1978). *Biochemistry* **17,** 2628.
Kam, C. M., Copher, J. C., and Powers, J. C. (1987). *J. Am. Chem. Soc.* **109,** 5044.
Kam, C. M., Fujikawa, K., and Powers, J. C. (1988). *Biochemistry* **27,** 2547.
Kemal, C., and Knowles, J. R. (1981). *Biochemistry* **20,** 3688.
Kitz, R., and Wilson, I. B. (1962). *JBC* **237,** 3245.
Klinman, J. P., and Krueger, M. (1982). *Biochemistry* **21,** 67.
Knowles, J. R. (1983). *Antibiotics (N.Y.)* **6,** 90.
Komives, E. A., and Ortiz de Montellano, P. R. (1987). *JBC* **262,** 9793.
Komives, E. A., Tew, D., Olmstead, M. M., and Ortiz de Montellano, P. R. (1988). *Inorg. Chem.* **27,** 3112.
Kuhn, H., Holzhutter, H.-G., Schewe, T., Hiebsch, C. M., and Rapoport, S. M. (1984). *EJB* **139,** 577.
Larsson, A., and Sjoberg, B.-M. (1986). *EMBO J.* **5,** 2037.
Latham, J. A., and Walsh, C. (1987). *J. Am. Chem. Soc.* **109,** 3421.
Lee, J. S., Jacobsen, N. E., and Ortiz de Montellano, P. R. (1988). *Biochemistry* **27,** 7703.
Likos, J. L., Ueno, H., Feldhaus, R. W., and Metzler, D. E. (1982). *Biochemistry* **21,** 4377.

Lippert, B., Metcalf, B. W., Jung, M. J., and Casara, P. (1977). *EJB* **74,** 441.
Lukton, D., Mackie, J. E., Lee, J. S., Marks, G. S., and Ortiz de Montellano, P. R. (1988). *Chem. Res. Toxicol.* **1,** 208.
McCluskey, S. A., Marks, G. S., Sutherland, E. P., Jacobsen, N., and Ortiz de Montellano, P. R. (1986). *Mol. Pharmacol.* **30,** 352.
Macdonald, T. L., Zirvi, K., Burka, L. T., Peyman, P., and Guengerich, F. P. (1982). *J. Am. Chem. Soc.* **104,** 2050.
Maeda, Y., and Ingold, K. U. (1980). *J. Am. Chem. Soc.* **102,** 328.
Mangold, J. B., and Klinman, J. P. (1984). *JBC* **259,** 7772.
Marcotte, P., and Walsh, C. (1975). *BBRC* **62,** 677.
Marnett, L. J., Weller, P., and Batista, J. R. (1986). *In* "Cytochrome *P*-450: Structure, Mechanism, and Biochemistry" (P. R. Ortiz de Montellano, ed.), pp. 29–76. Plenum, New York.
Matuszewska, B., and Borchardt, R. T. (1987). *JBC* **262,** 265.
Maycock, A. L., Suva, R. H., and Abeles, R. H. (1975). *J. Am. Chem. Soc.* **97,** 5613.
Maycock, A. L., Abeles, R. H., Salach, J. I., and Singer, T. P. (1976). *Biochemistry* **15,** 114.
Maycock, A. L., Aster, S. D., and Patchett, A. A. (1980). *Biochemistry* **19,** 709.
Metcalf, B. W., and Jung, M. J., (1979). *Mol. Pharmacol.* **16,** 539.
Metcalf, B. W., Bey, P., Danzin, C., Jung, M. J., Casara, P., and Vevert, J. P. (1978). *J. Am. Chem. Soc.* **100,** 2551.
Metcalf, B. W., Wright, C. L., Burkhart, J. P., and Johnston, J. O. (1981). *J. Am. Chem. Soc.* **103,** 3221.
Mincey, T., Bell, J. A., Mildvan, A. S., and Abeles, R. H. (1981). *Biochemistry* **20,** 7502.
Moorman, A. R., and Abeles, R. H. (1982). *J. Am. Chem. Soc.* **104,** 6785.
Morrison, J. F., and Walsh, C. T., (1988). *Adv. Enzymol. Relat. Areas Mol. Biol.* **61,** 210.
Nagahisa, A., Orme-Johnson, W. H., and Wilson, S. R. (1984). *J. Am. Chem. Soc.* **106,** 1166.
Neary, J. T., Soodak, M., and Maloof, F. (1984). "Methods in Enzymology," Vol. 107, p. 445.
Ohtaki, S., Nakagawa, H., Nakamura, S., Nakamura, M., and Yamazaki, I. (1985). *JBC* **260,** 441.
Olson, S. T., Massey, V., Ghisla, S., and Whitfield, C. D. (1979). *Biochemistry* **18,** 4724.
Ortiz de Montellano, P. R. (1987). *Acc. Chem. Res.* **20,** 289.
Ortiz de Montellano, P. R. (1988). *In* "Progress in Drug Metabolism" (G. G. Gibson, ed.), Vol. 11, pp. 99–148. Taylor & Francis, London.
Ortiz de Montellano, P. R., and Grab, L. A. (1986). *J. Am. Chem. Soc.* **108,** 5584.
Ortiz de Montellano, P. R., and Kunze, K. L. (1980a). *JBC* **255,** 5578.
Ortiz de Montellano, P. R., and Kunze, K. L. (1980b). *BBRC* **94,** 443.
Ortiz de Montellano, P. R., and Mathews, J. M. (1981a). *BJ* **195,** 761.
Ortiz de Montellano, P. R., and Mathews, J. M. (1981b). *Biochem. Pharmacol.* **30,** 1138.
Ortiz de Montellano, P. R., and Mico, B. A. (1980). *Mol. Pharmacol.* **18,** 128.
Ortiz de Montellano, P. R., and Reich, N. O. (1986). *In* "Cytochrome *P*-450: Structure, Mechanism, and Biochemistry" (P. R. Ortiz de Montellano, ed.), pp. 273–314. Plenum, New York.
Ortiz de Montellano, P. R., Mico, B. A., and Yost, G. S. (1978). *BBRC* **83,** 132.
Ortiz de Montellano, P. R., Beilan, H. S., and Kunze, K. L. (1981). *JBC* **256,** 6708.
Ortiz de Montellano, P. R., Kunze, K. L., and Augusto, O. (1982). *J. Am. Chem. Soc.* **104,** 3545.
Ortiz de Montellano, P. R., Augusto, O., Viola, F., and Kunze, K. L. (1983). *JBC* **258,** 8623.
Ortiz de Montellano, P. R., Mathews, J. M., and Langry, K. C. (1984). *Tetrahedron* **40,** 511.
Ortiz de Montellano, P. R., Choe, Y. S., DePillis, G., and Catalano, C. E. (1987). *JBC* **262,** 11641.
Ortiz de Montellano, P. R., David, S. K., Ator, M. A., and Tew, D. (1988). *Biochemistry* **27,** 5470.
Padgette, S. R., Wimalasena, K., Herman, H. H., Sirimanne, S. R., and May, S. W. (1985). *Biochemistry* **24,** 5826.
Paech, C., Salach, J. I., and Singer, T. P. (1980). *JBC* **255,** 2700.
Palfreyman, M. G., Bey, P., and Sjoerdsma, A. (1987). *Essays Biochem.* **23,** 28.

Palmer, J., and Abeles, R. H. (1979). *JBC* **254,** 1217.
Parkes, C., and Abeles, R. H. (1984). *Biochemistry* **23,** 6355.
Pascal, R. A., Jr., and Huang, D.-S. (1987). *J. Am. Chem. Soc.* **109,** 2854.
Pegg, A. E., McGovern, K. A., and Wiest, L. (1987). *BJ* **241,** 305.
Pellino, A. M., and Danenberg, P. V. (1985). *JBC* **260,** 10996.
Penning, T. M., and Talalay, P. (1981). *JBC* **256,** 6851.
Penning, T. M., Covey, D. F., and Talalay, P. (1981). *JBC* **256,** 6842.
Petry, T. W., and Eling, T. E. (1987). *JBC* **262,** 14112.
Pohl, L. R., and Krishna, G. (1978). *Biochem. Pharmacol.* **27,** 335.
Porter, D. J. T., and Bright, H. J. (1983). *JBC* **258,** 9913.
Poulos, T. L., Finzel, B. C., and Howard, A. J. (1987). *JMB* **195,** 687.
Que, L., Jr. (1983). *Coord. Chem. Rev.* **50,** 73.
Rajashekhar, B., Fitzpatrick, P. F., Colombo, G., and Villafranca, J. J. (1984). *JBC* **259,** 6925.
Rando, R. (1974). *Science* **185,** 320.
Rando, R. R. (1977). *Biochemistry* **21,** 4604.
Rando, R. R. (1984). *Pharm. Rev.* **36,** 111.
Rando, R. R., and Eigner, A. (1977). *Mol. Pharmacol.* **13,** 1005.
Reilly, P. E. B., and Ivey, D. E. (1979). *FEBS Lett.* **97,** 141.
Ringe, D., Petsko, G. A., Kerr, D. E., and Ortiz de Montellano, P. R. (1984). *Biochemistry* **23,** 2.
Ringe, D., Seaton, B. A., Gelb, M. H., and Abeles, R. H. (1985). *Biochemistry* **24,** 64.
Ringe, D., Mottonen, J. M., Gelb, M. H., and Abeles, R. H. (1986). *Biochemistry* **25,** 5633.
Roise, D., Soda, K., Yagi, T., and Walsh, C. T. (1984). *Biochemistry* **23,** 5195.
Salowe, S. P., Ator, M. A., and Stubbe, J. (1987). *Biochemistry* **26,** 3408.
Santi, D. V., McHenry, C. S., and Sommer, H. (1974). *Biochemistry* **13,** 471.
Schechter, P. J. (1986). *In* "New Anticonvulsant Drugs" (B. S. Meldrum, and R. J. Porter, eds.), pp. 265–275. John Libbey, London.
Schloss, J. V. (1988). *Acc. Chem. Res.* **21,** 348.
Schnackerz, K., Ehrlich, K., Geisemann, W., and Reed, T. (1979). *Biochemistry* **18,** 3557.
Schonbrunn, A., Abeles, R. H., Walsh, C. T., Ghisla, S., Ogata, H., and Massey, V. (1976). *Biochemistry* **15,** 1798.
Schwab, J. M., Ho, C-K., Li, W., Townsend, C. A., and Salituro, G. M. (1986). *J. Am. Chem. Soc.* **108,** 5309.
Silverman, R. B. (1983). *JBC* **258,** 14766.
Silverman, R. B. (1984). *Biochemistry* **23,** 5206.
Silverman, R. B. (1988). "Mechanism-Based Enzyme Inactivation: Chemistry and Enzymology." CRC Press, Boca Raton, Florida.
Silverman, R. B., and Géorge, C. (1988a). *Biochemistry* **27,** 3285.
Silverman, R. B., and George, C. (1988b). *BBRC* **150,** 942.
Silverman, R. B., and Hoffman, S. J. (1984). *Med. Res. Rev.* **4,** 415.
Silverman, R. B., and Invergo, B. J. (1986). *Biochemistry* **25,** 6817.
Silverman, R. B., and Levy, M. A. (1981). *Biochemistry* **20,** 1197.
Silverman, R. B., and Yamasaki, R. B. (1984). *Biochemistry* **23,** 1322.
Silverman, R. B., and Zieske, P. A. (1985). *Biochemistry* **24,** 2128.
Silverman, R. B., Hoffman, S. J., and Catus III, W. B. (1980). *J. Am. Chem. Soc.* **102,** 7126.
Silverman, R. B., Hiebert, C. K., and Vazquez, M. L. (1985). *JBC* **260,** 14648.
Silverman, R. B., Invergo, B. J., and Mathew, J. (1986). *J. Med. Chem.* **29,** 1840.
Singer, T. P., Salach, J. I., Castagnoli, N., Jr., and Trevor, A. (1986). *BJ* **235,** 785.
Sinnott, M. L. (1987). *In* "Enzyme Mechanisms" (M. I. Page and A. Williams, eds.), pp. 259–297. Royal Society of Chemistry, London.
Sinnott, M. L., and Smith, P. J. (1976). *J. Chem. Soc., Chem. Commun.*, p. 223.

Sinnott, M. L., and Smith, P. J. (1978). *BJ* **175,** 525.
Sinnott, M. L., Tzotzos, G. T., and Marshall, S. E. (1982). *J. Chem. Soc., Perkin Trans. 2,* p. 1665.
Sjoberg, B.-M., and Graslund, A. (1983). *Adv. Inorg. Biochem.* **5,** 87.
Sjoberg, B.-M., Graslund, A., and Eckstein, F. (1983). *JBC* **258,** 8060.
Sjoerdsma, A., and Schechter, P. J. (1984). *Clin. Pharmacol. Ther.* **35,** 287.
Sok, D.-E., Han, C.-Q., Pai, J.-K., and Sih, C. J. (1982). *BBRC* **107,** 101.
Soper, T. S., and Manning, J. M. (1982). *JBC* **257,** 13930.
Stearns, R. A., and Ortiz de Montellano, P. R. (1985). *J. Am. Chem. Soc.* **107,** 234.
Tang, S.-S., Trakman, P. C., and Kagan, H. M. (1983). *JBC* **258,** 4331.
Tantsunami, S., Yago, N., and Hosoe, M. (1981). *BBA* **662,** 226.
Thelander, L., Larsson, B., Hobbs, J., and Eckstein, F. (1976). *JBC* **251,** 1398.
Tsou, C.-L. (1988). *Adv. Enzymol. Relat. Areas Mol. Biol.* **61,** 381.
Ueno, H., Likos, J. J., and Metzler, D. E. (1982). *Biochemistry* **21,** 4387.
Veldink, G. A., and Vliegenthart, J. F. G. (1984). *Adv. Inorg. Biochem.* **6,** 139.
Waley, S. G. (1980). *BJ* **185,** 771.
Waley, S. G. (1985). *BJ* **227,** 843.
Walsh, C. (1977). *Horiz. Biochem. Biophys.* **3,** 36.
Walsh, C. (1979). "Enzymatic Reaction Mechanisms," pp. 777–827. Freeman, San Francisco, California.
Walsh, C. (1980). *Acc. Chem. Res.* **13,** 148.
Walsh, C. (1982). *Tetrahedron* **38,** 871.
Walsh, C. T. (1984). *Annu. Rev. Biochem.* **53,** 493.
Walsh, C. T., Schonbrunn, A., Lockridge, O., Massey, V., and Abeles, R. H. (1972). *JBC* **247,** 7858.
Wang, E., and Walsh, C. (1978). *Biochemistry* **17,** 1313.
Wang, E. A., and Walsh, C. T. (1981). *Biochemistry* **20,** 7539.
Washtien, W., and Abeles, R. H. (1977). *Biochemistry* **16,** 2485.
Weidmann, B., and Abeles, R. H. (1984). *Biochemistry* **23,** 2373.
Wenz, A., Thorpe, C., and Ghisla, S. (1981). *JBC* **256,** 9809.
Wenz, A., Ghisla, S., and Thorpe, C. (1985). *EJB* **147,** 553.
Westkaemper, R. B., and Abeles, R. H. (1983). *Biochemistry* **22,** 3256.
Williamson, P. R., and Kagan, H. M. (1987). *JBC* **262,** 8196.
Wiseman, J. S., Nichols, J. S., and Kolpak, M. X. (1982). *JBC* **257,** 6328.
Withers, S. G., Street, I. P., Bird, P., and Dolphin, D. H. (1987). *J. Am. Chem. Soc.* **109,** 7350.
Withers, S. G., Rupitz, K., and Street, I. P. (1988). *JBC* **263,** 7929.
Wolfson, G., Chisholm, J., Tashjian, A. H., Jr., Fish, S., and Abeles, R. H. (1986). *JBC* **261,** 4492.

6

Site-Specific Modification of Enzyme Sites

ROBERTA F. COLMAN
Department of Chemistry and Biochemistry
University of Delaware
Newark, Delaware 19716

I. Introduction

Identification of the amino acids which participate in catalytic and regulatory sites of enzymes and evaluation of their functional role are continuing goals of

THE ENZYMES, Vol. XIX

biochemists. Since the emergence of affinity labeling in the 1960s (e.g., *1, 2*), the application of this technique has expanded explosively until it is used widely in fields such as enzymology, immunology, receptor biochemistry, and pharmacology. In this approach a reagent is designed which is structurally similar to the natural substrate or regulator but with an additional reactive functional group capable of covalent attachment to many different amino acids. The expectation is that the affinity label will bind reversibly to a particular ligand site on the enzyme and, within the lifetime of the enzyme–reagent complex, will react irreversibly with amino acids accessible from the specific ligand site. The strategy of affinity labeling allows the identification of particular binding proteins within a complex biochemical mixture; it facilitates the location of polypeptide regions or subunits associated with specific binding sites in purified enzymes; it permits the experimental evaluation of enzyme binding sites predicted on the basis of so-called amino acid consensus sequences; and it provides an experimental approach for comparing operational ligand sites of a protein when in solution with those identified in a protein crystal by X-ray diffraction.

Many of the general principles of affinity labeling have been reviewed (*3–6*). More practical considerations regarding affinity labeling have been presented (*5, 7–9*), along with extensive examples. Certain specific aspects have been summarized including photoaffinity labeling of enzymes (*10*), affinity labeling of purine nucleotide sites in proteins (*11*), photoaffinity labeling of peptide hormone receptors (*12*), photoaffinity labeling using ATP analogs (*13*), and affinity labeling of hormone-specific proteins (*14*). In addition, two volumes have appeared recently on mechanism-based enzyme inactivation (*15*), which might be considered a special case of site-specific modification of enzyme sites; however, these enzyme-activated irreversible inhibitors are addressed separately in this volume (Chapter 5, Ator and Ortiz de Montellano). Rather than attempting to be comprehensive, this chapter focuses on the principles of affinity labeling, discusses several chemical classes of affinity reagents exhibiting different characteristic features, and highlights newer examples of the application of these types of reagents, as well as representative examples from the earlier literature.

II. General Characteristics of Site-Specific Modification

Because the goal of affinity labeling is to modify chemically amino acid residues within a specific binding site of an enzyme, the approach is best undertaken after a detailed examination of the structural requirements for binding to that site. These requirements can be ascertained by considering the range of acceptable alternate substrates for an enzyme or effective competitive inhibitors for the substrate or allosteric regulator. The choice or design of a new reagent will depend on a knowledge of which portions of the natural ligand can be altered without endangering recognition by the enzyme; these are the positions at which reactive substituents may be placed. The most desirable reagents are those which preserve the critical structural features of the natural ligand but which feature in addition a small substituent that reacts relatively indiscriminately with amino acids. Since the amino acid participants in ligand binding sites are often unknown prior to initiation of the affinity labeling experiments, the use of highly reactive, nonspecific substituents improves the probability of covalent reaction once the specificity of binding is assured by the structural fidelity of the rest of the site-directed reagent.

A characteristic of an affinity label (R) is that it forms reversibly an enzyme–reagent complex (ER) prior to irreversible covalent modification to yield modified enzyme (ER′), as follows:

$$E + R \underset{k_{-1}}{\overset{k_1}{\rightleftharpoons}} ER \xrightarrow{k_{max}} ER' \tag{1}$$

In many cases, a nonlinear dependence of the rate of modification on the concentration of affinity label (a "rate saturation" effect) is observed: the rate constant increases with increasing reagent concentration until the enzyme become saturated with reagent; thereafter, further increases in reagent concentration do not enhance the rate constant. In accordance with the simplified mechanism of reaction (1), the observed rate constant for modification (k_{obs}) can be expressed as

$$k_{obs} = \frac{k_{max}}{1 + (K_D/[R])} \tag{2}$$

where k_{obs} is the observed rate constant at a particular reagent concentration, k_{max} is the maximum rate constant for modification at saturating reagent concentrations, and $K_D = (k_{-1} + k_{max})/k_1$ and represents the concentration of reagent giving half the maximum inactivation rate (*16*). The reciprocal form of Eq. (2) is

$$\frac{1}{k_{obs}} = \frac{1}{k_{max}} + \frac{K_D}{k_{max}} \frac{1}{[R]} \tag{3}$$

A double reciprocal plot of $1/k_{obs}$ versus $1/[R]$ yields values for K_D and k_{max}. However, in some cases, examination of the reaction may be restricted to cases where $[R] \ll K_D$; this may occur, for example, because of limited solubility of the reagent or because of high rate constants that are at the limit of the measurable range for the analytical method used (*17*). In such cases, $k_{obs} \cong (k_{max}/K_D)[R]$ and is linearly dependent on [R] over the experimentally accessible reagent concentration range (*17*). Therefore, the failure to observe a "rate saturation" effect does not automatically exclude a compound as functioning as an affinity label.

An essential characteristic of an affinity label is that it exhibits limited incorporation (ideally, 1 mol/mol active site) in causing inactivation. Indeed, the amount of reagent incorporation is generally lower for the affinity label than for a reagent with the same functional group which lacks the structural features directing the affinity label to the ligand binding site. The importance of the native structure of an enzyme in limiting the amount of reagent incorporated is illustrated by the reaction of the substrate analog 3-bromo-2-ketoglutarate with pig heart NAD-dependent isocitrate dehydrogenase (*18*). This compound, when incubated under defined conditions for 90 min with the enzyme in the native state, is incorporated to the extent of 0.8 mol reagent/mol enzyme; in contrast, when the incubation is conducted under the same conditions with the exception that the enzyme is denatured in urea, as much as 5 mol of 3-bromo-2-ketoglutarate is incorporated per mole of enzyme subunit in the 90-min period (*18*). Although amino acid residues at the catalytic site are often considered to be unusually reactive, this is not always the case, and there is less justification for expecting the amino acid participants in an allosteric site to have remarkable reactivity. The binding of the affinity label to the native tertiary structure of the enzyme is primarily responsible for limiting the extent of chemical modification.

The substrate, regulatory compound, or other ligand which normally occupies the binding site is expected to compete with the affinity label, thereby decreasing the observed rate constant for modification. In fact, a specific effect of one of several natural ligands in decreasing the reaction rate can provide important evidence in identifying the target functional site of the affinity label (*7, 16*). The ability of increasing concentrations of the natural ligand to decrease the rate constant for modification by the affinity label allows the calculation of a dissociation constant for the enzyme–ligand complex at the site being attacked by the affinity label. This constant may then be compared with a K_d value for the enzyme–ligand complex determined independently (in the absence of the affinity label) (*18*).

The ability of an enzyme to catalyze its normal reaction with a covalently bound substrate, termed "catalytic competence," was proposed as another criterion of an affinity label (*19*). This is an extremely stringent criterion which can be satisfied only if the affinity label reacts covalently at a nonessential residue, if the incorporated reagent contains a functional group capable of undergoing reaction, and if the enzyme-bound affinity label can attain the required orientation within the active site. The product of modification of human placental estradiol 17β-dehydrogenase by 3-bromoacetoxyestrone meets this criterion (*19*): the covalently bound estrone is stereospecifically converted to estradiol in the presence of NADH. More recently, the criterion has successfully been applied to the modification of *Escherichia coli* RNA polymerase by several nucleotide analogs (*20*). However, catalytic competence of modified enzyme could not be demonstrated for NAD-dependent isocitrate dehydrogenase inactivated by 3-bromo-2-ketoglutarate (*18*) or for pyruvate kinase inactivated by 2-[(4-bromo-2,3-dioxobutyl)thio]-1,N^6-ethenoadenosine 5′-diphosphate (*21*), despite the fact that these reactions fulfill all other criteria of affinity labeling. Although the observation of catalytic competence can provide positive evidence for modification in the region of the active site, the lack of catalytic competence does not exclude active site modification since the formation of a covalent attachment may prevent the bound substrate from assuming the orientation required for enzyme catalysis.

Among the large number of site-specific reagents described in the literature, there are characteristics that are particularly desirable. In addition to structural similarity to the substrate or regulatory compound and the potential to react with many types of amino acids, these characteristics include water solubility, ease of synthesis, acceptable stability in aqueous solution over a moderate pH range, and the ability to form a stoichiometric, stable, isolable, and readily identifiable product from the reaction with enzyme. In this chapter, several chemical classes of affinity labels are considered, their notable features are evaluated in terms of these criteria, and selected examples of their application are presented.

III. Haloketones

The use of α-haloketones as reactive substituents dates back to the initial affinity labeling studies, as typified by the inactivation of chymotrypsin by *N*-tosylphenylalanyl chloromethyl ketone (*1*) and of 2-keto-3-deoxygluconate-6-phosphate aldolase by bromopyruvate (*22*). Many of the early investigations involving haloketone derivatives have been critically reviewed by Hartman (*23*). The effectiveness of haloketone derivatives is due, in part, to their high reactivity. For example, the pH-independent rate constant for reaction of 3-bromo-2-ketoglutarate with the thiol group of glutathione has been calculated as $2.2 \times 10^4\ M^{-1}\ \text{sec}^{-1}$ (*18*), about 4 orders of magnitude greater than the pH-independent rate constant for reaction of iodoacetate with glutathione ($3.2\ M^{-1}\ \text{sec}^{-1}$) (*24*).

A. Haloketone Derivatives of Nucleotides

In the late 1980s, haloketone derivatives of several nucleotides have been synthesized by coupling of 1,4-dibromo-2,3-butanedione with the thiol group of a coenzyme or nucleotide analog. Thus, *S*-(4-bromo-2,3-dioxobutyl)coenzyme A (BDB-CoA) was synthesized (*25*) and shown to inactivate several enzymes that bind acetyl-CoA (*25–29*), including citrate synthase and β-hydroxy-β-methylglutaryl-CoA reductase. In the case of citrate synthase, BDB-CoA appeared to function as an active site-directed irreversible inhibitor, with protection against inactivation being provided by benzoyl-CoA, a competitive inhibitor with respect to acetyl-CoA (*25*). Modified peptides have been isolated, and it was proposed that the active site-directed inactivation was due to reaction with either Lys-22 or Glu-363 located at or near the entrance to the active site (*29*).

The synthesis and detailed characterization of 4-bromo-2,3-dioxobutyl derivatives of mercaptopurine nucleotides have been described by Colman *et al.* The three compounds illustrated in Fig. 1 have the same reactive functional group located at distinctive positions of the purine ring: 2-(4-bromo-2,3-dioxo-butylthio)adenosine 5′-diphosphate (2-BDB-TA 5′-DP) (*30–32*), 6-(4-bromo-2,3-dioxobutylthio)adenosine 5′-diphosphate (6-BDB-TA 5′-DP) (*16, 33, 34*), and 8-(4-bromo-2,3-dioxobutylthio)adenosine 5′-diphosphate (8-BDB-TA 5′-DP) (*35*). In each case, either the 5′-monophosphates or 5′-triphosphates were also prepared. In addition, the 2- and 8-(4-bromo-2,3-dioxobutylthio) derivatives of cyclic AMP have been synthesized (*36*). These compounds are all closely related to the adenine nucleotides, and they are water-soluble and negatively charged at neutral pH. Typically, reactions are conducted between pH 6 and 7 at 25°C, using 100–600 μM reagent. The rate constant for decomposition of the BDB-nucleotides, as determined by release of free bromide, is pH dependent: for example, the $t_{1/2}$ for 6-BDB-TA 5′-DP was measured as 9.9 hr at pH 6.1 (*16*) and 61 min at pH 7.1 (*34*), whereas at pH 7.0 the $t_{1/2}$ for 2-BDB-TAMP was determined as 60 min (*30*). At pH values much above 7.1, the solvolysis reaction may be too rapid to allow these BDB derivatives to be useful as affinity labels of enzymes.

The bromoketo group is potentially reactive with most nucleophiles found in proteins, and the diketo moiety adds the possibility of reaction with arginine residues. Peptides containing cysteine (*21, 37*), histidine (*38*), tyrosine (*21*), glutamate (*39*), and aspartate (*40*) modified by BDB-nucleotides have actually been isolated and characterized. The placement of the reactive substituent at either the 2-, 6-, or 8-positions of the purine ring should allow the systematic probing of different subregions of a single purine-binding site in an enzyme or the labeling of distinct sites for enzymes with multiple nucleotide sites which

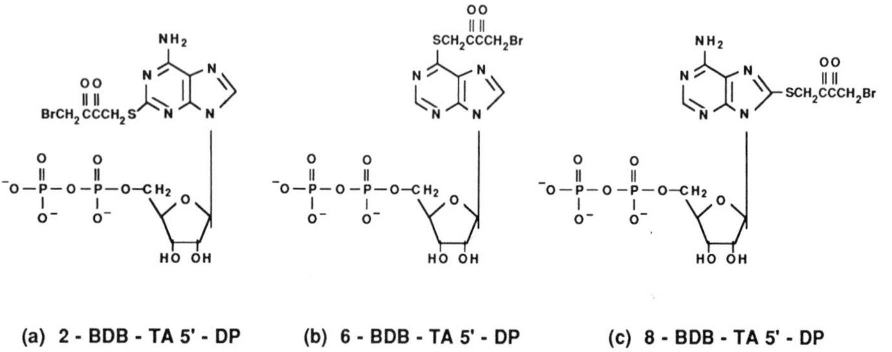

(a) 2 - BDB - TA 5' - DP (b) 6 - BDB - TA 5' - DP (c) 8 - BDB - TA 5' - DP

FIG. 1. Bromo-2,3-dioxobutyl nucleotide affinity labels. (a) 2-(4-Bromo-2,3-dioxobutylthio)-adenosine 5′-diphosphate; (b) 6-(4-bromo-2,3-dioxobutylthio)adenosine 5′-diphosphate; and (c) 8-(4-bromo-2,3-dioxobutylthio)adenosine 5′-diphosphate.

differ in structural specificity. For example, rabbit muscle pyruvate kinase is inactivated by 2-, 6-, and 8-BDB-nucleotides, but effective protection against inactivation by the 2- (*30*) and 6-substituted compounds (*33*) is provided by metal–nucleotide whereas phosphoenolpyruvate and metal ion maximally decrease the inactivation rate by 8-BDB-TA 5′-DP (*35*). All of these compounds appear to react irreversibly in the vicinity of the active site of pyruvate kinase but at different subsites. In contrast, the BDB derivatives thus far evaluated for their effect on bovine liver glutamate dehydrogenase react exclusively at regulatory sites, rather than at the active site: the 6-BDB-TADP reacts covalently at the NADH inhibitory site (*34*) whereas the 2-BDB-TADP acts as an affinity label of the ADP-activating site (*32*). In these studies, the amount of reagent incorporated into the proteins was measured either by analysis of the protein-bound organic phosphate or by using ^{32}P-labeled reagent. In isolating modified peptides, however, it has proved convenient to use nonlabeled reagent and to introduce a radioactive tag by reducing the dioxo groups of covalently bound reagent with [^3H]$NaBH_4$ (*21, 37–40*). The procedures used to isolate these modified peptides have sometimes taken advantage of the special ability of this *cis*-diol moiety of nucleotidyl peptides to complex with the dihydroxyboryl group during chromatography on phenyl boronate-agarose columns (*21, 37, 38*). When a tryptic digest is applied to the column in buffer at pH 8, the nucleotidyl peptides remain bound while the unmodified peptides are eluted; nucleotidyl peptides can then be eluted with distilled water.

The compounds shown in Fig. 2 all have a reactive bromoketo substituent at the 2-position, and, owing to the presence of ethenoadenosine, they are all fluorescent, with an emission peak at about 428 nm on excitation at 302 nm (*41*). These reagents provide a convenient approach to introduce a covalently bound fluorescent probe at a specific nucleotide-binding site of an enzyme. Because of its 2′-phosphate, 2-(4-bromo-2,3-dioxobutylthio)-1,N^6-ethenoadenosine 2′,5′-bisphosphate (2-BDB-TεA 2′,5′-DP, Fig. 2a) can first bind reversibly and then react covalently at $NADP^+$-binding sites (*39, 41*). A related affinity label featuring the free 6-amino group of the adenine ring, 2-(4-bromo-2,3-dioxobutylthio)adenosine 2′,5′-bisphosphate (2-BDB-TA 2′,5′-DP) has also been described (*42*). In contrast, with its 5′-diphosphate, 2-(4-bromo-2,3-dioxobutylthio)-1,N^6-ethenoadenosine 5′-diphosphate (2-BDB-TεA 5′-DP, Fig. 2b) would be expected to react at ADP or NAD sites in enzymes (*21*). The compound shown in Fig. 2c, 2-(3-bromo-2-oxopropylthio)-1,N^6-ethenoadenosine 5′-diphosphate (2-BOP-TεA 5′-DP), differs in having one less carbonyl group in the reactive side chain than 2-BDB-TεA 5′-DP. The monoketo compound does not enolize in aqueous solution, whereas the bromodioxobutyl group is extensively enolized (*31, 41*); thus, a comparison of the effects of 2-BDB with 2-BOP-TεA 5′-DP on an enzyme tests the importance of the enolate structure, the diketo groups, and/or the chain length in the affinity label.

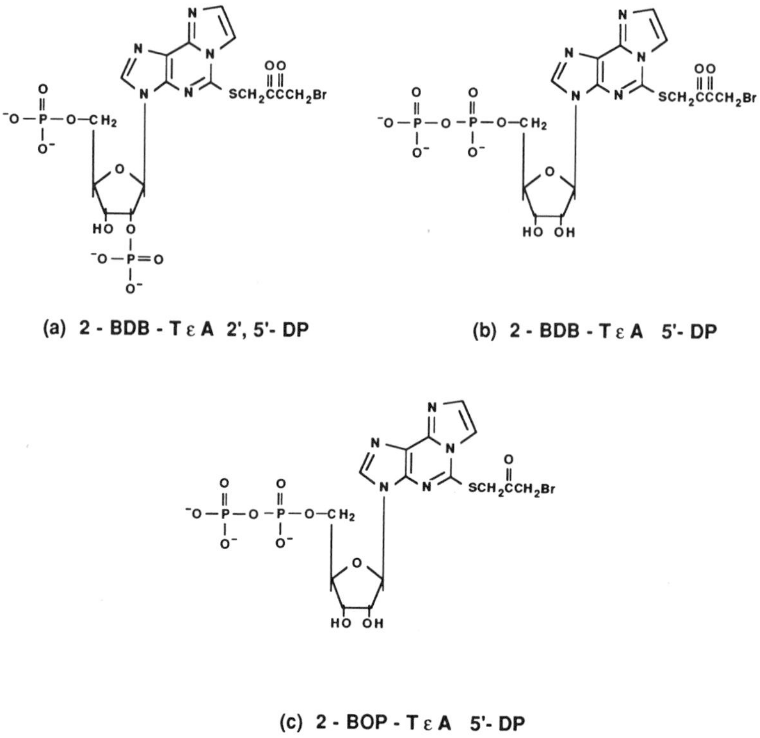

(a) 2 - BDB - TεA 2', 5'- DP

(b) 2 - BDB - TεA 5'- DP

(c) 2 - BOP - TεA 5'- DP

FIG. 2. Additional bromo-2,3-dioxobutyl nucleotide affinity labels. (a) 2-(4-Bromo-2,3-dioxobutylthio)-1,N^6-ethenoadenosine 2′,5′-diphosphate; (b) 2-(4-bromo-2,3-dioxobutylthio)-1,N^6-ethenoadenosine 5′-diphosphate; and (c) 2-(3-bromo-2-oxopropylthio)-1,N^6-ethenoadenosine 5′-diphosphate.

Pig heart $NADP^+$-specific isocitrate dehydrogenase was shown to be rapidly inactivated by 2-BDB-TεA 2′,5′-DP (Fig. 2a) with concomitant incorporation of 1 mol of reagent/mol of subunit as measured either by bound fluorescence or phosphorus (*41, 43*). Since protection against inactivation was provided by NADPH or NADP, but not by the substrate manganese–isocitrate, it was concluded that reaction had occurred at the coenzyme binding site. After digestion by trypsin, the nucleotidyl peptides were isolated by chromatography on DEAE-cellulose, followed by treatment with acid phosphatase (to decrease the negative charge by removing the phosphate groups from covalently bound reagent) and rechromatography on the same DEAE-cellulose column; the purification thus exploited the properties of the modified peptide introduced by the reagent (*43*). A single peptide correlating with the modification of the coenzyme site was isolated and subjected to gas phase sequencing: Asp-Leu-Ala-Gly-X-Ile-His-Gly-Leu-Ser-Asn-Val-Lys. Since no identifiable phenylthiohydantoin derivative was found at position 5, it was concluded that the amino acid at this position of the peptide was modified (*43*). Recent evidence obtained from the isolation of the corresponding peptide from unmodified enzyme indicates that X is cysteine (G. E. Smyth and R. F. Colman, unpublished data, 1990). The fluorescent enzyme-bound 2-BDB-TεA 2′,5′-DP at a coenzyme site has been used as an energy donor in resonance energy transfer experiments to estimate distances between coenzyme sites on the two subunits of the enzyme dimer and distances between the coenzyme and metal–isocitrate sites (*44*).

The same 2-BDB-TεA 2′,5′-DP was also found to inactivate the $NADP^+$-specific glutamate dehydrogenase from *Salmonella typhimurium*. The rate of inactivation exhibited a nonlinear dependence on the reagent concentration, indicative of reversible binding prior to irreversible modification (*39*). The presence of NADPH or $NADP^+$ in the reaction mixture completely prevented inactivation, suggesting that 2-BDB-TεA 2′,5′-DP also functioned as an affinity label of the coenzyme site in this $NADP^+$-specific dehydrogenase.

Rabbit muscle pyruvate kinase binds both ADP and phosphoenolpyruvate at its active site in catalyzing the formation of ATP and pyruvate. The fluorescent nucleotide analog 2-BDB-TεA 5′-DP (Fig. 2b) inactivated pyruvate kinase in a site-specific manner, but surprisingly the most effective protector against inactivation was phosphoenolpyruvate in the presence of mono- and divalent cations, rather than ADP (*21*). There is a marked structural similarity between phosphoenolpyruvate (PEP) and the enol form of the bromodioxobutyl group, suggesting that the enolized form of 2-BDB-TεA 5′-DP probably directs the reagent to the phosphoenolpyruvate site of the enzyme. In support of this conclusion, incubation of pyruvate kinase with the monoketone 2-BOP-TεA 5′-DP, which does not form an enolate, failed to cause appreciable inactivation (*21*). This result emphasizes the importance of examining the contributions of all structural features of a site-directed reagent before drawing conclusions about the nature of the target site. The compound 2-BDB-TεA 5′-DP may prove to be an affinity label of either ADP or phosphoenolpyruvate sites in various enzymes.

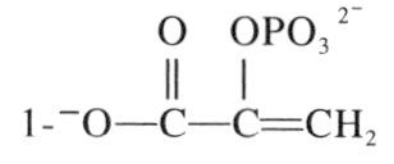

$$\begin{array}{c} \quad\ \ \text{O}\ \ \text{OPO}_3^{2-} \\ \quad\ \ \|\quad | \\ 1\text{-}^{-}\text{O—C—C=CH}_2 \end{array}$$

PEP

$$\begin{array}{c} \qquad\quad \text{O}\ \ \text{OH} \\ \qquad\quad \|\quad | \\ \text{Br—CH}_2\text{—C—C=CH—S—}\varepsilon\text{A—5}'\text{-DP} \end{array}$$

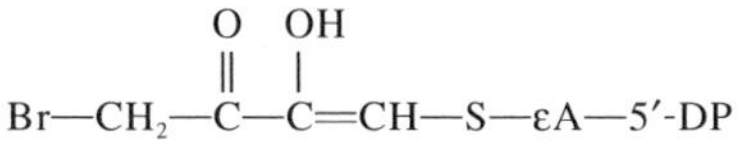

2-BDB-TεA 5′-DP

B. 3-Bromo-2-ketoglutarate

Bromopyruvate was first used as an affinity label of pyruvate binding sites by Meloche (*45*), and its application has been extended to a number of enzymes (*23*). By analogy, 3-bromo-2-ketoglutarate (BrKG) was synthesized as an affinity label of α-ketoglutarate binding sites by Mäntsälä and Zalkin (*46*) and shown to inactivate glutamate synthase. Hartman (*47*) demonstrated that BrKG was an inactivator of the pig heart $NADP^+$-dependent isocitrate dehydrogenase, which normally catalyzes the formation of α-ketoglutarate as the product of oxidative decarboxylation of isocitrate. More recently (*48*) it has been shown that BrKG is incorporated up to 1 mol of reagent per mole of enzyme dimer, concomitant with loss of 65% of the original activity. Apparently BrKG reacts and inactivates one subunit of the dimer, which decreases somewhat the activity of the second subunit. Complete protection against inactivation was provided by NADP or the NADP–α-ketoglutarate adduct but not by manganous isocitrate. BrKG modified only one peptide of the $NADP^+$-specific isocitrate dehydrogenase which was identical to the peptide labeled by the coenzyme analog 2-BDB-TεA 2′,5′-DP (*43, 48*). These data were interpreted to indicate that 3-bromo-2-ketoglutarate modifies an amino acid in the nicotinamide region of the coenzyme site proximal to the substrate site.

3-Bromo-2-ketoglutarate has also been demonstrated to be an affinity label of the NAD^+-specific isocitrate dehydrogenase of pig heart (*18*). In this case, the enzyme could be totally inactivated, and complete protection was provided by the substrate metal–isocitrate, rather than by coenzyme. It was considered that the substrate site was the functional target of BrKG in the NAD^+-specific enzyme, and the cysteine-modified peptide isolated from the enzyme [A. Saha, Y.-C. Huang, and R. F. Colman, (*48a*)] had no apparent sequence relationship to that purified from the BrKG-inactivated $NADP^+$-specific isocitrate dehydrogenase (*48*). In aqueous solution at pH 6.1, the decomposition of 3-bromo-2-ketoglutarate was shown to involve a complex series of reactions, the rates of which have been analyzed (*49*). Elimination of HBr occurs most rapidly to yield 3-ene-2-ketoglutarate, which can still add nucleophiles by a Michael-type addition to C-4 and thus can inactivate isocitrate dehydrogenase, albeit with different kinetic parameters than exhibited by BrKG (*48*). Since BrKG can generate more than one reactive species, it is apparent that, in any application of 3-bromo-2-ketoglutarate as an affinity label, the distribution of the several reactive species and the rates of their decomposition under the conditions used will have to be considered (*49*).

IV. Aldehyde Affinity Labels

Primary amines (such as the ε-amino group of lysine) react readily with aldehydes, often forming Schiff bases which can be stabilized by reduction with $NaBH_4$. Such a reaction forms the basis of the widespread use of pyridoxal phosphate to modify chemically the lysine residues of proteins. In recent years, in order to increase the specificity of labeling, aldehydes have been incorporated into more complex compounds which mimic the structures of the substrate or cofactor of the enzyme.

A. Periodate-Oxidized Nucleotides

The oxidation of ribonucleotides by periodate results in the cleavage of the bond between C-2′ and C-3′ of the ribose ring to yield the 2′,3′-dialdehyde derivative of the nucleotide (e.g., Fig. 3a). The compounds generated are structurally similar to the parent ribonucleotide, are capable of reacting with lysine residues, and (above all) are easy to prepare, thus accounting for the high popularity of their use as affinity labels of nucleotide sites in enzymes. Easterbrook-Smith *et al.* (*50*), in an often cited paper, demonstrated that the dialdehyde derivative of ATP (oATP) fulfilled the general criteria for an affinity label of pyruvate carboxylase. The magnesium complex of oATP functioned as a linear competitive inhibitor with respect to MgATP when evaluated in the absence of a reducing agent. On incubation of oATP with pyruvate carboxylase followed by addition of $NaBH_4$, irreversible inactivation was observed with 1.1 mol of reagent incorporated per mole of catalytic site at 100% inactivation. Protection by MgATP against inactivation indicated that oATP reacts at the metal–nucleotide site. Evidence was presented that a proteolytic digest of the enzyme modified with radioactive oATP, followed by reduction, contained a labeled product which comigrated chromatographically with a model adduct of oATP and lysine (*50*).

A large number of other enzymes have been tested for modification by dialde-

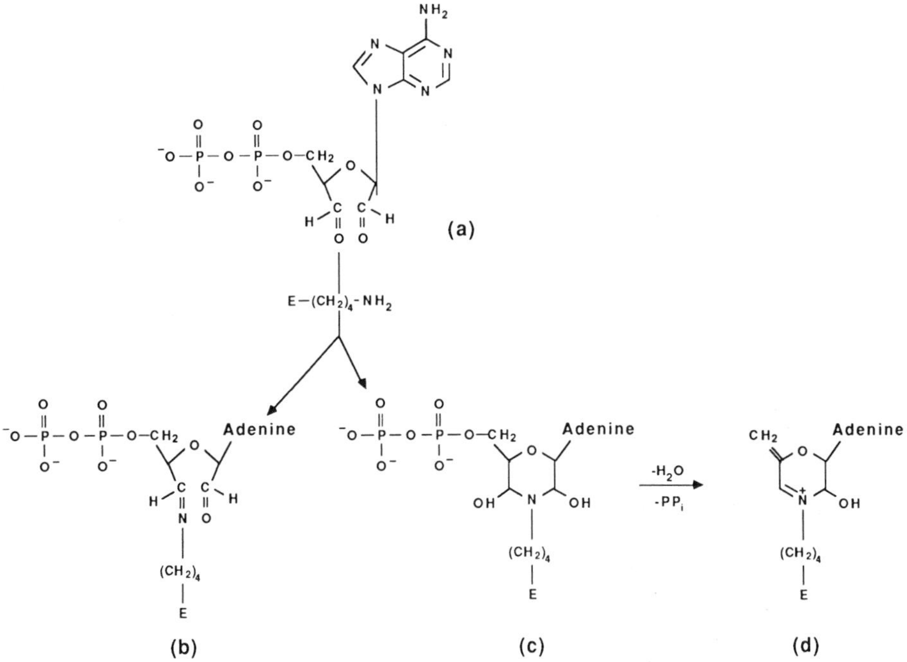

FIG. 3. Possible reaction products of 2′,3′-dialdehyde derivative of ADP with lysine in enzymes.

hyde derivatives of nucleotides. Some of this work has been reviewed (*4, 11*). Table I presents a partial list of enzymes in which affinity labeling experiments have been described using the dialdehyde derivatives of ATP, ADP, or AMP (*50–62*). In most of the cases, the catalytic site was the target of attack, but in the NAD$^+$-dependent isocitrate dehydrogenase an ADP allosteric site was modified (*60, 61*). Additional affinity labeling experiments have reported the use of the dialdehyde derivative of ethenoadenosine triphosphate for beef liver mitochondrial ATPase (*63*), of periodate-oxidized CDP for *Corynebacterium nephridii* ribonucleotide reductase (*64*), of dial-GDP or -GTP for rat liver succinyl-CoA synthetase (*65*) and rat liver elongation factor EF-2 (*66, 67*), and of the dialdehyde derivative of 8-azido-ATP for phosphorylase kinase (*68*).

Although the coenzyme NADP has two ribose moieties, because of the 2′-phosphate on the adenosine ribose, only the nicotinamide ribose is oxidized by periodate. The 2′,3′-dialdehyde derivative of NADP was first described by Rippa *et al.* (*69, 70*) and shown to inactivate the 6-phosphogluconate dehydrogenase of *Candida utilis*. NADP or NADPH protected the enzyme against inactivation, whereas ribose 5-phosphate or 6-phosphogluconate did not exhibit this protective effect, indicating that modification occurred at the coenzyme-binding site. The initial reaction was reversible on dilution, but became irreversible on the addition of $NaBH_4$, observations consistent with formation of a Schiff base as the initial product. The reaction product was identified from the reduced oNADP–enzyme complex, after hydrolysis with NaOH, by comparison with synthetic standards prepared from α-*N*-acetyllysine and D-glyceraldehyde (*71*).

TABLE I

ENZYMES COVALENTLY LABELED BY DIALDEHYDE DERIVATIVES OF ADENINE NUCLEOTIDES

Source and Enzyme	Ref.
Dialdehyde derivative of ATP	
Sheep liver pyruvate carboxylase	*50*
Rabbit muscle phosphorylase kinase	*51*
Bovine heart mitochondrial ATPase	*52*
Mycobacterium phlei ATPase	*53*
Bovine brain adenylate cyclase	*54*
Rabbit muscle phosphofructokinase	*55*
Saccharomyces cerevisiae phosphoenolpyruvate carboxykinase	*56*
Dialdehyde derivative of ADP	
Rabbit muscle pyruvate kinase	*57, 58*
Escherichia coli succinyl-CoA synthetase	*59*
Pig heart NAD$^+$-dependent isocitrate dehydrogenase	*60, 61*
Dialdehyde derivative of AMP	
Bacteroides symbiosus pyruvate phosphate dikinase	*62*

Periodate-oxidized NADP has been used as an affinity label for other NADP-specific enzymes including pigeon liver malic enzyme (*72, 73*), spinach ferredoxin–$NADP^+$ reductase (*74, 75*), and *Leuconostoc mesenteroides* glucose-6-phosphate dehydrogenase (*76*). Recently, a new fluorescent dialdehyde derivative has been reported as an affinity label of NADP sites, namely, periodate-oxidized 3-aminopyridine adenine dinucleotide phosphate (*77*).

Mas and Colman demonstrated that the 2′,3′-dialdehyde of NADPH could be generated enzymatically by reaction of oNADP with pig heart $NADP^+$-specific isocitrate dehydrogenase (*78*). Isolated oNADPH was further shown to function as an affinity label for the coenzyme site of the same isocitrate dehydrogenase (*78*). The 2′,3′-dialdehyde derivative of NADPH has been used to label a protein involved in the superoxide-generating oxidase activity of pig neutrophils (*79*) and human neutrophils (*80*). In addition, it has been shown to act as a coenzyme in the reductive amination of α-ketoglutarate catalyzed by bovine liver glutamate dehydrogenase and to function as a site-directed modifier of the reduced coenzyme-binding site of that enzyme (*81*).

For coenzymes with more than one ribose, it is desirable to restrict the oxidation to one ribose ring in order to yield a defined product at a single site on an enzyme. Direct treatment of NAD with periodate would lead to oxidation of both ribose moieties with formation of several reactive groups. Therefore, a reaction scheme was devised involving periodate treatment of NADP, followed by enzymic hydrolysis of the 2′-phosphate by alkaline phosphatase to yield a reactive NAD in which only the ribose adjacent to the nicotinamide is oxidized (*82*). The 2′-phosphate functions as a reversible protecting group of the adenosine ribose during the oxidation reaction. By analogy, oNADH was synthesized by enzymic reduction of oNADP to oNADPH by $NADP^+$-specific isocitrate dehydrogenase, followed by treatment with alkaline phosphatase to remove the 2′-phosphate "blocking group" (*82*). The 2′,3′-dialdehyde derivatives of NAD and NADH both react specifically with pig heart NAD^+-dependent isocitrate dehydrogenase.

In many of the examples cited of affinity labeling of enzymes by periodate-oxidized nucleotides, it has been assumed that the reaction involved formation of a Schiff base with an enzymic lysine, as in Fig. 3a,b. However, in very few papers has direct evidence been presented supporting the existence of a Schiff base intermediate. Lowe and Beechey (*83*), after examining in detail the structure of periodate-oxidized ATP, concluded that in aqueous solution there is little free aldehyde; rather, the compound exists predominantly as an equilibrium mixture of three dialdehyde monohydrates (cyclic hemiacetals) and a dihydrate. The presence of cyclic hemiacetals may account for the ability of periodate-oxidized NADP and NADPH to function as coenzymes in several enzymic reactions (e.g., *78, 81*). In many cases, the product of the covalent reaction of an enzyme and periodate-oxidized nucleotide may be a dihydroxymorpholino derivative (Fig. 3c), which is similar to the cyclic hemiacetals observed in aqueous solutions of oATP (*83*). Such a dihydroxymorpholino derivative, which would not be affected by $NaBH_4$, was proposed by Gregory and Kaiser (*55*) to account for the partial reactivation on dialysis of oATP-inactivated phosphofructokinase even after treatment with $NaBH_4$. It is relevant that in the affinity labeling of $NADP^+$-dependent isocitrate dehydrogenase by [^{14}C]oNADPH (*78*), of glucose-6-phosphate dehydrogenase by [^{14}C]oNADP (*76*), and of NAD^+-specific isocitrate dehydrogenase by [^{14}C]ADP (*60*), no significant specific incorporation of tritium was found after treatment of the modified enzymes with [^3H]$NaBH_4$, despite the stoichiometric incorporation of the ^{14}C-labeled reagents. These observations are consistent with formation of a dihydroxymorpholino derivative.

Another reaction complicating analysis of the products of reaction of enzymes and periodate-oxidized nucleotides is the well-known β-elimination which can be catalyzed by primary amines (Fig. 3c,d) (*84*). Such a β-elimination mechanism could account for the greater incorporation of [2,8-^3H]oATP than of [α-^{32}P]oATP during the inactivation of mitochondrial ATPase (*52*). Similarly, the greater incorporation of [8-^{14}C]oADP than of [α-^{32}P]oADP into the activator site of NAD^+-dependent isocitrate dehydrogenase can be attributed to elimination of the pyrophosphate (*60*). Evidence has been presented that the periodate-oxidized 3-aminopyridine adenine dinucleotide phosphate (used as an affinity label for malic enzyme) is readily cleaved into separate 3-aminopyridine and adenosine moieties, but the location of the phosphoryl groups was not ascertained (*77*). Clearly, in order to interpret affinity labeling experiments using periodate-oxidized nucleotides, consideration must be devoted to the position of the radiolabel and to the structural integrity of the nucleotide during the reaction as well as once bound to the enzyme. The simplicity of preparation of the reactive compounds should not be taken as representative of the ease of their application as affinity labels.

A peptide from pyruvate kinase labeled with oADP (*58*) and one from ferredoxin–$NADP^+$ reductase labeled with oNADP (*75*) have been isolated and characterized. These are exceptions. Despite the large number of papers describing the kinetics of affinity labeling by periodate-oxidized nucleotides, there are very few reports of the identification of particular amino acids labeled by these reagents within determined peptide sequences. For enzyme products other than a Schiff base reducible by $NaBH_4$, the instability of the products in the proteolytic digests of modified enzymes under conditions of peptide purification has precluded isolation of labeled peptides in most cases.

B. Nucleoside Polyphosphopyridoxals

Pyridoxal 5′-phosphate has long been used as a group-specific reagent to modify lysine residues in proteins (*85*). As a monoaldehyde, it has been demonstrated to form Schiff bases with the ε-amino group of lysine residues in proteins, and there are many documented examples of reduction of these Schiff bases and isolation of the stabilized modified lysines. In 1985 and 1986, two laboratories designed and synthesized a new class of affinity label aimed at lysines in nucleotide sites of enzymes by linking pyridoxal phosphate (PLP) to a nucleotide. Tagaya *et al.* (*86, 87*) reported on the synthesis of uridine diphosphopyridoxal (UP_2-PL) and Tagaya and Fukui (*87, 88*), as well as Tamura *et al.* (*89*), described the preparation of adenosine diphosphopyridoxal (AP_2-PL) (Fig. 4a,b). Glycogen synthase was inactivated by UP_2-PL, with the reaction rate exhibiting a nonlinear dependence on the concentration of UP_2-PL indicative of reversible binding of enzyme–reagent complex (K_i 2.5 μM) prior to complete inactivation (*86*). Protection against inactivation was provided by UDPglucose and UDP, but not by the allosteric activator glucose 6-phosphate. After stabilization of the enzyme product with $NaBH_4$ and digestion by chymotrypsin, a modified peptide was purified by following the pyridoxyl absorbance or fluorescence. The sequence of the peptide was determined and the altered residue identified as lysine.

The AP_2-PL was initially shown to inactivate lactate dehydrogenase (*88*), hexokinase, adenylate kinase, and 3-phosphoglycerate kinase (*89*). Yeast hexokinase binds AP_2-PL, with a K_d value of 23 μM (*90*), prior to inactivation.

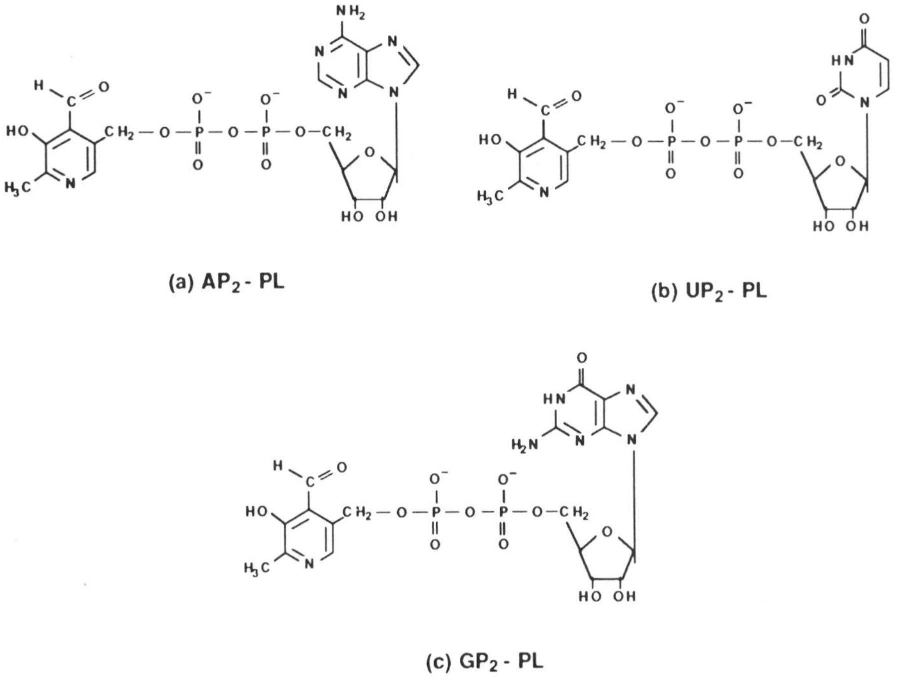

(a) AP_2 - PL

(b) UP_2 - PL

(c) GP_2 - PL

FIG. 4. Nucleoside polyphosphopyridoxals. (a) Adenosine diphosphopyridoxal; (b) uridine diphosphopyridoxal; and (c) guanosine diphosphopyridoxal.

Glucose enhances the binding of the reagent, whereas ATP competes with PLP-AMP, suggesting that modification occurs in the ATP-binding site. Reduction with [^{3}H]$NaBH_4$ yielded a stable product, a single tryptic peptide was isolated and its modified residue identified as lysine (*90*). Affinity labeling of adenylate kinase by AP_2-PL also yielded, after reduction, a unique modified peptide with lysine-21 as the target site (*91*). The same lysine was modified by adenosine tri- and tetraphosphopyridoxals, which was interpreted as indicating that either the affinity label or the lysine-21 region is capable of considerable mobility (*92*). In contrast, *E. coli* F_1-ATPase is more effectively inactivated by AP_3-PL and AP_4-PL than by AP_2-PL (*93*). Additional enzymes which are inactivated by adenosine polyphosphopyridoxals include the *E. coli* transcription termination factor ρ (*94*), the γ subunit of phosphorylase kinase (*95*), pyridoxal kinase (*96*), and aldose reductase (aldehyde reductase) (*97*). These nucleoside polyphosphopyridoxals apparently have the potential to modify ATP, ADP, NAD(P), and pyridoxal sites in enzymes. The guanosine analogs GP_2-PL (Fig. 4c) and GP_3-PL have also been synthesized and shown to react with the protein product, p21, of the *ras* gene (*98*).

As a class of affinity labels, the nucleoside polyphosphopyridoxals exhibit many of the favorable features without many of the drawbacks of the periodate-oxidized nucleotides. The nucleoside polyphosphopyridoxals are structurally close to the natural coenzymes, are water soluble and negatively charged at neutral pH, and exhibit high affinity for the enzymes with which they have been tested. Their targets have thus far been limited to lysine residues, but they react to form Schiff bases which can be stabilized by reduction with $NaBH_4$. Once reduced, the products are stable throughout the peptide purification procedures, and the sequences of the modified peptides can readily be determined.

C. Additional Aldehyde Derivatives

A novel approach to site-specific modification has involved the incubation of DNA-dependent RNA polymerase with 4-[*N*-(β-hydroxyethyl)-*N*-methyl]benzaldehyde esters of AMP, ADP, or ATP:

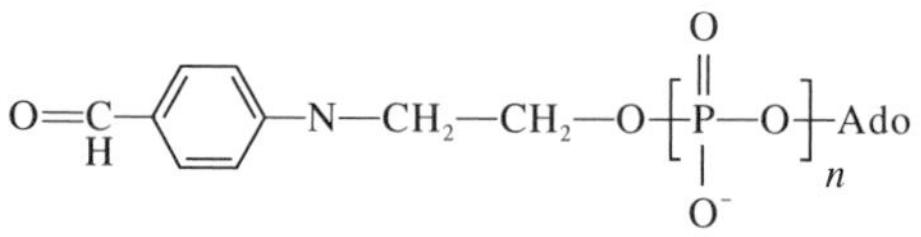

$$O{=}\underset{H}{C}{-}C_6H_4{-}N{-}CH_2{-}CH_2{-}O{\left[\overset{O}{\overset{\|}{P}}\underset{O^-}{\underset{|}{}}{-}O\right]_n}{-}Ado \qquad (n = 1, 2, 3)$$

followed by reduction by $NaBH_4$ (*20, 99, 100*). The benzaldehydes react with free amino groups of the protein at or near the active site, and the resultant Schiff bases are stabilized by reduction. These esters bound to the polymerase are positioned to participate in subsequent phosphodiester formation in the presence of template DNA. When the covalently labeled polymerase is incubated with radioactive nucleoside triphosphate as the incoming substrate nucleotide, the enzyme becomes radiolabeled as a result of the phosphodiester formation catalyzed by the competent polymerase. The radiolabeled nucleotide which is selected depends on the DNA template used and the nucleotide sequence of its particular promoter. This strategy has been used by Grachev *et al.* to label RNA polymerase from wheat germ (*99*) and *E. coli* (*20*) and by Schaffner *et al.* for T7 RNA polymerase (*100*).

Affinity labels with reactive aldehyde substituents have not been limited to nucleotide derivatives. Johanson and Henkin (*101*) synthesized a new antifolate to deliver a glyoxal group near an arginine residue predicted by X-ray crystallographic studies to interact with the *p*-aminobenzoylcarbonyl of methotrexate in the active site of dihydrofolate reductase. The compound they designed was diaminopteridine glyoxal (DAP-glyoxal), which they demonstrated to react rapidly with polyarginine in the presence of borate buffer. DAP-glyoxal, in the absence of borate, was shown to inhibit *E. coli* and *L. casei* dihydrofolate reductase competitively with the substrate dihydrofolate. In borate buffer, dihydrofolate reductase was inactivated by DAP-glyoxal, with a rate constant that exhibited saturation kinetics with respect to [DAP-glyoxal]. Inactivation was enhanced by the coenzyme NADPH and prevented by dihydrofolate, as might be expected for an affinity label of this enzyme (*101*). The reagent remained covalently bound to denatured enzyme at low pH, although it dissociated at neutral pH. This behavior allowed the cleavage of the modified protein by CNBr and purification of large fragments. Although the target amino acid could not be definitively identified because of instability of the reaction product, it was localized to a region of the protein including Arg-52 of *E. coli* dihydrofolate reductase, the amino acid designated by the X-ray studies to be close to the benzoyl carbonyl. This study illustrates the potential of affinity labeling to evaluate, for an enzyme in solution, the interactions between substrate and active site based on the crystal structure of enzyme–inhibitor complexes.

V. Fluorosulfonylbenzoyl Nucleosides

The fluorosulfonylbenzoyl nucleosides (Fig. 5) are reactive derivatives which have been used as site-specific labels of nucleotide sites of a large number of proteins. These compounds retain the purine and ribose structures while substituting a reactive sulfonyl fluoride in the phosphoryl region of the natural nucleotides. 5′-*p*-Fluorosulfonylbenzoyladenosine (5′-FSBA), the first of this group to be synthesized (*102, 103*), might be considered as an analog of ADP, ATP, or NADH. The carbonyl adjacent to the 5′ position is structurally equivalent to the α-phosphoryl group, and, if the molecule is arranged in an extended conformation, the sulfonyl fluoride may be located in a position corresponding to the

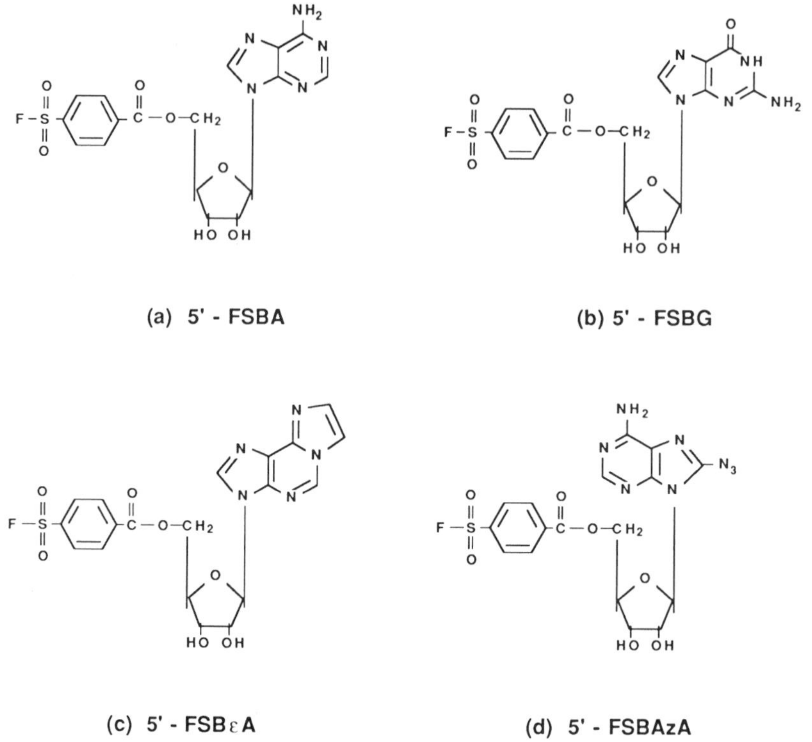

(a) 5' - FSBA

(b) 5' - FSBG

(c) 5' - FSBεA

(d) 5' - FSBAzA

FIG. 5. Fluorosulfonylbenzoyl nucleoside affinity labels. (a) 5′-*p*-Fluorosulfonylbenzoyladenosine; (b) 5′-*p*-fluorosulfonylbenzoylguanosine; (c) 5′-*p*-fluorosulfonylbenzoyl-1,N^6-ethenoadenosine; and (d) 5′-*p*-fluorosulfonylbenzoyl-8-azidoadenosine.

γ-phosphoryl of ATP or to the nicotinamide ribose of NADH. The sulfonyl fluoride is an electrophilic agent capable of covalent reaction with tyrosine and lysine (*104*), histidine (*105–107*), cysteine (*105, 108–112*), and serine (*113*) residues in proteins. Stable derivatives of tyrosine and lysine result from reaction with the fluorosulfonylbenzoyl nucleosides. On acid hydrolysis, the ester linkages between the benzoyl and nucleoside groups are hydrolyzed, and the acid-stable products are carboxybenzenesulfonyl tyrosine (CBS-Tyr) and carboxybenzenesulfonyl lysine (CBS-Lys), which have been separately synthesized (*104*). These products can be distinguished from each other and from other natural amino acids on a standard amino acid analyzer (*104*). CBS-Tyr and CBS-Lys are the products of protein modification by 5′-FSBA which have most commonly been identified in enzymes, and these have often been located within a defined amino acid sequence.

In several cases, reaction of 5′-FSBA with cysteine has been postulated to involve initial formation of a thiolsulfonate derivative of 5′-FSBA, followed by a rapid displacement of the sulfinic acid moiety by an adjacent cysteine to generate a disulfide bond (*108*). Such a reaction sequence accounts for the lack of correlation between reagent incorporation and loss of measurable sulfhydryl groups, as well as for the reactivation by dithiothreitol. Because the reagent has been displaced from the target site on the enzyme, the location of the specific cysteines is not straightforward. For *S*-adenosylhomocysteinase, Gomi *et al.* (*112*) have determined the amino acid sequences of peptides containing cysteines modified by 5′-FSBA by the following strategy: 5′-FSBA-inactivated enzyme was treated with nonradioactive iodoacetate to block all available thiol groups. The 5′-FSBA-generated enzyme cystine was subsequently reduced with mercaptoethanol to reactivate the enzyme, and the newly exposed cysteines were radiolabeled with iodo[^{14}C]acetate (*112*). Proteolytic digestion yielded labeled peptides which were purified and characterized.

The 5′-*p*-fluorosulfonylbenzoyladenosine has specifically modified NAD-binding sites in several dehydrogenases and reductases, as summarized in Table II (*114–126*), and of ADP/ATP sites in a large number of kinases (Table III) (*127–152*). In addition, 5′-FSBA provides affinity labeling of adenine nucleotide or adenosine sites in such diverse proteins as ATPases, synthetases, myosin, DNA polymerase, glycine methyltransferase, and aggregin (an ADP receptor protein of human platelet membranes); these proteins are listed, along with selected references, in Table IV (*153–186*). The general characteristics of the fluorosulfonylbenzoyl nucleosides, as well as the best conditions for their reaction with enzymes, have been discussed (*11*).

In many cases, 5′-FSBA reacts at active sites; however, in bovine liver glutamate dehydrogenase, all the fluorosulfonylbenzoyl nucleosides react at regulatory sites. This enzyme is a hexamer of identical subunits, each of which has two regulatory sites for ADP, two for GTP, and two for NADH (one catalytic

TABLE II

ENZYMES COVALENTLY MODIFIED AT NAD(P)H SITES BY 5′-*p*-FLUOROSULFONYLBENZOYLADENOSINE

Enzyme	Ref.
Glutamate dehydrogenase	*102, 104, 114*
Malate dehydrogenase	*115*
Xanthine dehydrogenase	*116*
Nicotinamide-nucleotide transhydrogenase	*117*
NADH–cytochrome-b_5 reductase	*118*
Aldehyde reductase	*119*
DT-Diaphorase	*120*
3α,20β-Hydroxysteroid dehydrogenase	*121, 122*
Estradiol 17β-dehydrogenase	*123, 124*
3β-Hydroxy-5-ene steroid dehydrogenase	*125, 126*

TABLE III

KINASES COVALENTLY LABELED BY 5′-*p*-FLUOROSULFONYLBENZOYLADENOSINE

Enzyme	Ref.
Pyruvate kinase	*108, 127–130*
Phosphofructokinase	*131–135*
Glycerol kinase	*136*
Ribulose-5-phosphate kinase	*137, 138*
6-Phosphofructo-2-kinase/fructose-2,6-bisphosphatase	*139*
cAMP-Dependent protein kinase	*140–144*
cGMP-Dependent protein kinase	*145, 146*
Casein kinase II	*147*
3-Hydroxy-3-methylglutaryl-CoA reductase kinase	*148*
Epidermal growth factor-stimulated protein kinase	*149–152*

and one regulatory) (*187*). Incubation of glutamate dehydrogenase with 0.3 m*M* 5′-FSBA at 30°C and pH 8 yields enzyme which retains its catalytic activity, activation by ADP, and inhibition by GTP, but which has lost its normal inhibition by high concentrations of NADH (*102, 104*). In contrast to native enzyme, modified enzyme containing 3 mol –SBA/mol hexameric enzyme exhibits normal Michaelis–Menten kinetics in a plot of velocity versus NADH concentration

TABLE IV

OTHER ATP/ADP AND ADENOSINE BINDING PROTEINS COVALENTLY LABELED BY 5′-*p*-FLUOROSULFONYLBENZOYLADENOSINE

Enzyme	Ref.
F_1-ATPase	*106, 153–158*
Na^+,K^+-Transport ATPase	*159*
Na^+,Mg^{2+}-ATPase	*160*
Chloroplast ATPase	*161*
Carbamoyl-phosphate synthase	*162–165*
Phosphoribosylpyrophosphate synthetase	*166*
Glutamine synthetase	*167, 168*
5-Oxo-L-prolinase	*169*
Luciferase	*170*
Myosin	*110, 171, 172*
Actin	*172*
DNA polymerase I	*173*
recA Protein	*174*
Acetyl-CoA carboxylase	*175*
Glycine methyltransferase	*176, 177*
S-Adenosylhomocysteinase	*178, 179*
Aggregin (platelet ADP receptor protein)	*180–186*

(*104*). Equal amounts of CBS-Tyr and CBS-Lys were determined as the reaction products of 5′-FSBA with glutamate dehydrogenase. It was concluded that both lysine and tyrosine are present in the region of the NADH regulatory site; both are close to the sulfonyl fluoride of the enzyme-bound reagent and have an equal probability of reacting. Covalent modification of either residue on three of the six subunits of the active enzyme is sufficient to eliminate NADH inhibition. Two nucleosidyl peptides were isolated from 5′-FSBA-modified enzyme by chromatography on phenyl boronate-agarose followed by HPLC. These were shown to contain Tyr-190 and Lys-420 as the only reaction sites with 5′-FSBA (*114*). Although these residues are widely separated in the linear sequence, the affinity labeling experiments indicated that they are close to each other in the native structure of the enzyme.

Not only does 5′-FSBA act as an affinity label of specific sites in purified enzymes, but it also can react with a single protein in such a complex system as an intact cell. ADP induces a change in the shape of human platelets from disks to spiculated spheres, followed by platelet aggregation and granule secretion (*188*). ADP has been postulated to bind to a receptor protein on the exterior surface of the platelet membranes. Incubation of radioactive 5′-FSBA with intact platelets in plasma led to the time-dependent inhibition of ADP-induced shape changes concomitantly with labeling of a 100,000-dalton protein isolated from the membrane fraction (*180*). This protein is protected by ADP against labeling by 5′-FSBA and has been proposed to be an ADP receptor protein, designated aggregin, responsible for shape change and subsequent aggregation of platelets. 5′-FSBA has been used as a covalent probe of the ADP site of aggregin, reflecting the effect that ADP exerts on platelet function when bound to its receptor; these studies have recently been reviewed (*186*).

Additional fluorosulfonylbenzoyl analogs of nucleosides have been synthesized, some of which are shown in Fig. 5. The 3′-*p*-fluorosulfonylbenzoyladenosine (3′-FSBA) has been prepared and characterized (*103, 189*). Although it is not as close an analog of ATP, ADP, and NAD as is 5′-FSBA, 3′-FSBA may react with amino acids immediately adjacent to a natural coenzyme site and thereby help to identify these proximal residues. The compound 5′-*p*-fluorosulfonylbenzoylguanosine (5′-FSBG, Fig. 5b) was synthesized by reaction of guanosine hydrochloride with *p*-fluorosulfonylbenzoyl chloride (*190*) and might be expected to be specifically directed toward GTP sites in proteins. In glutamate dehydrogenase, 5′-FSBG reacts covalently, causing elimination of one of the GTP regulatory sites and decreased sensitivity to GTP inhibition (*190, 191*). Rabbit muscle pyruvate kinase is relatively indiscriminate in its structural requirements for nucleotide and uses GDP in addition to ADP as coenzyme. Accordingly, it is not surprising that pyruvate kinase is inactivated by reaction of 5′-FSBG in the region of the active site (*109*), although the reaction characteristics are not identical to those of 5′-FSBA with the same enzyme (*129*). In contrast, rat liver cytosolic phosphoenolpyruvate carboxykinase, which requires guanosine or inosine nucleotides for catalysis, is specifically inactivated by 5′-FSBG, whereas 5′-FSBA causes inactivation at only 1% of the rate of 5′-FSBG under the same conditions (*192*). 5′-FSBG also reacts at guanine nucleotide sites on isolated platelet membranes, a result that has been interpreted as reaction with a GTP-binding protein that modulates α_2-adrenergic receptor interactions and mediates inhibition of adenylate cyclase (*193*). Additionally, GTP-binding sites of tubulin are covalently modified by 5′-FSBG (*194, 195*).

The compound 5′-*p*-fluorosulfonylbenzoylinosine (5′-FSBI) was first described by Williamson and Meister (*169*). This compound has been found to react with 5-oxo-L-prolinase (*169*) and with bovine heart mitochondrial F_1-ATPase (*196*), both of which can bind ITP. Furthermore, the synthesis of several new 5′-*p*-fluorosulfonylbenzoylarabinosyl nucleoside analogs has recently been reported (*197*); these include 5′-FSB derivatives of arabinosylcytosine, arabinosyladenine, arabinosylguanine, arabinosyluracil, and thymidine.

The structure illustrated in Fig. 5c is 5′-*p*-fluorosulfonylbenzoyl-1,N^6-ethenoadenosine (5′-FSBεA), a fluorescent nucleotide analog with an emission peak at 412 nm and an excitation maximum of 308 nm (*198*). This compound provides a convenient means of introducing a covalently bound fluorescent probe at nucleotide sites in proteins. 5′-FSBεA reacts as an affinity label of the active site of pyruvate kinase (*105, 198*). Bovine liver glutamate dehydrogenase is reversibly inhibited by 1,N^6-ethenoadenosine 5′-triphosphate, and the enzyme is irreversibly modified by 5′-FSBεA at a regulatory site that can also be occupied by GTP (*199*). The target amino acid of 5′-FSBεA has been identified as Tyr-262 of glutamate dehydrogenase (*200*). The fluorescent properties of this covalently bound nucleotide have been utilized in resonance energy transfer experiments in which an estimate was made of the distance between the GTP site covalently labeled by 5′-FSBεA (energy donor) and (as energy acceptor) an ADP site occupied reversibly by 2′(3′)-*O*-(2,4,6-trinitrophenyl)adenosine 5′-diphosphate (TNP-ADP) (*201*). The fluorescence of bound –SBεA is quenched by increasing concentrations of TNP-ADP, allowing an estimated average distance of 18 Å between the allosteric GTP and ADP sites. Rabbit muscle phosphofructokinase (*202*), fructose-1,6-bisphosphate aldolase (*203*), and phosphoglycerate kinase (*204*) all react covalently with 5′-FSBεA with limited reagent incorporation.

A related fluorescent nucleotide affinity label, 5′-*p*-fluorosulfonylbenzoyl-2-aza-1,N^6-ethenoadenosine (5′-FSBaεA), in which a nitrogen replaces the carbon atom at the 2-position of 5′-FSBεA, has also been prepared (*205, 206*). 5′-FSBaεA has different spectral properties, with an emission peak at 490 nm and an excitation maximum at 356 nm. Thus, it can be used in a complementary manner to 5′-FSBεA as a covalently bound chromophore. 5′-FSBaεA reacts at a GTP regulatory site of glutamate dehydrogenase, and energy transfer experiments have led to an estimated distance between the catalytic and GTP sites in this enzyme (*205*). The cAMP site of phosphofructokinase has been labeled by 5′-FSBaεA, and this probe has yielded distance estimates between this site and particular reactive cysteines on the enzyme (*206*).

Radioactive fluorosulfonylbenzoyl nucleosides have generally been synthesized by reaction of the labeled nucleoside with nonradioactive *p*-fluorosulfonylbenzoyl chloride (e.g., *11, 103*). This synthetic route is satisfactory for many applications of site-specific modification, particularly for reactions and subsequent procedures conducted at neutral pH. However, the ester linkage between the benzoyl and nucleoside moieties has limited stability below pH 6 and above pH 9, and any gel electrophoresis experiments or isolation conditions at these pH extremes (including acid hydrolysis prior to amino acid composition determination) may lead to loss of the radioactive tag. An alternate synthesis strategy for 5′-FSBA has been reported by Esch and Allison (*207*) in which the radioactive label is derived from *p*-amino[*carboxy*-^{14}C]benzoic acid which is converted to *p*-fluorosulfonyl[^{14}C]benzoyl chloride prior to coupling with adenosine. This strategy, which can be generalized to synthesis of all the fluorosulfonylbenzoyl nucleosides, allows the isolation of radioactive CBS-Lys and CBS-Tyr following acid hydrolysis of labeled peptides in which lysine or tyrosine is modified.

Figure 5 pictures all of the fluorosulfonylbenzoyl nucleosides in an extended conformation. However, evidence from proton magnetic resonance as well as from fluorescence spectroscopy indicates that 5′-FSBA, 5′-FSBG, and 5′-FSBεA exist in solution partially in a conformation in which the purine ring is intramolecularly stacked on the benzoyl ring (*208*). These three compounds (Fig. 5a–c) differ in their sensitivity to disruption of stacking (*208*). The observed differences in solution conformation (between stacked and extended forms) of these nucleotide analogs may account for their reactions with amino acids within either the pyrophosphate or the purine portion of the nucleotide-binding site.

Figure 5d shows a new bifunctional affinity label, 5′-*p*-fluorosulfonylbenzoyl-8-azidoadenosine (5′-FSBAzA), which contains both an electrophilic fluorosulfonyl moiety and a photoactivatable azido group (*209*). Following stoichiometric incorporation of reagent through the fluorosulfonyl at a specific site, photolysis of the tethered molecule leads to a higher efficiency of photoincorporation than is normally observed when azidonucleotides are free in solution (see Section VI). Furthermore, this compound should reveal the amino acids which are adjacent to the residue which is initially labeled and thus help to elucidate the tertiary structure of the enzyme in the region of the nucleotide site. Reaction of 5′-FSBAzA at the NADH regulatory site of bovine liver glutamate dehydrogenase has been described (*209*).

As a class of site-directed modifying agents, the fluorosulfonylbenzoyl nucleosides offer several favorable features: they are structurally related to nucleotides, even though they lack phosphoryl groups and require the addition of a small amount of organic solvent to maintain stability in aqueous solutions. The fluorosulfonyl hydrolyzes sufficiently slowly at neutral pH [$t_{1/2}$ 43 min at pH 8.0; $t_{1/2}$ 223 min at pH 7.65 (*11*)] to allow analysis of the reaction rates with enzymes. The reactive group is capable of covalent reaction with a large number of amino acids, and experience is accumulating (Tables II–IV) in the isolation of stable reaction products either for determining the stoichiometry of incorporation or for isolating and evaluating the amino acid sequence of labeled peptides.

VI. Photoreactive Compounds

A. General Considerations

Unique features distinguish the site-specific modification of enzymes by photoreactive compounds from reactions by other classes of affinity labels. First, these compounds form covalent bonds with enzymes only when deliberately activated by light; therefore, it is possible to characterize the reversible interaction between enzyme and reagent if light is excluded. Second, photoreactive compounds are less discriminating than electrophiles in their reactions with amino acids. They can react by a variety of mechanisms including insertion into C–H, O–H, or N–H bonds, cycloaddition to multiple bonds, and abstraction of a hydrogen atom from a C–H bond followed by coupling of two monoradicals; thus, they are not restricted to reaction with nucleophilic amino acids but rather have the potential to react also with the hydrophobic amino acids that are often in the binding sites of substrates or coenzymes. Photoreactive compounds exhibit certain negative characteristics as compared to other classes of site-directed modifiers. Often the photoactivated compound reacts rapidly with solvent, causing the efficiency of photoincorporation into enzyme to be low and making it difficult to ascertain the characteristics of a stoichiometrically labeled enzyme. Furthermore, few of the reaction products of photoaffinity labeling are structurally defined, and they are often unstable and difficult to isolate; thus, relatively few photoaffinity labeled peptides have been isolated and sequenced. Consideration must also be given to the optimal wavelength for photoactivation: irradiation in the spectral region characteristic of protein absorption often leads to nonspecific inactivation; therefore, irradiation at wavelengths greater than 300 nm is desirable. Many of the general principles of photoaffinity labeled have been discussed (*10, 210*).

Three major classes of photoreactive compounds have been used to label specific sites of proteins: nitrenes, carbenes, and free radicals. Carbenes, generated by photolysis of diazo derivatives, are highly reactive and relatively nonselective in their target sites. Nitrenes, formed by photoactivation of azido derivatives, are less reactive than carbenes and more electrophilic, exhibiting a preference for attack of O–H as compared to C–H bonds. Free radicals, formed by irradiation of α,β-unsaturated ketones, purines, or pyrimidines ("direct photoaffinity labeling") exhibit a preference for targeting C–H bonds, but the chemistry of these reactions is not well defined.

B. Azido Derivatives

In photoaffinity labeling experiments in the 1980s, the most commonly used of the classes of photoreactive compounds are the azido derivatives, particularly those including an 8-azidopurine group. 8-Azido-ATP (8-N_3-ATP) and 8-azido-GTP (8-N_3-GTP) were first synthesized by B. E. Haley and co-workers, and the techniques involved in their application have been reviewed (*211*). A few representative examples follow. The ATP hydrolytic site of the *recA* protein from *E. coli* has been covalently modified by 8-N_3-ATP (*212–214*). Although at most 0.15 mol of reagent was bound per mole of protein (*212*), a peptide containing modified tyrosine-264 was isolated as the major site of labeling (*213*). It is interesting that the 8-azido group reacted with the same tyrosine residue in this protein as did 5′-*p*-fluorosulfonylbenzoyladenosine, suggesting to these authors that 8-N_3-ATP binds to the *recA* protein in the anti rather than the syn conformation which is thought to be the predominant form in solution. Fructose-1,6-bisphosphatase is inactivated by photoaffinity labeling of the allosteric AMP site by 8-N_3-AMP (*215*); in this case, up to about 0.5 mol reagent was incorporated per enzyme subunit.

Cyclic AMP-binding sites have been labeled by 8-N_3-cAMP in a variety of preparations, ranging from unfractionated red cell membranes (*216*) and partially purified protein kinase (*217*) to the homogeneous regulatory subunit of cAMP-dependent protein kinase (*218*). In the latter case, the affinity of the reagent is extremely high, and the stoichiometry approaches 1 mol of 8-N_3-cAMP/mol of subunit; a modified tyrosine has been identified. Recently, bovine pancreatic ribonuclease A has been found to photoincorporate 8-N_3-cAMP, with indirect evidence suggesting that one of two threonines is modified (*219*).

The difficulties that are often encountered in attempting to structurally characterize the products of azidonucleotide reactions are illustrated by the photoaffinity labeling of ADPglucose synthetase by 8-N_3-ADP-glucose (*220*). Seven different radioactive peptides were observed, and the target amino acids within the identified peptides could not be designated with any certainty. The photolabeling of terminal deoxynucleotidyl transferase by 8-N_3-2′-deoxyadenosine 5′-triphosphate was used to assess the subunit location of the deoxynucleoside triphosphate sites (*221*). Since α and β subunits were both photolabeled, the authors concluded that the nucleotide-binding domain involves contributions from both polypeptide chains. The synthesis of a new fluorescent photoaffinity label, 2′,3′-*O*-(2,4,6-trinitrophenyl)-8-azido-ATP, and its photoincorporation into Ca^{2+}-ATPase have been reported (*222*). An additional novel investigation involved the incorporation of 8-N_3-adenosine into tRNA, which was then incubated with ribosomes and photoactivated to cross-link the tRNA to particular identifiable ribosomal proteins (*223*).

Some proteins specific for GTP will accept 8-N_3-GTP, but not 8-N_3-ATP, as a specific photoprobe. Tubulin provides an example: in the dark, 8-N_3-GTP substitutes for GTP in promoting tubulin polymerization, whereas photolysis leads to incorporation into the β subunit of the heterodimeric protein (*224, 225*). Protection against photoincorporation of 8-N_3-GTP was provided specifically by GTP but not ATP, indicating that the decreased incorporation could not be attributed to the trivial effect of light absorption by the added nucleotides. Photoaffinity labeling of the G_α and G_β protein subunits of retinal rod outer segments was also effected by 8-N_3-GTP (*226*).

The photolabile 2-azido-ADP was synthesized and found to undergo rearrangement, promoted by alkali, to the photostable tetrazole form (*227*). Although 2-N_3-ADP mimics ADP in promoting platelet aggregation, it does not function as a specific photoaffinity label of the ADP receptor protein (*227*); the authors postulate that the 2-N_3-ADP is bound to the receptor protein in an orientation favoring reaction of the nitrene with solvent. In contrast, 2-N_3-ADP did photolabel nucleotide-binding sites of the ADP/ATP carrier of beef heart mitochondria (*228*). The photoincorporated reagent was localized in large peptide fragments of the protein, but specifically labeled amino acids were not convincingly identified. The chloroplast F_1-ATPase has also been photolabeled by 2-N_3-ADP or 2-N_3-ATP during repeated or prolonged irradiation to allow time for reequilibration between the tetrazole and azido species (*229*).

5-Azido-2′-deoxyuridine 5′-triphosphate (5-N_3-dUTP) has been enzymically converted by *E. coli* DNA polymerase (in the dark) to a photoreactive DNA which was then used to photolabel the *lac* repressor (*230*). Alternatively, when 5-N_3-dUTP was irradiated in the presence of DNA polymerase, it functioned as a photoaffinity label of the enzyme (*231*). The syntheses of several photoreactive 3′-arylazido derivatives of ATP (*232*) as well as of 9-(3′-azido-3′-deoxy-β-D-xylofuranosyl)adenosine and -guanosine 5′-triphosphates (*233*) have been described.

Other azido derivatives of nucleotides have been designed as specific photoaffinity labels of coenzyme sites. Arylazido-β-alanine-NAD, with the reactive substituent at the 3′-position of the adenosine ribose, can function as a probe of the NAD-binding region of dehydrogenases and reductases, although the reactive group is relatively large and at a distance from the natural structural components of NAD (*234–236*). Two photolabels of coenzyme A have been described: *p*-azidobenzoyl-CoA (*237*) and *N*-(3-iodo-4-azidophenylpropionamide)cysteinyl-5-(2′-thiopyridylcysteine)-CoA (*238*). A bifunctional azidonucleotide, 3′-arylazido-8-azidoadenosine 5′-triphosphate, was synthesized as a photosensitive cross-linking agent; however, the extent of cross-linking of the F_1-ATPase of *Micrococcus luteus* was extremely limited (*239*). The relatively small, chromophoric compounds 2-[(4-azido-2-nitrophenyl)amino]ethyl di- and triphosphates (*240, 241*) and azidonitrophenylphosphate (*242*) are photoreactive analogs of phosphate or pyrophosphate which may bind and react at nucleotide-binding sites in enzymes.

Although the great majority of azido derivatives have been modeled on nucleotide structures, a few have been aimed at other chemical classes of binding sites. One such example is N^{α}-acetylasparaginyllysyl-(N^{ε}-azidobenzoyl)threoninamide, a photoreactive azido tripeptide which mimics the asparagine acceptor site of an oligosaccharide on a potential N-linked glycoprotein (*243*). This compound was used to label the oligosaccharyltransferase that catalyzes the synthesis of such glycoproteins. Another example is *N*-(4-azidosalicyl)galactosamine, a photoaffinity label of a carbohydrate (*N*-acetylgalactosamine) binding site in the protein discoidin I, a lectin of *Dictyostelium discoideum* (*244*). A third is *N*-(*p*-azido-*m*-iodophenethylamidoisobutyl)norepinephrine, used to label the β-adrenergic receptor in guinea pig lung membranes (*245*).

C. Carbenes and Free Radicals

Despite the fact that they can be much more reactive than nitrenes, carbene derivatives are less frequently cited in reports on photoaffinity labels. Two such derivatives of thyroid hormone have been described in which the amino group of the alanine side chain of 3,5,3′-triiodo-L-thyronine or 3,5,3′,5′-tetraiodo-L-thyronine was derivatized with 2-diazo-3,3,3-trifluoropropionate (*246*). Irradiation at 254 nm of these compounds with several cell lines yielded low incorporation of reagent (0.5–13.5%) into two forms of thyroid hormone receptor proteins. Low incorporation was also characteristic of early reports of photoaffinity labeling (e.g., *247*) in which $O^{2'}$-ethyl-2-diazomalonyl-cAMP and $N^6,O^{2'}$-di(ethyl-2-diazomalonyl)-cAMP were used to generate carbenes which modify rabbit muscle phosphofructokinase. Similarly, only 3–14% of the potential sites of lactate dehydrogenase were labeled by the carbene derived from 3-(3*H*-diazirino)pyridine adenine dinucleotide (*248*).

3′-*O*-(4-Benzoyl)benzoyl-ATP is an α,β-unsaturated ketone capable of being excited to a diradical triplet state by a 10-min irradiation with UV light of wavelengths exceeding 300 nm (*249*). The compound, in the dark, is a substrate for mitochondrial F_1-ATPase and on irradiation causes 70% inactivation of the enzyme, with incorporation of the γ-^{32}P-labeled reagent exclusively into the β subunit (*249*). This photoreactive benzophenone has also been used to photolabel myosin after trapping it at the active site by cross-linking two proximate cysteine thiol groups (*250, 251*). This technique allowed the recovery of photolabeled protein in approximately 50–60% yields. Cremo and Yount also synthesized two fluorescent photoaffinity labels, 2′-deoxy-3′-*O*-(4-benzoylbenzoyl)- and 3′(2′)-*O*-(4-benzoylbenzoyl)-1,N^6-ethenoadenosine 5′-phosphate (*252*); they used myosin as a test system to demonstrate the efficiency of incorporation and the potential applications as fluorescent probes. The authors propose that the carbonyl of benzophenone is converted to a tertiary alcohol linked covalently to the protein (*252*). Several steroids have been used as photoaffinity labels, including Δ^6-testerosterone-agarose for Δ^5-3-ketosteroid isomerase (steroid Δ-isomerase) (*253*) and triamcinolone for liver glucocorticoid receptor (*254*). In the case of the purified glucocorticoid receptor, the modified enzyme was enzymically cleaved and the radioactive reagent localized to two amino acids.

Direct photoaffinity labeling by nucleotides irradiated by low-wavelength UV light in the presence of enzymes has been decreasing in use in the last few years. Examples include the effect of irradiation of dATP with ribonucleotide reductase (*255*), of NAD with diphtheria toxin fragment A (*256*), of GTP with *ras* p21 (*257*), of dNTP with DNA polymerase (*258*), and of dTTP with ribonucleotide reductase (*259*). The major goal is identification of a ligand-binding domain in a purified protein or designation of a particular protein with a nucleotide-binding site in a mixture of proteins. In most of these studies, irradiation is continued for relatively long time periods (30–60 min), and there is no correlation between activity change and reagent incorporation; thus, it is difficult to ascertain whether the modification of the residues is responsible for loss of function or if these amino acids become exposed when the enzyme is denatured.

VII. Haloacyl and Alkyl Halide Derivatives

Among the chemical classes of affinity labels, the haloacyl and alkyl halide group is perhaps the most varied in the range of natural compounds which they are designed to mimic. Representative examples of haloacyl derivatives are listed in Table V (*260–279*) and include steroids, nucleotides, coenzymes, and peptides. Although the pH-independent reaction rates of bromoacetate with thiol compounds (*24*) are about 4 orders of magnitude lower than those of haloketones (e.g., *18*), haloacyl substituents can readily be linked to amines and alcohols. Thus, derivatives of biologically functional molecules can often be prepared directly. The stability of the affinity labels under the usual solvent conditions may compensate for the relatively low reaction rates, frequently leading to a specific modification of a particular site of an enzyme despite the fact that the reaction period required may be long.

In addition, the reaction product of an amino acid with an affinity label of the haloacetyl type is converted by acid hydrolysis to a carboxymethylamino acid derivative; many carboxymethylamino acids are well characterized and therefore readily identifiable in modified enzymes. Haloacetyl derivatives of cysteine, lysine, histidine, methionine, glutamate, aspartate, and tyrosine have been re-

TABLE V

HALOACYL DERIVATIVES USED AS AFFINITY LABELS OF ENZYMES

Class and compound	Enzyme	Ref.
Steroids		
16α-Bromoacetoxyestradiol 3-methyl ether	Estradiol 17β-dehydrogenase	*260*
12β-Hydroxy-4-estrene-3,17-dione 12-bromoacetate	Estradiol 17β-dehydrogenase	*261*
3-Methoxyestriol 16-bromoacetate	Estradiol 17β-dehydrogenase	*261*
5α-Dihydrotestosterone 17-bromoacetate	3α,20β-Hydroxysteroid dehydrogenase	*262*
17β-(Bromoacetoxy)progesterone	3α,20β-Hydroxysteroid dehydrogenase	*262*
17β-(Bromoacetoxy)progesterone	Cytochrome *P*-450	*263*
17β-(Bromoacetoxy)-5α-dihydro-testosterone	3α-Hydroxysteroid dehydrogenase	*264*
21-Bromoacetoxydesoxycorticosterone	3α-Hydroxysteroid dehydrogenase	*264*
17α-Estradiol 17-(bromoacetate)	Estradiol 17β-dehydrogenase	*265*
17β-Estradiol 17-(bromoacetate)	Estradiol 17β-dehydrogenase	*265*
Coenzyme A derivatives		
Bromoacetyl-thioester coenzyme A	Thiolase	*266*
3-Chloropropionyl coenzyme A	3-Hydroxy-3-methylglutaryl-CoA synthase	*267*
3-Chloropropionyl coenzyme A	Fatty acid synthase	*268*
Nucleotide derivatives		
p-(Bromoacetamido)phenyluridyl pyrophosphate	UDPGalactose 4-epimerase	*269*
5′-Bromoacetamido-5′-deoxyadenosine	3α,20β-Hydroxysteroid dehydrogenase	*270*
(Bromoacetamido)nucleoside	Ribonuclease A	*271*
Adenosine 5′-(2,3-dibromohydrogen succinate)	Pyruvate kinase, myokinase	*272*
N^6-(*p*-Bromoacetaminobenzyl)-AMP	Glyceraldehyde-3-phosphate dehydrogenase	*273*
	Myokinase	*273*
	Phosphoglycerate kinase	*274*
Miscellaneous analogs		
N-Bromoacetylpyridoxamine	Aspartate aminotransferase	*275*
Bromoacetyl peptide	cAMP-Dependent protein kinase	*276*
3-Bromoacetylchloramphenicol	Chloramphenicol acetyltransferase	*277*
N-Bromoacetyl-3,3′,5-triiodothyronine	Iodothyronine deiodase (iodide peroxidase)	*278*
2-(4-Bromoacetamido)anilino-2-deoxy-pentitol 1,5-bisphosphate	Ribulose-bisphosphate carboxylase/oxygenase	*279*

ported. For example, by comparison with carboxymethylhistidine standards, it was established that histidine is the target of 16α-bromoacetoxyestradiol 3-methyl ether in estradiol 17β-dehydrogenase (*260*), and peptides containing critical modified histidines have been identified (*261*). The recent demonstration that the same histidine in estadiol 17β-dehydrogenase is labeled by affinity labels containing the reactive bromoacetyl group at position 17 in the D ring of the steroid and at position 3 in the A ring was interpreted as indicating that the 17-substituted compounds bind to the catalytic site in the "wrong way" relative to the orientation required for catalysis (*265*); it is obvious that even if a natural ligand affords total protection against a reactive structural analog, one cannot assume that the two compounds bind in the same orientation. Furthermore, without direct demonstration, it should not be assumed that tightly bound radioactive reagent is necessarily covalently bound to protein: the inactivation of UDPgalactose 4-epimerase by the active site-directed *p*-(bromoacetamido)phenyluridyl pyrophosphate proved to be due to alkylation of the adenine ring of enzyme-bound NAD (*280*).

Extreme caution must be used in interpreting the experimental data obtained from studies of the site-specific modification of enzyme sites. For an enzyme whose structure has been determined by X-ray crystallography, such as bovine pancreatic ribonuclease, the results of affinity labeling by a reactive nucleoside can be compared with the crystal structures of various enzyme–ligand complexes (*271*). The general characteristics of the haloacetyl class of affinity labels have been summarized (*281*).

Alkyl halides are even less reactive than acyl halides, as indicated by the compilation of reaction rates of thiolate anions with various types of alkyl halides (*282*). Nevertheless, potentially useful affinity labels have been synthesized with alkyl halide substituents and have been shown to specifically inactivate several enzymes, albeit slowly; the low reactivity of the alkyl halides may minimize nonspecific reaction. Adenosine 5′-(2-bromoethyl)phosphate has been characterized and reported to inactivate NAD^+-dependent isocitrate dehydrogenase (*283*). The 2′- and 3′-(2-bromoethyl)-AMP labels have also been synthesized, and model reactions of the bromoethyl-AMPs with cysteine, lysine, histidine, and tyrosine have been studied (*284*). More recently, esters of adenosine 5′-monophosphate have been prepared with ethyl, propyl, or hexyl moieties and bromo or chloro substituents at the ω position (*285*). Yeast alcohol dehydrogenase exhibited enhanced inactivation by the hexyl derivative, but inactivation rates of other dehydrogenases were unremarkable. Two iodopropyl derivatives of cAMP have been described, namely, 1,N^6-(3-iodopropyleno)adenosine 3′,5′-cyclic monophosphate and 3′-*O*-(2-iodo-3-hydroxypropyl)adenosine 3′,5′-cyclic monophosphate; the latter inactivates cAMP phosphodiesterase from human platelets, with a pseudo-first-order rate constant of 0.147 hr^{-1} (*286*).

Several other purine nucleotide derivatives with alkyl halide substituents have appeared in the literature, including adenosine 5′-chloromethane phosphonate, adenosine 5′-chloromethylpyrophosphate, and adenosine 5′-(β-bromoethane phosphonate) (*287–291*). These compounds have been evaluated as affinity labels of such nucleotide-binding enzymes as leucyl- and tryptophanyl-tRNA synthetases (leucine– and tryptophan–tRNA ligases), phosphorylase *b*, and cAMP-dependent protein kinase. The use of alkyl halides as electrophilic agents is not confined to nucleotide affinity labels: muscarinic acetylcholine receptors have been specifically tagged with *N*-(2-chloroethyl)-*N*-(2′,3′-propyl)-2-aminoethyl-benzilate, an analog of the muscarinic antagonist, propylbenzilylcholine (*292*).

VIII. Miscellaneous Affinity Labels

Although most of the reactive compounds designed to modify specific sites on enzymes can be included in the general chemical classes of Sections III–VII, there are a few noteworthy exceptions. The reagent 7-chloro-4-nitro-2,1,3-benzoxadiazole (NBD-Cl) exhibits "saturation kinetics" in its inactivation of histidinol dehydrogenase, suggesting reversible binding prior to covalent modification (*293*). The authors have pointed out the structural similarity between the substrate histidinol and NBD-Cl, with its benzoxadiazole ring. Indeed, substantial protection against inactivation was provided by L-histidinol as well as by the competitive inhibitors histamine and imidazole, and substrate protection of one of the two major modified cysteine-containing peptides was observed. However, in other enzymes, such as oat root ATPase, NBD-Cl appears to function as an adenine analog, since inactivation of this ATPase can be prevented by MgADP or MgATP (*294*).

The design of inhibitors of proteases has paralleled the development of the technique of site-specific modification of enzymes, much of which has been reviewed (*1*). Among the new types of affinity labels for cysteine proteases is peptidyldiazomethane, in which the specificity is determined by the amino acid sequence (*295*). The carboxylic–phosphoric mixed anhydride of mestitoyl-AMP has yielded effective and specific modification of nucleotide sites in such enzymes as pyruvate kinase (*296*) and tryptophanyl-tRNA synthetase (*297*). For the latter enzyme, protection against inactivation by both tryptophan and ATP, in addition to the higher affinity of the reagent as compared with ATP, is consistent with reaction at the tryptophanyl adenylate site (*297*). A coenzyme A analog with a thioarsenite group (*S*-dimethylarsino-CoA) was synthesized and shown to react covalently with the thiol group of cysteine in a model reaction (*298*); this is a highly discriminating reagent that irreversibly inactivates phosphotransacetylase.

IX. Conclusions

The application of site-specific modification of enzymes and other proteins has become increasingly common, and a wide range of chemical classes is available from which to select or design a reagent for exploring a particular enzyme. Structural similarity to the natural ligand is always desirable to ensure target specificity, but the goal of the investigation must be clear to permit a rational choice of the nature of the reactive group.

For studies aimed at identifying amino acid participants in active or regulatory sites, or for those aimed at evaluating the kinetic properties, equilibrium binding, spectral characteristics, or X-ray structure of a stoichiometrically modified enzyme, it is important to choose a reagent type with the potential for forming quantitatively a stable, covalent bond between enzyme and affinity label. The haloketone, fluorosulfonylbenzoyl, and haloacyl derivatives may be most appropriate for these applications since they generally yield products of enzyme nucleophiles which can be isolated and identified within an amino acid sequence. Affinity labels of this type can be used to provide information regarding the effect of occupying the particular site on the conformation of the enzyme, on the reactivity of other sites, or on subunit interactions. They can provide a rational basis for the design of site-directed mutagenesis studies.

Reagents with distinctive absorption or fluorescence spectra can be used to introduce chromophores into specific sites on enzymes in order to monitor conformational changes or to estimate distances between known functional sites on enzymes. Different characteristics in a reagent may be desirable or acceptable for studies with the goal of distinguishing a specific receptor on an intact cell, a particular ligand-binding protein in an unfractionated cell extract, or one among several subunits of a pure protein which has the specialized function of recognizing a given regulatory compound. In these cases, rapid indiscriminate reaction with amino acids within the specific ligand site may be the most important property of the affinity label. Periodate-oxidized nucleotides and photoreactive compounds can be suitable reagents for this type of study despite their limited record in yielding stable products which can be purified and located within a primary sequence. The overall purpose of site-directed modification of an enzyme is to reveal new insights into the function or structure of that protein, and deliberate consideration of the most appropriate affinity label can optimize the accomplishment of this goal.

Acknowledgments

This work was supported by U.S. Public Health Service Grants DK 37000 and DK 39075 and by National Science Foundation Grant DMB-88-04706.

References

1. Shaw, E. (1970). "The Enzymes," 3rd Ed., Vol. 1, p. 91.
2. Baker, B. R. (1976). "Design of Active Site-Directed Irreversible Enzyme Inhibitors." Wiley, New York.

3. Plapp, B. V. (1982). "Methods in Enzymology," Vol. 87, p. 469.
4. Bazaes, S. (1987). *In* "Chemical Modification of Enzymes" (J. Eyzaguirre, ed.), Chap. 3, p. 35. Horwood, Chichester, England.
5. Jakoby, W. B., and Wilchek, M. (eds.) (1977). "Methods in Enzymology," Vol. 46.
6. Bell, J. E., and Bell, E. T. (1988). *In* "Proteins and Enzymes," Chap. 8, p. 184. Prentice-Hall, Englewood Cliffs, New Jersey.
7. Colman, R. F. (1989). *In* "Protein Function: A Practical Approach" (T. Creighton, ed.), Chap. 4, p. 77. IRL, Oxford.
8. Glazer, A. N., De Lange, R. J., and Sigman, D. S. (1975). *In* "Chemical Modification of Proteins, Selected Methods and Analytical Procedures," Chap. 5, p. 135. American Elsevier, New York.
9. Lundblad, R. L., and Noyes, C. M. (1984). *In* "Chemical Reagents for Protein Modification," Vol. 2, Chap. 6, p. 141. CRC Press, Boca Raton, Florida.
10. Schaffer, H.-J. (1987). *In* "Chemical Modification of Enzymes" (J. Eyzaguirre, ed.), Chap. 4, p. 45. Horwood, Chichester, England.
11. Colman, R. F. (1983). *Annu. Rev. Biochem.* **52,** 67.
12. Eberle, A. N., and de Graan, P. N. E. (1985). "Methods in Enzymology," Vol. 109, p. 129.
13. Schoner, W., and Scheiner-Bobis, G. (1988). "Methods in Enzymology," Vol. 156, p. 312.
14. Sweet, F., and Murdock, G. L. (1987). *Endocr. Rev.* **8,** 154.
15. Silverman, R. B. (1988). "Mechanism-Based Enzyme Inactivation: Chemistry and Enzymology, Vols. 1 and 2. CRC Press, Boca Raton, Florida.
16. Huang, Y.-C., and Colman, R. F. (1984). *JBC* **259,** 12481.
17. Kettner, C., and Shaw, E. (1981). "Methods in Enzymology," Vol. 80, p. 826.
18. Bednar, R. A., Hartman, F., and Colman, R. F. (1982). *Biochemistry* **21,** 3681.
19. Groman, E. V., Schultz, R. M., and Engel, L. L. (1975). *JBC* **250,** 5450.
20. Grachev, M. A., Kolocheva, T. I., Lukhtanov, E. A., and Mustaev, A. A. (1987). *EJB* **163,** 113.
21. DeCamp, D. L., and Colman, R. F. (1989). *JBC* **264,** 8430.
22. Meloche, H. P. (1970). *Biochemistry* **9,** 5050.
23. Hartman, F. C. (1977). "Methods in Enzymology," Vol. 46, p. 130.
24. Dickens, F. (1933). *BJ* **27,** 1142.
25. Owens, M. S., and Barden, R. E. (1978). *ABB* **187,** 299.
26. Clements, P. R., Barden, R. E., Ahmad, P. M., and Ahmad, F. (1979). *BBRC* **86,** 278.
27. Katiyar, S. S., Pan, O., and Porter, J. W. (1982). *BBRC* **104,** 517.
28. Dugan, R. E., and Katiyar, S. S. (1986). *BBRC* **141,** 278.
29. Hammond, D. C., Kruggel, W. G., Lewis, R. V., and Barden, R. E. (1986). *JBC* **261,** 8424.
30. Kapetanovic, E., Bailey, J. M., and Colman, R. F. (1985). *Biochemistry* **24,** 7586.
31. Huang, Y.-C., Bailey, J. M., and Colman, R. F. (1986). *JBC* **261,** 14100.
32. Batra, S. P., and Colman, R. F. (1986). *JBC* **261,** 15565.
33. Colman, R. F., Huang, Y.-C., King, M. M., and Erb, M. (1984). *Biochemistry* **23,** 3281.
34. Batra, S. P., and Colman, R. F. (1984). *Biochemistry* **23,** 4940.
35. DeCamp, D. L., Lim, S., and Colman, R. F. (1988). *Biochemistry* **27,** 7651.
36. Grant, P. G., DeCamp, D. L., Bailey, J. M., Colman, R. W., and Colman, R. F. (1990). *Biochemisty* **29,** 887.
37. Batra, S. P., and Colman, R. F. (1986). *Biochemistry* **25,** 3508.
38. Batra, S. P., Lark, R. H., and Colman, R. F. (1989). *ABB* **270,** 277.
39. Bansal, A., Dayton, M. A., Zalkin, H., and Colman, R. F. (1989). *JBC* **264,** 9827.
40. Huang, Y.-C., and Colman, R. F. (1989). *JBC* **264,** 12208.
41. Bailey, J. M., and Colman, R. F. (1985). *Biochemistry* **24,** 5367.
42. Bailey, J. M., and Colman, R. F. (1987). *Biochemistry* **26,** 6858.
43. Bailey, J. M., and Colman, R. F. (1987). *JBC* **262,** 12620.

44. Bailey, J. M., and Colman, R. F. (1987). *Biochemistry* **26,** 4893.
45. Meloche, H. P. (1965). *BBRC* **18,** 277.
46. Mäntsälä, P., and Zalkin, H. (1976). *JBC* **251,** 3294.
47. Hartman, F. C. (1981). *Biochemistry* **20,** 894.
48. Ehrlich, R. S., and Colman, R. F. (1987). *JBC* **262,** 12614.
48a. Saha, A., Huang, Y-C., and Colman, R. F. (1989). *Biochemistry* **28,** 8425.
49. Bednar, R. A., Hartman, F. C., and Colman, R. F. (1982). *Biochemistry* **21,** 3690.
50. Easterbrook-Smith, S. B., Wallace, J. C., and Keech, D. B. (1976). *EJB* **62,** 125.
51. King, M. M., and Carlson, G. M. (1981). *Biochemistry* **20,** 4382.
52. Lowe, P. N., and Beechey, R. B. (1982). *Biochemistry* **21,** 4073.
53. Kumar, G., Kalra, V. K., and Brodie, A. F. (1979). *JBC* **254,** 1964.
54. Westcott, K. R., Olwin, B. B., and Storm, D. R. (1980). *JBC* **225,** 8767.
55. Gregory, M. R., and Kaiser, E. T. (1979). *ABB* **196,** 199.
56. Saavedra, C., Araneda, S., and Cardemil, E. (1988). *ABB* **267,** 38.
57. Hinrichs, M. V., and Eyzaguirre, J. (1982). *BBA* **704,** 177.
58. Bezares, G., Eyzaguirre, J., Hinrichs, M. V., Heinrickson, R. L., Reardon, I., Kemp, R. G., Latshaw, S. P., and Bazaes, S. (1987). *ABB* **253,** 133.
59. Nishimura, J. S., Mitchell, T., Collier, G. E., Matula, J. M., and Ball, D. J. (1983). *EJB* **136,** 83.
60. King, M. M., and Colman, R. F. (1983). *Biochemistry* **22,** 1656.
61. Ehrlich, R. S., and Colman, R. F. (1984). *JBC* **259,** 11936.
62. Evans, C. T., Goss, N. H., and Wood, H. G. (1980). *Biochemistry* **19,** 5809.
63. Wakagi, T., and Ohta, T. (1982). *J. Biochem. (Tokyo)* **92,** 1403.
64. Tsai, P. K., and Hogenkamp, H. P. C. (1983). *ABB* **226,** 276.
65. Ball, D. J., and Nishimura, J. S. (1980). *JBC* **255,** 10805.
66. Nilsson, L., and Nygard, O. (1985). *EJB* **148,** 299.
67. Nilsson, L., and Nygard, O. (1988). *EJB* **171,** 293.
68. King, M. M., Carlson, G. M., and Haley, B. E. (1982). *JBC* **257,** 14058.
69. Rippa, M., Signorini, M., Signori, R., and Dallocchio, F. (1975). *FEBS Lett.* **51,** 281.
70. Rippa, M., Bellini, T., Signorini, M., and Dallocchio, F. (1979). *ABB* **196,** 619.
71. Dallocchio, F., Negrini, R., Signorini, M., and Rippa, M. (1976). *ABB* **429,** 629.
72. Chang, G. G., and Huang, T. M. (1979). *BBRC* **86,** 829.
73. Chang, G. G., Chang, T. C., and Huang, T. M. (1982). *Int. J. Biochem.* **14,** 621.
74. Chan, R. L., and Carrillo, N. (1984). *ABB* **229,** 340.
75. Chan, R. L., Carrillo, N., and Vallejos, R. H. (1985). *ABB* **240,** 172.
76. White, B. J., and Levy, H. R. (1987). *JBC* **262,** 1223.
77. Chang, G. G., Shiao, M.-S., Liaw, J. G., and Lee, H. J. (1989). *JBC* **264,** 280.
78. Mas, M. T., and Colman, R. F. (1983). *JBC* **258,** 9332.
79. Umei, T., Takeshige, K., and Minakami, S. (1986). *JBC* **261,** 5229.
80. Smith, R. M., Curnulte, J. T., and Babior, B. M. (1989). *JBC* **264,** 1958.
81. Lark, R. H., and Colman, R. F. (1986). *JBC* **261,** 10659.
82. Saha, A., and Colman, R. F. (1988). *ABB* **264,** 665.
83. Lowe, P. N., and Beechey, R. B. (1982). *Bioorg. Chem.* **11,** 55.
84. Schwartz, D. E., and Gilham, P. T. (1972). *J. Am. Chem. Soc.* **94,** 8921.
85. Eyzaguirre, J. (1987). *In* "Chemical Modification of Enzymes" (J. Eyzaguirre, ed.), Chap. 1, p. 9. Horwood, Chichester, England.
86. Tagaya, M., Nakano, K., and Fukui, T. (1985). *JBC* **260,** 6670.
87. Tagaya, M., and Fukui, T. (1986). *J. Protein Chem.* **5,** 129.
88. Tagaya, M., and Fukui, T. (1986). *Biochemistry* **25,** 2958.
89. Tamura, J. K., Rakov, R. D., and Cross, R. L. (1986). *JBC* **261,** 4126.

90. Tamura, J. K., LaDine, J. R., and Cross, R. L. (1988). *JBC* **263,** 7907.
91. Tagaya, M., Yagami, T., and Fukui, T. (1987). *JBC* **262,** 8257.
92. Yagami, T., Tagaya, M., and Fukui, T. (1988). *FEBS Lett.* **229,** 261.
93. Noumi, T., Tagaya, M., Miki-Takeda, K., Maeda, M., Fukui, T., and Futui, M. (1987). *JBC* **262,** 7686.
94. Dombroski, A. J., LaDine, J. R., Cross, R. L., and Platt, T. (1988). *JBC* **263,** 18810.
95. Tagaya, M., Hayakawa, Y., and Fukui, T. (1988). *JBC* **263,** 10219.
96. Dominici, P., Scholz, G., Kwok, F., and Churchich, J. E. (1988). *JBC* **263,** 14712.
97. Morjana, N. A., Lyons, C., and Flynn, T. G. (1989). *JBC* **264,** 2912.
98. Ohmi, N., Hoshino, M., Tagaya, M., Fukui, T., Kawakita, M., and Hattori, S. (1988). *JBC* **263,** 14261.
99. Grachev, M. A., Hartman, G. R., Maximova, T. G., Mustaev, A. A., Schaffner, A. R., Sieber, H., and Zaychikov, E. F. (1986). *FEBS Lett.* **200,** 287.
100. Schaffner, A. R., Jorgensen, E. D., McAllister, W. T., and Hartman, G. R. (1987). *Nucleic Acids Res.* **15,** 8773.
101. Johanson, R. A., and Henkin, J. (1986). *JBC* **260,** 1465.
102. Pal, P. K., Wechter, W. J., and Colman, R. F. (1975). *JBC* **250,** 8140.
103. Colman, R. F., Pal, P. K., and Wyatt, J. L. (1977). "Methods in Enzymology," Vol. 46, p. 240.
104. Saradambal, K. V., Bednar, R. A., and Colman, R. F. (1981). *JBC* **256,** 11866.
105. Tomich, J. M., and Colman, R. F. (1985). *BBA* **827,** 344.
106. Bullough, D. A., and Allison, W. S. (1986). *JBC* **261,** 5722.
107. Liao, T.-H. (1984). *Fed. Proc., Fed. Am. Soc. Exp. Biol.* **43,** 1774.
108. Annamalai, A. E., and Colman, R. F. (1981). *JBC* **256,** 10276.
109. Tomich, J. M., Marti, C., and Colman, R. F. (1981). *Biochemistry* **20,** 6711.
110. Togashi, C. T., and Reisler, E. (1982). *JBC* **257,** 10112.
111. Marshall, M., and Fahien, L. A. (1985). *ABB* **241,** 200.
112. Gomi, T., Ogawa, H., and Fujioka, M. (1986). *JBC* **261,** 13422.
113. Paulos, R. L., and Price, P. A. (1974). *JBC* **249,** 1453.
114. Schmidt, J. A., and Colman, R. F. (1984). *JBC* **259,** 14515.
115. Roy, S., and Colman, R. F. (1979). *Biochemistry* **18,** 4683.
116. Nishino, T., and Nishino, T. (1987). *Biochemistry* **26,** 3068.
117. Phelps, D. C., and Hatefi, Y. (1985). *Biochemistry* **24,** 3503.
118. Chen, S., Haniu, M., Iyanagi, T., and Shively, J. E. (1986). *J. Protein Chem.* **5,** 133.
119. Flynn, T. G., and Cronin, C. N. (1985). *Fed. Proc., Fed. Am. Soc. Exp. Biol.* **44,** 1619.
120. Liu, X.-F., Chen, S., Yuan, H., Haniu, M., Iyanagi, T., and Shively, J. E. (1988). *Fed. Proc., Fed. Am. Soc. Exp. Biol.* **2,** A1024.
121. Sweet, F., and Samant, B. R. (1980). *Steroids* **36,** 2675.
122. Sweet, F., and Samant, B. R. (1981). *Biochemistry* **20,** 5170.
123. Tobias, B., and Strickler, R. C. (1981). *Biochemistry* **20,** 5546.
124. Inano, H., and Tamaoki, B-I. (1985). *J. Steroid Biochem.* **22,** 681.
125. Ishi-Ohba, H., Inano, H., and Tamaoki, B-I. (1986). *J. Steroid Biochem.* **25,** 555.
126. Ishi-Ohba, H., Saiki, N., Inano, H., and Tamaoki, B-I. (1986). *J. Steroid Biochem.* **24,** 753.
127. Wyatt, J. L., and Colman, R. F. (1977). *Biochemistry* **16,** 1333.
128. Likos, J. J., Hess, B., and Colman, R. F. (1980). *JBC* **255,** 9388.
129. Annamalai, A. E., Tomich, J. M., Mas, M. T., and Colman, R. F. (1982). *ABB* **219,** 47.
130. DeCamp, D. L., and Colman, R. F. (1986). *JBC* **261,** 4499.
131. Mansour, T. E., and Colman, R. F. (1978). *BBRC* **81,** 1370.
132. Pettigrew, D. W., and Frieden, C. (1978). *JBC* **253,** 3623.
133. Weng, L., Henrickson, R. L., and Mansour, T. E. (1980). *JBC* **255,** 1492.

134. Ogilvie, J. W. (1983). *Biochemistry* **22,** 5908.
135. Ogilvie, J. W. (1985). *Biochemistry* **24,** 317.
136. Pettigrew, D. W. (1987). *Biochemistry* **26,** 1723.
137. Krieger, T. J., and Miziorko, H. M. (1986). *Biochemistry* **25,** 3496.
138. Krieger, T. J., Mende-Mueller, L., and Mizioko, H. M. (1987). *BBA* **915,** 112.
139. El-Maghrabi, M., Pate, T. M., D'Angelo, G., Correia, J. J., Lively, M. O., and Pilkis, S. J. (1987). *JBC* **262,** 11714.
140. Zoller, M. J., and Taylor, S. S. (1979). *JBC* **254,** 8363.
141. Zoller, M. J., Nelson, N. C., and Taylor, S. S. (1981). *JBC* **256,** 10837.
142. Hixson, C. S., and Krebs, E. G. (1979). *JBC* **254,** 7509.
143. Bhatnagar, D., Hartl, F. T., Roskoski, R., Jr., Lessor, R. A., and Leonard, N. J. (1984). *Biochemistry* **23,** 4350.
144. Hagiwara, M., Inagaki, M., and Hidaka, H. (1987). *Mol. Pharmacol.* **31,** 523.
145. Hixson, C. S., and Krebs, E. G. (1981). *JBC* **256,** 1122.
146. Hashimoto, E., Takio, K., and Krebs, E. G. (1982). *JBC* **257,** 727.
147. Hathaway, G. M., Zoller, M. J., and Traugh, J. A. (1981). *JBC* **256,** 11442.
148. Ferrer, A., Caelles, C., Nassot, N., and Hegardt, F. G. (1987). *JBC* **262,** 13507.
149. Buhrow, S. A., Cohen, S., and Staros, J. V. (1982). *JBC* **257,** 4019.
150. Buhrow, S. A., Cohen, S., Garbers, D. L., and Staros, J. V. (1983). *JBC* **258,** 7824.
151. Russo, M. W., Lukas, T. J., Cohen, S., and Staros, J. V. (1985). *JBC* **260,** 5205.
152. Buhrow, S. A., and Staros, J. V. (1985). "Methods in Enzymology," Vol. 109, p. 816.
153. Esch, F. S., and Allison, W. S. (1978). *JBC* **253,** 6100.
154. DiPietro, A., Godinot, C., Martin, J.-C., and Gautheron, D. C. (1979). *Biochemistry* **18,** 1738.
155. DiPietro, A., Godinot, C., and Gautheron, D. C. (1981). *Biochemistry* **20,** 6312.
156. Bitar, K. G. (1982). *BBRC* **109,** 30.
157. Bullough, D. A., Yoshida, M., and Allison, W. S. (1986). *ABB* **244,** 865.
158. Bullough, D. A., Verburg, J. G., Yoshida, M., and Allison, W. S. (1987). *JBC* **262,** 11675.
159. Cooper, J. B., and Winter, C. G. (1980). *J. Supramol. Struct.* **13,** 165.
160. Lewis, R. N. A. H., George, R., and McElhaney, R. N. (1986). *ABB* **247,** 201.
161. DeBenedetti, E., and Jagendorf, A. (1979). *BBRC* **96,** 440.
162. Boettcher, B. R., and Meister, A. (1980). *JBC* **255,** 7129.
163. Powers, S. G., Muller, G. W., and Kafka, N. (1983). *JBC* **258,** 7545.
164. Marshall, M., and Fahien, L. A. (1985). *ABB* **241,** 200.
165. Kim, H., Wan, H., and Evans, D. R. (1988). *Fed. Proc., Fed. Am. Soc. Exp. Biol.* **2,** A1549.
166. Harlow, K. W., and Switzer, R. L. (1985). *Biochemistry* **24,** 3360.
167. Foster, W. B., Griffith, M. J., and Kingdon, H. S. (1981). *JBC* **256,** 882.
168. Pinkovsky, H. B., Ginsburg, A., Reardon, I., and Heinrikson, R. F. (1984). *JBC* **259,** 9616.
169. Williamson, J. M., and Meister, A. (1982). *JBC* **257,** 9161.
170. Lee, Y., Esch, F. S., and DeLuca, M. A. (1981). *Biochemistry* **20,** 1253.
171. Saitoh, M., Ishikawa, T., Matsushima, S., Naku, M., and Hidaka, H. (1987). *JBC* **262,** 7796.
172. Bennett, J. S., Vilaire, G., Colman, R. F., and Colman, R. W. (1981). *JBC* **256,** 1185.
173. Pandey, V. N., and Modak, M. J. (1988). *JBC* **263,** 6068.
174. Knight, K. L., and McEntee, K. (1985). *JBC* **260,** 10177.
175. Chen, S., and Kim, K.-H. (1982). *JBC* **257,** 9953.
176. Fujioka, M., and Ishiguro, Y. (1986). *JBC* **261,** 6346.
177. Fujioka, M., Takata, Y., Konishi, K., and Ogawa, H. (1987). *Biochemistry* **26,** 5696.
178. Takata, Y., and Fujioka, M. (1984). *Biochemistry* **23,** 4357.
179. Gomi, T., Ogawa, H., and Fujioka, M. (1986). *JBC* **261,** 13422.
180. Bennett, J. S., Colman, R. F., and Colman, R. W. (1978). *JBC* **253,** 7346.

181. Figures, W. R., Niewiarowski, S., Morinelli, T. A., Colman, R. F., and Colman, R. W. (1981). *JBC* **256,** 7789.
182. Colman, R. W., and Figures, W. R. (1984). *Mol. Cell. Biochem.* **59,** 101.
183. Mills, D. C. B., Figures, W. R., Scearce, L. M., Stewart, G. J., Colman, R. F., and Colman, R. W. (1985). *JBC* **260,** 8078.
184. Figures, W. R., Scearce, L. M., Wachtfogel, Y., Chen, J., Colman, R. F., and Colman, R. W. (1986). *JBC* **261,** 5981.
185. Colman, R. W., Figures, W. R., Wu, Q.-X., Chung, S.-Y., Morinelli, T. A., Tuszynski, G. P., Colman, R. F., and Niewiarowski, S. (1988). *ABB* **262,** 298.
186. Colman, R. W., Puri, R. N., Zhou, F., and Colman, R. F. (1988). *In* "Platelet Membrane Receptors" (G. A. Jamison, ed.), p. 263. Alan R. Liss, New York.
187. Colman, R. F. (1990). *In* "A Study of Enzymes, Volume III: Mechanism of Enzyme Action" (S. A. Kuby, ed.), in press. CRC Press, Boca Raton, Florida.
188. Born, G. V. R. (1965). *Nature (London)* **206,** 1121.
189. Pal, P. K., Wechter, W. J., and Colman, R. F. (1975). *Biochemistry* **14,** 707.
190. Pal, P. K., Reischer, R. J., Wechter, W. J., and Colman, R. F. (1978). *JBC* **253,** 6644.
191. Pal, P. K., and Colman, R. F. (1979). *Biochemistry* **18,** 838.
192. Jadus, M., Hanson, R. W., and Colman, R. F. (1981). *BBRC* **101,** 884.
193. Limbird, L. E., Buhrow, S. A., Speck, J. L., and Staros, J. V. (1983). *JBC* **258,** 10289.
194. Steiner, M. (1984). *BBRC* **123,** 92.
195. Prasad, A. R. S., and Luduena, R. F. (1984). *Fed. Proc., Fed. Am. Soc. Exp. Biol.* **43,** 1705.
196. Bullough, D. A., and Allison, W. A. (1986). *JBC* **261,** 14171.
197. Novotny, L., and Plunkett, W. (1987). *Nucleic Acids Res. Symp. Series* **18,** 81.
198. Likos, J. J., and Colman, R. F. (1981). *Biochemistry* **20,** 491.
199. Jacobson, M. A., and Colman, R. F. (1982). *Biochemistry* **21,** 2177.
200. Jacobson, M. A., and Colman, R. F. (1984). *Biochemistry* **23,** 6377.
201. Jacobson, M. A., and Colman, R. F. (1983). *Biochemistry* **22,** 4247.
202. Ogilvie, J. W. (1983). *Biochemistry* **22,** 5915.
203. Palczewski, K., Hargrave, P. A., Folta, E. J., and Kochman, R. (1985). *EJB* **146,** 309.
204. Wiksell, E., and Larsson-Raznikiewicz, M. (1987). *JBC* **262,** 14472.
205. Jacobson, M. A., and Colman, R. F. (1984). *Biochemistry* **23,** 3789.
206. Craig, D. W., and Hammes, G. G. (1980). *Biochemistry* **19,** 330.
207. Esch, F. S., and Allison, W. S. (1978). *Anal. Biochem.* **84,** 642.
208. Jacobson, M. A., and Colman, R. F. (1984). *JBC* **259,** 1454.
209. Dombrowski, K. E., and Colman, R. F. (1989). *Arch. Biochem. Biophys.* **275,** 302.
210. Bayley, H., and Knowles, J. R. (1977). "Methods in Enzymology," Vol. 46, p. 69.
211. Potter, R. L., and Haley, B. E. (1983). "Methods in Enzymology," Vol. 91, p. 613.
212. Knight, K. L., and McEntee, K. (1985). *JBC* **260,** 867.
213. Knight, K. L., and McEntee, K. (1985). *JBC* **260,** 10185.
214. Knight, K. L., and McEntee, K. (1986). *PNAS* **83,** 9289.
215. Marcus, F., and Haley, B. E. (1979). *JBC* **254,** 259.
216. Haley, B. E. (1975). *Biochemistry* **14,** 3852.
217. Pomerantz, A. H., Rudolph, S. A., Haley, B. E., and Greengard, P. (1975). *Biochemistry* **14,** 3858.
218. Ringheim, G. E., Saraswat, L. D., Bubis, J., and Taylor, S. S. (1988). *JBC* **263,** 18247.
219. Wower, J., Aymie, M., Hixson, S. S., and Zimmerman, R. A. (1989). *Biochemistry* **28,** 1563.
220. Lee, Y. M., and Preiss, J. (1986). *JBC* **261,** 1058.
221. Evans, R. K., and Coleman, M. S. (1989). *Biochemistry* **28,** 707.
222. Seebregts, C. J., and McIntosh, D. B. (1989). *JBC* **264,** 2043.
223. Wower, J., Hixson, S. S., and Zimmerman, R. A. (1988). *Biochemistry* **27,** 8114.

224. Geahlen, R. L., and Haley, B. E. (1977). *PNAS* **74,** 4375.
225. Geahlen, R. L., and Haley, B. E. (1979). *JBC* **254,** 11982.
226. Kohnken, R. E., and McConnell, D. G. (1985). *Biochemistry* **24,** 3803.
227. Macfarlane, D. E., Mills, D. C. B., and Srivastava, P. C. (1982). *Biochemistry* **21,** 544.
228. Dalbon, P., Brandelin, G., Bonlay, F., Hoppe, J., and Vignais, P. V. (1988). *Biochemistry* **27,** 5141.
229. Xue, Z., Zhou, J.-M., Melese, T., Cross, R. L., and Boyer, P. D. (1987). *Biochemistry* **26,** 3749.
230. Evans, R. K., Johnson, J. D., and Haley, B. E. (1986). *PNAS* **83,** 5382.
231. Evans, R. K., and Haley, B. E. (1987). *Biochemistry* **26,** 269.
232. Jeng, S. J., and Guillory, R. J. (1975). *J. Supramol. Struct.* **3,** 448.
233. Panka, D., and Dennis, D. (1984). *JBC* **259,** 8384.
234. Chen, S., and Guillory, R. J. (1977). *JBC* **252,** 8990.
235. Yamaguchi, M., Chen, S., and Hatefi, Y. (1986). *Biochemistry* **25,** 4864.
236. Chen, S., Lee, T. D., Legesse, K., and Shively, J. E. (1986). *Biochemistry* **25,** 5391.
237. Lau, E. P., Haley, B. E., and Barden, R. E. (1977). *Biochemistry* **16,** 2581.
238. Ruoho, A. E., Woldegiorgis, G., Kobayashi, C., and Shrago, E. (1989). *JBC* **264,** 4168.
239. Schafer, H.-J., and Dose, K. (1984). *JBC* **259,** 15301.
240. Okamoto, Y., and Yount, R. G. (1985). *PNAS* **82,** 1575.
241. Nakamaye, K. L., Wells, J. A., Bridenbaugh, R. L., Okamoto, Y., and Yount, R. G. (1985). *Biochemistry* **24,** 5226.
242. Garvin, J., Michel, L., Dupuis, A., Issartel, J.-P., Lunardi, J., Hoppe, J., and Vignais, P. (1989). *Biochemistry* **28,** 1442.
243. Welply, J. K., Shenbugamurthi, P., Naider, F., Park, H. P., and Lennarz, W. J. (1985). *JBC* **260,** 6459.
244. Kohnken, R. E., and Berger, E. A. (1987). *Biochemistry* **26,** 8727.
245. Resek, J. F., and Ruoho, A. E. (1988). *JBC* **263,** 14410.
246. Horowitz, Z. D., Sahnoun, H., Pascual, A., Casanova, J., and Samuels, H. H. (1988). *JBC* **263,** 6636.
247. Brunswick, D. J., and Cooperman, B. S. (1971). *PNAS* **68,** 1801.
248. Standring, D. N., and Knowles, J. R. (1988). *Biochemistry* **19,** 2811.
249. Williams, N., and Coleman, P. S. (1982). *JBC* **257,** 2834.
250. Mahmood, R., and Yount, R. G. (1984). *JBC* **259,** 12956.
251. Mahmood, R., Cremo, C., Nakamaye, K. L., and Yount, R. G. (1987). *JBC* **262,** 14479.
252. Cremo, C. R., and Yount, R. G. (1987). *Biochemistry* **26,** 7524.
253. Hearne, M., and Benisek, W. F. (1985). *Biochemistry* **24,** 7111.
254. Carlstedt-Duke, J., Stromstedt, P. E., Persson, B., Cederlund, E., Gustafsson, J.-A., and Jornvall, H. (1988). *Biochemistry* **263,** 6842.
255. Caras, J. W., and Martin, D. W., Jr. (1982). *JBC* **257,** 9508.
256. Carroll, S. F., McCloskey, J. A., Crain, P. F., Oppenheimer, N. J., Marschner, T. M., and Collier, R. J. (1985). *PNAS* **82,** 7237.
257. Basu, A., and Modak, M. J. (1987). *JBC* **262,** 2369.
258. Pandey, V. N., Williams, K. R., Stone, K. L., and Modak, M. J. (1987). *Biochemistry* **26,** 7744.
259. Kierdaszuk, B., and Eriksson, S. (1988). *Biochemistry* **27,** 4952.
260. Chin, C-C., and Warren, J. C. (1975). *JBC* **250,** 7682.
261. Murdock, G. L., Chin, C-C., and Warren, J. C. (1986). *Biochemistry* **25,** 641.
262. Sweet, F., and Samant, B. R. (1980). *Biochemistry* **19,** 978.
263. Onoda, M., Haniu, M., Yanagibashi, K., Sweet, F., Shively, J. E., and Hall, P. F. (1987). *Biochemistry* **26,** 657.

264. Penning, T. M., Carlson, K. E., and Sharp, R. B. (1987). *BJ* **245,** 269.
265. Murdock, G. L., Warren, J. C., and Sweet, F. (1988). *Biochemistry* **27,** 4452.
266. Davis, J. T., Chen, H-H., Moore, R., Nishitani, Y., Masamune, S., Sinskey, A. J., and Walsh, C. T. (1987). *JBC* **262,** 90.
267. Miziorko, H. M., and Behnke, C. (1985). *JBC* **260,** 13516.
268. Miziorko, H. M., Behnke, C., Ahmad, P. M., and Ahmad, F. (1986). *Biochemistry* **25,** 468.
269. Wong, Y-H. H., Winer, F. B., and Frey, P. A. (1979). *Biochemistry* **18,** 5332.
270. Samant, B. R., and Sweet, F. (1983). *JBC* **258,** 12779.
271. Hummel, C. F., Pincas, M. R., Brandt-Rauf, P. W., Frei, G. M., and Carty, R. P. (1987). *Biochemistry* **26,** 135.
272. Berghauser, J., and Geller, A. (1974). *FEBS Lett.* **38,** 254.
273. Suzuki, K., Eguchi, C., and Imahori, K. (1977). *J. Biochem. (Tokyo)* **81,** 1147.
274. Suzuki, K., Eguchi, C., and Imahori, K. (1977). *J. Biochem. (Tokyo)* **81,** 1393.
275. Mattingly, J. R., Jr., Farach, H. A., and Martinez-Carrion, M. (1983). *JBC* **258,** 6243.
276. Mobashery, S., and Kaiser, E. T. (1988). *Biochemistry* **27,** 3691.
277. Kleanthous, C., Cullis, P. M., and Shaw, W. V. (1985). *Biochemistry* **24,** 5307.
278. Mol, J. A., Docter, R., Kaptein, E., Jansen, G., Hennemann, G., and Visser, T. J. (1984). *BBRC* **124,** 475.
279. Herndon, C. S., and Hartman, F. C. (1984). *JBC* **259,** 3102.
280. Wong, Y-H., and Frey, P. A. (1979). *Biochemistry* **18,** 5337.
281. Wilchek, M., and Givol, D. (1977). "Methods in Enzymology," Vol. 46, p. 153.
282. Dahl, K. H., and McKinley-McKee, J. S. (1981). *Bioorg. Chem.* **10,** 329.
283. Roy, S., and Colman, R. F. (1980). *JBC* **255,** 7517.
284. Bednar, R. A., and Colman, R. F. (1982). *J. Protein Chem.* **1,** 203.
285. Fries, R. W., Bohlken, D. P., Murch, B. P., Lerdal, K. G., and Plapp, B. V. (1983). *ABB* **225,** 110.
286. Reimann, J. E., Grant, P. G., Colman, R. W., and Colman, R. F. (1983). *J. Protein Chem.* **2,** 113.
287. Kovaleva, G. K., Ivanov, L. L., Madoyan, I. A., Favorova, O. O., Severin, E. S., Gulyaev, N. N., Baranova, L. A., Shabarova, Z. A., Sokolova, N. I., and Kiselev, L. L. (1978). *Biochemistry (Engl. Transl.)* **43,** 525.
288. Krauspe, R., Kovaleva, G. K., Gulyaev, N. N., Baranova, L. A., Agalarova, M. B., Severin, E. S., Sokolova, N. I., Shabarova, Z. A., and Kiselev, L. L. (1978). *Biochemistry (Engl. Transl.)* **43,** 656.
289. Skolysheva, L. K., Vul'fson, P. L., Gulyaev, N. N., and Severin, E. S. (1978). *Biochemistry (Engl. Transl.)* **43,** 1914.
290. Mikhailova, L. I., Vul'fson, P. L., Skolysheva, L. K., Agalarova, M. B., and Severin, E. S. (1978). *Biochemistry (Engl. Transl.)* **43,** 2016.
291. Grivennikov, I. A., Bulargina, T. V., Khropov, Y. V., Gulyaev, N. N., and Severin, E. S. (1979). *Biochemistry (Engl. Transl.)* **44,** 771.
292. Curtis, C. A. M., Wheatley, M., Bansal, S., Birdsall, M. J. M., Eveleigh, P., Peddler, E. K., Poyner, D., and Hulme, E. C. (1989). *JBC* **264,** 489.
293. Grubmeyer, C. T., and Gray, W. R. (1986). *Biochemistry* **25,** 4778.
294. Randall, S. K., and Sze, H. (1987). *JBC* **262,** 7135.
295. Crawford, C., Mason, R. W., Wikstrom, P., and Shaw, E. (1988). *BJ* **253,** 751.
296. Hampton, A., Harper, P. J., Sasaki, S., Howgate, P., and Preston, R. K. (1975). *BBRC* **65,** 945.
297. Madoyan, I. A., Favorova, O. O., Kovaleva, G. K., Sokolova, N. I., Shabarova, Z. A., and Kiselev, L. L. (1981). *FEBS Lett.* **123,** 156.
298. Duhr, E. F., Owens, M. S., and Barden, R. E. (1983). *BBA* **749,** 84.

7

Stereochemistry of Enzyme-Catalyzed Reactions at Carbon

DONALD J. CREIGHTON • NUNNA S. R. K. MURTHY
Department of Chemistry and Biochemistry
University of Maryland Baltimore County
Baltimore, Maryland 21228

THE ENZYMES, Vol. XIX

I. Introduction

The general topic of enzyme reaction stereochemistry was last reviewed in this treatise by Popják in 1970 (*1*). At that time, excitement in the area was clearly due to a proliferation of methods using all three isotopes of hydrogen (^{1}H, protium; ^{2}H, deuterium; and ^{3}H, tritium) to elucidate the most subtle stereochemical aspects of enzyme-catalyzed reactions. The clever use of these isotopes had successfully revealed the stereochemical features of several of the enzymes involved in glycolysis, the citric acid cycle, and lipid metabolism. With respect to the nicotinamide adenine dinucleotide (phosphate) [$NAD(P)^{+}$]-dependent dehydrogenases, the stereochemistry of hydrogen transfer at C-4 of the nicotinamide ring revealed a nearly equal abundance in nature of both A- and B-side specific dehydrogenases; all of this work was based on the pioneering studies of Westheimer, Vennesland, and co-workers (*2*). Perhaps the most exciting development at the time came from the laboratories of Cornforth (*3*) and Arigoni (*4*) in which the experimental and conceptual foundation was laid for the stereochemical analysis of chiral [^{1}H,^{2}H,^{3}H]methyl groups produced or consumed during enzymic catalysis. Finally, Popják's review recounted what continues to be one of the most elegant and comprehensive stereochemical analyses of a metabolic pathway, namely, that for the biosynthesis of cholesterol from mevalonate.

The intervening years since the earlier review by Popják have seen continued advances in the methods of stereochemical analysis. Numerous sophisticated uses of chiral [^{1}H,^{2}H,^{3}H]methyl groups as probes of enzyme mechanisms have been published. An important advance in recent years involves the synthesis and stereochemical analysis of chiral [^{16}O,^{17}O,^{18}O]organophosphates and -organothiophosphates as probes of enzyme-catalyzed phosphoryl transfer reactions, a topic reviewed elsewhere in this treatise. The difficult problem of determining the stereochemistry of hydrogen transfer to the two diastereotopic faces of the isoalloxazine ring of the flavin adenine dinucleotide (FAD)-dependent dehydrogenases has been solved with the use of exchange-inert deazaflavin analogs of

FAD. Aside from advances in the methods of analysis, the dramatic increase in the number of enzymes of known stereochemistry has provided additional fuel for the continuing debate over the relative importance of ancestral versus mechanistic explanations for stereochemical trends among enzymes of a given reaction type.

A. Aims and Past Reviews

Because of these advances, enzyme reaction stereochemistry has grown to such proportions that no single book chapter (because of limited space) or single author (because of limited knowledge) can hope for a thorough coverage of both the experimental and conceptual aspects of this topic. Therefore, this chapter emphasizes the interpretation of enzyme reaction stereochemistry rather than methods. Examples have been selected from the literature in order to illustrate broad principles that may be applicable to large classes of enzymes. Particular emphasis is given to enzyme systems for which X-ray crystallographic measurements provide a structural context in which to evaluate possible mechanistic interpretations of stereochemical trends. The methods of stereochemical analysis are only briefly summarized in cases where existing reviews can be referenced for experimental details.

For readers less familiar with this general topic, the review by Popják provides an excellent historical perspective on enzyme reaction stereochemistry as well as a concise description of pertinent stereochemical nomenclature (*1*). This chapter uses the Cahn–Ingold–Prelog convention for assigning absolute configuration to chiral centers (*5–7*). Chemically identical substituents at prochiral centers are designated according to Hanson's convention (*8*). Previous general reviews of enzyme stereochemistry published since 1969 may be found elsewhere (*9–17*). Recent specialized reviews of the stereochemistry of enzyme-catalyzed methyl (*18–20*) and phosphoryl (*21–25*) group transfers are also available. Critical discussions of the possible interpretations of stereochemical trends among different enzyme-catalyzed reactions have been published (*16, 26*).

B. Physicochemical Basis of Stereoselectivity

Enzyme-catalyzed reactions are generally, though not uniformly, characterized by either the stereospecific or stereoselective interconversion of selected stereoisomers of substrates and products via single, well-defined stereochemical pathways. The glycolytic enzyme L-lactate dehydrogenase (LDH) from pig heart illustrates this phenomenon; the stereochemical fidelity of hydride transfer between the (*S*)-enantiomer of lactate and the *re* face (A side) of the nicotinamide ring of bound NAD^+ is over 99.999998% (*27*) [Eq. (1)]:

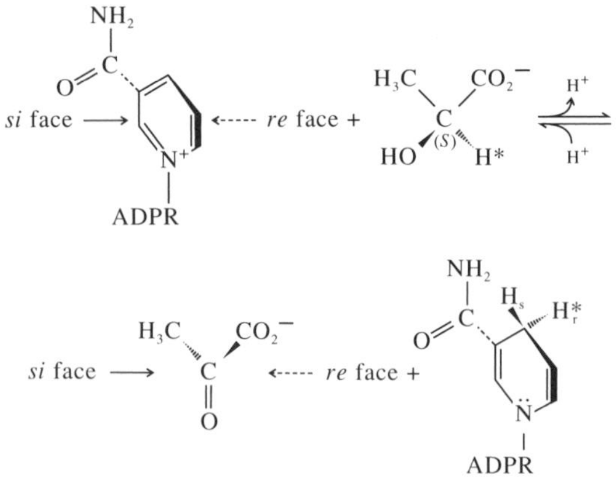

NH_2 C O
si face ⟶ N^+ ⇠ *re* face + H_3C CO_2^- C (*S*) HO H* ⇌ H^+ H^+
ADPR

NH_2 C O H_s H_r^*
si face ⟶ H_3C CO_2^- C O ⇠ *re* face + N ADPR (1)

where ADPR is adenosine diphosphoribose. In the reverse direction, hydrogen is directly transferred from the *pro*-(*R*) position at C-4 of the dihydronicotinamide ring to the *re* face of pyruvate. In 1966 Popják and Cornforth formulated the general physicochemical requirements for an enzyme to distinguish among chiral molecules when they suggested that the "dissymmetric treatment of a substrate by an enzyme can occur whenever the enzyme imposes, whether actively by binding or passively by obstruction, a particular orientation on the substrate at the site of action" (*28*). Recent high-resolution X-ray crystallographic measurements on the NAD^+-dependent dehydrogenases provide some of the clearest experimental support for this statement.

In the case of LDH, a structural explanation for the observed stereochemistry with normal substrates can be deduced from the X-ray structure of the binary complex formed between pig heart LDH and the catalytically competent coenzyme–substrate analog (3*S*)-5-(3-carboxy-3-hydroxypropyl)NAD^+ (**1**) (*29*) (Fig. 1).

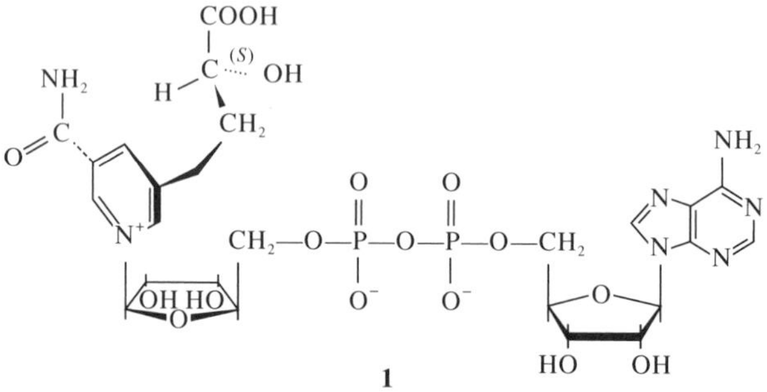

COOH (*S*) H C OH CH_2 NH_2 C O N^+ OH HO O
CH_2—O—P(=O)(O⁻)—O—P(=O)(O⁻)—O—CH_2 O HO OH NH_2 N N N N

1

Thus, specificity for (*S*)-lactate can be explained on the basis of strong polar interactions between the carboxyl and hydroxyl functions of the (*S*)-lactyl moiety and Arg-171, His-195, and, possibly, Arg-109, thereby orienting the hydrogen at the chiral carbon of the lactyl function for direct transfer to the *re* face of the

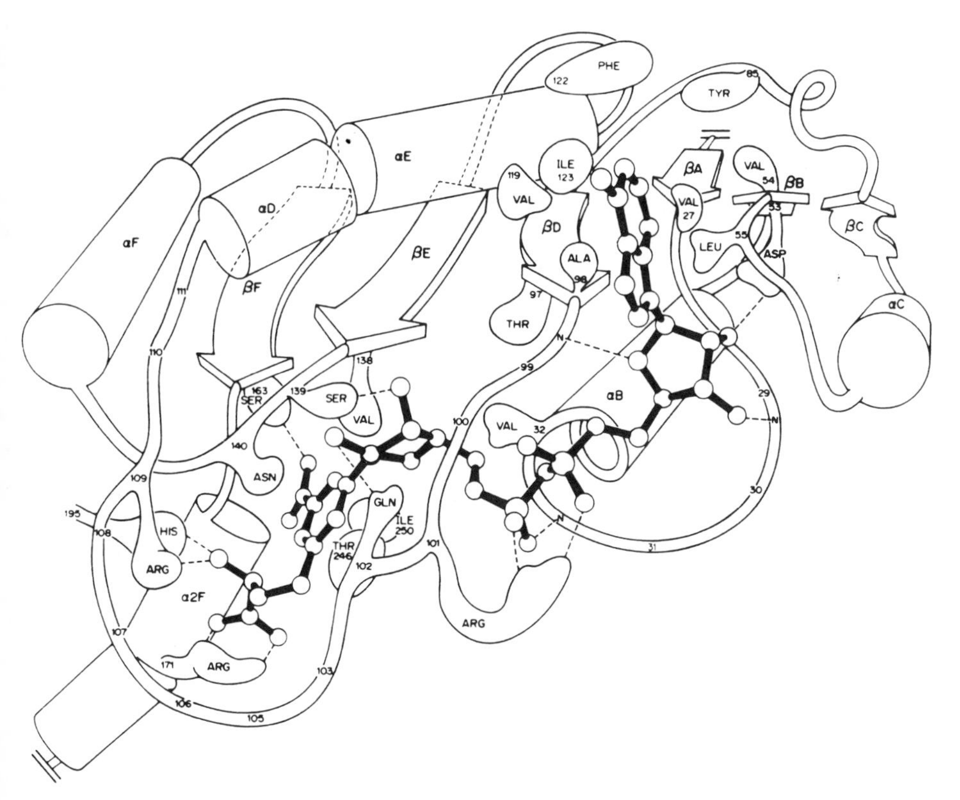

FIG. 1. Diagrammatic representation of the binary complex formed between (3*S*)-5(3-carboxy-3-hydroxypropyl)NAD^+ (**1**) and lactate dehydrogenase (pig heart), deduced on the basis of X-ray crystallographic measurements. The C-2 hydrogen of the (*S*)-lactyl moiety is positioned above the *re* face of the nicotinamide ring. [From Ref. (*29*), with permission.]

nicotinamide ring. The nicotinamide ring appears to be fixed in one orientation in the active site owing to a potential hydrogen-bonding interaction between Ser-163 and the amido function of the ring. Steric interactions with hydrophobic residues on the *si* face (B side) of the nicotinamide may further stabilize the ring in its observed orientation.

The importance of "obstruction" (steric hindrance) in determining the stereochemical course of enzymic reactions is further illustrated by a comparison of the stereospecificities of the alcohol dehydrogenases from yeast and from horse liver. The enzymes from both sources are A-side (*re* face) specific dehydrogenases wherein the *pro-(R)* hydrogen at C-1 of the alcohol is in the transferring position [Eq. (2)]:

$$NAD^+ + \mathrm{C}(H_r^*)(H_s)(R)(OH) \underset{H^+}{\overset{H^+}{\rightleftharpoons}} R\text{—}CHO + NADH^* \quad (2)$$

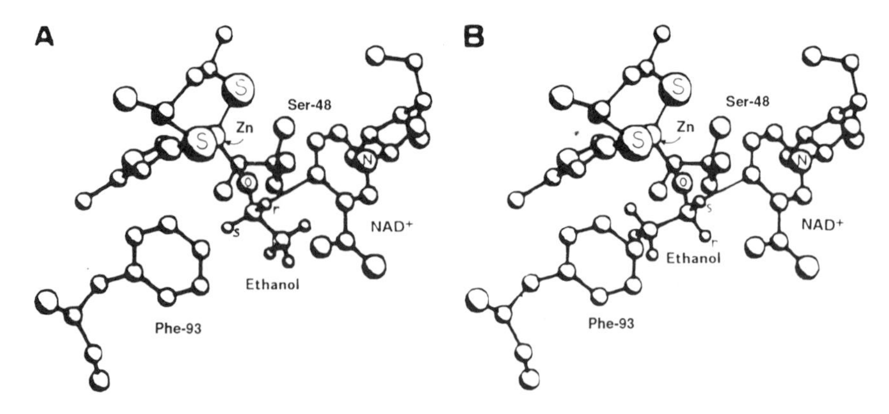

FIG. 2. Reasonable structural explanation for the stereospecificity of hydrogen transfer catalyzed by horse liver alcohol dehydrogenase. The binary complexes were obtained by building ethanol into the X-ray structure of the holoenzyme, on the basis of the assumption that the ethanolic oxygen is directly coordinated to the active site zinc ion. (A) When the *pro*-(*R*) proton of ethanol is directed at C-4 of the nicotinamide ring, the methyl function is favorably positioned in the active site; (B) when the *pro*-(*S*) proton is directed at C-4, the methyl function interacts sterically with Phe-93. [From Ref. (*30*), with permission.]

A possible structural explanation for the *pro*-(*R*) specificity of the horse liver enzyme can be obtained by building the ethanol molecule into a model of the active site obtained from X-ray measurements on the binary NAD^+·enzyme complex (*30*) (Fig. 2). When the *pro*-(*R*) hydrogen of ethanol is positioned for transfer to the coenzyme, the methyl group fits well into a pocket between Ser-48 and Phe-93; the complex in which the *pro*-(*S*) hydrogen is positioned for transfer appears to be disfavored by steric interactions between the methyl function of substrate and the aromatic ring of Phe-93. However, the postulated steric restriction may not be absolute, since the *pro*-(*S*) hydrogen of the alcohol is also transferred to bound NAD^+ at a rate that is approximately 2–4% of that for the hydrogen at the *pro*-(*R*) position (*31*). In contrast, yeast alcohol dehydrogenase stereospecifically transfers only the *pro*-(*R*) hydrogen of bound ethanol, perhaps reflecting increased steric restrictions owing to the presence of bulkier residues at positions 48 (Thr) and 93 (Trp) in this enzyme (*31–33*). Clearly, the stereochemistry of enzyme-catalyzed reactions must reflect the highly defined organization of residues within active sites.

C. MECHANISTIC VERSUS ANCESTRAL EXPLANATIONS FOR STEREOCHEMICAL TRENDS

An important conceptual challenge is to understand why a particular organization of active site residues (stereochemistry) has been retained during biological evolution (*16, 26*). This question relates to the general issue of how well enzyme structure and function are adapted to a particular catalytic role within cells.

Three general explanations have been envisioned for an observed stereochemistry. (1) The stereochemistry may reflect an arrangement of active site functional groups that for mechanistic reasons imparts greater catalytic efficiency to the enzyme than is possible on the basis of other arrangements that could have evolved. If greater catalytic efficiency has survival value to the host organism, the stereochemistry will tend to be conserved by natural selection. (2) The stereochemistry may be due to one of several mechanistically equivalent arrangements of functional groups inherited by chance from a preexisting enzyme catalyzing an analogous reaction, an ancestral explanation for the observed stereochemistry. (3) The stereochemistry may reflect a *nonoptimal* arrangement of functional groups because insufficient time has passed for the natural selection of an optimal arrangement, because an optimal arrangement has no survival value, because an optimal arrangement is constrained from evolving since it is tied to some other property of the enzyme that has greater survival value (e.g., protein stability, allosteric control), or because to evolve to a more efficient catalyst it must unavoidably pass through an inefficient enzyme.

Functional explanations for stereochemistry are most easily identified in some cases of substrate stereospecificity where a particular stereoisomer of substrate/product that is used/produced by an enzyme in a metabolic pathway must be in register with those of the neighboring enzymes in the pathway, in order to maintain metabolic coupling. In contrast, the stereochemical pathway followed during catalytic interconversion of bound substrate and product is often more difficult to explain. For example, the A-side specificity of L-lactate dehydrogenase cannot be explained on the basis of metabolic usage, since the same enantiomer of product would arise from transfer of either the A-side [C-4 *pro*-(*R*)] or B-side [C-4 *pro*-(*S*)] hydrogen of bound NADH to the *re* face of bound pyruvate [Eq. (1)]. This kind of "cryptic" stereochemistry frequently occurs among enzyme-catalyzed reactions. An important challenge is to identify cases in which the cryptic stereochemistry of an enzyme-catalyzed reaction may have a basis in mechanistic advantage.

The following generalizations and tentative conclusions emerge from a survey of the reaction stereochemistries and related physicochemical properties of the enzymes reviewed in this chapter:

1. *NAD(P)$^+$- and FAD-dependent dehydrogenases*. Numerous attempts have been made to discern a pattern in the stereochemistries of hydrogen transfer with respect to C-4 of the nicotinamide ring among different NAD(P)$^+$-dependent dehydrogenases. Although several provocative empirical generalizations have

been proposed, there has yet to be a single, cogent explanation for these stereochemistries on the basis of either catalytic mechanism or evolutionary relatedness. The same conclusion applies to the FAD-dependent dehydrogenases with respect to the stereochemistries of hydrogen transfer to the two diastereotopic faces of the isoalloxazine ring of FAD.

2. *Isomerases and pyridoxal phosphate (PLP)-dependent transaminases (aminotransferases).* The proton transfer reactions associated with isomerases and PLP-dependent transaminases are generally suprafacial processes. This may reflect a mechanistic advantage of a single active site base functioning in both proton abstraction and readdition at the same diastereotopic face of enediol(ate), dienol(ate), or enamine intermediates.

3. *Aldolases, PLP- and pyruvyl-dependent amino acid decarboxylases, and metal ion-dependent β-keto acid decarboxylases.* Carbon–carbon lyases generally (although not uniformly) involve retention of configuration at the carbon atom involved in heterolytic bond cleavage when there is a readily identifiable electrophilic component to catalysis, either an intermediate Schiff base or a catalytically essential divalent metal ion that could function to stabilize an intermediate carbanion. This empirical generalization implies that for reactions involving discrete intermediates there may be an intrinsic mechanistic advantage of a stereochemical mechanism in which a single active site structure can be used to accommodate the two structurally similar transition states associated with retention of configuration.

4. *Enzyme-catalyzed Claisen-type condensations, and metal ion-independent decarboxylases.* Carbon–carbon lyases generally involve inversion of configuration in cases where there is no readily identifiable electrophilic component to catalysis. This implies that catalysis of these reactions requires either the concerted or tightly coupled action of two or more active site functional groups, in order to trap (or avoid) unstable intermediates formed along the reaction pathway.

5. *β-Elimination reactions.* Enzyme-catalyzed elimination reactions β to the carbonyl functions of ketones or thioesters, or those involving enamine intermediates, generally proceed by syn processes. anti-Eliminations are generally observed in the remaining cases. As with the carbon–carbon lyases, there appears to be a relationship between reaction stereochemistry and the stability of reaction intermediates.

6. *Methyl group transfer reactions.* Methyl transfers from active biological methylating agents containing sulfonium ions generally involve inversion of configuration of the methyl group, consistent with a one-step S_N2 process. Retention of configuration is generally observed in cases where the methyl group donor is much less chemically reactive, implying the formation of a reactive methyl–enzyme intermediate.

II. Nicotinamide-Dependent Dehydrogenases: Hydride Transfer to Carbon

The NAD(P)$^+$-dependent dehydrogenases constitute the single most extensively studied class of enzymes with respect to reaction stereochemistry. In the most recent compilation of 157 dehydrogenases, 77 are A-side specific and 80 are B-side specific (*34*); examples are given in Table I. Whether the observed stereochemistries of the dehydrogenases have a basis in mechanism or evolutionary relatedness is currently a matter of debate (*2*).

A. History and Methods

Vennesland, Westheimer, and co-workers were the first to demonstrate that the two hydrogens at C-4 of the dihydronicotinamide ring of coenzyme are enzymatically nonequivalent, when they showed that monodeuterio-NADH results from an equilibrium mixture composed of yeast alcohol dehydrogenase (YADH), NAD$^+$, and excess [1-$^2H_2$]ethanol in water solvent (*35*) [Eq. (3)]:

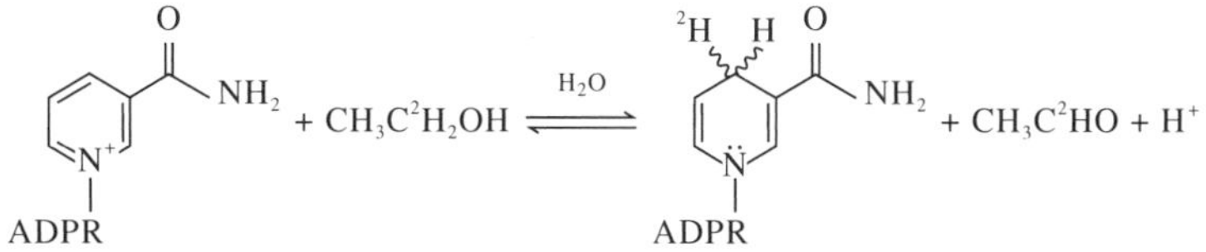

$$\text{NAD}^+\ (\text{ADPR}) + CH_3C^2H_2OH \overset{H_2O}{\rightleftharpoons} {}^2H,H\text{-NADH}\ (\text{ADPR}) + CH_3C^2HO + H^+ \qquad (3)$$

Had the enzyme not been able to distinguish between the two prochiral hydrogens of the coenzyme, dideuterio-NADH would have been isolated. Subsequent studies by Vennesland and co-workers demonstrated that the hydrogen transferred by YADH is not the same as that transferred by glyceraldehyde-3-phosphate dehydrogenase (GAPDH) and selected other dehydrogenases (*36–41*). This observation gave rise to Vennesland's convention in which "A-side specific" dehydrogenases are defined as those that transfer the same hydrogen of the coenzyme as that transferred by YADH, and "B-side specific" dehydrogenases are defined as those that transfer the alternate prochiral hydrogen as does GAPDH. The absolute stereochemistry of hydrogen transfer was later established by chemical degradation of monodeuterio-NADH, obtained from the action of an A-side dehydrogenase, to chiral [2-^{2}H]succinate having the same optical rotation as that of authentic (−) (2*R*)-[2-^{2}H]succinate (**2**) (*42*) [Eq. (4)]:

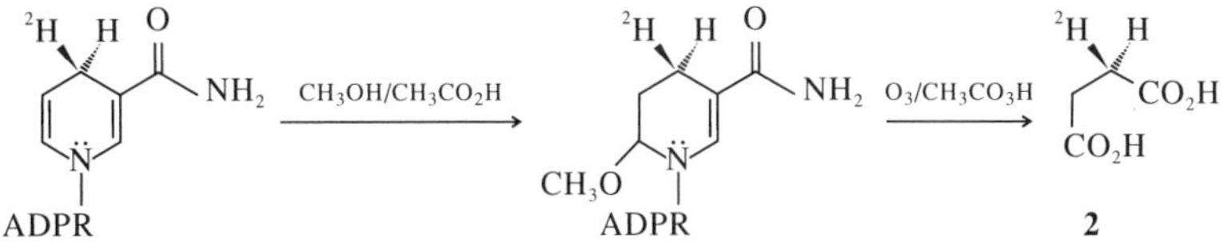

$$\text{NADH}\ ({}^2H,H;\ \text{ADPR}) \xrightarrow{CH_3OH/CH_3CO_2H} \text{6-}CH_3O\text{-NADH}\ (\text{ADPR}) \xrightarrow{O_3/CH_3CO_3H} {}^2H,H\text{-}C(CO_2H)CH_2CO_2H\ (\mathbf{2}) \qquad (4)$$

TABLE I

EXAMPLES OF A- AND B-SIDE DEHYDROGENASES[a]

A-side specific	B-side specific
Alcohol dehydrogenase (yeast)	Homoserine dehydrogenase
Alcohol dehydrogenase (liver)	Glycerol-3-phosphate dehydrogenase
Glycerol dehydrogenase	D-Xylulose reductase
Aldose reductase	L-Xylulose reductase
Histidinol dehydrogenase	Mannitol-1-phosphate dehydrogenase
Shikimate dehydrogenase	*myo*-Inositol 2-dehydrogenase
Glyoxylate reductase	UDPglucose dehydrogenase
L-Lactate dehydrogenase	3-Hydroxybutyrate dehydrogenase
D-Lactate dehydrogenase	3-Hydroxylacyl-CoA dehydrogenase
Glycerate dehydrogenase	Phosphogluconate dehydrogenase (decarboxylating)
Mevaldate reductase	Glucose dehydrogenase
Hydroxymethylglutaryl-CoA reductase (rat, yeast)	Galactose dehydrogenase
Malate dehydrogenase	Glucose-6-phosphate dehydrogenase
Isocitrate dehydrogenase	3α-Hydroxysteroid dehydrogenase (*Pseudomonas testosteroni*, rooster comb)
3α-Hydroxysteroid dehydrogenase (rat liver)	β-Hydroxysteroid dehydrogenase
Phosphoglycerate dehydrogenase	20β-Hydroxysteroid dehydrogenase
Sorbose dehydrogenase	Ribitol dehydrogenase
20α-Hydroxysteroid dehydrogenase	Testosterone 17β-dehydrogenase
Formate dehydrogenase	Carnitine dehydrogenase
Acetaldehyde dehydrogenase (acylating)	L-Fucose dehydrogenase
Succinate-semialdehyde dehydrogenase	D-Sorbitol-6-phosphate dehydrogenase
Acyl-CoA dehydrogenase	15-Hydroxyprostaglandin dehydrogenase
Orotate reductase	Aspartate semialdehyde dehydrogenase
2-Hexadecenal reductase	Glyceraldehyde phosphate dehydrogenase
Alanine dehydrogenase	Xanthine dehydrogenase
Tetrahydrofolate dehydrogenase	Cortisone α-reductase
Methylenetetrahydrofolate dehydrogenase	Glutamate dehydrogenase
Cytochrome-b_5 reductase	Glutamate dehydrogenase ($NADP^+$)
Nitrate reductase	Leucine dehydrogenase
Ferredoxin–$NADP^+$ reductase	Glutathione reductase
NADPH dehydrogenase	Lipoamide reductase
17-Ketostearic acid reducing enzyme (to L-17-hydroxystearic acid)	

[a] The range of different reactions catalyzed by the dehydrogenases is illustrated. [From Ref. (*2*), with permission.]

Thus, A-side dehydrogenases transfer the *pro-(R)* hydrogen at C-4 of the dihydronicotinamide ring and B-side dehydrogenases transfer the *pro-(S)* hydrogen. Early experiments also demonstrated that dehydrogenases can distinguish between the diastereotopic or enantiotopic faces of the carbonyl substrates, as well as those of coenzyme (*43, 44*). Lactate dehydrogenase [Eq. (1)] and alcohol dehydrogenase [Eq. (2)] are two such examples.

The seminal work of Westheimer and Vennesland laid the foundation for the methods currently used to determine the stereochemistry of hydrogen transfer with respect to the nicotinamide ring of the coenzyme. The experimental strategy commonly employed is to use authentic (4*R*)- or (4*S*)-[4-^{2}H/^{3}H]NADH as a substrate for the dehydrogenase of unknown stereochemistry, and then to determine whether deuterium or tritium is lost from or retained in the resultant oxidized coenzyme. If tritium is used as a tracer isotope, radiolabel counting can be used to decide whether tritium is present or absent in the oxidized coenzyme isolated from the reaction mixture. Alternatively, if deuterium is the tracer isotope, ^{1}H-NMR spectroscopy can be used to detect the presence of deuterium at C-4 of the oxidized coenzyme (Fig. 3) (*45*). The NMR method has the advantage of being able to monitor deuterium content at the exact site of action of the dehydrogenase on the bound coenzyme. Mass spectrometry has also been used

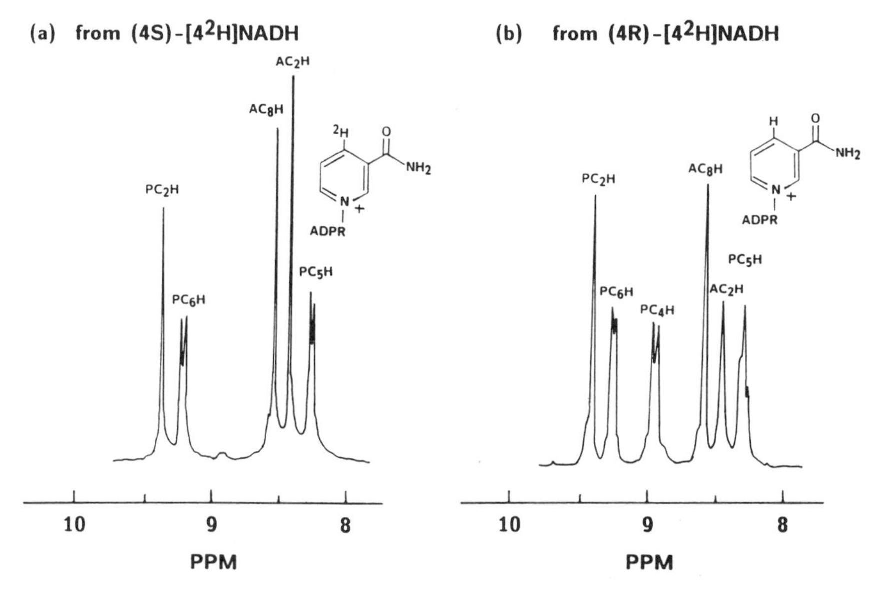

FIG. 3. ^{1}H-NMR spectra (220 MHz) of NAD$^+$ samples produced by salicylate 1-monooxygenase in the presence of salicylate, O_2, and (a) (4*S*)-[4-^{2}H]NADH or (b) (4*R*)-[4-^{2}H]NADH. Chemical shifts are versus trimethylsilyl [*U*-^{2}H]propionate. The chemical shift assignments are indicated by the prefix letters P and A, referring to the pyridine and adenine rings, respectively. The absence of the PC_4H resonance in the left-hand spectrum (a) and its presence in the right-hand spectrum (b) indicate that the enzyme is A-side specific. [From Ref. (*45*), with permission.]

to detect the presence of deuterium in the parent ion of nicotinamide generated by pyrolysis of samples of NAD^+ as small as 1 μg (*46*),

$$\left[\text{H or }^2\text{H} \text{ nicotinamide } (C(=O)NH_2) \right]^{+\cdot}$$

H, m/z 122; ^{2}H, m/z 123

B. Empirical Generalizations

Numerous attempts have been made to discern a pattern in the different stereochemistries of the dehydrogenases (Table I). One of the earliest empirical generalizations, for which there are few exceptions, are those of Bentley (*47*): "(1) The stereospecificity of a particular reaction is fixed and does not depend on the source of the enzyme. (2) The stereospecificity of a particular reaction is the same in those cases in which both NAD^+ and $NADP^+$ can be used as coenzymes. (3) If an enzyme utilizes a range of substrates, the stereospecificity with respect to hydrogen transfer [is the same for all of them]." The first generalization suggests that the stereospecificity with respect to the nicotinamide ring is a highly conserved property, once the dehydrogenase evolves to be specific for a particular substrate. The second and third generalizations indicate that the stereochemistry is largely fixed by specific interactions between the active site residues and the nicotinamide ring of bound coenzyme, for example, as indicated by X-ray measurements on lactate dehydrogenase (Fig. 1). Numerous additional generalizations have been proposed that imply either an evolutionary or mechanistic explanation for the stereochemistries of the dehydrogenases (*34*). Some of the more provocative explanations are summarized below.

1. *Consecutive Dehydrogenases*

Davies *et al.* observed that consecutive dehydrogenases in a metabolic pathway tend to have the same stereochemistry, in accordance with the possibility that some metabolic pathways may have evolved in a sequential fashion, wherein the genes for the first enzymes in a pathway gave rise to the genes for subsequent enzymes in the pathway (*48*). Two examples that fit this generalization are found in the metabolism of ethanol (involving two A-side dehydrogenases) and D-xylulose (involving two B-side dehydrogenases):

$$\begin{array}{l} \text{Ethanol} \xrightarrow{\text{alcohol dehydrogenase}} \text{acetaldehyde} \xrightarrow{\text{aldehyde dehydrogenase}} \text{acetate} \\ \text{D-Xylulose} \xrightarrow{\text{D-xylulose reductase}} \text{xylitol} \xrightarrow{\text{L-xylulose reductase}} \text{L-xylulose} \end{array} \quad (5)$$

Five additional pathways conform to this generalization: two involve A-side dehydrogenases and three, B-side dehydrogenases. Apparent exceptions to the rule are the sequential nitrate (A-side) and nitrite (B-side) reductases from *Candida utilis* (*49*) and the sequential alcohol dehydrogenase and aldehyde reductase involved in the conversion of cinnamyl alcohol to cinnamic acid in plants (*50*).

2. *Dehydrogenases Using Small Unphosphorylated Substrates*

In a review of the literature in 1978, You *et al.* noted the broad generalization that dehydrogenases which use unphosphorylated substrates containing three carbon atoms or less are A-side specific (*51*). These authors suggest that A-side dehydrogenases may have arisen before B-side dehydrogenases, on the basis of the assertion that small unphosphorylated substrate molecules may have served as the predominant energy source in the prebiotic chemical environment. The question of why B-side dehydrogenases ever arose is not addressed in their hypothesis.

3. *Direct Coenzyme Transfer*

An intriguing generalization was recently proposed by Srivastava and Bernhard who suggested that pairs of different dehydrogenases capable of transferring coenzyme directly between their active sites via an intermediate ternary complex are of opposite stereochemistries (*52*) [Eq. (6)]:

$$E_1{\cdot}NADH + E_2 \rightleftharpoons E_1{\cdot}NADH{\cdot}E_2 \rightleftharpoons E_1 + E_2{\cdot}NADH \tag{6}$$

where E_1 is the A-side dehydrogenase and E_2, the B-side dehydrogenase. This kind of "direct metabolite transfer" may be a general phenomenon in intermediary metabolism (*53–56*). The essential observations in support of the above generalization are as follows: (1) The inclusion of excess apolactate dehydrogenase ([LDH] $\cong 1.5 \times 10^{-4}$ *M*) in reaction mixtures separately composed of the A-side dehydrogenases for malate, alanine, or sorbitol and NADH ($\sim 2.5 \times 10^{-5}$ *M*) produces a dramatic decrease in the rate of product formation consistent with sequestration of NADH by apo-LDH. (2) However, the inclusion of excess apo-LDH in a reaction mixture composed of the B-side-specific glutathione reductase and oxidized glutathione produces only a minor decrease in rate, suggesting that NADH bound to apo-LDH is directly accessible to the active site of glutathione reductase via a ternary complex. (3) The corollary experiments with glyceraldehyde-3-phosphate dehydrogenase (B side) suggest that this enzyme is able to provide coenzyme directly to the active sites of A-side dehydrogenases but not to glutathione reductase. These observations imply a functional explanation for the stereochemistries of some dehydrogenases that is somehow based on the mechanics of coenzyme transfer between the active sites of dehydrogenases having opposing stereochemistries (*57*).

However, a word of caution is appropriate at this point. The kinetic analysis that forms the basis of the generalization proposed by Srivastava and Bernhard is neither experimentally nor conceptually straightforward. A recent reexamination of the kinetics of NADH transfer between lactate dehydrogenase and α-glycerol-3-phosphate dehydrogenase fully supports a free-diffusion mechanism *not* involving a kinetically significant ternary complex (*58*), although in a previous kinetic study by Srivastava and Bernhard this system was concluded to involve direct metabolite transfer (*59*). Thus, it may be premature to fully accept the basic premise of the generalization proposed by Srivastava and Bernhard.

4. *Stereoelectronic Control*

Probably the most provocative generalization is that recently proposed by Benner which forms the basis for an equally provocative mechanistic explanation for stereochemical trends among the dehydrogenases (*60*). This generalization is based on the observation that when the equilibrium constant for a dehydrogenase reaction $\{K_{eq} = [\text{–C(=O)–}][\text{NAD(P)H}][\text{H}^+] \div [\text{-CH(OH)-}][\text{NAD(P)}^+]\}$ with natural substrates is less than 10^{-12} *M*, the dehydrogenase is usually A-side specific; for reactions having equilibrium constants greater than 10^{-10} *M*, the dehydrogenase is generally B-side specific. Both A- and B-side dehydrogenases are found to catalyze reactions for which K_{eq} is within 1 order of magnitude of 10^{-11} *M*. Benner reports that of the 130 dehydrogenases for which both the stereochemistry and equilibrium constants are known, 120 appear to fit the generalization and 5 might not (*61*). Of the 120 enzymes that fit the generalization, 77 have equilibrium constants removed from the region in which both A- and B-side dehydrogenases occur. Nevertheless, the validity of Benner's generalization has been controversial, owing in part to confusion over the natural substrates for some of the dehydrogenases (*61, 62*). An extensive review of Benner's generalization can be found elsewhere (*2*).

Benner proposes that the empirical generalization can be rationalized on the basis of mechanistic principles using the following postulates. First, dehydrogenases are assumed to have evolved to optimal catalytic efficiency, in part because of the ability to perturb the equilibrium constant for bound substrates and products toward unity, independent of the overall equilibrium constant for the reaction. Enzymes capable of this kind of balancing of free energies of bound species include lactate dehydrogenase and the yeast and horse liver alcohol dehydrogenases; the theoretical significance of this phenomenon has been discussed elsewhere (*63, 64*).

Second, A-side dehydrogenases are postulated to have evolved to bind a conformation of the coenzyme in which the nicotinamide ring has an anti-orientation with respect to the ribose ring; B-side dehydrogenases evolved to bind the conformation having the syn-orientation:

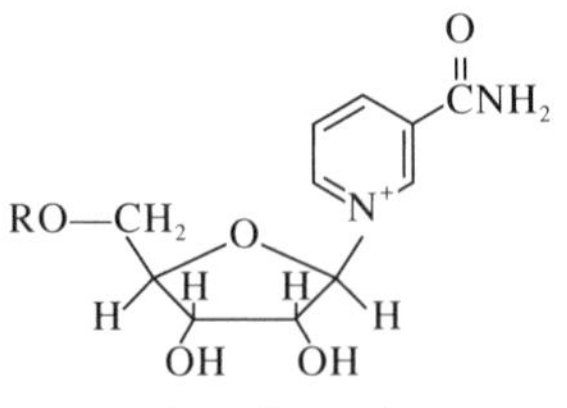

anti-conformation

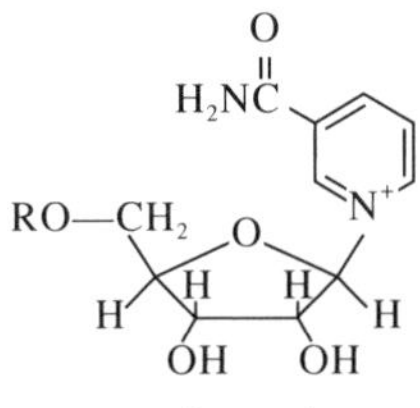

syn-conformation

That the conformation of bound coenzyme may correlate generally with the stereochemistry of a dehydrogenase was first suggested by Rossmann *et al.* (*65*) and later by Kaplan and co-workers (*51*) and Walsh (*66*). Five different A-side dehydrogenases have been demonstrated to bind the anti-conformation of coenzyme and four B-side dehydrogenases to bind the syn-conformation, on the basis of X-ray and nuclear Overhauser effect measurements (Table II) (*67–74*). Glutathione reductase may be an exception to this generalization, since this B-side dehydrogenase reportedly binds the anti-conformation (*74*). However, the extent to which this enzyme deviates from the generalization is unclear, as

TABLE II

CONFORMATIONS OF NICOTINAMIDE COENZYMES BOUND TO ACTIVE SITES OF A- AND B-SIDE DEHYDROGENASES/REDUCTASES, AS DETERMINED BY X-RAY CRYSTALLOGRAPHY AND NUCLEAR OVERHAUSER EFFECT MEASUREMENTS

Enzyme (source)	Conformation of bound coenzyme	Method[a]	Ref.
A-side			
Alcohol dehydrogenase			
Horse liver	*anti*-NAD^+	X-Ray	*67*
Horse liver	*anti*-NAD^+	NOE	*68*
Yeast	*anti*-NAD^+	NOE	*68*
Malate dehydrogenase			
Pig	*anti*-NAD^+	X-Ray	*67*
Lactate dehydrogenase			
Dogfish	*anti*-NAD^+	X-Ray	*67*
Beef heart	*anti*-NAD^+	NOE	*69*
Dihydrofolate reductase			
Lactobacillus casei	*anti*-NADH	X-Ray	*70*
L. casei	*anti*-$NADP^+$	NOE	*71*
Sorbitol dehydrogenase			
Sheep liver	*anti*-NAD^+	NOE	*72*
B-side			
Glyceraldehyde-3-phosphate dehydrogenase			
Lobster muscle	*syn*-NAD^+	X-Ray	*67*
Bacillus stearothermophilis	*syn*-NAD^+	X-Ray	*73*
Glutathione reductase			
Human erythrocytes	*anti*-NAD^+	X-Ray	*74*
Glucose-6-phosphate dehydrogenase			
Leuconostoc mesenteroides	*syn*-$NAD(P)^+$	NOE	*69*
Glutamate dehydrogenase			
Rat liver	*syn*-$NAD(P)^+$	NOE	*69*

[a] X-Ray, X-Ray crystallography; NOE, nuclear Overhauser effect.

the crystallographic coordinates have not yet been reported for the enzyme–coenzyme complex.

Third, the syn-conformation of the coenzyme is postulated to be a stronger reducing species than the anti-conformation. This postulate is indirectly supported by NMR measurements of the equilibrium distribution of the two conformers of the oxidized and reduced forms of nicotinamide mononucleotide, indicating that the syn-conformation is a stronger reducing agent than the anti-conformation by roughly 1.3 kcal/mol (*75, 76*). Thus, binding of the anti-conformation of the coenzyme by A-side dehydrogenases is proposed to provide one means by which the enzyme can perturb the ratio of bound substrates and products toward unity for reactions that are highly favorable in the direction of reduction of the carbonyl substrate. Alternatively, binding of the syn-conformation by B-side dehydrogenases has the same advantage for substrates not so easily reduced. For reactions for which K_{eq} is within 1 order of magnitude of 10^{-11} *M*, both A- and B-side dehydrogenases are found since neither conformation of NADH would be preferred. The reason why the break occurs at a K_{eq} value of 10^{-11} *M*, or an effective equilibrium constant of 10^{-4} at pH 7, presumably depends on the nature of the reaction mechanism defining the relationship between the equilibrium constants for the bound and free substrate and product molecules.

Finally, the syn- and anti-conformations of bound coenzyme are postulated to preferentially transfer the *pro-*(*S*) and *pro-*(*R*) hydrogens, respectively. This is based on the assumption that the dihydronicotinamide ring is stabilized in the boat conformation, owing to orbital overlap between the lone pair of electrons on the nitrogen and the anti-bonding σ^*(C–O) orbital of the ribosyl moiety:

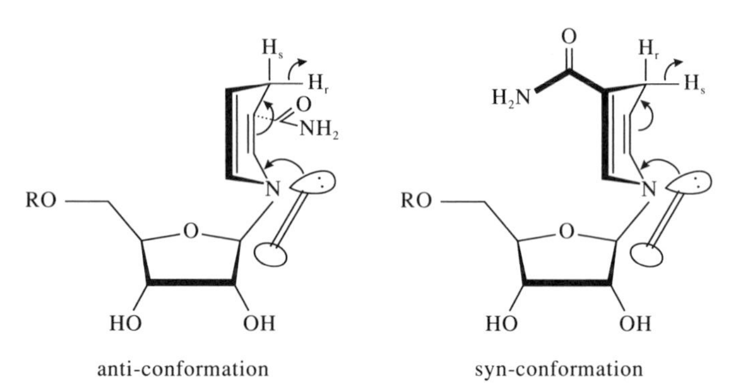

Thus, for reasons of orbital overlap in the transition state, the pseudoaxial hydrogens are most easily transferred. Evidence has not, as yet, been found in support of a boat conformation for coenzyme bound to enzymes. However, a boat conformation in the transition state is consistent with the results of recent computational studies (*77*).

Finally, Benner's theory is noteworthy not because its correctness is assured, but because it emphasizes that a mechanistic explanation for the stereochemistries of the dehydrogenases is possible on the basis of the different physicochemical properties of the diastereotopic hydrogens of the dihydronicotinamide ring.

III. Flavin-Dependent Oxidoreductases: Hydride Transfer to Nitrogen

Flavoenzymes catalyze a diverse set of oxidation–reduction reactions involving tightly bound FAD or flavin mononucleotide (FMN) (*78, 79*) [Eq. (7)]:

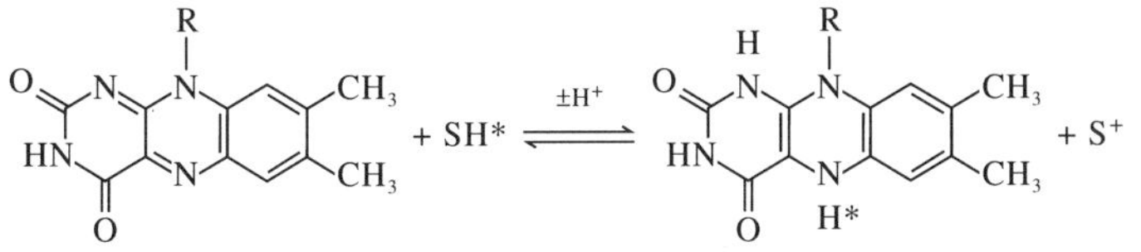

$$\text{FAD/FMN} + SH^* \overset{\pm H^+}{\rightleftharpoons} \text{FADH}_2\text{/FMNH}_2 + S^+ \qquad (7)$$

FAD/FMN

FADH$_2$/FMNH$_2$

FAD/FADH$_2$, R = $CH_2(CHOH)_3CH_2OPOPOCH_2$–(adenosine; O O, O$^-$ O$^-$; NH$_2$; HO OH)

FMN/FMNH$_2$, R = $CH_2(CHOH)_3CH_2OPO_3H^-$

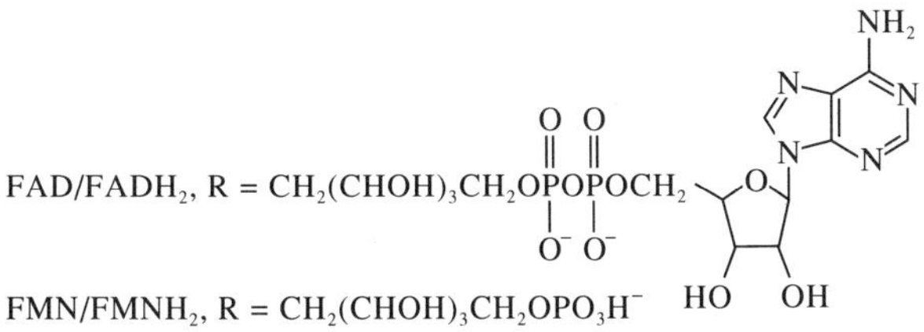

In analogy to the NAD(P)$^+$-dependent dehydrogenases, hydrogen transfer can take place to either the *re* or the *si* face of the isoalloxazine ring system:

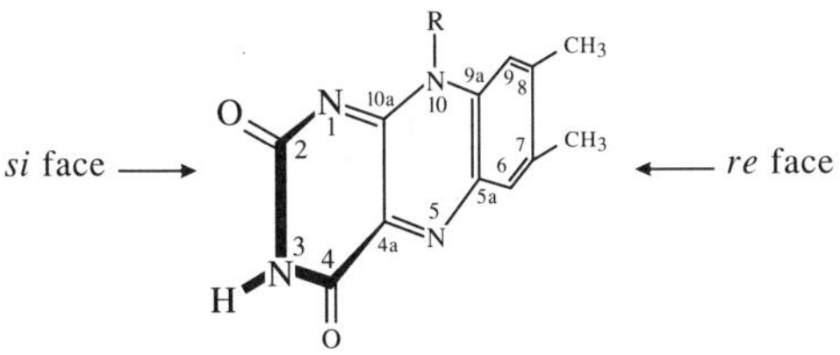

si face ⟶ [isoalloxazine ring, positions 1, 2, 3, 4, 4a, 5, 5a, 6, 7, 8, 9, 9a, 10, 10a; R, CH3, CH3, H, O] ⟵ *re* face

Unfortunately, isotope transfer methods cannot be used experimentally to determine the stereochemistry of hydrogen transfer, because the hydrogen at N-5 of FADH$_2$/FMNH$_2$ rapidly exchanges with solvent protons. An important experimental achievement in recent years has been the development of a general method for determining this stereochemistry with the use of exchange-inert 5-

deazaflavins that have been demonstrated to serve as catalytically competent substrate analogs for many flavoenzymes (*80–84*). This has allowed division of the FAD-dependent oxidoreductases into two stereochemically distinct classes (Table III) (*85–89*).

Glutathione reductase is the best studied member of the FAD-dependent oxidoreductases and serves to illustrate the stereochemical and mechanistic problems associated with this general class of enzymes. This enzyme functions to

TABLE III

FLAVOENZYME REACTIONS INVOLVING HYDROGEN TRANSFER BETWEEN SUBSTRATE AND EITHER THE *re* OR *si* FACES OF BOUND FLAVIN

Enzyme	Source	Substrate	Side of flavin interacting with substrate	Ref.
Reductases/dehydrogenases				
Glutathione reductase	Human erythrocytes	NADPH	*re*	*74*
NAD^+:FMN oxidoreductase	*Beneckea harveyi*	NADH	*re*	*85, 86*
Thioredoxin reductase	*Escherichia coli*	NADPH	*re*	*87*
Mercuric reductase	*Pseudomonas aeruginosa*	NADPH	*re*	*87*
General acyl-CoA dehydrogenase	Pig kidney	Acyl-CoA	*re*	*87*
D-Lactate dehydrogenase	*Megasphaera elsdenii*	Pyruvate	*si*	*88*
Oxidases				
Glucose oxidase	*Aspergillus niger*	Glucose	*re*	*87*
D-Amino-acid oxidase	Pig kidney	Pyruvate/NH_4^+	*re*	*88*
L-Lactate oxidase	*Mycobacterium smegmatis*	Pyruvate	*si*	*89*
Glycolate oxidase	Spinach	Glycolate	*si*	*89*
Oxygenases/hydroxylases				
p-Hydroxybenzoate hydroxylase	*Pseudomonas fluorescens*	NADPH	*re*	*87*
Melilotate hydroxylase	*Pseudomonas*	NADH	*re*	*87*
Anthranilate hydroxylase	*Trichosporum cutaneum*	NADPH	*re*	*87*
Cyclohexanone monooxygenase	*Acinetobacter* NCIB	$AcPyADP^+$ [a]	*re*	*88*
2-Methyl-3-hydroxy-pyridine-5-carboxylic acid oxygenase	*Pseudomonas* MA-1	$AcPyAD^+$ [b]	*re*	*88*

[a] Oxidized 3-acetylpyridine adenine dinucleotide phosphate.
[b] Oxidized 3-acetylpyridine adenine dinucleotide.

maintain high concentrations of reduced glutathione (GSH) in cells. Catalysis is due to two distinct partial reactions involving the cyclical oxidation–reduction of two catalytically essential cysteinyl residues (*90*) [Eqs. (8) and (9)]:

$$H^+ + NADPH + E(S{-}S) \longrightarrow NADP^+ + E(SH)(SH) \quad (8)$$

$$E(SH)(SH) + GSSG \longrightarrow E(S{-}S) + 2GSH \quad (9)$$

In a seminal study, Fisher and Walsh indirectly demonstrated that the glutathione reductase reaction likely involves direct hydrogen transfer from bound NADPH to N-5 of bound FAD, on the basis of the finding that [5-^{3}H]dihydro-5-deazaflavin (**3H**) results from incubating [4-^{3}H]NADPH with glutathione reductase reconstituted with 5-deazaflavin (**3**) (*82*) [Eq. (10)]:

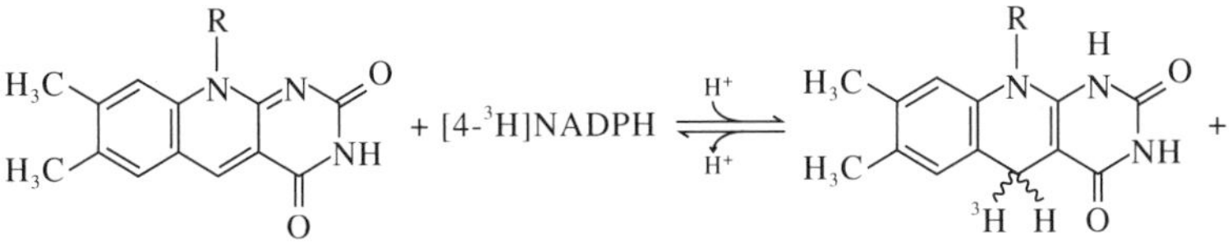

$$\mathbf{3} + [4\text{-}^3H]NADPH \underset{H^+}{\overset{H^+}{\rightleftharpoons}} \mathbf{3H} + NADP^+ \quad (10)$$

However, this approach could not be used to establish the absolute stereochemistry of hydrogen transfer because the reduced and oxidized forms of 5-deazaflavin, once released from the enzyme, undergo a rapid scrambling reaction that results in racemization at C-5 of the reduced species (*83*). Nevertheless, subsequent X-ray crystallographic measurements on the stable ternary enzyme·FAD·NADP$^+$ complex demonstrated that the *re* and *si* faces of the isoalloxazine and nicotinamide rings (respectively) are next to one another (*74*):

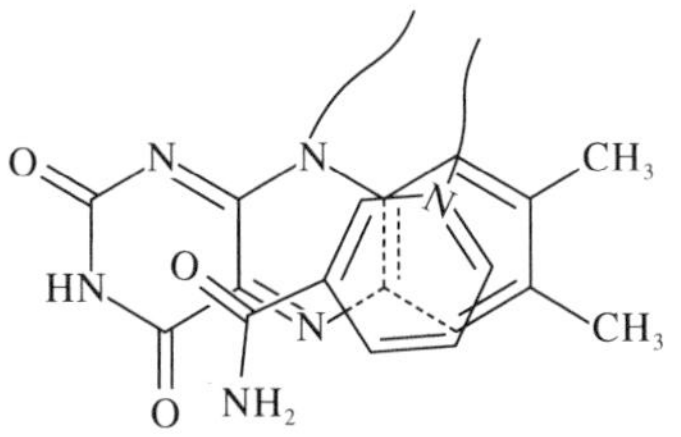

Moreover, the catalytically essential protein sulfhydryl groups, arising from Cys-58 and Cys-63, are located near the *si* face of the isoalloxazine ring. Thus, catalysis by glutathione reductase appears to involve hydrogen transfer reactions at both diastereotopic faces of the isoalloxazine ring system.

A. Methods

The development of a general method for determining the absolute stereochemistry of hydrogen transfer among different flavoenzyme reactions was inspired by early stereochemical studies on the enzymes associated with methane-producing bacteria that use as a naturally occurring cofactor 7,8-didemethyl-8-hydroxy-5-deazaflavin (**4**) (coenzyme F_{420}):

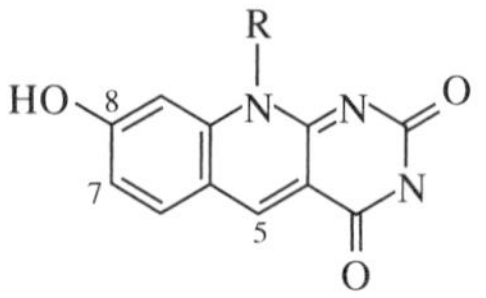

4

$$R = CH_2(CHOH)_3CH_2OP(=O)(O^-)\text{–O–}CH(CH_3)C(=O)NHCH(CO_2^-)(CH_2)_2C(=O)NHCH(CO_2^-)(CH_2)_2CO_2^-$$

Unlike the synthetic deazaflavin (**3H**) employed with glutathione reductase [Eq. (10)], the reduced form of F_{420} (**4H**) is much less susceptible to the kind of nonenzymic scrambling reaction that leads to racemization at C-5. This fact allowed the determination of the stereochemistries associated with several F_{420}-dependent enzymes, including the selenium-containing hydrogenase from *Methanococcus vannielii,* using the substrate analog 7,8-didemethyl-8-hydroxy-5-[5-^{2}H]deazariboflavin (**5**) (*85*) [Eq. (11)]:

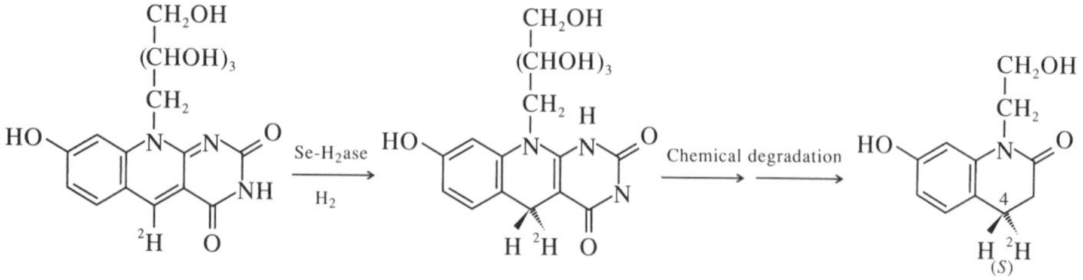

5 $\xrightarrow[H_2]{\text{Se-H}_2\text{ase}}$ **5H** $\xrightarrow{\text{Chemical degradation}}$ **6H** (*S*) (11)

The addition of hydrogen to the *si* face of **5** was established by chemical degradation of **5H** to the 3,4-[4-^{2}H]dihydrolactam (**6H**), shown to have the (*S*)-configuration at C-4, owing to the correspondence between its optical rotatory dispersion curve and that of authentic **6H.** Moreover, this assignment indirectly established that the NAD^+ :FMN oxidoreductase from *Beneckea harveyi* likely involves hydrogen transfer to the *re* face of FMN, on the basis of an earlier study showing that **5** is stereospecifically reduced by this enzyme (*86*).

Pai and co-workers recently developed a general strategy for determining the stereochemistries of the flavoenzymes that is fundamentally based on the stereochemical features of the glutathione reductase reaction (*87, 88*). Their experimental approach relies on the observations that (1) the reduced forms of the 8-demethyl-8-hydroxy-5-deazaflavins (**7**) are not susceptible to nonenzymic racemization,

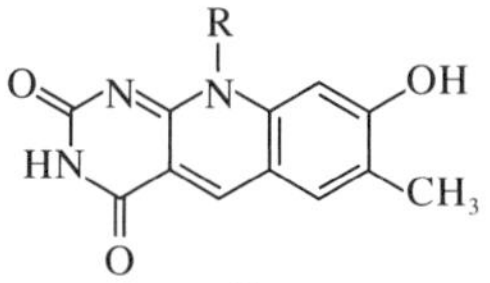

7

(2) the apoproteins of numerous flavoenzymes can be successfully reconstituted with these analogs, and (3) the glutathione reductase reaction involves hydrogen transfer from NADPH to the *re* face of the isoalloxazine ring (*74*). The analysis of an enzyme of unknown stereochemistry first involves reconstitution of the enzyme with **7,** followed by reaction with the reduced form of the second substrate. The **7H** is then desorbed from the enzyme and reconstituted with the apoprotein of glutathione reductase. Finally, treatment with $NADP^+$ results in transfer of the *pro*-(*R*) hydrogen at C-5 of bound **7H** to $NADP^+$. Thus, if isotopically labeled substrate (SH*) is used with the enzyme (E_1) of unknown stereochemistry and hydrogen transfer is to the *pro*-(*S*) position at C-5 of bound **7** [Eq. (12)],

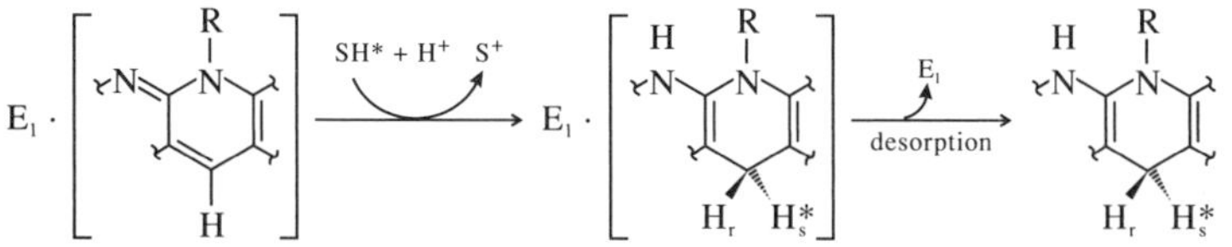

$$E_1 \cdot [\ldots] \xrightarrow{SH^* + H^+ \quad S^+} E_1 \cdot [\ldots H_r\ H_s^*] \xrightarrow[\text{desorption}]{E_1} \ldots H_r\ H_s^* \qquad (12)$$

the isotope will be retained in the analog after analysis with glutathione reductase (E_2) [Eq. (13)].

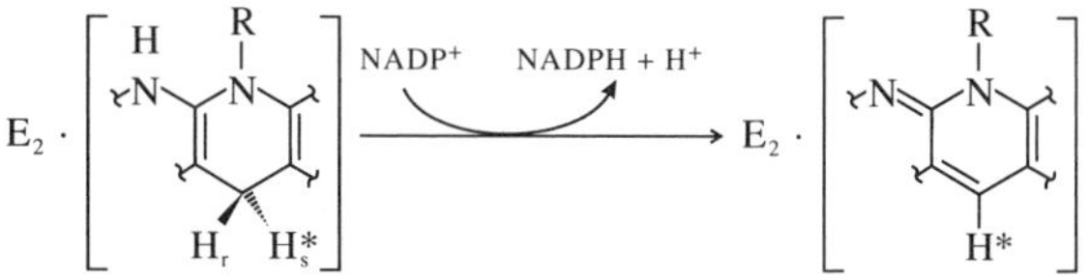

$$E_2 \cdot [\ldots H_r\ H_s^*] \xrightarrow{NADP^+ \quad NADPH + H^+} E_2 \cdot [\ldots H^*] \qquad (13)$$

Alternatively, if hydrogen transfer is to the *pro*-(*R*) position during the first step [Eq. (12)], the isotope will be transferred to $NADP^+$ during the analysis step [Eq. (13)].

The stereochemistry of hydrogen transfer in the direction of reduction of substrate can be determined with the use of **7H,** enriched with tritium in the *pro*-(*R*) position at C-5. This compound is prepared by NaB^3H_4 treatment of general acyl-CoA dehydrogenase reconstituted with **7,** which has been established to involve stereoselective reduction at the *re* face of the bound flavin. The stereoselectively reduced flavin is then released from the dehydrogenase and reconstituted into the flavoenzyme of unknown stereochemistry. On treatment of the enzyme with oxidized substrate, most of the tritium will either be retained in the flavoenzyme or transferred to substrate, depending on whether hydrogen transfer involves the *pro*-(*S*) or *pro*-(*R*) positions at C-5 (respectively) of the flavin analog.

B. Trends

The stereospecificities of 15 different flavoenzymes have so far been determined by a combination of solution techniques and X-ray crystallography (Table III). In analogy with the $NAD(P)^+$-dependent dehydrogenases, hydrogen transfer to the *re* and *si* faces of bound flavin are both significant stereochemical options. Pai and co-workers have identified two tentative correlations between stereospecificity and substrate structure: (1) flavoenzymes using pyridine nucleotides uniformly involve hydrogen transfer to the *re* face of the flavin ring, and (2) flavoenzymes using α-hydroxy acids involve hydrogen transfer to the *si* face of the flavin ring (*88*). These two trends are, perhaps, most easily explained by the conservation of a defined arrangement of active site residues brought forward during biological evolution from common ancestral proteins. Nevertheless, a hidden mechanistic explanation for these trends cannot be excluded.

IV. Isomerases Involving Proton Transfer: Importance of Single-Base Mechanisms

Numerous enzyme-catalyzed isomerization reactions involve either 1,2-hydrogen transfers [Eq. (14)]

$$-\underset{\underset{H^*}{|}}{\overset{\overset{HO}{|}}{C}}-\overset{\overset{O}{\|}}{C}- \rightleftharpoons -\overset{\overset{O}{\|}}{C}-\underset{\underset{H^*}{|}}{\overset{\overset{OH}{|}}{C}}- \qquad (14)$$

or 1,3-allylic hydrogen transfers [Eq. (15)].

$$>C=\overset{\overset{H}{|}}{C}-\underset{|}{\overset{\overset{H^*}{|}}{C}}- \rightleftharpoons -\underset{|}{\overset{\overset{H^*}{|}}{C}}-\overset{\overset{H}{|}}{C}=C< \qquad (15)$$

A substantial body of evidence indicates that hydrogen migration is generally a suprafacial process mediated by a single active site base. Single-base mechanisms of this type may have fundamental advantages over mechanisms involving antarafacial hydrogen transfer requiring two active site bases, as first suggested by Hanson and Rose (*16*).

A. 1,2-Proton Transfer Reactions

The aldose–ketose isomerases constitute the best studied class of enzymes catalyzing 1,2-proton transfer reactions (Tables IV and V). Isomerization generally involves significant intramolecular hydrogen transfer with variable amounts of solvent proton exchange (*16*). This argues for the formation of an enediol(ate) intermediate facilitated by a single active site base partially shielded from solvent [Eq. (16)].

$$\mathrm{R{-}\underset{\underset{\ddot{B}}{H}}{\overset{HO}{C}}{-}\overset{O}{\overset{\|}{C}}{-}H} \rightleftharpoons \mathrm{R{-}\overset{HO}{C}{=}\overset{O^-}{C}{-}H \;\; \overset{H}{B}} \rightleftharpoons \mathrm{R{-}\overset{O}{\overset{\|}{C}}{-}\underset{\underset{\ddot{B}}{H}}{\overset{OH}{C}}{-}H} \qquad (16)$$

In the case of the xylose isomerase reaction, for which solvent proton exchange has not been detected, an intramolecular hydride transfer mechanism cannot be excluded (*91*).

Remarkably, the configuration of the putative enediol intermediates for all of the aldose–ketose isomerases is uniformly cis, although suprafacial hydrogen transfer is observed at both faces of the enediol(ate). Isomerases that operate on (2*R*)-aldoses transfer hydrogen at the *re–re* face of the enediol(ate) to the *pro*-(*R*) position at C-1 of the product ketose (Table IV); isomerases that operate on (2*S*)-aldoses transfer hydrogen at the *si–si* face to the *pro*-(*S*) position of the ketose (Table V). Hanson and Rose speculate that the observed stereochemistries imply the use of a minimum number of catalytic groups in order to facilitate catalysis and that this may represent an intrinsic mechanistic advantage over other stereochemical alternatives (*16*). A *cis*-enediol intermediate would allow a single electrophilic group in the active site to accommodate both oxygen atoms of the enediol; suprafacial hydrogen transfer can be accomplished with a single active site base. This kind of economic use of catalytic groups may have advantages over a mechanism involving a *trans*-enediol(ate) intermediate requiring a minimum of two electrophilic groups and antarafacial hydrogen transfer involving a catalytic base at each of the two diastereotopic faces of the enediol intermediate. In any event, the stereochemical uniformities observed among the aldose–ketose isomerases are not easily explained on the basis of divergent evolution from a common ancestral enzyme, given the absence of evidence for se-

TABLE IV

ALDOSE-KETOSE ISOMERASES THAT USE (2*R*)-ALDOSES AS SUBSTRATES[a]

cis-Enediolate

Enzyme	Substrate
Triose-phosphate isomerase (EC 5.3.1.1)	D-Glyceraldehyde 3-phosphate ($R = CH_2OPO_3^{2-}$)
Ribose-5-phosphate isomerase (EC 5.3.1.6)	D-Ribose 5-phosphate [$R = (CHOH)_2CH_2OPO_3^{2-}$]
Glucose-6-phosphate isomerase (EC 5.3.1.9)	D-Glucose 6-phosphate [$R = (CHOH)_3CH_2OPO_3^{2-}$]
Glucosamine-6-phosphate isomerase (deaminating) (EC 5.3.1.10)	2-Amino-2-deoxy-D-glucose 6-phosphate [replace OH by NH_2; $R = (CHOH)_3CH_2OPO_3^{2-}$]
L-Arabinose isomerase (EC 5.3.1.4)	L-Arabinose [$R = (CHOH)_2CH_2OH$]
D-Xylose isomerase (EC 5.3.1.5)	D-Xylose [$R = (CHOH)_2CH_2OH$]

[a]The reactions involve suprafacial hydrogen transfer at the *re–re* face of the intermediate enediol(ate) (*16*).

quence similarities among the pentose isomerases (*92*) and among the phospho sugar isomerases (*93*).

1. *Triose-Phosphate Isomerase*

Triose-phosphate isomerase provides the clearest experimental support for a single-base mechanism. The active site base involved in the proton transfer reaction is almost surely a single carboxyl group arising from Glu-165. Early affinity labeling studies identified this residue in the active site (*66*). In support of an essential catalytic role for this residue, site-directed mutagenesis of Glu-165 to Asp-165 diminishes k_{cat} by several orders of magnitude in both the forward and reverse directions (*94*). Model building of the dihydroxyacetone phosphate (DHAP) molecule into the X-ray structure of the isomerase from chicken muscle positions the carboxyl oxygens of Glu-165 above the C-1 and C-2 carbons of substrate, an arrangement well suited for catalyzing a suprafacial 1,2-hydrogen transfer (*95, 96*) (Fig. 4). This orientation of Glu-165 with respect to bound

TABLE V

ALDOSE–KETOSE ISOMERASES THAT USE (2*S*)-ALDOSES AS SUBSTRATES[a]

cis-enediolate

Enzyme	Substrate
D-Mannose-6-phosphate isomerase (EC 5.3.1.8)	D-Mannose 6-phosphate [R = $(CHOH)_3CH_2OH$]
D-Arabinose isomerase (EC 5.3.1.3)	D-Arabinose [R = $(CHOH)_2CH_2OH$]

[a]The reactions involve suprafacial hydrogen transfer at the *si–si* face of the intermediate enediol(ate) (*16*).

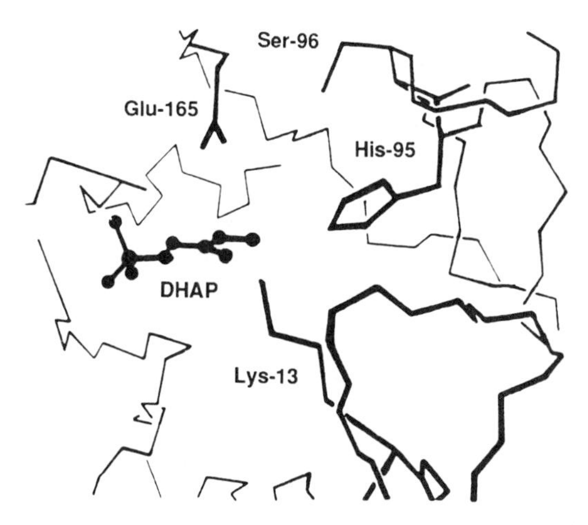

FIG. 4. Hypothetical binary complex formed between dihydroxyacetone phosphate and triose-phosphate isomerase constructed from published X-ray coordinates for the chicken muscle enzyme (*96*). [From Ref. (*95*), with permission.]

substrate has been largely confirmed on the basis of X-ray measurements on the actual binary complexes formed from the crystalline isomerase from yeast with DHAP and with the transition state analog, phosphoglycolohydroxamate (*97*). Moreover, the positively charged ε-amino of a single active site lysine appears to be in a position to interact with both the C-1 and C-2 oxygens of bound substrate, although other groups in the active site (e.g., His-95) are also in a position to play an electrophilic role in catalysis.

2. *Glucose-Phosphate Isomerase*

Glucose-phosphate isomerase is one of the best studied enzymes catalyzing the interconversion of aldo- and ketohexose phosphates. An active site carboxyl group is a possible candidate for the base catalyzing the intramolecular proton transfer reaction. The affinity label 1,2-anhydro-D-mannitol 6-phosphate (**8**) inactivates the enzyme by forming an ester linkage between C-1 of the affinity label and an active site carboxyl of a glutamic acid residue (*98*).

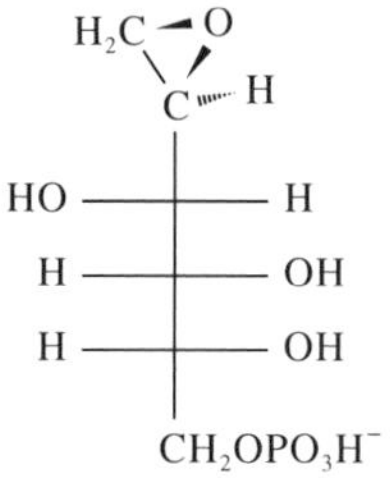

H_2C—O, C—H, HO—H, H—OH, H—OH, $CH_2OPO_3H^-$

8

The glutamic acid residue is probably catalytically important because the pH dependence of the rate of inactivation is similar to the pH dependence of V_{max} with the normal substrates for the enzyme. Moreover, the presence of an active site lysine has been postulated on the basis of pH–rate profiles and on the basis of the loss of activity in the presence of both pyridoxal phosphate and BH_4^- (*99*).

The substrate specificity of glucose-phosphate isomerase illuminates additional stereochemical subtleties of the isomerase reaction (*100*). In the aldose to ketose direction, the enzyme potentially operates on an equilibrium mixture of substrate forms composed of two cyclic hemiacetals (the α- and β-anomers of glucose 6-phosphate) and trace quantities of the acyclic aldehyde form [Eq. (17)]:

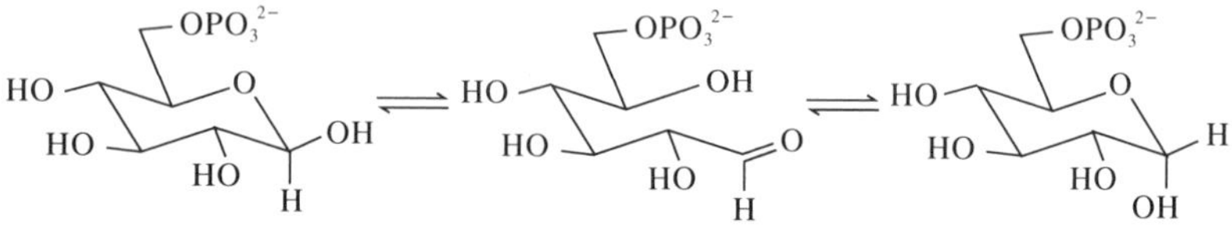

(17)

β-D-Glucopyranose 6-phosphate (62%) ⇌ free aldehyde (trace) ⇌ α-D-glucopyranose 6-phosphate (38%)

The half-life for nonenzymic interconversion of the anomers is approximately 15 sec under physiological conditions. In the ketose to aldose direction, fructose 6-phosphate is composed of two cyclic ketals and trace quantities of the acyclic ketone [Eq. (18)]:

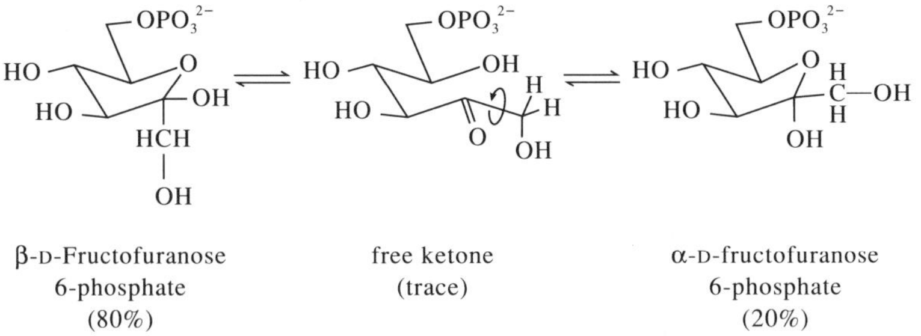

(18)

β-D-Fructofuranose 6-phosphate (80%) ⇌ free ketone (trace) ⇌ α-D-fructofuranose 6-phosphate (20%)

The half-life for nonenzymic interconversion of the anomers is about 0.4 sec. Stopped-flow kinetic measurements, using the isomerase from yeast, indicate that both the α- and β-anomers of glucose 6-phosphate are used directly by the enzyme to give the α- and β-anomers of fructose 6-phosphate, although the α-anomer is consumed at least 20-fold faster than the β-anomer. In addition, the enzyme is capable of catalyzing the interconversion (anomerization) of the α- and β-anomers. With the α-anomer of glucose 6-phosphate, the anomerization reaction is approximately twice as fast as the isomerization reaction.

These observations can be rationalized by a mechanism in which the enzyme can bind either the α- or β-anomers and catalyze their interconversion via the acyclic species (Scheme I).

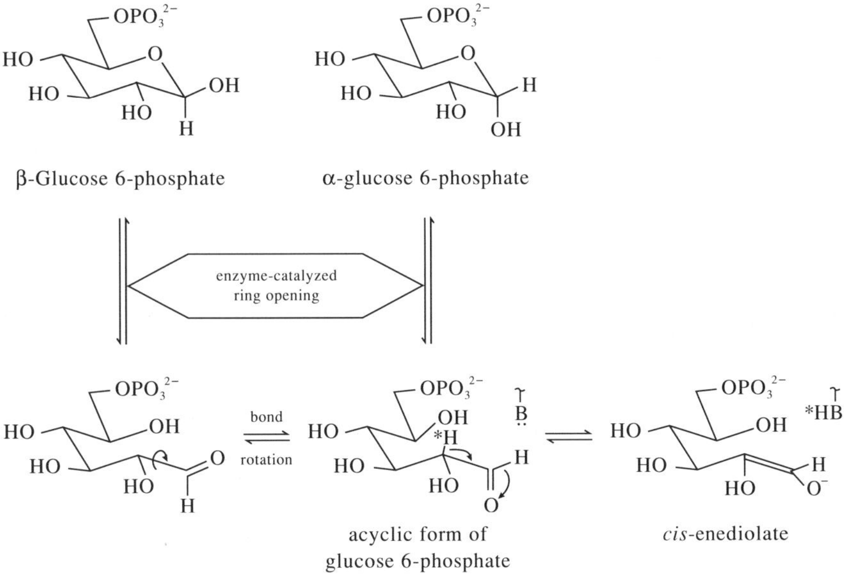

β-Glucose 6-phosphate α-glucose 6-phosphate

enzyme-catalyzed ring opening

bond rotation

acyclic form of glucose 6-phosphate

cis-enediolate

SCHEME I

Rotation about the bond axis between C-1 and C-2 of bound glucose 6-phosphate would allow reversible addition of the C-5 hydroxyl to either face of the aldehydic carbonyl function. Preference for use of the α-anomer over the β-anomer may be a consequence of the cisoid arrangement of the C-1 and C-2 hydroxyls in both the chair and boat conformations of the sugar, an arrangement well suited for formation of a *cis*-enediol intermediate [Eq. (19)]:

$$\rightleftharpoons \quad \rightleftharpoons \textit{cis}\text{-enediolate} \qquad (19)$$

Consistent with this explanation for anomeric specificity, mannose-phosphate isomerase (the substrate for which is the C-2 epimer of glucose 6-phosphate) uses the β-anomer of substrate [Eq. (20)]:

$$\rightleftharpoons \quad \rightleftharpoons \textit{cis}\text{-enediolate} \qquad (20)$$

It should be emphasized that the anomerization reaction is not a necessary side reaction of all of the isomerases. For example, mannose-phosphate isomerase has no detectable anomerase activity. Perhaps the anomerase activity of glucose-phosphate isomerase is physiologically significant, preventing the slow non-enzymic rate of interconversion of the anomers from limiting the isomerization reaction under physiological conditions. The relationship between active site functional groups involved in the enediol proton transfer reaction and in catalyzed ring opening of glucose 6-phosphate is a topic of considerable interest (*101*).

3. *Glyoxalase I*

Glyoxalase I (lactoylglutathione lyase) has several stereochemical features in common with the aldose–ketose isomerases. This zinc metalloenzyme catalyzes the GSH-dependent conversion of a variety of aromatic and aliphatic α-ketoaldehydes to their corresponding α-hydroxythio esters having the (*R*)-configuration at C-2 (*102, 103*) [Eq. (21)]:

$$RC(=O)CHO + GSH \rightleftharpoons R\text{-}C^{(R)}(OH)(H)\text{-}C(=O)SG \qquad (21)$$

A putative physiological role for this enzyme is to function together with the thioester hydrolase glyoxalase II (hydroxyacylglutathione hydrolase) to chemically remove from cells cytotoxic methylglyoxal (R = CH_3) as D-lactate (*104*).

Glyoxalase I operates on an equilibrium mixture of diastereomeric thiohemiacetals [(1*R*)-**9** and (1*S*)-**9**] formed by preequilibrium addition of GSH to the *re* or *si* faces of the aldehydic carbonyl group of the α-ketoaldehyde [Eq. (22)]:

$$\underset{(1R)\text{-}\mathbf{9}}{\mathrm{RC(=O)}\text{—}\overset{(R)}{\mathrm{C}}\mathrm{H(SG)(OH)}} \rightleftharpoons \mathrm{RC(=O)CHO + GSH} \rightleftharpoons \underset{(1S)\text{-}\mathbf{9}}{\mathrm{RC(=O)}\text{—}\overset{(S)}{\mathrm{C}}\mathrm{H(OH)(SG)}} \qquad (22)$$

Vander Jagt and co-workers were the first to present clear evidence that either one or both of the thiohemiacetals are the preferred substrates over the free aldehyde and GSH, on the basis of stopped-flow kinetic measurements (*103*). Subsequent pulse–chase isotope trapping experiments showed that both (1*R*)-**9** and (1*S*)-**9** (R = C_6H_5) are directly used by the enzyme with similar k_{cat}/K_m efficiencies (*105*). In analogy with glucose-phosphate isomerase, the ability of glyoxalase I to use both diastereomers directly as substrates provides a mechanism for preventing the nonenzymic rate of interconversion of the diastereomers from limiting the rate of removal of methylglyoxal from cells (*106*).

An enediol proton transfer mechanism for glyoxalase I, mediated by a single active site base, is now well supported on the basis of solvent isotope exchange studies (*107*), intermediate trapping experiments (*108*), and partitioning studies using fluoromethylglyoxal (*109, 110*) [Eq. (23)]:

$$\mathrm{CH_3C(=O)\text{—}C(OH)(H)\text{—}SG}\ \cdots\ \ddot{\mathrm{B}} \rightleftharpoons \mathrm{CH_3C(O^-)=C(OH)\text{—}SG}\ \ \mathrm{H^+B} \rightleftharpoons \mathrm{CH_3C(OH)(H)\text{—}C(=O)\text{—}SG}\ \cdots\ \ddot{\mathrm{B}} \qquad (23)$$

Kozarich and Chari suggest that the enzymic reaction most likely involves a *cis*-enediol intermediate, on the basis of two observations. First, the enzyme has the surprising ability to catalyze the stereospecific conversion of the thiohemiacetal (**11**), formed from (glutathiomethyl)glyoxal (**10**) and β-mercaptoethanol, to the thioester **12,** established to have the (*S*)-configuration at C-2 on the basis of chemical degradation to L-lactate (*111*) [Eq. (24)]:

$$\underset{\mathbf{10}}{\mathrm{GSCH_2C(=O)\text{—}CH(=O)}} \overset{\mathrm{RSH}}{\rightleftharpoons} \underset{\mathbf{11}}{\mathrm{GSCH_2C(=O)\text{—}CH(OH)\text{—}SR}} \xrightarrow{\mathrm{GXI}} \underset{\mathbf{12}}{\mathrm{GSCH_2\text{—}\overset{(S)}{C}H(OH)\text{—}C(=O)\text{—}SR}} \qquad (24)$$

Second, glyoxalase I catalyzes stereospecific exchange of the *pro*-(*S*) hydrogen at C-3 of glutathiohydroxyacetone (**13**) with solvent ^2H_2O (*105*).

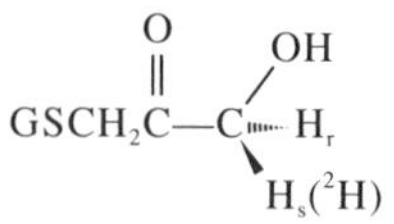

$GSCH_2C(=O)-C(OH)H_rH_s(^2H)$

13

These observations can be rationalized by a model of the binary enzyme·**11** complex in which the oxygen atoms at C-2 and C-3 are constrained to be cisoid, perhaps owing to the proximity of the catalytically essential site zinc ion in the active site (*113, 114*):

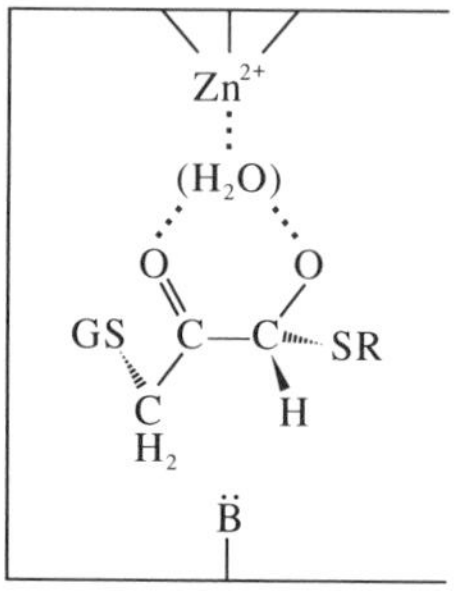

Zn^{2+} (H_2O) O O GS C—C SR C H H_2 B̈

Thus, the formation of **12** is consistent with a catalytic base above the *re* face of the C-2 carbonyl group. If **13** binds to the active site in a fashion analogous to that of **12,** and the *same* catalytic base is involved in catalyzed solvent exchange, the intermediate enediol(ate) *must* have the cis configuration. Extrapolating to the normal substrate for the enzyme, an analogous binary complex can be envisioned:

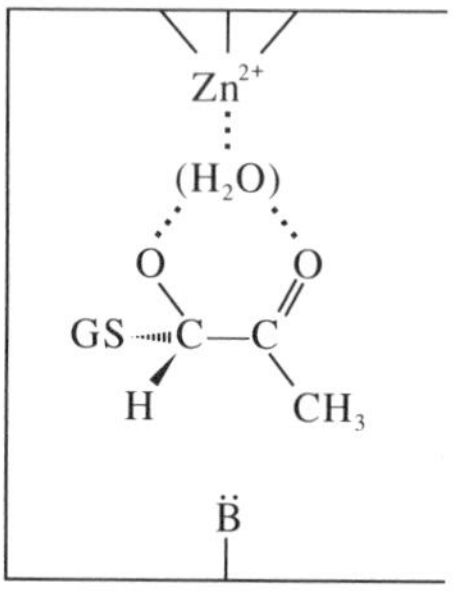

Zn^{2+} (H_2O) O O GS C—C H CH_3 B̈

Thus, glyoxalase I may well conform to the stereochemical generalization first established for the aldose–ketose isomerases.

The mechanism by which the enzyme is able to use both (1*R*)-**9** and (1*S*)-**9** as substrates is unclear. Perhaps the enzyme catalyzes interconversion of the diastereomers prior to the conversion of bound (1*S*)-**9** to product, in a fashion analogous to glucose-phosphate isomerase (Scheme I).

B. 1,3-ALLYLIC PROTON TRANSFER REACTIONS

Isomerases involving 1,3-allylic proton transfers [Eq. (15)] include Δ^5-3-ketosteroid isomerase and the aconitate and vinylacetyl-CoA isomerases. Early isotope exchange studies on these enzymes demonstrated that hydrogen migration is a suprafacial process, apparently arising from a single active site base. The reaction catalyzed by Δ^5-3-ketosteroid isomerase (*Pseudomonas testosteroni*) in ^2H_2O solvent involves substantial intramolecular hydrogen transfer (*115–117*) between the C-4β and C-6β positions of substrate (*118*), suggesting that a catalytic base partially shielded from solvent is positioned above the β face of bound steroid [Eq. (25)]:

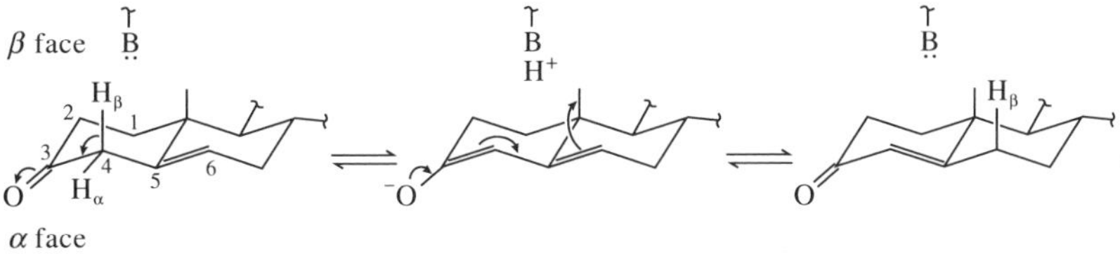

(25)

The reaction catalyzed by aconitate isomerase (*Pseudomonas putida*) in ^2H_2O solvent involves 4% tritium transfer between the *pro*-(*S*) positions of *cis*- and *trans*-aconitate (*119*) [Eq. (26)]:

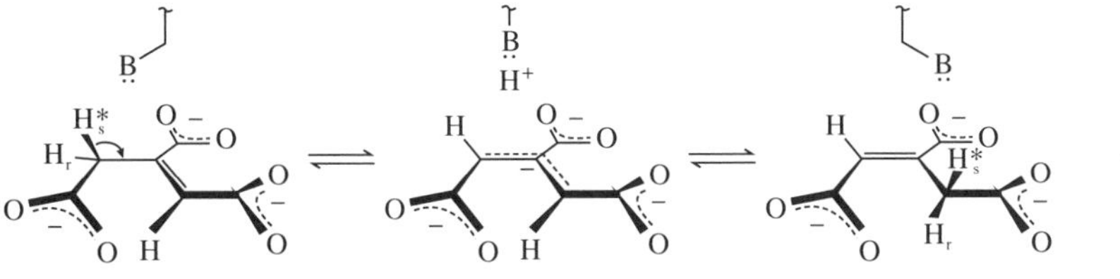

(26)

Vinylacetyl-CoA isomerase (ox liver) catalyzes the interconversion of β-methylvinylacetyl-CoA and β-methylcrotonyl-CoA without exchange with solvent tritium, consistent with intramolecular proton transfer (*120, 121*) [Eq. (27)]:

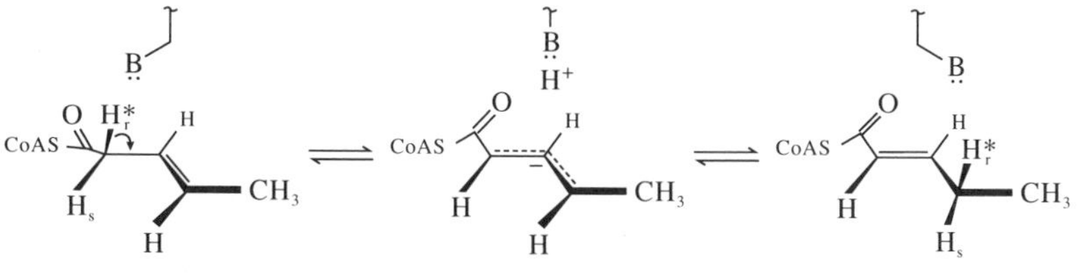

(27)

Of the three isomerases, Δ^5-3-ketosteroid isomerase (*P. testosteroni*) has received the greatest experimental attention in recent years, revealing some intriguing stereochemical features of this enzyme (*122*). This isomerase has been demonstrated to use a variety of Δ^4- and Δ^5-3-ketosteroids as substrates, for example, 5-androstene-3,17-dione (**14**) to form 4-androstene-3,17-dione (**15**):

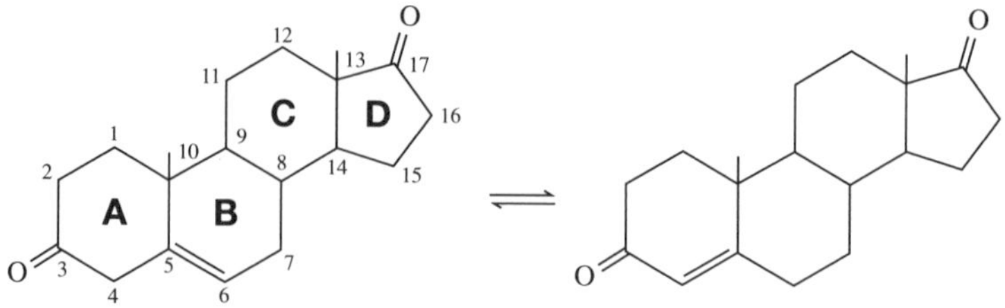

(28)

14 **15**

The nature of the catalytic base involved in the intramolecular proton transfer between the C-4β and C-6β positions of substrate has not been unequivocally identified. However, the carboxyl group of Asp-38 is a reasonable candidate. Photodecarboxylation of this residue, in the presence of 19-nortestosterone, to give Ala-38 results in complete loss of enzyme activity (*123–125*). Moreover, mutagenesis of Asp-38 to Asn-38 leads to a $10^{5.6}$-fold decrease in k_{cat} (*126*). Kuliopulos *et al.* have formulated a tentative model of the productive enzyme–substrate complex in which Asp-38 is proposed to be the catalytic base above the β face of bound steroid (*127*) (Fig. 5). Their model building studies are based on the finding that the spin-labeled substrate analog, spiro[doxyl-2,3′-5′α-androstan]-17′β-ol (**16**), binds at the active site of the isomerase.

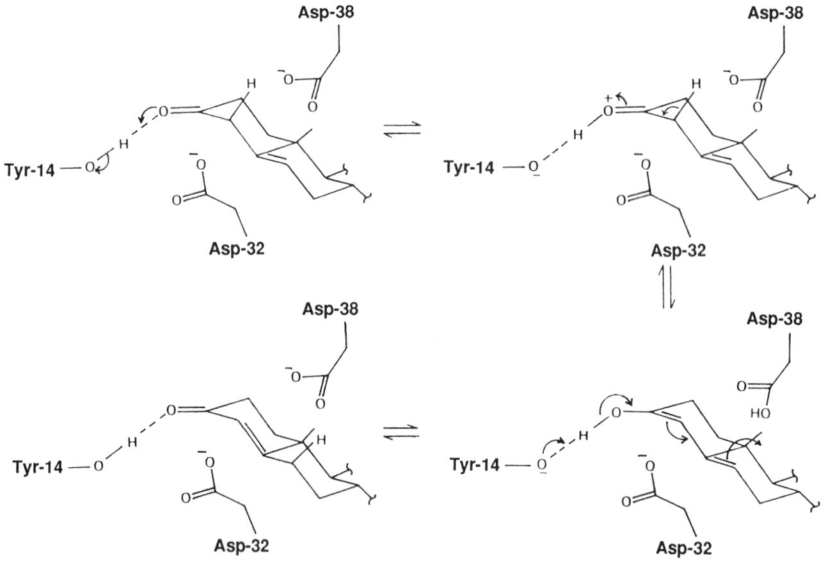

FIG. 5. Proposed mechanism for Δ^5-3-ketosteroid isomerase consistent with NMR, X-ray, and stereochemical data, as well as with the results of site-directed mutagenesis (*126, 127*).

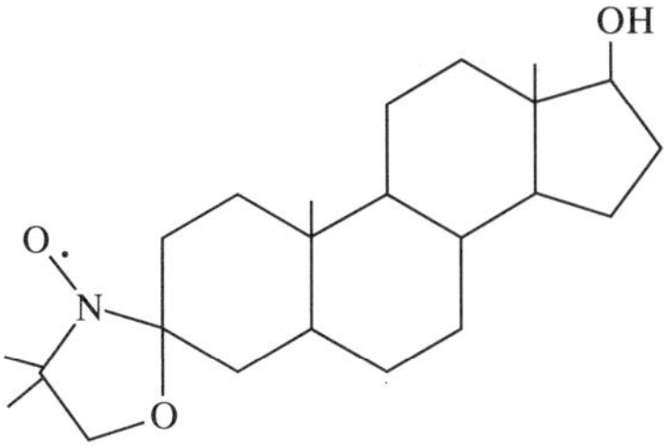

OH

O
N
O

16

Distances between the nitroxide of bound steroid and several assigned protons in or near the active site could be determined on the basis of the paramagnetic effects of the nitroxide on the relaxation rates of the active site protons. These distances could then be used to dock the structure of the spin-labeled steroid into a partially refined 2.5-Å resolution X-ray structure of the isomerase (*128*). However, in order to rationalize the chemical evidence that Asp-38 is involved in the catalytic mechanism, it was necessary to assume that steroid substrates and the spin-labeled steroid (**16**) are bound not only with reversal of the C-3 and C-17 positions, but also with reversal of the planes of the steroid ring systems ("upsidedown" binding), positioning Asp-38 above the C-4β and C-6β protons.

That steroids may bind to the isomerase in more than one orientation was first demonstrated by Pollack and co-workers when they showed that both 3β- and 17β-spirooxiranyl steroids covalently modify Asp-38 (*129, 130*). This suggests that steroids can bind to the active site with either the A or the D rings in close proximity to Asp-38. An intriguing observation made by these workers is that active site modification involves nucleophilic attack of the carboxyl group of Asp-38 at either the methylene or spiro carbons of the oxirane from the α faces of both the 3β- and 17β-oxiranyl steroids (*129–132*). This was demonstrated in the case of the 3β-oxiranyl steroid (**17**) by using ^{18}O-labeled steroid to inactivate the isomerase. Tryptic digestion of the reaction mixture resulted in two steroid-labeled peptides (**18a** and **18b**) that, upon base hydrolysis, gave two diastereomeric steroid molecules containing ^{18}O (**19a** and **19b**) (Scheme II).

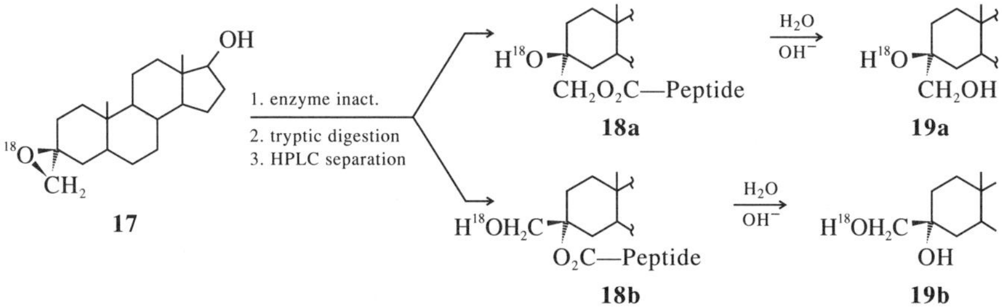

OH
^{18}O
CH_2
17

1. enzyme inact.
2. tryptic digestion
3. HPLC separation

$H^{18}O$
CH_2O_2C—Peptide
18a

H_2O
OH^-

$H^{18}O$
CH_2OH
19a

$H^{18}OH_2C$
O_2C—Peptide
18b

H_2O
OH^-

$H^{18}OH_2C$
OH
19b

SCHEME II

The distribution of ^{18}O between the hydroxyl and hydroxymethylene functions of **19a** and **19b,** determined by the fragmentation pattern of these two species during mass spectrometry, could only result from addition of the carboxyl of the enzyme to the α face of the steroid. Had the carboxyl added to the spiro carbon from the β face, the ^{18}O would have been located in the hydroxymethylene function of **19a.** Clearly, if the binding of the oxiranyl steroids is analogous to that of steroid substrates, Asp-38 could not be involved in the normal intramolecular proton transfer associated with the isomerization reaction. However, as recently emphasized by Pollack *et al.*, the oxiranyl steroids could conceivably bind to the active site in an upsidedown orientation in comparison to that of the substrate steroids (*133*). In this case, Asp-38 would still be the most probable base involved in the intramolecular proton transfer reaction. Perhaps upsidedown binding accounts for the report that the C-4α hydrogen of 5-androstene-3,17-dione undergoes slow labilization in the presence of the isomerase (*134, 135*).

V. Enzyme-Catalyzed Aldol- and Claisen-Type Condensations; β-Keto Acid Decarboxylases: Stereochemistry versus Stability of Reaction Intermediates

Enzymes involved in aldol- and Claisen-type condensations and β-keto acid decarboxylases catalyze electrophilic substitution reactions at the α-carbon of carbonyl substrates involving either retention or inversion of configuration at C_α [Eq. (29)]:

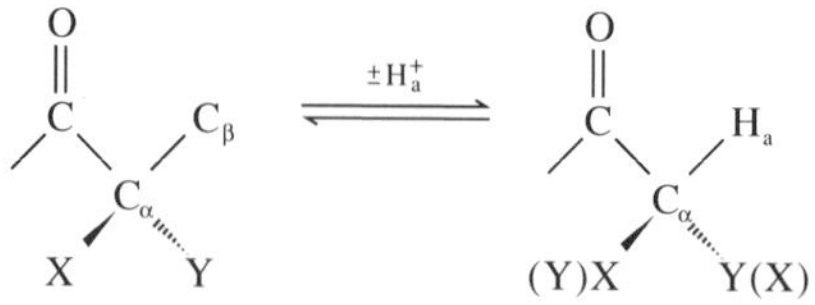

$$\underset{X \quad Y}{(O{=}C)(C_\beta)C_\alpha} \xrightleftharpoons{\pm H_a^+} \underset{(Y)X \quad Y(X)}{(O{=}C)(H_a)C_\alpha} \tag{29}$$

The following empirical generalization applies to many, although perhaps not all, of these enzymes: In cases where catalysis involves the intermediate formation of a Schiff base or a catalytically essential divalent metal ion appears to play an electrophilic role in catalysis, retention of configuration is generally observed; in cases in which neither Schiff bases nor metal ions are directly involved in catalysis, inversion of configuration is usually observed. This generalization, when viewed within the context of model chemistry, implies a relationship between the lifetime of the carbanion intermediate generated along the reaction pathway and reaction stereochemistry, as discussed at the end of this section.

A. Stereochemical Analysis Using Chiral [^1H,^2H,^3H]Methyl Groups

Many of the enzymic reactions that form the basis of the stereochemical generalizations discussed in this and subsequent sections of this chapter involve the interconversion of methyl and methylene functions [Eq. (29), where X = H_b and Y = H_c]. An essential achievement over the last 20 years has been the development of methods for analyzing the stereochemistry of such interconversions on the basis of (1) synthesis of substrate molecules containing chiral [^1H,^2H,^3H]methyl groups of known configuration, and (2) methods for analyzing the configuration of chiral [^1H,^2H,^3H]methyl groups in product molecules (*136–138*).

From an historical perspective, the laboratories of Cornforth (*3*) and Arigoni (*4*) were the first to synthesize chiral [^1H,^2H,^3H]acetate that was converted to chiral [^1H,^2H,^3H]acetyl-CoA [acetate kinase/phosphotransacetylase (phosphate acetyltransferase)] for the purpose of determining the stereochemistry of the malate synthase reaction. The synthesis of chiral acetate originally put forth by Cornforth is still noteworthy for its chemical elegance (Fig. 6). The successful

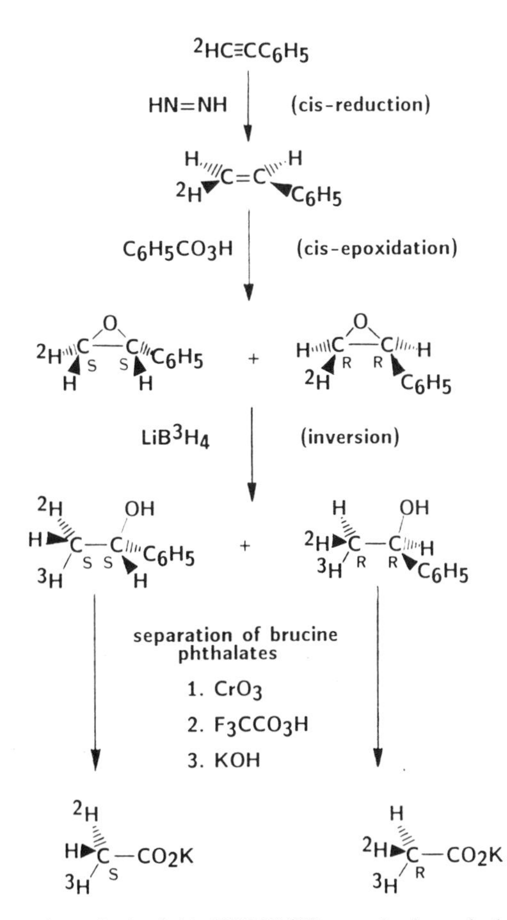

FIG. 6. Asymmetric synthesis of chiral [^1H,^2H,^3H]acetate by the method of Cornforth *et al.* (*3*).

resolution of the stereochemistry of the malate synthase reaction (inversion) depended on significant discrimination against the loss of deuterium versus protium from chiral [^{1}H,^{2}H,^{3}H]acetyl-CoA [k_H/k_D = 3.7–3.8 (*139*)] during the formation of (*S*)-malate. Thus, pure (*R*-methyl)-[*methyl*-^{1}H,^{2}H,^{3}H]acetyl-CoA results in the formation of excess (2*S*,3*S*)-[3-^{2}H,^{3}H]malate versus (2*S*,3*R*)-[3-^{1}H,^{3}H]malate, on the basis of the observation that incubation of the isolated (*S*)-malate with fumarase (fumarate hydratase) (catalyzing the anti-elimination of the elements of water) results in the retention of approximately 79% of the tritium in the fumarate and loss of the remaining tritium to water (*139, 140*) (Scheme III).

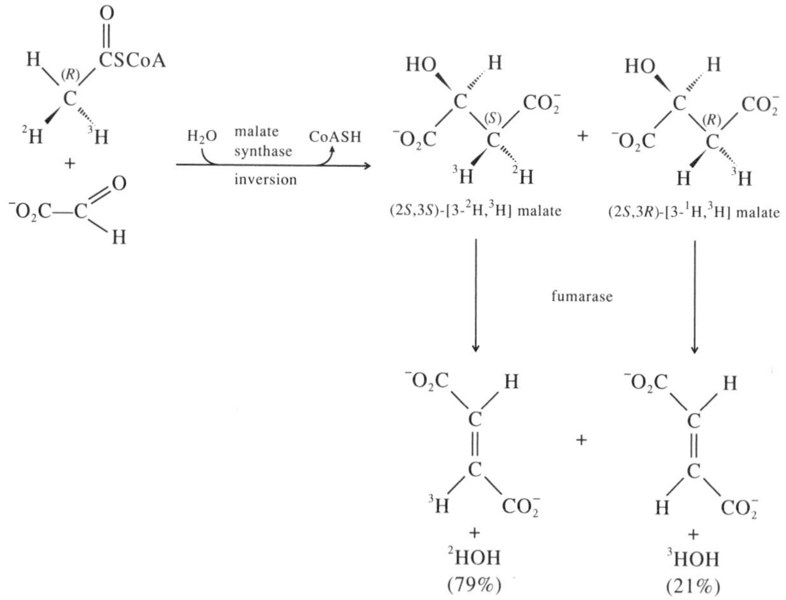

SCHEME III

Conversely, using pure (*S*-methyl)-[*methyl*-^{1}H,^{2}H,^{3}H]acetyl-CoA as substrate results in about 21% retention of tritium in the fumarate and loss of the remaining tritium to water.

Since these early studies, numerous additional procedures have been developed for the synthesis of chiral labeled acetic acid and other chiral labeled molecules by chemical (*141–151*) and enzymic methods (*152–155*). The malate synthase/fumarase system described above continues to be the most widely used method for the stereochemical analysis of chiral labeled acetic acid resulting from the degradation of product molecules containing chiral methyl groups.

B. Aldol Condensations

The general class of enzymes catalyzing aldol or retroaldol condensation reactions are the aldolases (Table VI) (*156–161*). In principle, there are four possible stereochemical routes to product depending on whether C–C bond formation involves retention or inversion of configuration at the methyl or methylene carbon atom α to the ketonic carbonyl [Eq. (30)]:

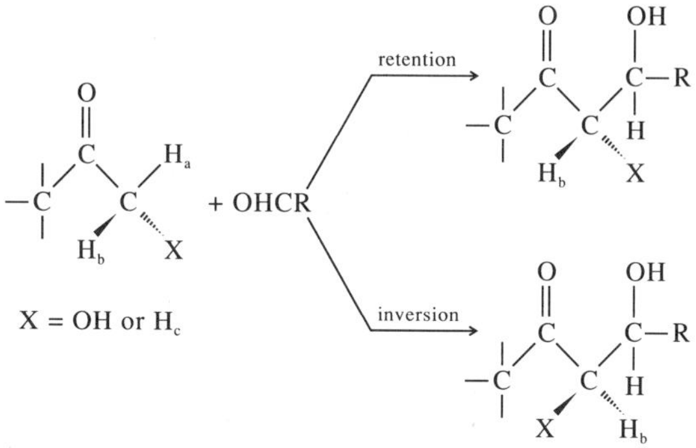

+ OHCR — retention → ; inversion →

$X = OH$ or H_c

(30)

and whether C–C bond formation takes place at the *re* or the *si* face of the aldehydic carbonyl group [Eq. (31)]:

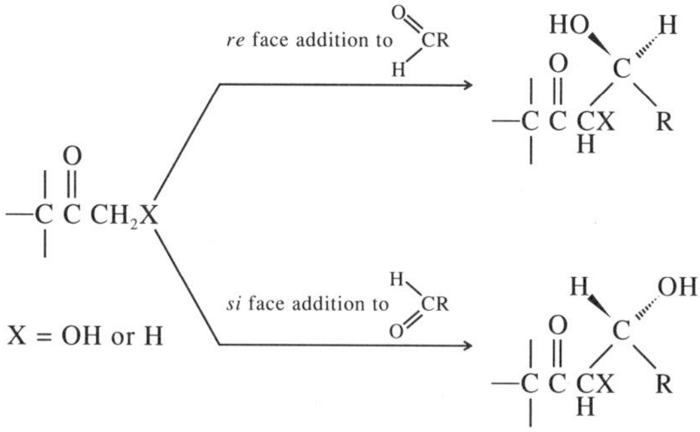

re face addition to O=CR(H) →

si face addition to H(O=)CR →

$X = OH$ or H

(31)

However, only two of the four possible stereochemical pathways have been observed among the enzymes so far examined: retention of configuration involving bond formation to either the *re* or the *si* face of the aldehydic substrate (Table VI). Mechanistically, the aldolases are divided into two general classes: those involving covalent catalysis owing to the intermediate formation of a Schiff base within the active site (Class I) and those that are metal activated (Class II).

1. *Class I Adolases*

The higher eukaryotes contain aldolases with mechanisms of action involving the formation of a ketimine intermediate between substrate and an essential ac-

TABLE VI

ENZYME-CATALYZED ALDOL CONDENSATIONS INVOLVING RETENTION OF CONFIGURATION

Enzyme	Source (class)	Method of stereochemical analysis	Nucleophilic addition to	Ref.
D-Fructose-1,6-bisphosphate aldolase (EC 4.1.2.13)	Muscle (I), yeast (II)	[P_iOH_2C–C(=O)–C(H_s)(*H_r)(OH) + H–C(=O)–$C_{(R)}$(H)(HO)–CH_2OP_i ⇌ P_iOH_2C–C(=O)–$C_{(S)}$(*H)(OH)–$C_{(R)}$(H)(OH)–$C_{(R)}$(OH)(H)–CH_2OP_i] [a]	*si* face	*156*
L-Rhamnulose-1-phosphate aldolase (EC 4.1.2.19)	*E. coli* (II)	[P_iOH_2C–C(=O)–C(H_r)(HO)(H_s*) + H–C(=O)–$C_{(S)}$(H)(OH)–CH_3 ⇌ P_iOH_2C–C(=O)–$C_{(R)}$(HO)(H*)–$C_{(S)}$(H)(OH)–$C_{(S)}$(H)(OH)–CH_3] [b]	*re* face	*157*
2-Keto-3-deoxy-6-phospho-D-gluconate aldolase (EC 4.1.2.14)	*P. putida* (I), *Pseudomonas saccharophila* (I)	[^-O_2C–C(=O)–C(H_a)(H_c)(H_b) + H–C(=O)–$C_{(R)}$(HO)(H)–CH_2OP_i ⇌ ^-O_2C–C(=O)–C(H_c)(H_b)–$C_{(S)}$(H)(OH)–$C_{(R)}$(OH)(H)–CH_2OP_i] [c,d]	*si* face	*158*, *159*

Enzyme	Source	Reaction		Stereochemistry	Ref.
2-Keto-3-deoxy-6-phospho-D-galactonate aldolase (EC 4.1.2.21)	*P. saccharophila* (?)	$^{-}O_2C{-}C(=O){-}CH_aH_bH_c$ + $O{=}CH{-}C^{(R)}H(OH){-}CH_2OP_i$ ⇌ $^{-}O_2C{-}C(=O){-}CH_bH_c{-}C^{(R)}H(OH){-}C^{(R)}H(OH){-}CH_2OP_i$	c,d	*re* face	*159*
2-Keto-4-hydroxyglutarate aldolase (EC 4.1.3.16)	Liver (I)	$^{-}O_2C{-}C(=O){-}CH_aH_bH_c$ + $O{=}CH{-}CO_2^-$ ⇌ $^{-}O_2C{-}C(=O){-}CH_bH_c{-}C^{(R+S)}H(OH){-}CO_2^-$	c,e	*si* + *re* faces	*160*
2-Keto-3-deoxyhexarate aldolase	*E. coli* (II)	$^{-}O_2C{-}C(=O){-}CH_aH_bH_c$ + $O{=}CH{-}CO_2^-$ ⇌ $^{-}O_2C{-}C(=O){-}CH_bH_c{-}C^{(R+S)}H(OH){-}CO_2^-$	c,e	*si* + *re* faces	*160*

(*continued*)

TABLE VI (*continued*)

Enzyme	Source (class)	Method of stereochemical analysis	Nucleophilic addition to	Ref.
myo-Inositol-1-phosphate synthase (EC 5.5.1.4)	Testis (?), pollen (?)	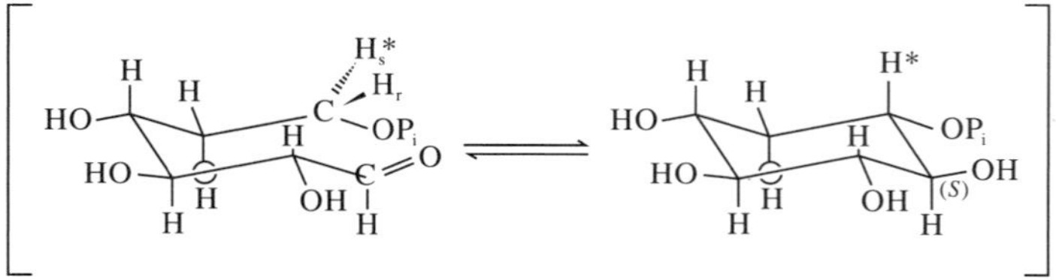 [f]	*re* face	*161*

[a]Retention of configuration is based on the observation that the enzyme stereospecifically catalyzes solvent tritium exchange with the *pro*-(*S*) proton at C-3 of dihydroxyacetone phosphate, the same position to which (*R*)-glyceraldehyde 3-phosphate must add during the condensation reaction in order to generate the (*S*)-configuration at C-3 of D-fructose 1,6-bisphosphate (*156*).

[b]Retention of configuration is based on the observation that the enzyme stereospecifically catalyzes solvent tritium exchange with the *pro*-(*R*) proton at C-3 of dihydroxyacetone phosphate, the same position to which (*S*)-2-hydroxypropanal must add during the condensation reaction in order to generate the (*R*)-configuration at C-3 of L-rhamnulose 1-phosphate (*157*).

[c]$H_a = {}^1H$; when $H_b = {}^2H$, $H_c = {}^3H$; when $H_b = {}^3H$, $H_c = {}^2H$.

[d]Configurational analysis of the product involved chemical conversion [(a) $NaBH_4$, (b) $NaIO_4$, (c) Ag_2O/NH_4OH] to [3-^{3}H,^{2}H]malate, followed by chiral analysis using fumarase.

[e]Configurational analysis of the product involved oxidative decarboxylation (H_2O_2) to [3-^{3}H,^{2}H]malate, followed by chiral analysis using fumarase.

[f]Retention of configuration is based on the observations that ring closure of (6*R*)-D-[1-^{14}C,6-^{3}H]glucose 6-phosphate results in expulsion of tritium to solvent whereas ring closure of (6*S*)-D-[1-^{14}C,6-^{3}H]glucose 6-phosphate results in retention of tritium in the product (*161*).

tive site lysine residue (*162–164*). The fructose-1,6-bisphosphate aldolase from rabbit muscle is one of the best studied members of this class. The mechanism of Fig. 7 is consistent with the stereochemical and mechanistic information currently available on this enzyme. In the direction of cleavage, the enzyme uses the β-anomer and acyclic forms of fructose 1,6-bisphosphate (*165, 166*). The early demonstration that the enzyme catalyzes stereospecific exchange with solvent of the *pro*-(*S*) proton at C-3 of dihydroxyacetone phosphate at a rate at least as fast as the cleavage reaction suggests that formation of the enamine precedes the condensation reaction with glyceraldehyde 3-phosphate (*167, 168*). The C–H bond to be cleaved during formation of the enamine is presumably perpendicular to the plane of the ketimine in order to ensure maximal orbital overlap in the transition state. This is in accordance with model systems demonstrating stereoselective loss of the axial α-protons from the iminium ions derived from β-hydroxy- and β-acetoxy-*trans*-decalones (*169*).

The observation of net retention of configuration by the muscle enzyme, as well as the other aldolases, suggests that a single active site base could alternatively function in a general base and general acid capacity in the exchange and condensation reactions, respectively (Fig. 7) (*16*). However, the stereochemistry does not absolutely require a single-base mechanism. For example, Hupe and co-workers propose a two-base mechanism in which the phosphate at C-1 of the

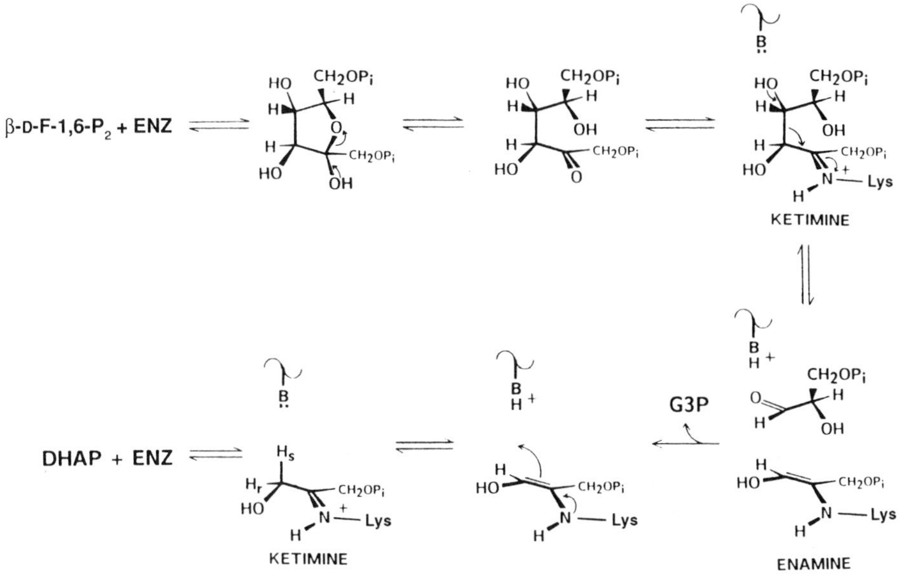

β-D-F-1,6-P$_2$ + ENZ

HO CH$_2$OP$_i$ H H O H CH$_2$OP$_i$ HO OH

HO CH$_2$OP$_i$ H H OH H CH$_2$OP$_i$ HO O

B HO CH$_2$OP$_i$ H H OH H CH$_2$OP$_i$ HO N — Lys H

KETIMINE

B H+ CH$_2$OP$_i$ O H H OH

G3P

H HO CH$_2$OP$_i$ N — Lys H

ENAMINE

B H+ H HO CH$_2$OP$_i$ N — Lys H

B H$_s$ H$_r$ HO CH$_2$OP$_i$ N — Lys H

KETIMINE

DHAP + ENZ

FIG. 7. Mechanism of fructose-1,6-bisphosphate aldolase (rabbit muscle) consistent with stereochemical data.

ketimine catalyzes the exchange reaction in a cyclic transition state and an active site group (BH) functions as a general acid in the condensation reaction (*170*) [Eq. (32)]. Their model was formulated, in part,

(32)

on the basis of the finding that dihydroxyacetone *O*-sulfate forms a ketimine with the rabbit muscle enzyme but does not undergo subsequent α-proton abstraction to form the enamine (*171*). The lower pK_a of the oxygens on the sulfate could account for the inability to catalyze proton abstraction at C-1. However, their model appears less likely in view of the earlier observation that the muscle enzyme also catalyzes stereospecific exchange of the *pro*-(*S*) hydrogen at C-1 of dihydroxyacetone phosphate, D-fructose 1,6-bisphosphate, and D-fructose 1-phosphate, although at somewhat slower rates than the *pro*-(*S*) hydrogen at C-3 of dihydroxyacetone phosphate (*172*). This implies that the C-1–H_s bond is orthogonal to the plane of the ketimine (at least some of the time), an arrangement that precludes the cyclic transition state of Eq. (32). Possible candidates for the base(s) involved in the exchange and/or condensation reactions include Asp-33, Glu-189, and Glu-187, all located near the Schiff base lysine (229) in the three-dimensional X-ray structure (2.7-Å resolution) of the rabbit muscle enzyme (*173*). Corresponding X-ray measurements on 2-keto-3-deoxy-6-phospho-D-gluconate aldolase from *P. putida* (2.8-Å resolution) show Glu-149, Glu-131, and His-63 near the Schiff base lysine (Lys-144) (*174*). Early affinity labeling experiments on this enzyme using bromopyruvate suggested Glu-56 as the most likely base involved in the exchange reaction (*175, 176*). However, Glu-56 is located 25 Å from Lys-144, too distant for direct participation in the catalytic mechanism.

The stereochemistry of Schiff base formation between fructose 1,6-bisphosphate and the aldolase from liver has also been addressed (*177*). Suggestive evidence for the intermediate formation of a (2*R*)-carbinolamine is based on the observation that BH_4^- reduction of substrate on the enzyme followed by acid hydrolysis of the protein gives exclusively glucitollysine and not mannitollysine. This indicates that the *re* face of the ketimine is exposed to solvent, and it implies that OH left from the same direction in other to form the ketimine. On this basis, the ε-amino of the lysine must add to the *si* face of the substrate carbonyl [Eq. (33)]:

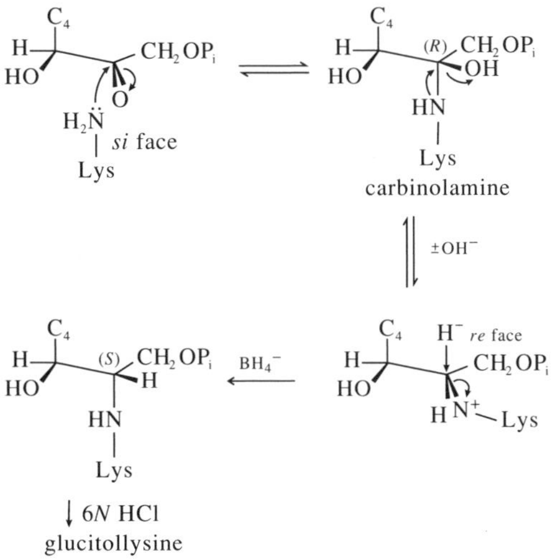

C_4 H HO CH_2OP_i O $H_2\ddot{N}$ *si* face Lys ⇌ C_4 H HO (*R*) CH_2OP_i OH HN Lys carbinolamine

$\pm OH^-$

C_4 H HO (*S*) CH_2OP_i H HN Lys ← BH_4^- ← C_4 H HO H^- *re* face CH_2OP_i HN^+—Lys (33)

↓ 6*N* HCl
glucitollysine

The analogous reduction experiment carried out with pyruvate plus 2-keto-3-deoxy-6-phospho-D-gluconate aldolase results in almost equal reduction at both diastereotopic faces of the ketimine (*178*). Apparently, solvent accessibility to the ketimine is not uniformly restricted among all aldolases.

2. *Class II Aldolases*

Class II aldolases, which are generally found in bacteria and yeast, require a tightly bound divalent metal ion for activity. Figure 8 depicts a stereomechanistic model for fructose-1,6-bisphosphate aldolase from yeast, the best studied member of the class II aldolases. The finding that the apoenzyme has anomerase activity indicates that the enzyme can bind both anomers of substrate and catalyze their interconversion (*166*), perhaps by a mechanism analogous to that envisioned for phosphoglucose isomerase (Scheme I). As emphasized previously, net retention of configuration is consistent with (but does not prove) a single-base mechanism.

That an active site Zn^{2+} ion plays an electrophilic role in catalysis is supported by several observations. Each subunit contains a single tightly bound Zn^{2+} ion that is essential for activity. Rutter and co-workers first suggested that the Zn^{2+} functions catalytically by coordinating to and polarizing the carbonyl function of bound substrate (*179, 180*). However, recent NMR measurements using the Mn^{2+}–holoenzyme indicate that bound dihydroxyacetone phosphate is too distant from the metal ion for direct coordination (*181*). On the other hand, the distances are consistent with an outer sphere complex that could permit transmission of the electronic effects of the metal ion through an intervening base (AH), perhaps an imidazole. That the carbonyl functions of bound dihydroxy-

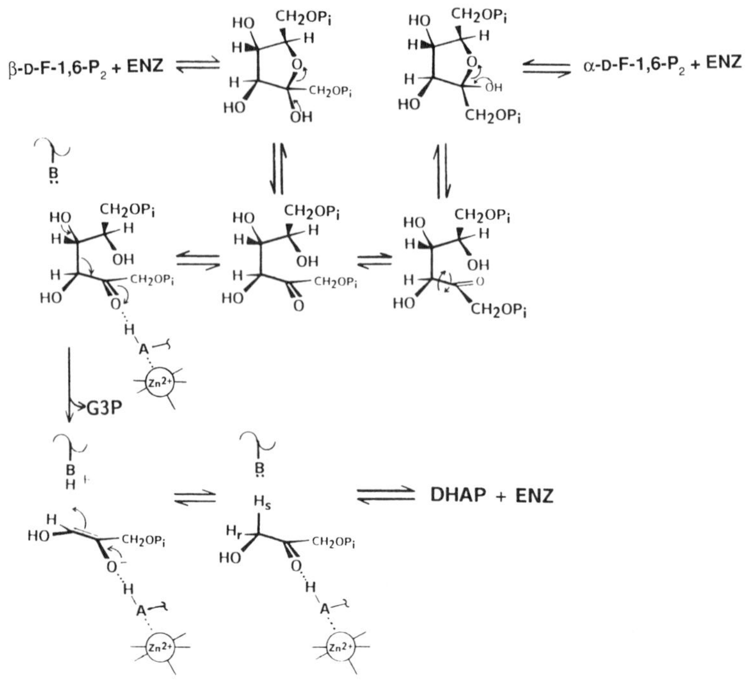

FIG. 8. Mechanism of fructose-1,6-bisphosphate aldolase (yeast) consistent with stereochemical and NMR data.

acetone phosphate and fructose 1,6-bisphosphate are polarized by the active site divalent metal ion is consistent with the results of kinetic studies using different metal-substituted yeast aldolases (*182*). Thus, the function of the active site Zn^{2+} ion appears to be mechanistically analogous to the function of the iminium ion form of the Schiff base in the active site of the class I aldolases.

The nature of the putative active site base involved in the proton exchange and condensation reactions has not been clearly established. However, Fourier transform infrared spectrophotometeric measurements suggest that the aldehydic carbonyl of bound glyceraldehyde-3-phosphate is significantly polarized by the enzyme, possibly due to the protonated form of an active site carboxyl group (*183*).

3. *myo-Inositol-1-Phosphate Synthase*

myo-Inositol-1-phosphate synthase deserves additional comment because of detailed differences between its reaction mechanism and those of the better known aldolases described above. The enzyme occurs in microorganisms, plants, and animals and requires NAD^+ as an essential cofactor. The aldol condensation reaction occurs subsequent to the NAD^+-dependent oxidation of the C-5 hydroxyl of bound D-glucose 6-phosphate; NADH-dependent reduction of the bound intermediate generates the final product (*184–187*) [Eq. (34)].

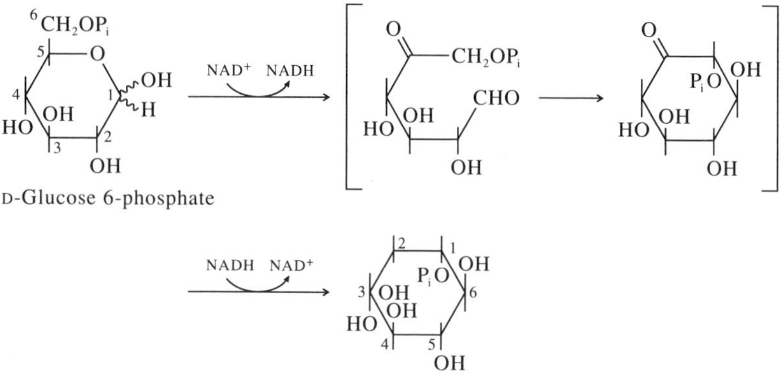

NAD^+ NADH

D-Glucose 6-phosphate

NADH NAD^+

(34)

L-*myo*-inositol 1-phosphate

Floss and co-workers established net retention of configuration at C-6 of substrate using the enzyme from both bull testis and pollen in a carefully executed double-radiolabeling experiment in which they compared the extent of tritium retention in L-*myo*-inositol 1-phosphate using either (6*R*)- or (6*S*)-D-[1-^{14}C,6-^{3}H]glucose 6-phosphate as substrates (*188*). Prior to these experiments, Byun *et al.*, using the synthase from rat testis, concluded that cyclization involved loss of the *pro*-(*S*) proton at C-6 of substrate, requiring net inversion of configuration (*189*). However, their experimental protocol involved the use of only one isomer of the labeled substrate and did not employ double-radiolabel counting to assess the extent of tritium retention in the product. Since it seems unlikely that the enzymes from bull testis and rat testis involve opposing stereochemical modes, the original report of Byun *et al.* appears problematic. As discussed by Floss and co-workers, the mechanistic classification of the synthase as a Class I or Class II aldolase is unclear (*188*).

4. *Anomeric Specificity and Metabolic Coupling*

The different anomeric specificities of the aldolases can be rationalized in terms of their different metabolic functions within cells (*190–193*). For example, fructose-1,6-bisphosphate aldolase from muscle uses the β-anomer and acyclic forms of substrate, but not the α-anomer. In fact, the α-anomer is a tight-binding inhibitor of the enzyme (K_i 1.3 μM). Since muscle tissue is primarily glycolytic, the capacity of the aldolase to use the β-anomer as substrate allows direct metabolic coupling with phosphofructokinase that produces the β-anomer as product. In contrast, the corresponding aldolase from yeast can use/produce both the α- and β-anomers of fructose 1,6-bisphosphate, owing to the anomerase activity of the enzyme. Since yeast cells carry out both glycolysis and gluconeogenesis, the yeast enzyme allows metabolic coupling to either phosphofructokinase (during glycolysis) or fructose bisphosphatase (specific for the α-anomer during gluconeogenesis). The ability of liver aldolase to use the β-anomer of fructose 1,6-bisphosphate and *not* be inhibited by the α-anomer can also be rationalized in terms of more efficient metabolic coupling during gluconeogenesis in liver tissue (*194*). Finally, the bacterial enzyme 2-keto-3-deoxy-6-phospho-D-gluconate aldolase has been demonstrated to use the acyclic form of substrate, thus allowing direct coupling to the preceding enzyme, gluconate-6-phosphate dehydratase, that produces the acyclic species as product (*195*).

C. Claisen-Type Condensations

Enzyme-catalyzed Claisen-type condensations are formally similar to enzyme-catalyzed aldol condensations, with the exception that the nucleophilic substrate is an ester or thioester [Eq. (35)]:

$$-X-\overset{\overset{O}{\|}}{C}-\underset{|}{\overset{|}{C}}-H + -\overset{\overset{O}{\|}}{C}- \rightleftharpoons -X-\overset{\overset{O}{\|}}{C}-\underset{|}{\overset{|}{C}}-\underset{|}{\overset{|}{C}}-OH \qquad (35)$$

$X = S \text{ or } O$

There are, however, clear stereomechanistic differences between these two classes of enzyme-catalyzed reactions. The Claisen-type condensations uniformly involve inversion of configuration at the α-carbon of the esteratic substrate, involving C–C bond formation at either the *re* or the *si* face of the ketonic or aldehydic substrate (Table VII) (*196–211*). Moreover, neither Schiff bases nor metal ions have been directly implicated in the catalytic mechanisms of these enzymes. Unlike the aldolases, these enzymes do not catalyze rapid enolization of the nucleophilic substrate in the absence of the second substrate. Inversion of configuration suggests that at least two catalytic groups, perhaps operating in concert, facilitate C–C bond formation. Physicochemical measurements on citrate synthase are consistent with this interpretation of inversion of configuration.

Citrate synthase catalyzes the only C–C bond-forming reaction in the citric acid cycle. As a general rule, the enzyme from plants, animals, and bacteria catalyzes the addition of acetyl-CoA to the *si* face of oxaloacetate so that the added acetyl function contributes the *pro*-(*S*) carboxymethylene arm of citrate. However, the citrate synthase from *Clostridium acidiurici* is an exception to the rule in that acetyl-CoA contributes the *pro*-(*R*) carboxymethylene arm of citrate, owing to the addition of acetyl-CoA to the *re* face of oxaloacetate. In spite of this variation, three independent research groups demonstrated that both *re*- and *si*-face-specific citrate synthases involve inversion of configuration at the methyl group of acetyl-CoA (*197, 198, 200*). For example, Eggerer *et al.* demonstrated inversion for both enzymes by employing (R-methyl)-[*methyl*-$^1H,^2H,^3H$]acetyl-CoA in the direction of synthesis (*197*) [Eq. (36)]:

$$(CoA)S\overset{O}{\overset{\|}{C}}-\underset{^3H}{C}(R)(H)(^2H) + {}^-O_2CCH_2\overset{O}{\overset{\|}{C}}CO_2^- \xrightarrow[-CoASH]{H_2O,\ \text{citrate }(si)\text{-synthase}} \quad ; \quad \xrightarrow[-CoASH]{H_2O,\ \text{citrate }(re)\text{-synthase}}$$

$$^-O_2C-CH_2-C(OH)(CO_2^-)-C(S)H(^2H)(^3H)-CO_2^- \xrightarrow[H_2O]{\text{citrate lyase}} {}^-O_2C-C(R)H(^2H)(^3H) + \text{oxaloacetate}$$

$$^-O_2C-C(S)H(^2H)(^3H)-C(OH)(CO_2^-)-CH_2-CO_2^- \xrightarrow[H_2O]{\text{citrate lyase, malate dehydrogenase (+NADH)}} {}^-O_2C-C(S)H(^2H)(^3H)-C(OH)(H)-CO_2^- + \text{acetate} \qquad (36)$$

The *si*-citrate synthase must involve inversion of configuration because treatment of the product [2-$^3H,^2H$]citrate with citrate lyase (previously established to involve inversion of configuration) gives (R)-[$^1H,^2H,^3H$]acetate. The *re*-citrate synthase must also involve inversion of configuration because treatment of the [2-$^3H,^2H$]citrate with citrate lyase and malate dehydrogenase gives ($2S,3S$)-[3-$^2H,^3H$]malate, since treatment with fumarase leads to retention of tritium in fumarate.

Inversion of configuration suggests that more than one active site group is involved in the proton transfer reactions associated with citrate formation. An active site base appears to facilitate abstraction of protons from the methyl group of bound acetyl-CoA, once the appropriate conformation of the active site is induced by binding of the second substrate, oxaloacetate. This conclusion is indirectly supported by the finding that citrate synthase is unable to catalyze the exchange of the methyl protons of acetyl-CoA with solvent, except in the presence of L-malate, a structural analog of oxaloacetate (*212*). Polarization of the ketonic carbonyl of oxaloacetate, presumably by a second active site catalytic

TABLE VII

ENZYME-CATALYZED CLAISEN-TYPE CONDENSATIONS INVOLVING INVERSION OF CONFIGURATION

Enzyme	Source	Method of stereochemical analysis		Nucleophilic addition to	Ref.
Malate synthase (EC 4.1.3.2)	Yeast	$(CoA)S\text{-}C(=O)\text{-}CH_aH_bH_c + H\text{-}C(=O)\text{-}CO_2^- \xrightarrow[H_2O]{(CoA)SH} {}^-O_2C\text{-}CH(OH)\text{-}C^{(S)}H_bH_c\text{-}CO_2^-$	*a,b*	*si* face	*3, 4, 196*
Citrate lyase (EC 4.1.3.6)	*A. aerogenes*	${}^-O_2C\text{-}C(OH)(CO_2^-)\text{-}CH_bH_c\text{-}CH_2\text{-}CO_2^- \xrightarrow{H_aOH_a} {}^-O_2C\text{-}CH_aH_bH_c + {}^-O_2C\text{-}C(=O)\text{-}CH_2\text{-}CO_2^-$	*c,d*	*si* face	*197–199*
Citrate (*si*)-synthase (EC 4.1.3.7)	Heart	$(CoA)S\text{-}C(=O)\text{-}CH_aH_bH_c + {}^-O_2C\text{-}C(=O)\text{-}CH_2\text{-}CO_2^- \xrightarrow[H_2O]{(CoA)SH} {}^-O_2C\text{-}CH_bH_c\text{-}C(OH)(CO_2^-)\text{-}CH_2\text{-}CO_2^-$	*a,e,d*	*si* face	*197, 198, 200–202*

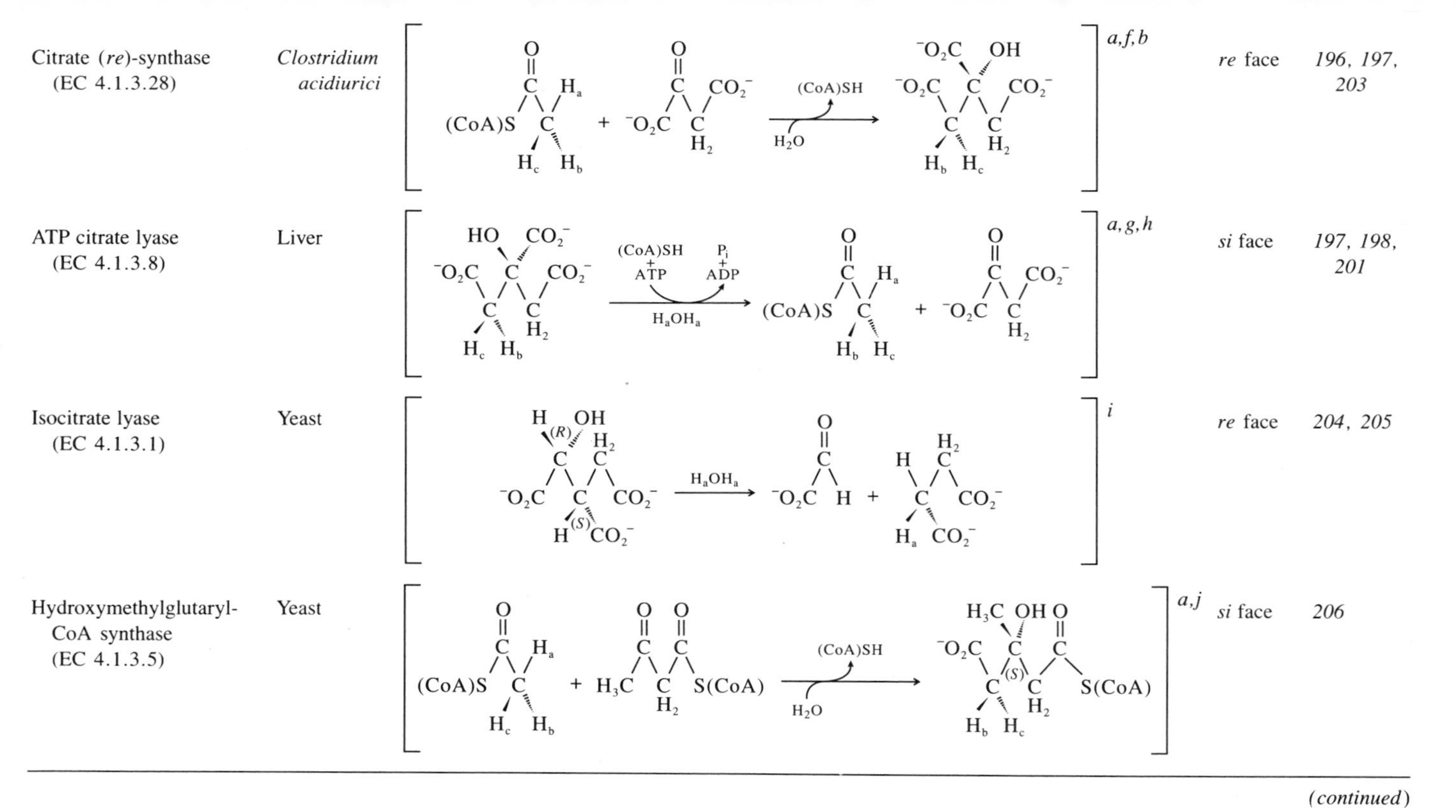

Enzyme	Source	Reaction	Notes	Face	References
Citrate (*re*)-synthase (EC 4.1.3.28)	*Clostridium acidiurici*		*a,f,b*	*re* face	*196, 197, 203*
ATP citrate lyase (EC 4.1.3.8)	Liver		*a,g,h*	*si* face	*197, 198, 201*
Isocitrate lyase (EC 4.1.3.1)	Yeast		*i*	*re* face	*204, 205*
Hydroxymethylglutaryl-CoA synthase (EC 4.1.3.5)	Yeast		*a,j*	*si* face	*206*

(*continued*)

TABLE VII (*continued*)

Enzyme	Source (class)	Method of stereochemical analysis	Nucleophilic addition to	Ref.
Hydroxymethylglutaryl-CoA lyase (EC 4.1.3.4)	Liver	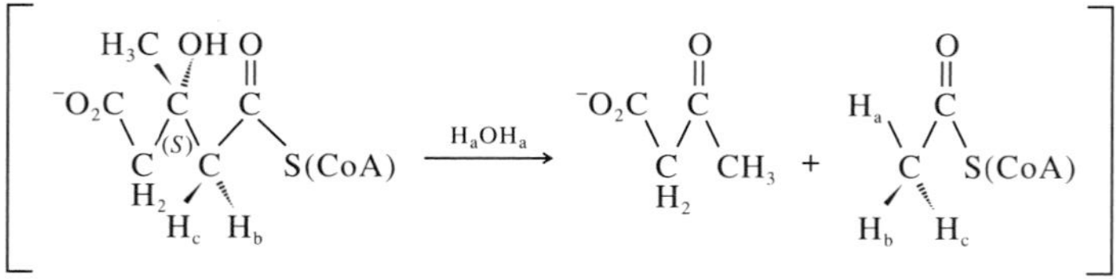 $[{}^-O_2C-CH_2-C^{(S)}(CH_3)(OH)-CH_bH_c-C(=O)-S(CoA) \xrightarrow{H_aOH_a} {}^-O_2C-CH_2-C(=O)-CH_3 + H_a-CH_bH_c-C(=O)-S(CoA)]$ [k,h]	*re* face	*207*
Citramalate lyase (EC 4.2.1.56; itaconyl-CoA hydratase)	*Clostridium tetanomorphum*	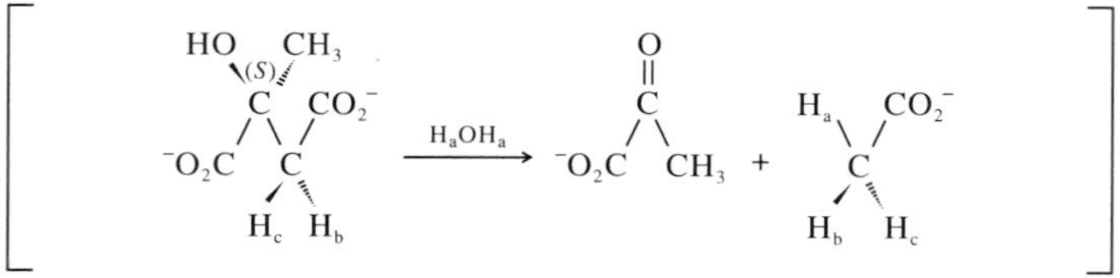 $[{}^-O_2C-C^{(S)}(OH)(CH_3)-CH_cH_b-CO_2^- \xrightarrow{H_aOH_a} {}^-O_2C-C(=O)-CH_3 + H_a-CH_bH_c-CO_2^-]$ [l,d]	*si* face	*208, 209*
Oxaloacetase (EC 3.7.1.1)	*A. niger*	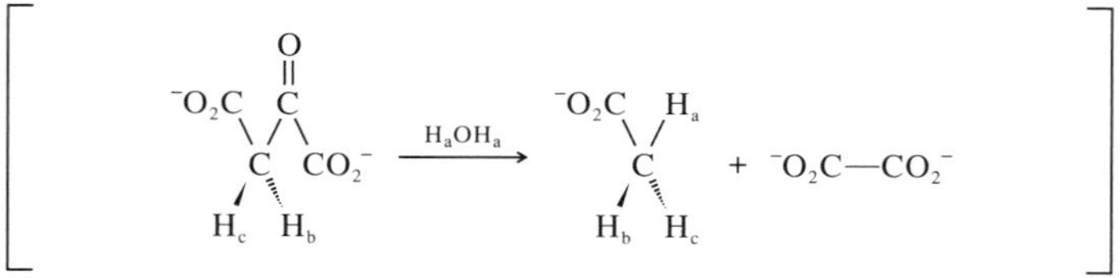 $[{}^-O_2C-CH_cH_b-C(=O)-CO_2^- \xrightarrow{H_aOH_a} {}^-O_2C-CH_aH_bH_c + {}^-O_2C-CO_2^-]$ [m,d]	?	*210*

Acetyl-CoA acetyl-transferase (thiolase) (EC 2.3.1.9)	Heart	$\left[H_3C{-}C(=O){-}C(H_cH_b){-}C(=O){-}S(CoA) \xrightarrow[H_aOH_a]{(CoA)SH} H_3C{-}C(=O){-}S(CoA) + H_a{-}C(H_bH_c){-}C(=O){-}S(CoA) \right]$	[a,n,h]	?	*211*

[a] $H_a = {}^1H$; when $H_b = {}^2H$, $H_c = {}^3H$; when $H_b = {}^3H$, $H_c = {}^2H$.

[b] Configurational analysis at C-3 of (2*S*, 3*R* or 3*S*)-[3-^{3}H,^{2}H]malate employed fumarase (see text) (*3, 4, 196*).

[c] $H_a = {}^2H$. The [^{3}H,^{1}H]citrate was synthesized from either [2,3-$^3H_2$]fumarate in H_2O ($H_c = {}^1H$, $H_b = {}^3H$) or fumarate in ^{3}HOH ($H_c = {}^3H$, $H_b = {}^1H$) by the sequential use of fumarase, L-malate dehydrogenase, and citrate (*re*)-synthase (*197*).

[d] Configurational analysis of the [^{3}H,^{2}H,^{1}H]acetate first involved conversion to [^{3}H,^{2}H,^{1}H]acetyl-CoA followed by the sequential use of malate synthase and fumarase (see text).

[e] Configurational analysis of [^{3}H,^{2}H]citrate involved conversion to [^{3}H,^{2}H,^{1}H]acetate using citrate lyase (*197*).

[f] Configurational analysis of the [^{3}H,^{2}H]citrate involved conversion to [^{3}H,^{2}H]malate by the sequential use of citrate lyase and L-malate dehydrogenase (*197*).

[g] The [^{3}H,^{2}H]citrate was synthesized from [^{3}H,^{2}H,^{1}H]acetate and oxaloacetate using citrate (*si*)-synthase (*197*).

[h] Configurational analysis of the [^{3}H,^{2}H,^{1}H]acetyl-CoA involved the sequential use of malate synthase and fumarase (see text).

[i] $H_a = {}^2H$. The formation of (+)(2*S*)-[2-^{2}H,^{1}H]succinate was established on the basis of optical rotation measurements (*205*).

[j] Configurational analysis at C-4 of the (3*S*)-[4-^{3}H,^{2}H]-3-hydroxy-3-methylglutaryl-CoA first involved chemical reduction to the corresponding mevalonates. The configuration at C-2 of the labeled mevalonates was established by conversion to androst-1,4-diene-3,17-dione (using a preparation of rat liver and a strain of *Mycobacterium phlei*), independently determined to involve loss of the *pro*-(*S*) hydrogen at C2 of mevalonate (*206*).

[k] $H_a = {}^2H$; when $H_c = {}^3H$, $H_b = {}^1H$; when $H_c = {}^1H$, $H_b = {}^3H$. The (3*S*)-[2-^{3}H,^{1}H]-3-hydroxy-3-methylglutaryl-CoA was synthesized by chemical methods from the [4-^{3}H]mevalonates (*207*).

[l] When $H_a = {}^1H$, $H_c = {}^2H$ and $H_b = {}^3H$; when $H_a = {}^2H$, $H_c = {}^1H$ and $H_b = {}^3H$. The [^{3}H,^{2}H(or ^{1}H)]citramalate was synthesized via the mesaconase reaction (*208*).

[m] When $H_a = {}^3H$, $H_c = {}^1H$ and $H_b = {}^2H$; when $H_a = {}^1H$, $H_c = {}^3H$ and $H_b = {}^2H$. The [3-^{2}H,^{3}H(or ^{1}H)]oxaloacetate was synthesized from (2*S*)-[3-$^2H_2$]malate by the sequential use of fumarase and malate dehydrogenase in either H_2O or ^{3}HOH (*210*).

[n] The [^{3}H,^{2}H]acetoacetyl-CoA species were synthesized from (3*S*)-3-hydroxy[2-$^3H_2$]butyryl-CoA and (3*S*)-3-hydroxy[2-$^2H_2$]butyryl-CoA by the sequential use of enoyl-CoA hydratase (in ^2H_2O or ^{3}HOH, respectively) and (3*S*)-specific 3-hydroxyacyl-CoA dehydrogenase (*211*).

group, is strongly indicated by the chemical shift change of the C-13 resonance arising from C-2 of oxaloacetate on binding to citrate synthase (*213*). Additional stereomechanistic information is available from the crystal structure of the ternary citrate synthase (chicken heart)·oxaloacetate·S-carboxymethyl-CoA complex recently solved to a resolution of 1.9 Å by Remington and co-workers (*214*). Model building of acetyl-CoA into a binary enzyme·oxaloacetate complex suggests that the condensation reaction might proceed in two discrete stages involving the direct participation of three key catalytic residues (Fig. 9). First, Asp-375 and His-274 are positioned to participate in the general acid–base-catalyzed formation of the neutral enol of acetyl-CoA. Second, His-320 functions as a general acid catalyst during the condensation of the enol at the *si* face of bound oxaloacetate. This mechanism emphasizes how several catalytic residues might be brought to bear on bound substrates in a reaction mechanism involving what appears to be an unstable intermediate.

D. β-Keto Acid Decarboxylases

β-Keto acid decarboxylases catalyze the irreversible decarboxylation of β-keto acids that are either used directly as substrates or are produced as initial reaction intermediates during the oxidative decarboxylation of β-hydroxy acids [Eq. (37)] (Table VIII) (*215–223*).

$$-\overset{\overset{\displaystyle O}{\|}}{C}-\underset{\underset{\displaystyle R}{|}}{\overset{\overset{\displaystyle H(OH)}{|}}{C}}-CO_2^- \xrightarrow{H^+ \quad CO_2} -\overset{\overset{\displaystyle O}{\|}}{C}-\underset{\underset{\displaystyle R}{|}}{\overset{\overset{\displaystyle H(OH)}{|}}{C}}-H \qquad (37)$$

Uridine diphosphoglucuronate decarboxylase also belongs in this category of enzymes (*221*). This enzyme requires catalytic amounts of NAD^+ for activity. Bound substrate is initially oxidized at C-4 to give a β-keto acid intermediate; decarboxylation followed by reduction at C-4 by bound NADH completes the catalytic cycle. As a class, the β-keto acid decarboxylases do not exhibit the kind of stereochemical uniformity observed among the aldolases (retention) and enzyme-catalyzed Claisen-type condensations (inversion); both stereochemical modes are observed among the decarboxylases (Table VIII). Nevertheless, most of these enzymes appear to conform to the empirical generalization given at the beginning of this section (*223*).

Those decarboxylases involving net retention of configuration generally require catalytically essential divalent metal ions that may be directly involved in stabilizing enol intermediates generated from the decarboxylation step. In the case of malic enzyme, an active site divalent metal ion (Mn^{2+}) has been implicated to be near bound substrate, on the basis of NMR paramagnetic relaxation

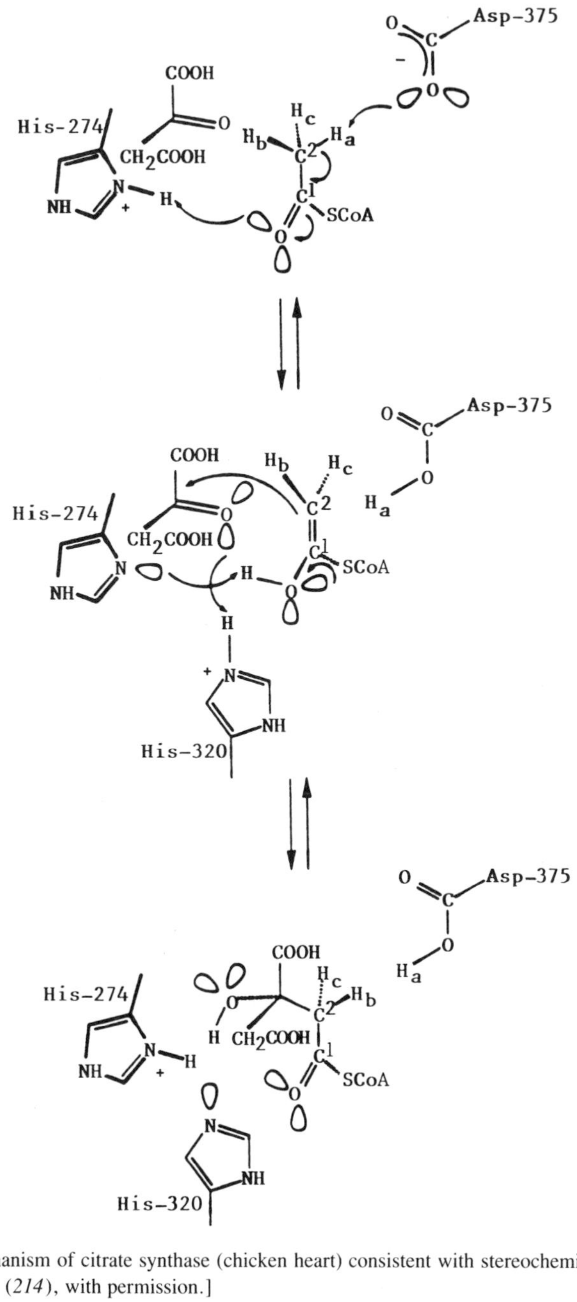

FIG. 9. Mechanism of citrate synthase (chicken heart) consistent with stereochemical and X-ray data. [From Ref. (*214*), with permission.]

TABLE VIII

β-KETO ACID DECARBOXYLASES INVOLVE BOTH NET RETENTION AND INVERSION OF CONFIGURATION

Enzyme	Source	Method of stereochemical analysis	Schiff base or metal ion requirement	Ref.
Isocitrate dehydrogenase ($NADP^+$) (EC 1.1.1.42)	Heart	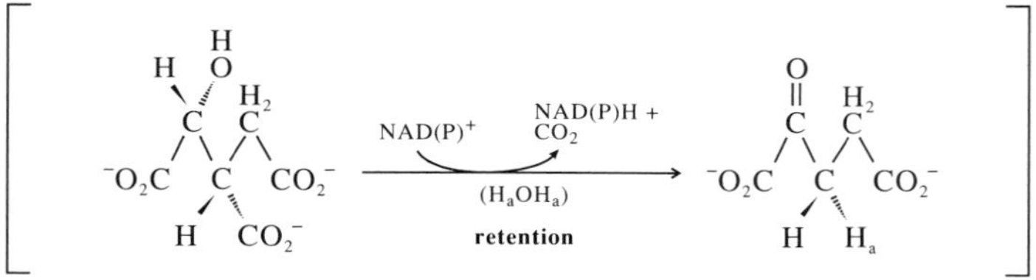 $NAD(P)^+ \rightarrow NAD(P)H + CO_2$ (H_aOH_a) **retention** [a]	Mg^{2+}	*215, 216*
Isocitrate dehydrogenase (NAD^+) (EC 1.1.1.41)	Heart	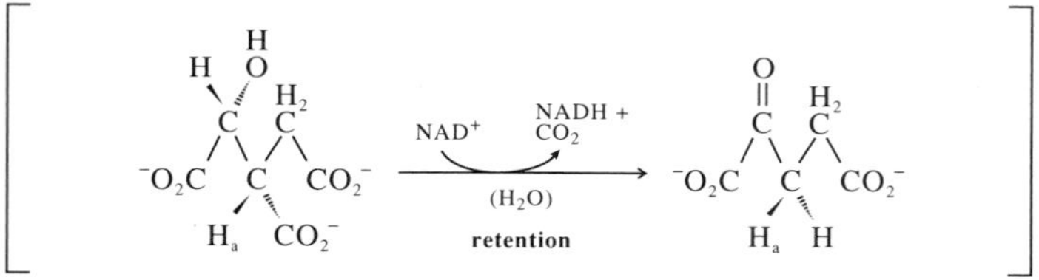 $NAD^+ \rightarrow NADH + CO_2$ (H_2O) **retention** [b]	Mg^{2+}	*217*
Malic enzyme (EC 1.1.1.40)	*E. coli*	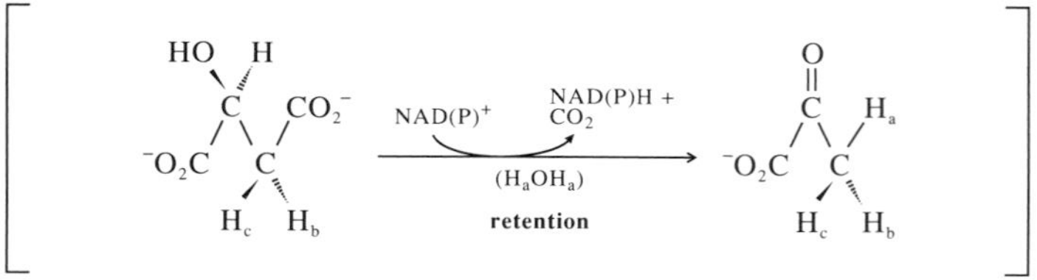 $NAD(P)^+ \rightarrow NAD(P)H + CO_2$ (H_aOH_a) **retention** [c]	Mn^{2+} or Mg^{2+}	*218*

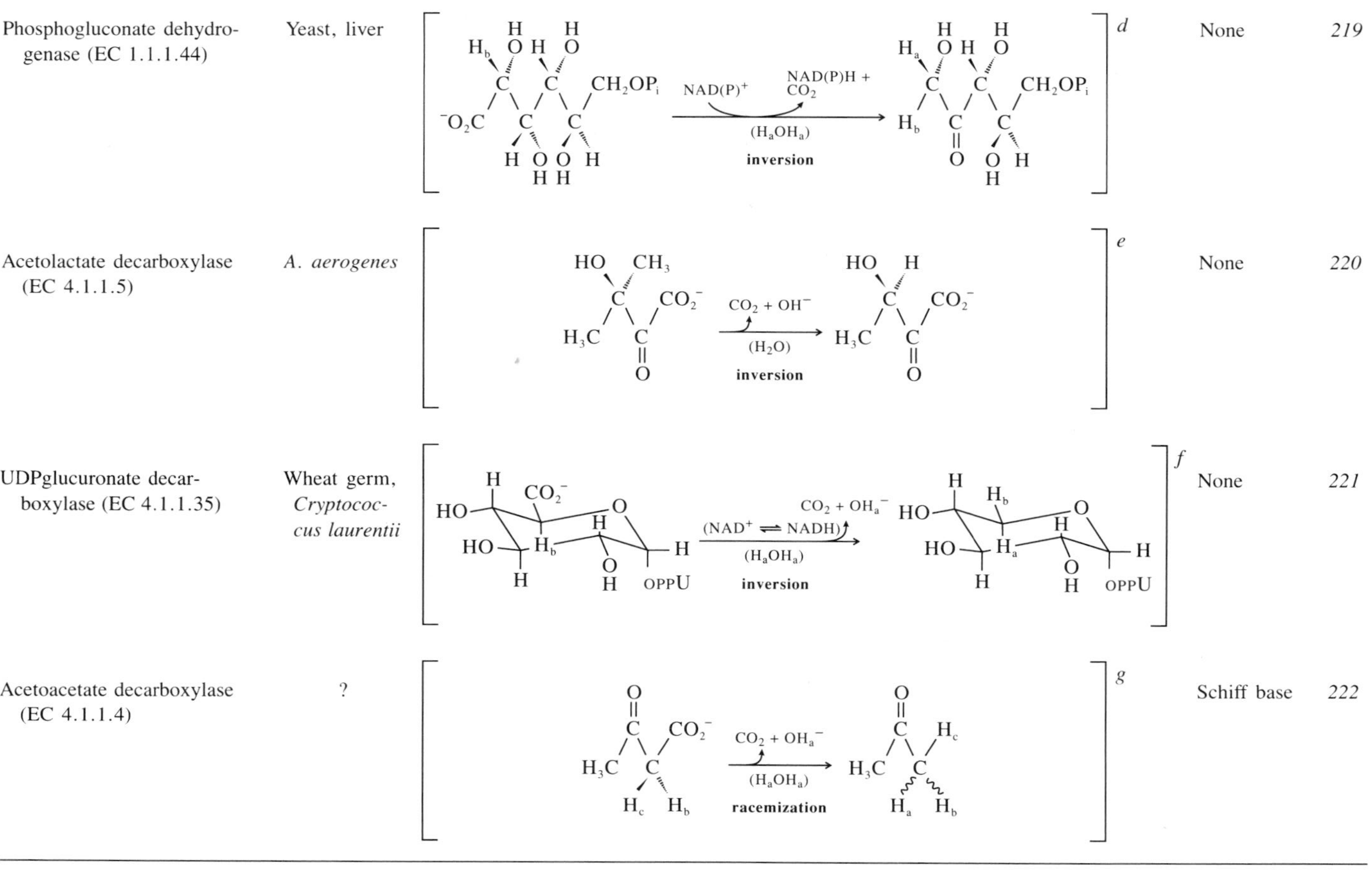

Enzyme	Source	Reaction		Cofactor	Reference
Phosphogluconate dehydrogenase (EC 1.1.1.44)	Yeast, liver	$NAD(P)^+ \rightarrow NAD(P)H + CO_2$ (H_aOH_a) **inversion**	*d*	None	*219*
Acetolactate decarboxylase (EC 4.1.1.5)	*A. aerogenes*	$\rightarrow CO_2 + OH^-$ (H_2O) **inversion**	*e*	None	*220*
UDPglucuronate decarboxylase (EC 4.1.1.35)	Wheat germ, *Cryptococcus laurentii*	($NAD^+ \rightleftharpoons NADH$) $\rightarrow CO_2 + OH_a^-$ (H_aOH_a) **inversion**	*f*	None	*221*
Acetoacetate decarboxylase (EC 4.1.1.4)	?	$\rightarrow CO_2 + OH_a^-$ (H_aOH_a) **racemization**	*g*	Schiff base	222

(*continued*)

TABLE VIII (*continued*)

Enzyme	Source	Method of stereochemical analysis	Schiff base or metal ion requirement	Ref.
Oxaloacetate decarboxylase (EC 4.1.1.3)	*P. putida*	$\left[{}^{-}O_2C{-}C(=O){-}C(H_c)(H_b){-}CO_2^{-} \xrightarrow[(H_aOH_a)\ \textbf{inversion}]{CO_2 + OH_a^{-}} {}^{-}O_2C{-}C(=O){-}C(H_c)(H_a)(H_b) \right]^h$	Me^{2+}	*223*

[a] $H_a = {}^3H$, Configurational analysis of [3-^{3}H,^{1}H]-α-ketoglutarate involved chemical conversion to [3-^{3}H,^{1}H]malate, followed by treatment with fumarase (see text) (216).

[b] $H_a = {}^3H$. [^{3}H]Citrate was synthesized from *cis*-aconitate in ^{3}HOH using aconitase. The formation of (3*R*)-[3-^{3}H]-α-ketoglutarate was established on the basis of the observation that the NADP$^+$-dependent isocitrate dihydrogenase [known to catalyze solvent proton exchange with the *pro-(S)* hydrogen at C-3 of α-ketoglutarate (*184*)] did not catalyze tritium loss from the [^{3}H]-α-ketoglutarate (*217*).

[c] $H_a = {}^1H$; when $H_c = {}^2H$, $H_b = {}^3H$; when $H_c = {}^3H$, $H_b = {}^2H$. The (3*R*)- and (3*S*)-[3-^{3}H,^{2}H]malates were synthesized from (3*R*)- and (3*S*)-[^{3}H,^{2}H,^{1}H]pyruvate using pyruvate carboxylase (retention of configuration). Configurational analysis of the product [^{3}H,^{2}H,^{1}H]pyruvate involved the sequential use of pyruvate carboxylase, malate dehydrogenase, and fumarase (*218*).

[d] When $H_a = {}^1H$, $H_b = {}^3H$; when $H_a = {}^3H$, $H_b = {}^1H$. Configurational analysis at C-1 of [1-^{3}H,^{1}H]ribulose 5-phosphate involved periodate oxidation to [^{3}H]glycolic acid followed by treatment with glycolic acid oxidase [catalyzing the loss of the *pro-(R)* hydrogen] (*219*).

[e] The product was demonstrated to be (−)(3*R*)-acetoin on the basis of optical rotation measurements (*220*).

[f] $H_a = {}^1H$, $H_b = {}^3H$. Configurational analysis at C-5 of [5-^{3}H,^{1}H]uridine diphospho-D-xylose involved conversion to [^{3}H]glycolic acid (from C-4 and C-5), using chemical and enzymic methods, followed by treatment with glycolic acid oxidase [catalyzing the loss of the *pro-(R)* hydrogen] (*221*).

[g] $H_a = {}^2H$; when $H_c = {}^3H$, $H_b = {}^1H$; when $H_c = {}^1H$, $H_b = {}^3H$. The (2*R*)- and (2*S*)-[2-^{3}H,^{1}H]acetoacetates were synthesized from chemically produced (2*R*, 3*R*)- and (2*S*, 3*R*)-3-[2,3-$^3H_2$]hydroxybutyrate using 3-hydroxybutyrate dehydrogenase. Configurational analysis of the [^{3}H,^{2}H,^{1}H]acetone involved oxidation to [^{3}H,^{2}H,^{1}H]acetate, conversion to [^{3}H,^{2}H,^{1}H]acetyl-CoA, and the sequential use of malate synthase and fumarase (see text) (*222*).

[h] When $H_a = {}^1H$, $H_c = {}^3H$ and $H_b = {}^2H$; when $H_a = {}^2H$, $H_c = {}^3H$ and $H_b = {}^1H$. The (3*S*)-[3-^{3}H,^{2}H]- and (3*S*)-[3-^{3}H,^{1}H]oxaloacetates were enzymically synthesized from stereospecifically labeled samples of aspartate using glutamate–oxaloacetate transaminase. Configurational analysis of the samples of [^{3}H,^{2}H^{1}H] pyruvate involved enzymic conversion to chiral labeled acetate (lactate dehydrogenase/lactate oxidase), and the sequential use of malate synthase and fumarase (see text) (*223*).

measurements (*224*). The metal ion may facilitate decarboxylation by polarizing the carbonyl group of bound oxaloacetate (*225*). In support of an enolpyruvate intermediate, the enzyme has been demonstrated to catalyze rapid solvent proton exchange with pyruvate (*226*). Similarly, isocitrate dehydrogenase ($NADP^+$) catalyzes stereospecific exchange of the *pro-(S)* hydrogen at C-3 of α-ketoglutarate with solvent, suggesting the presence of a catalytically important base in the active site (*216*).

In contrast, decarboxylases that do not involve either divalent metal ions or Schiff base intermediates generally involve inversion of configuration (Table VIII). This suggests that, in the absence of a ready means of stabilizing carbanion intermediates, the concerted or tightly coupled action of at least two catalytic groups is required, perhaps involving a transition state in which there is significant negative charge development at the α-carbon:

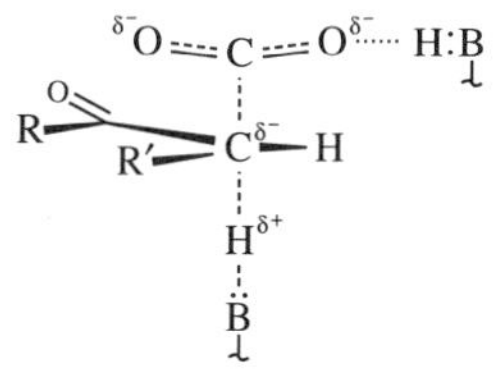

Such a transition state might be significantly stabilized by the neighboring carbonyl function, and it would explain the requirement for an α-keto acid intermediate in the uridine diphosphoglucuronate decarboxylase reaction.

Two *apparent* exceptions to the above generalization have been recently identified by Benner and co-workers (*222, 223*) (Table VIII). The first exception is the bacterial enzyme oxaloacetate decarboxylase. This enzyme requires divalent metal ions for activity and would be predicted to involve retention of configuration, provided that the metal ion plays an electrophilic role in catalysis, consistent with model chemistry (*227, 228*). However, this enzyme has been demonstrated to involve inversion of configuration (*223*). An experimental test for the occurrence of an obligatory enolpyruvate intermediate along the reaction pathway for this enzyme would be most useful. The second apparent exception is the reaction catalyzed by the Schiff base enzyme acetoacetate decarboxylase, also predicted to go by retention of configuration. Remarkably, this enzyme reaction proceeds with virtually complete racemization, on the basis of a recently published study by Rozzell and Benner (*222*). These workers suggest that stereorandom addition of solvent protons to the two diastereotopic faces of the intermediate enamine might be comprehensible if protonation of the enamine is simply not catalyzed by an active site base. An alternative possibility is that rotation about the C-2–N bond of the enamine allows both faces of the enamine to be exposed to a single active site base [Eq. (38)].

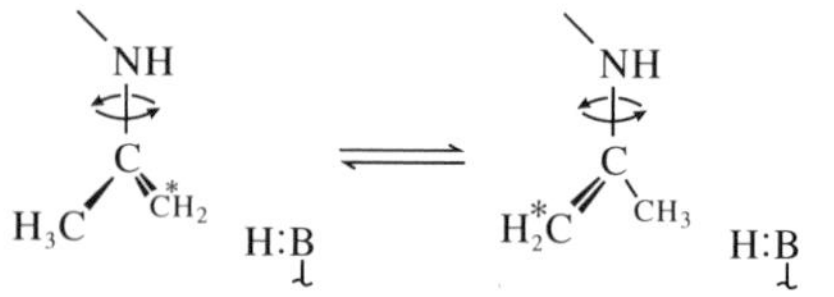

(38)

Whatever the explanation, the stereochemistry of the acetoacetate decarboxylase reaction is controlled by factors not common to the other β-keto acid decarboxylases.

E. Mechanistic Implications

The empirical generalization emphasized at the beginning of this section suggests that in many cases reaction stereochemistry is related to the stability of the carbanions/enamines generated along the reaction pathway. Reactions involving the intermediate formation of Schiff bases or metal-stabilized enol(ate)s generally go by retention of configuration; reactions for which there is no evidence for kinetically stable intermediates generally go by inversion of configuration.

These trends may have an underlying mechanistic explanation. For reactions involving stable intermediates, retention of configuration may reflect the advantage implicit in the use of a similar disposition of active site functional groups to stabilize two transition states of similar structure, for example,

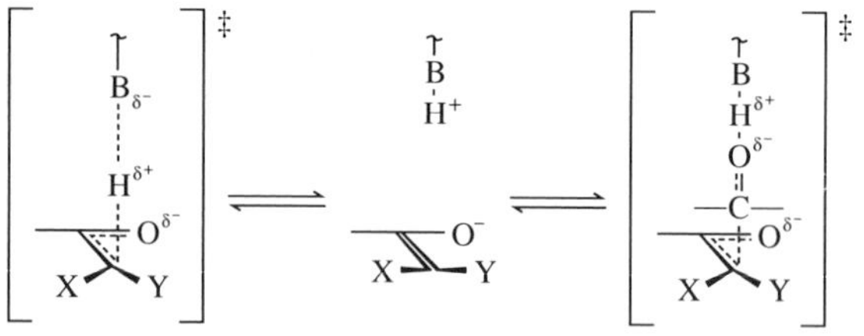

(39)

This is a partial restatement of the "minimum number [of active site bases] rule" proposed by Hansen and Rose which implies that there are mechanistic advantages to the reuse of active site bases for more than one purpose during catalysis versus the alternative use of multiple bases (*16*). For reactions involving kinetically unstable intermediates, the advantages implicit in the retention stereochemical mode appear to have been overridden by the need for the concerted or tightly coupled action of two or more active site functional groups that function to trap the unstable intermediate, thus, requiring an invertive stereochemical mode.

Thibblin and Jencks (*229*) succinctly summarized this notion for simple chemical model systems when they suggested that "the lowest energy pathway for a reaction that proceeds through an unstable intermediate complex containing the elements of three molecules involves a preassociation of the reactants prior to the first covalent step when the intermediate reverts to reactants (k'_{-1}) faster than the final reactant can diffuse away (k_{-a})." (Fig. 10). For example, a preassociation/concerted mechanism of this type appears to apply to the general acid-catalyzed cleavage of the anion of 1-phenylcyclopropanol to 1-phenylpropanone (*230*) [Eq. (40)]:

$$BH^+ + \text{>C}\overset{CH_2}{\triangle}\text{C}(O^-)(C_6H_5) \rightleftharpoons \left[B \cdot \overset{\delta+}{H} \cdots \overset{\delta-}{C} \cdots CH_2 \cdots C(C_6H_5) \overset{\delta-}{\cdots} O \right]^{\ddagger} \rightleftharpoons B + HC{-}CH_2{-}C(C_6H_5){=}O \quad (40)$$

Structure–activity correlations indicate an open transition state with significant negative charge development at C-3. Moreover, this mechanism is consistent with inversion of configuration observed during the base-catalyzed cleavage of

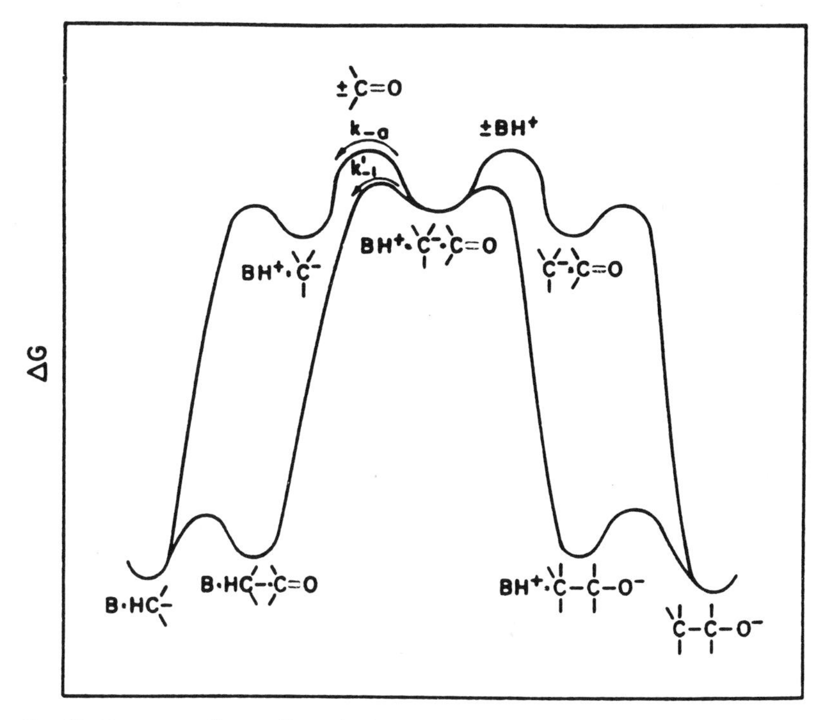

FIG. 10. Free energy diagram illustrating how a preassociation mechanism (lower curve) provides the lowest energy pathway for a reversible carboligation reaction when the intermediate carbanion reacts more rapidly with BH^+ (k'_{-1}) than with the carbonyl function (k_{-a}). [From Ref. (*229*), with permission.]

trans-1-methyl-2-phenylcyclopropanol in aqueous dioxane (*229, 230*). Thus, the invertive stereochemical mode observed among many enzymes may represent an adaptation to the lowest free-energy pathway for reactions involving carbanion intermediates with very short (or nonexistent) lifetimes.

VI. Pyridoxal Phosphate- and Pyruvyl-Dependent Enzymes: Flexible Response to Different Chemical Challenges

The pyridoxal phosphate (PLP)-dependent enzymes catalyze a diverse set of chemical transformations of amino acids. These include transamination, decarboxylation, and racemization reactions [Eqs. (41–43)],

$$^{+}H_3N-C(CO_2^-)(R)-H \xrightarrow[\text{transamination}]{R'C(=O)CO_2^-} H_3\overset{+}{N}-C(CO_2^-)(R')-H + RC(=O)CO_2 \quad (41)$$

$$^{+}H_3N-C(CO_2^-)(R)-H \xrightarrow[\text{decarboxylation}]{H^+} H_3\overset{+}{N}CH_2R + CO_2 \quad (42)$$

$$^{+}H_3N-C(CO_2^-)(R)-H \xrightarrow[\text{racemization}]{} H-C(CO_2^-)(R)-NH_3^+ \quad (43)$$

as well as substitution reactions at the β- and γ-carbons [Eq. (44)].

$$-C(X)-C(Y)-C(CO_2^-)(H)-NH_3^+ \xrightarrow{Z} -C(Z)-C(Y)-C(CO_2^-)(H)-NH_3^+ + X \quad \text{or} \quad -C(X)-C(Z)-C(CO_2^-)(H)-NH_3^+ + Y \quad (44)$$

However, extensive stereochemical studies suggest that catalysis by all of these enzymes is achieved by fundamentally similar spacial relationships among bound substrates, PLP, and active site residues, as emphasized in a review by Floss and Vederas (*231*). These enzymes exemplify the economy implicit in single-base mechanisms. The PLP-dependent decarboxylases generally, although not uniformly, conform to the generalization of retention of configuration for carbon–carbon lyases involving stable intermediates.

Common to all of the PLP-dependent enzymes so far examined is an initial transaldimination reaction between bound amino acid and an internal aldimine formed between PLP and an active site lysine (*232, 233*) [Eq. (45)]:

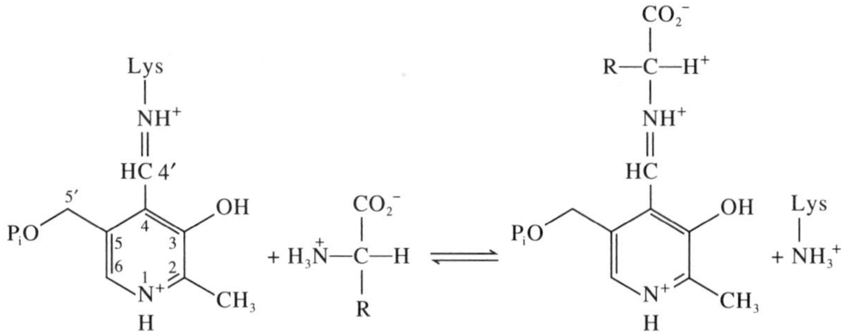

(45)

Heterolytic bond cleavage at the α-carbon of the resultant external aldimine results in the formation of a resonance-stabilized quinoid intermediate [Eq. (46)]:

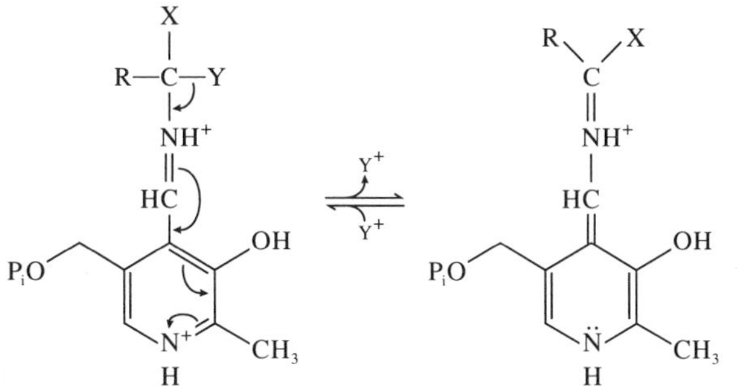

(46)

where X is CO_2^- and Y is H for transaminases and racemases, and X is H and Y is CO_2^- for decarboxylases. Dunathan was the first to suggest that selection of the bond to be cleaved (X versus Y) is achieved by stabilizing a conformation of the external aldimine in which the leaving group is orthogonal to the plane of the ring system, thus, ensuring maximum orbital overlap with the π system in the transition state (**20**) (*234*).

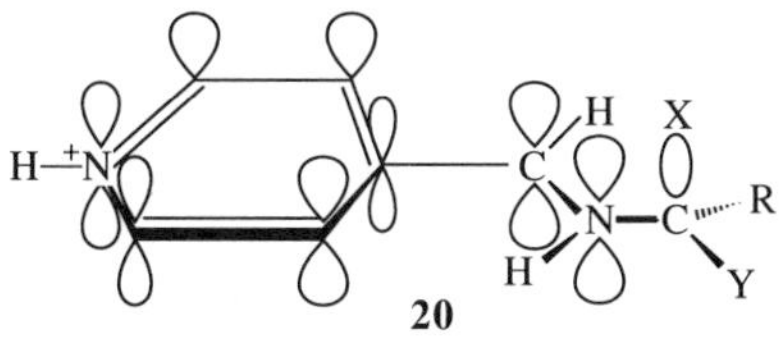

20

This reasoning is supported by model studies in which Schiff base complexes between PLP and amino acids undergo nonenzymic racemization and hydrogen exchange at rates that are proportional to the conformers in which the C_α–H_α bond is orthogonal to the plane of the pyridine ring of PLP(*235*).

A. Transaminases

Stereochemical and X-ray crystallographic studies on the transaminases (aminotransferases) have given rise to the clearest picture of the spatial relationship between bound substrates and catalytic residues that may apply generally to the PLP-dependent enzymes. Transamination involves the reversible interconversion of L-α-amino acids and α-keto acids by a double-displacement mechanism involving transiently formed pyridoxamine monophosphate (PMP) [Eq. (47)]:

(47)

In order to maintain coplanarity between the pyridine ring and the aldimine function, the bulky substituents about the aldimine must be trans; a cis-configuration is precluded for steric reasons.

That catalytic interconversion of the aldimine and ketimine involves suprafacial 1,3-hydrogen transfer between the α-carbon and C-4′ at the *si* face of the aldimine is indicated by two observations. First, the *pro*-(*S*) hydrogen at C-4′ of pyridoxamine is in the transferring position because the apo form of aspartate transaminase catalyzes stereospecific exchange of deuterium from (4′*S*)-[4′-^{2}H]pyridoxamine with solvent in the presence of oxaloacetate (*236*) [Eq. (48)]:

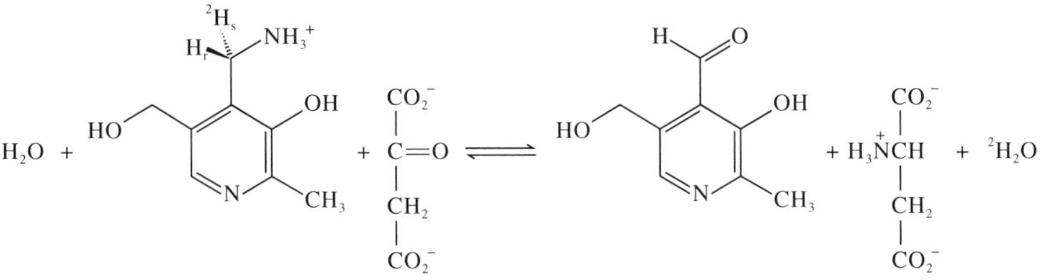

$$H_2O + \ldots + {}^{-}O_2C{-}C({=}O){-}CH_2{-}CO_2^{-} \rightleftharpoons \ldots + H_3\overset{+}{N}CH(CO_2^{-}){-}CH_2{-}CO_2^{-} + {}^2H_2O \qquad (48)$$

The same conclusion applies for L-alanine/pyruvate (*237*), pyridoxamine/pyruvate (*238*), and dialkylamino acid transaminase (*239*). Second, small but significant amounts of direct hydrogen transfer (as either ^2H or ^3H) between the α-carbon and C-4′ have been detected during catalysis by the latter two enzymes (*238, 239*). Thus, the clear implication of these results is that a single active site base operating at the *si* face of the aldimine facilitates proton transfer between the α-carbon and C-4′ [Eq. (47)].

These conclusions are consistent with the results of recent high-resolution (2.8 Å) X-ray crystallographic measurements on the holoenzyme form of mitochondrial aspartate transaminase (aspartate aminotransferase) and its complexes with substrate analogs (*240, 241*). On the basis of these measurements, Kirsch *et al.* formulated a catalytic mechanism for aspartate transaminase that describes the probable spacial features of the active site complexes that form at different stages of transamination (*242*) (Fig. 11). The initial transaldimination reaction involves attack of the α-amino group of the bound amino acid on the *re* face of the internal aldimine formed from PLP and Lys-258. The resultant tetrahedral intermediate then decomposes to form the external aldimine, positioning the ε-amino of Lys-258 above the *si* face of this intermediate. Lys-258 appears to be the best candidate for the catalytic base involved in the suprafacial 1,3-proton transfer to form the ketimine intermediate. Hydrolysis of the ketimine completes the first cycle of catalysis wherein the transferred proton ends up in the *pro-*(*S*) position at C-4′ of PMP, consistent with the previously described stereochemical results. The X-ray measurements also provide compelling evidence for reorientation of the coenzyme during catalysis, a possibility first suggested by Ivanov and Karpeisky on the basis of spectroscopic and stereochemical data (*243*). As approximately indicated in Fig. 11, the pyridine ring of the cofactor has limited rotational freedom during catalysis. Conversion of the internal aldimine to the external aldimine results in a reorientation of the ring, primarily owing to an approximately 30° rotation about the bond axis between C-5 and C-5′. Linear dichroism measurements on single crystals of aspartate transaminase also suggest that the pyridine ring undergoes reorientation during catalysis (*244*). More-

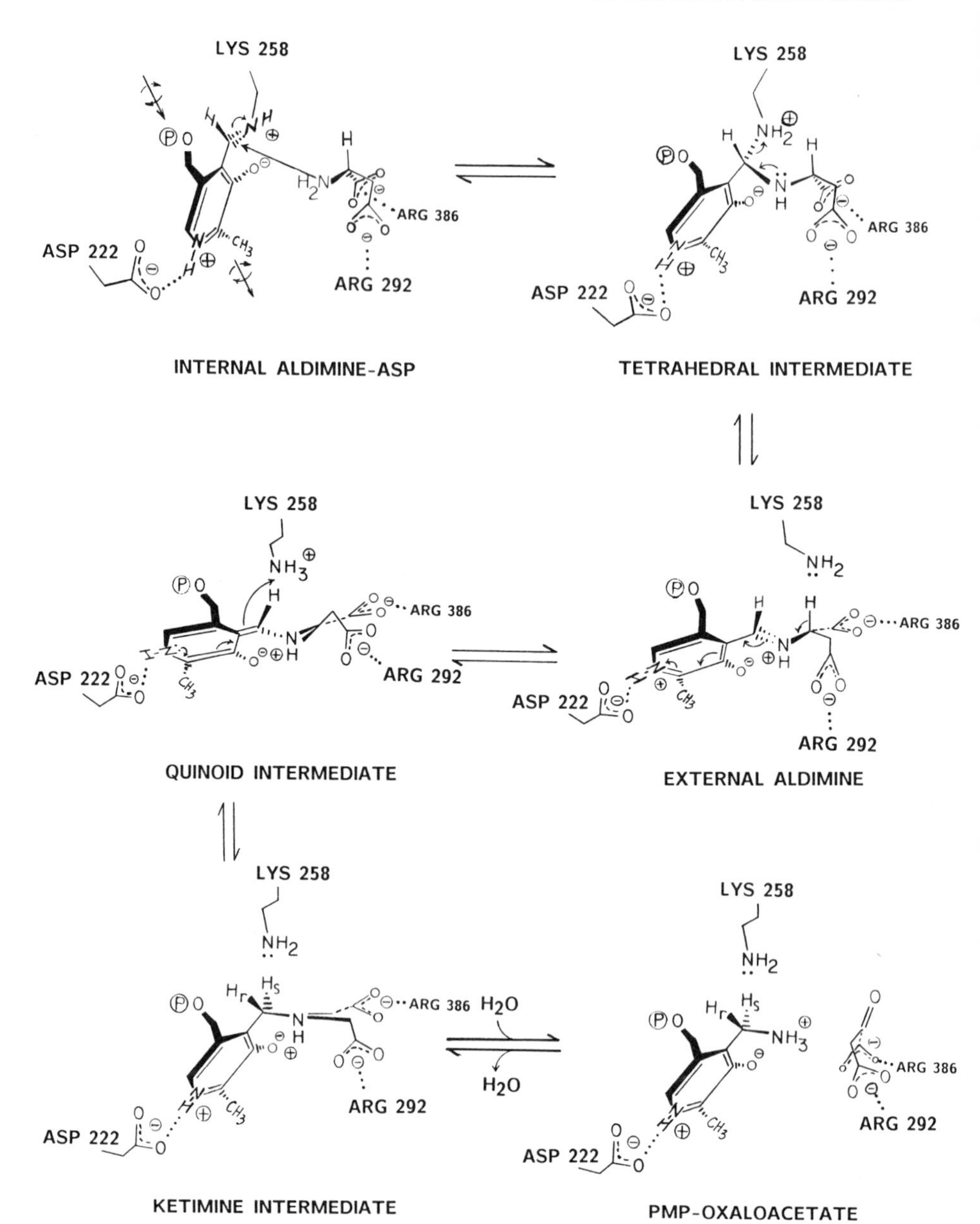

FIG. 11. Proposed catalytic mechanism for aspartate aminotransferase deduced on the basis of the crystallographic structures of the holoenzyme (mitochondria) and its complexes with substrate analogs (*242*). Model building was used to formulate the most probable structures formed along the reaction pathway with normal substrates.

over, reorientation of the ring offers a reasonable explanation for the finding that BH_4^- reduction of the holoenzyme (internal aldimine) primarily takes place at the *re* face of C=N, whereas BH_4^- reduction of the external aldimine formed with aspartate takes place primarily at the *si* face (*237*). A similar observation has been made in the case of holotryptophanase (*245*). Apparently, alternate faces of PLP are exposed to solvent during interconversion of the aldimines.

B. Pyridoxal Phosphate-Dependent Decarboxylases

The members of the PLP-dependent decarboxylases have several stereochemical features in common, the best documented of which is net retention of configuration at the α-carbon during the decarboxylation of α-L-amino acids (Table IX). Belleau and Burba were the first to establish the overall stereochemistry of a PLP-dependent decarboxylase, namely, that for the tyrosine decarboxylase from *Streptococcus faecalis* (*246*). Perhaps even more remarkable than their experimental achievement was their prediction that enzyme-catalyzed decarboxylation would proceed by retention of configuration: "It might be expected . . . that the transition state for the release of carbon dioxide should resemble that for protonation of the α-carbon [of the quinonoid intermediate] since in all probability the same active site accommodates the R group [at the α-carbon] in both transition states. Accordingly, overall retention of configuration may be expected in the enzymic decarboxylation of amino acids." This prediction was confirmed by conducting the enzymic decarboxylation of L-tyrosine in ^2H_2O solvent and demonstrating that the resultant [α-^{2}H]tyramine exhibits a significant deuterium kinetic isotope effect (k_H/k_D = 2) during catalytic oxidation by monoamine oxidase; the oxidase was independently shown to exhibit an isotope effect of similar magnitude with synthetic (αR)-[α-^{2}H]tyramine (k_H/k_D = 2.3), whereas a smaller isotope effect was found with synthetic (αS)-[α-^{2}H]tyramine (k_H/k_D = 1.25) [Eq. (49)]:

$$\text{HO–C}_6\text{H}_4\text{–CH}_2\text{C}(\text{CO}_2^-)(\text{H})(\text{NH}_3^+) \xrightarrow[\text{decarboxylase}]{^2\text{H}^+ \quad \text{CO}_2} \text{HO–C}_6\text{H}_4\text{–CH}_2\text{C}(^2\text{H})(\text{H})(\text{NH}_3^+)$$

$$\xrightarrow[\text{monoamine oxidase/O}_2\text{/H}_2\text{O}]{k_H/k_D = 2} \text{HO–C}_6\text{H}_4\text{–CH}_2\text{C}(=\text{O})\text{H} + {}^2\text{H}_2\text{O}_2 + \text{NH}_4^+ \qquad (49)$$

Since this early experiment, few exceptions have been found to the general rule of retention of configuration among PLP-dependent decarboxylases operating on a range of different amino acids (Table IX) (*247–260*).

Decarboxylation appears to take place at the same diastereotopic face of the cofactor as that employed by the transaminases. For example, glutamate decar-

TABLE IX

NET REACTION STEREOCHEMISTRIES OF THE PYRIDOXAL PHOSPHATE-DEPENDENT DECARBOXYLASES

Enzyme	Source	Substrate(s)	Stereochemistry	Ref.
Tyrosine decarboxylase	*Streptococcus faecalis*	L-Tyrosine	Retention	*246, 247*
		L-Tryptophan	Retention	
Aromatic-amino-acid decarboxylase	Human	3-(3,4-Dihydroxyphenyl)-2-methyl-L-alanine	Retention	*248*
Aromatic-amino-acid decarboxylase	Hog kidney	L-Tyrosine	Retention	*247, 231*
		L-Tryptophan	Retention	
Lysine decarboxylase	*Bacillus cadaveris, E. coli*	L-Lysine	Retention	*249–251*
Glutamate decarboxylase	*E. coli*	L-Glutamate	Retention	*252, 253*
		α-*Methyl*-L-glutamate	Retention	
Glutamate decarboxylase	Mammalian	L-Glutamate	Retention	*254*
Histidine decarboxylase	Mammalian	L-Histidine	Retention	*255*
Ornithine decarboxylase	Kidney, *E. coli, B. cadaveris*	L-Ornithine	Retention	*251, 256–258*
Arginine decarboxylase	*E. coli*	L-Arginine	Retention	*251, 257*
Meso-α, ε-diaminopimelate decarboxylase	*Bacillus sphaericus*, wheat germ	DL-Diaminopimelate	Inversion	*259, 260*

boxylase catalyzes a slow abortive transamination reaction in the presence of L-glutamate or L-α-methylglutamate that results in solvent proton incorporation at the *pro*-(*S*) position at C-4′ of PMP (*261*) (Fig. 12). This suggests that the α-carboxyl group is above the *si* face of the external aldimine, provided that the same active site group (BH) is involved in both abortive transamination and decarboxylation. As with the transaminases, the *si* face of the external aldimine

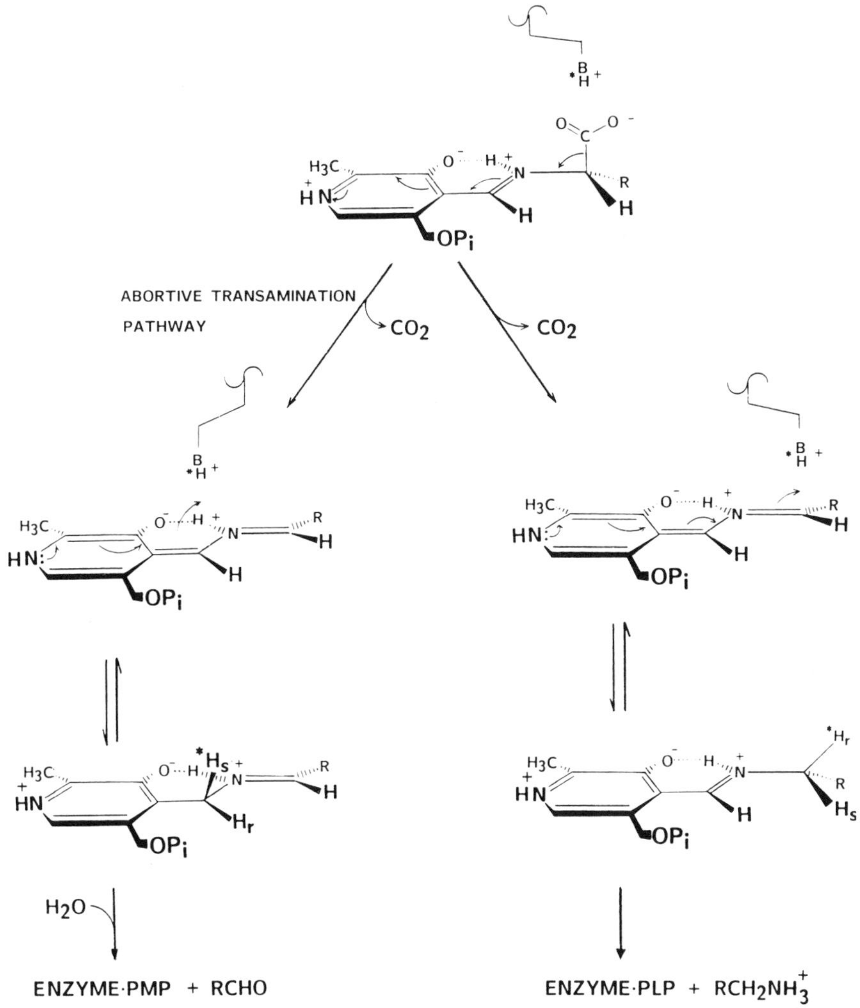

FIG. 12. Hypothetical stereochemical mechanism for abortive transamination during the glutamate decarboxylase reaction (*261*). The same active site base (BH^+) is envisioned to be involved in both the normal decarboxylation reaction and abortive transamination.

of tyrosine decarboxylase is most likely exposed to solvent since treatment of the enzyme with NaB^3H_4 in the presence of tyrosine or tryptamine results in tritium incorporation (72–77%) at the *pro-(S)* position at C-4′ of PMP (*262*). Borohydride reduction of the holoenzyme form of tyrosine decarboxylase takes place primarily at the *re* face (78–98%) of the internal aldimine (*262*). All of these observations taken together suggest a strong mechanistic and, perhaps, evolutionary relationship between the transaminases and decarboxylases (*263*). In addition, the PLP-dependent decarboxylases are an extension of the empirical generalization that carbon–carbon lyase reactions involving stable intermediates usually involve retention of configuration.

Nevertheless, an exception to the general rule of retention has recently been discovered in the form of *meso*-α,ε-diaminopimelate decarboxylase from *Bacillus sphaericus* (*259*). This PLP-dependent enzyme, which catalyzes the final step in lysine biosynthesis, is the only known amino acid decarboxylase to operate on an α-carbon having the D-configuration (*264*). Inversion of configuration was demonstrated for the enzyme from *Bacillus sphaericus* by conducting the decarboxylation reaction in ^2H_2O solvent and isolating as product (6*R*)-L-[6-^{2}H]lysine [Eq. (50)]:

$$\begin{matrix} {}^-O_2C \\ {}^+H_3\overset{+}{N} \\ H \end{matrix} \overset{(S)}{>} C(CH_2)_3C \overset{(R)}{<} \begin{matrix} CO_2^- \\ NH_3^+ \\ H \end{matrix} \xrightarrow{{}^2H^+ \quad CO_2} \underset{\mathbf{21}}{\begin{matrix} {}^-O_2C \\ H_3\overset{+}{N} \\ H \end{matrix} \overset{(S)}{>} C(CH_2)_3C \overset{(R)}{<} \begin{matrix} H_s \\ NH_3^+ \\ {}^2H_r \end{matrix}} \qquad (50)$$

The absolute configuration of **21** was established by two methods. First, **21** was converted to 5-phthalimido[5-^{2}H]valerate by the use of chemical and enzymic methods and shown to have the same optical rotatory properties as those of authentic (5*R*)-5-phthalimido[5-^{2}H]valerate produced from L-glutamate by an established stereochemical route. Second, **21** was converted to [1-^{2}H]cadavarine with L-lysine decarboxylase, followed by treatment with diamine oxidase to form pelletierine. Retention of all of the deuterium in the pelletierine demonstrated that the deuterium must be in the *pro-(R)* position, since the oxidase reaction is known to labilize hydrogen at the *pro-(S)* position [Eq. (51)]:

$$\mathbf{21} \xrightarrow[\text{decarboxylase}]{\text{L-lysine}} H_3\overset{+}{N}{-}CH_2(CH_2)_3{-}C \begin{matrix} H_s \\ NH_3^+ \\ {}^2H_r \end{matrix} \xrightarrow[\text{oxidase}]{\text{diamine}} \begin{matrix} C^2HO \\ | \\ (CH_2)_3 \\ | \\ CH_2 \\ | \\ NH_3^+ \end{matrix} + \begin{matrix} NH_3^+ \\ | \\ CH^2H \\ | \\ (CH_2)_3 \\ | \\ CHO \end{matrix} \qquad (51)$$

The stereochemistry of *meso*-α,ε-diaminopimelate decarboxylase from a eukaryotic source (wheat germ) also involves inversion of configuration (*260*). As suggested by Floss and Vederas inversion might be comprehensible if the enzyme evolved from a preexisting L-amino acid decarboxylase in which the dispositions of the coenzyme, active site base, and the R group of the amino acid moiety were largely preserved, thus requiring the α-CO_2^- to be above the *re* face of the external aldimine (*231*) [Eq. (52)]:

(52)

C. Pyruvyl-Dependent Amino Acid Decarboxylases

Pyruvyl-dependent amino acid decarboxylases are mechanistically analogous to the PLP-dependent amino acid decarboxylases wherein a pyruvyl group in amido linkage to the amino terminus of the protein functions in place of PLP. The formation of a Schiff base linkage between the α-amino function of the amino acid and the ketonic carbonyl of the pyruvyl moiety is supported by the results of borohydride trapping experiments with L-histidine decarboxylase (*Lactobacillus* 30a) in the presence of substrate (*265*). Evidence could not be found for a reducible internal aldimine in the absence of substrate.

Like most of the PLP-dependent decarboxylases, these enzymes involve retention of configuration at the α-carbon (Fig. 13). Chang and Snell first observed that histidine decarboxylase catalyzes the conversion of L-histidine to histamine in ^2H_2O solvent with the stereospecific incorporation of one deuterium atom from ^2H_2O solvent (*266*). Retention of configuration was tentatively suggested on the basis of a comparison of the optical rotation properties of the deuterated histamine with those of model compounds. This conclusion was later confirmed for histidine decarboxylase from both *Lactobacillus* 30a and *Clostridium welchii* by a method that employed diamine oxidase for the configurational analysis of the (αS)-[α-^{3}H]histamine resulting from enzymic decarboxylation of (αS)-[α-^{3}H]histidine in H_2O; diamine oxidase catalyzes stereospecific removal of the *pro*-(*S*) hydrogen at the α-methylene of histamine (*267, 268*) [Eq. (53)]:

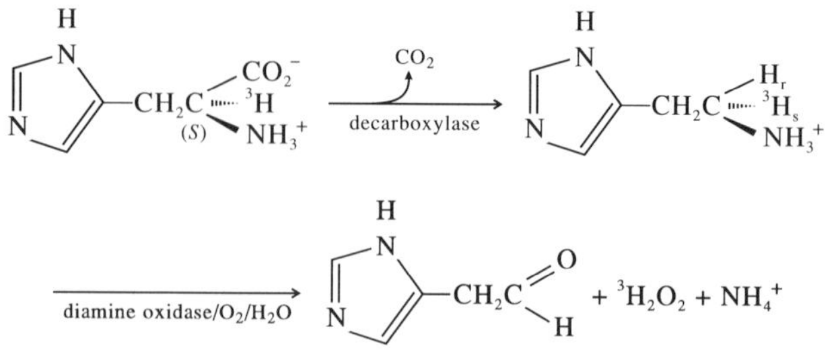

$$\text{Im-CH}_2\text{C}(^3\text{H})(\text{CO}_2^-)(\text{NH}_3^+)\ (S) \xrightarrow[\text{decarboxylase}]{\text{CO}_2} \text{Im-CH}_2\text{C}(\text{H}_r)(^3\text{H}_s)(\text{NH}_3^+) \xrightarrow{\text{diamine oxidase/O}_2\text{/H}_2\text{O}} \text{Im-CH}_2\text{C(=O)H} + {}^3\text{H}_2\text{O}_2 + \text{NH}_4^+ \quad (53)$$

Using a similar experimental strategy, the pyruvyl-dependent *S*-adenosylmethionine decarboxylase (*Escherichia coli*) was also demonstrated to involve retention of configuration at the α-carbon of the methionyl moiety (*269*). The

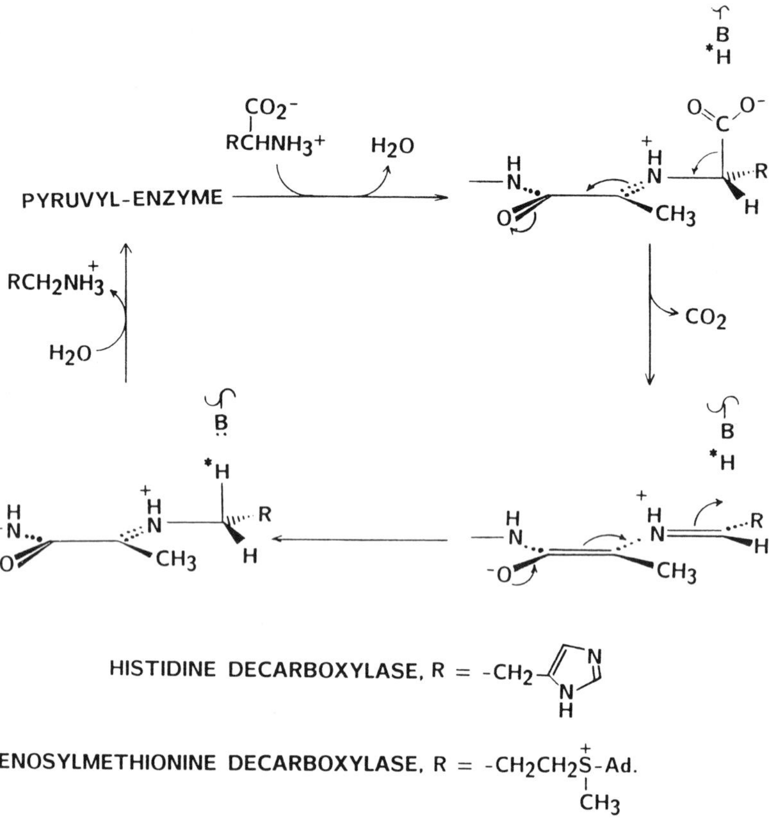

HISTIDINE DECARBOXYLASE, R = $-CH_2$-(imidazolyl)

S-ADENOSYLMETHIONINE DECARBOXYLASE, R = $-CH_2CH_2\overset{+}{S}(CH_3)$-Ad.

FIG. 13. Hypothetical stereochemical mechanism for the pyruvyl-containing amino acid decarboxylases consistent with retention of configuration at the α-carbon of the amino acid. The α-carboxyl function is arbitrarily positioned above the *si* face of the initial ketimine intermediate.

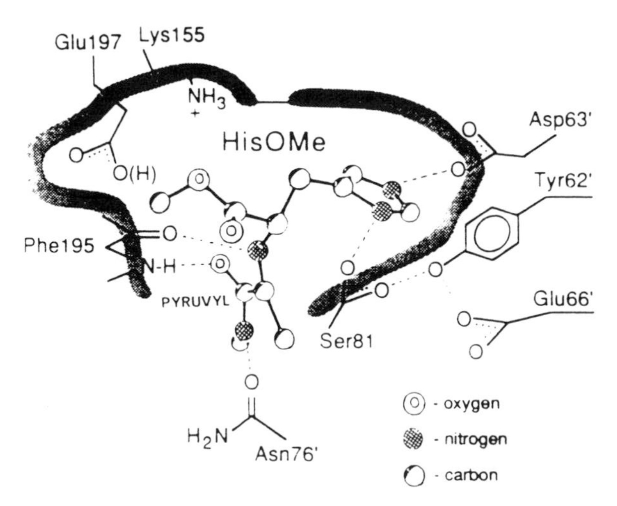

FIG. 14. Diagrammatic representation of the binary complex formed between histidine decarboxylase and histidine *O*-methyl ester (HisOMe) deduced from X-ray crystallographic measurements. [From Ref. (*270*), with permission.]

implication here is that the two pyruvyl-dependent decarboxylases employ a common catalytic mechanism, possibly mediated by a single base in the active site. This appears to be supported by X-ray measurements at 2.5-Å resolution on the crystalline binary complex formed from histidine decarboxylase (*Lactobacillus* 30a) and the substrate analog, histidine *O*-methyl ester (*270*) (Fig. 14). The C_α–CO_2H bond of the analog is orthogonal to the plane of the ketamine, an orientation appropriate for decarboxylation on the basis of stereoelectronic principles. The carboxyl group of Glu-197 (if protonated) is appropriately positioned to potentially bind the α-carboxyl of the normal substrate and subsequently transfer a proton to the *pro*-(*R*) position of histamine once CO_2 has left the active site.

D. ELIMINATION–SUBSTITUTION REACTIONS

The stereochemistry of the PLP-dependent elimination/substitution reactions involving the β- and γ-carbons of amino acid substrates generally conforms to the trends observed among the PLP-dependent transamination and decarboxylation reactions at the α-carbon [Eq. (44)]. All of the enzymes so far studied that catalyze nucleophilic replacement reactions at the β-carbon (five examples; e.g., tryptophan synthase) and/or α,β-elimination reactions (seven examples; e.g., tryptophanase) involve retention of configuration at the β-carbon (*231*) [Eq. (54)]:

$$\text{XH} + \text{Y-C(H}_a\text{)(H}_b\text{)-C(H)(H}_3\overset{+}{\text{N}})\text{-CO}_2^- \xleftarrow[\text{substitution}]{\text{YH}} \text{X-C(H}_a\text{)(H}_b\text{)-C(H)(H}_3\overset{+}{\text{N}})\text{-CO}_2^- \xrightarrow[\text{elimination}]{\text{H}_c\text{OH}_c} \text{H}_c\text{-C(H}_a\text{)(H}_b\text{)-C(=O)-CO}_2^- + \text{NH}_4^+ + \text{XH} \quad (54)$$

Catalysis by some, if not all, of these enzymes appears to involve intramolecular proton transfers mediated by a single active site base. In the case of the tryptophanase reaction, an intramolecular 1,3-proton transfer takes place between C-2 of tryptophan and C-3 of the indol leaving group, on the basis of significant tritium transfer between these two carbons using (2*S*)-[2-^{3}H]tryptophan as substrate (*271*). This is consistent with a single active site base catalyzing a suprafacial proton transfer with respect to the putative aminoacrylate intermediates [Eq. (55)]:

$$\text{(pyridoxal-tryptophan aldimine, B:, } \text{H}_c\text{)} \longrightarrow \text{(quinonoid, B-H}_c^+\text{)} \xrightarrow{*\text{H}^+ \;\; (\text{H}_c)\text{indole}} \text{(aminoacrylate aldimine, B-*H}^+\text{)} \longrightarrow \text{(quinonoid, *H, B:)} \longrightarrow \text{pyruvate} + \text{NH}_3 \quad (55)$$

Two enzymes that catalyze electrophilic replacement reactions at C_β are kynureninase and aspartate β-decarboxylase. In the case of the former enzyme, the replacement reaction at C_β involves retention of configuration, and intramolecular tritium transfer is observed between C_α of kynurenine and the methyl group of the product alanine, consistent with a single-base mechanism (*272*) [Eq. (56)]:

$$\longrightarrow \longrightarrow \text{alanine} \qquad (56)$$

where R is *o*-aminophenyl. In marked contrast to these observations, the aspartate β-decarboxylase reaction involves inversion of configuration at C_β, and different active site bases are involved in the proton transfer reactions at C_α and C_β (*273*). The evolutionary and/or mechanistic significance of this stereochemical mode for aspartate β-decarboxylase is unclear.

Cystathionine γ-synthase is the best studied enzyme catalyzing both γ-replacement and β,γ-elimination reactions. The enzyme is found in plants and bacteria and normally functions to catalyze the formation of cystathionine from *O*-acylhomoserine and cysteine during the biosynthesis of methionine (*66*) [Eq. (57)]:

$$H_3\overset{+}{N}\underset{H}{\overset{CO_2^-}{C}}CH_2CH_2OAcyl + H_3\overset{+}{N}\underset{H}{\overset{CO_2^-}{C}}CH_2SH \longrightarrow H_3\overset{+}{N}\underset{H}{\overset{CO_2^-}{C}}CH_2CH_2SCH_2\underset{H}{\overset{CO_2^-}{C}}\overset{+}{N}H_3 + HOacyl \qquad (57)$$

The enzyme also catalyzes a β,γ-elimination reaction with homoserine (in the absence of cysteine) to form α-ketobutyrate, a reaction having no apparent physiological significance. The results of various isotope-labeling studies on both the normal substitution and aberrant elimination reactions have given rise to a reasonably clear picture of the stereochemical features of this interesting reaction. Experiments on the enzyme from *Salmonella* by Flavin and co-workers, as well as others, demonstrated that the β,γ-elimination reaction involves significant intramolecular proton transfer from both the C_α and the *pro*-(*R*) position at C_β of *O*-succinylhomoserine to the C_γ methyl group of α-ketobutyrate (*274–279*). This process appears to involve retention of configuration at C_γ (*231*). These observations have been rationalized in terms of a mechanism involving a single (possibly polyprotic) base that facilitates a suprafacial proton transfer with respect to a vinylglycine–PLP intermediate having a cisoid conformation [Eq. (58)]:

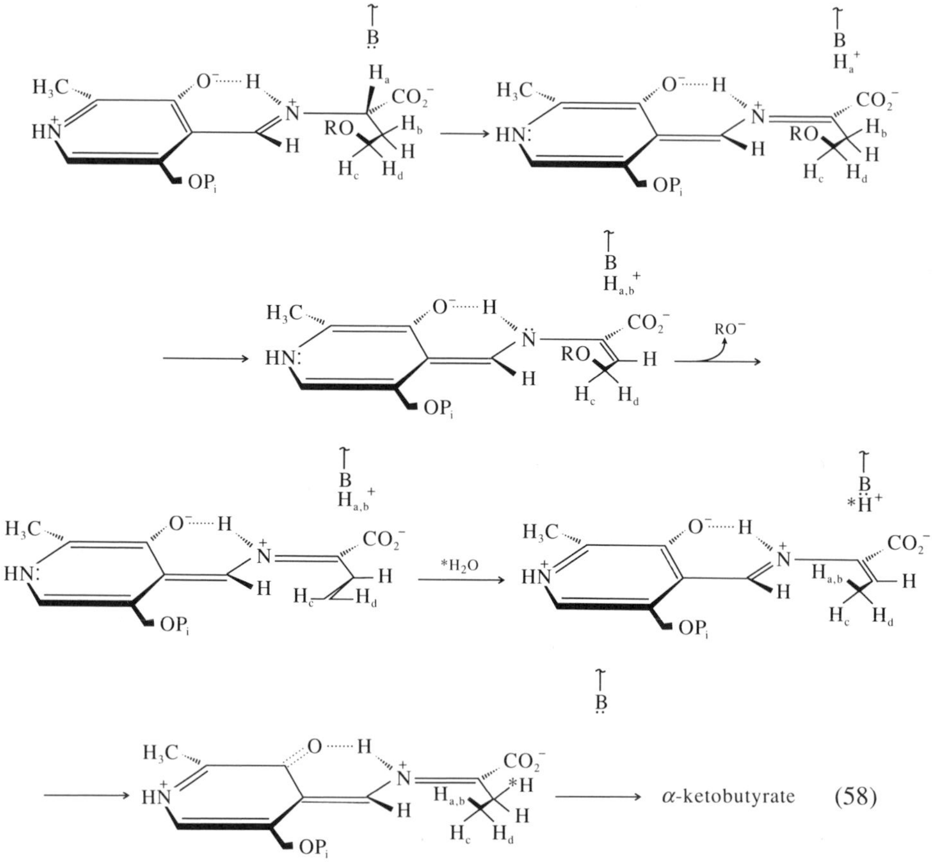

$\longrightarrow$ α-ketobutyrate (58)

In order to determine the stereochemical course at C_γ during the normal replacement reaction, Chang and Walsh converted (*Z*)-DL-[4-^{2}H]vinylglycine and (*E*)-DL-[3,4-$^2H_2$]vinylglycine to cystathionine in presence of cystathionine γ-synthase (*Salmonella*) and cysteine (*280*). Stereospecific chemical and enzymic degradation of the resultant samples of cystathionine to homoserine followed by comparison of the NMR spectra with those of authentic (4*R*)- and (4*S*)-L-[4-^{2}H]homoserine established that cysteine must add to the *si* face of the vinylglycine–PLP complex. Net retention of configuration at C_γ was established by converting authentic (4*R*)- and (4*S*)-L-[4-^{2}H]-*O*-succinylhomoserine to the corresponding cystathionines and comparing their NMR spectra with those of the original cystathionines obtained from the stereospecifically labeled vinylglycines. Taken together these observations suggest the following geometry for the intermediate adduct complex containing the active site base that is involved in the proton transfer reactions associated with catalysis of the normal replacement reaction (**22**):

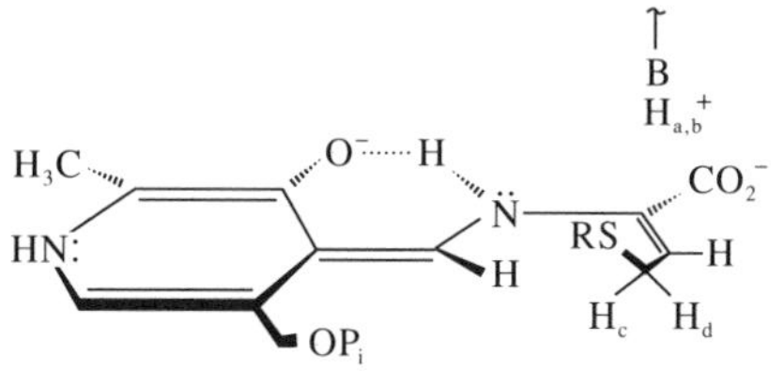

22

where RS is cysteinyl.

VII. β-Elimination–Addition Reactions: E2 versus E1cB Mechanisms

Aside from selected PLP-dependent enzymes, a range of different enzymes catalyze either syn-eliminations, in which the leaving groups are eclipsed (Table X) (*281–290*), or anti-eliminations, in which the leaving groups are antiperiplanar (Table XI) (*291–310*). In the vast majority of cases these two options are stereochemically "cryptic" (giving the same product), since one of the carbons involved in the elimination reaction is a prochiral methylene carbon [e.g., Eq. (59)]:

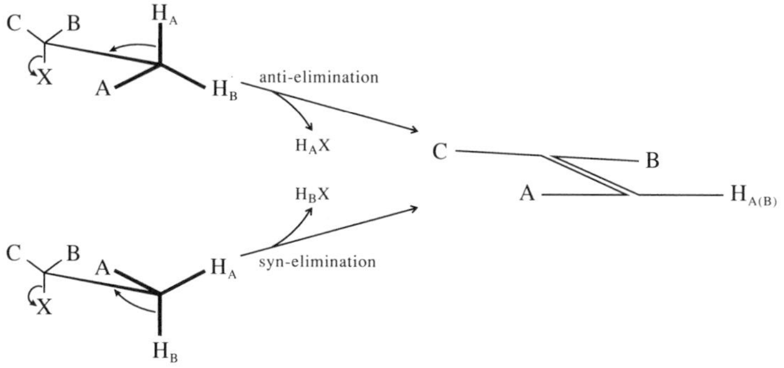

(59)

The question of central importance is whether these two stereochemical options have a basis in mechanistic advantage or whether they were inherited by chance

TABLE X

ENZYME-CATALYZED 1,2-ELIMINATION–ADDITION REACTIONS INVOLVING syn-STEREOCHEMISTRY

Enzyme	Source	Chemistry of elimination step	Ref.
β-Hydroxydecanoylthioester dehydratase (EC 4.2.1.60)	*E. coli*	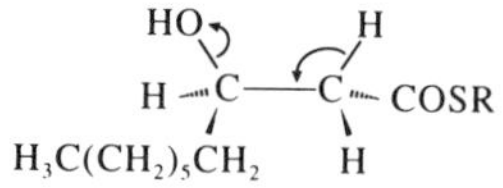 HO, H; H–C–C–COSR; $H_3C(CH_2)_5CH_2$, H	*281*
β-Hydroxybutyryl-CoA dehydratase (EC 4.2.1.55)	*Saccharomyces cerevisiae*	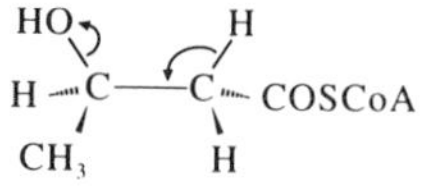 HO, H; H–C–C–COSCoA; CH_3, H	*282*
Methylglutaconyl-CoA hydratase (EC 4.2.1.18)	Liver	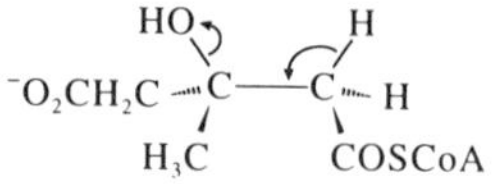 HO, H; $^-O_2CH_2C$–C–C–H; H_3C, COSCoA	*283*
Dehydroquinate dehydratase (EC 4.2.1.10)	*A aerogenes, E. coli*	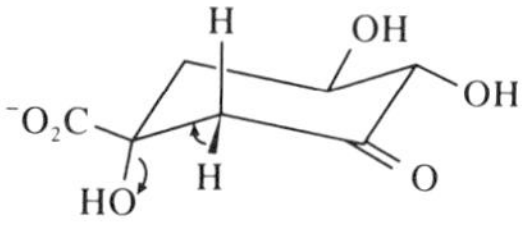 H, OH, OH; ^-O_2C, H, O; HO	*284*
UV Endonuclease V	Bacteriophage T_4	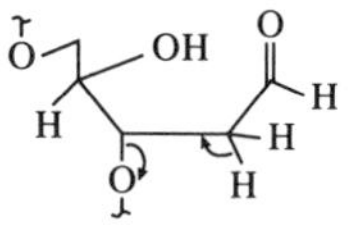 O, OH, O; H, H, H; O, H	*285*
S-Adenosylhomocysteine hydrolase (EC 3.3.1.1)	Liver	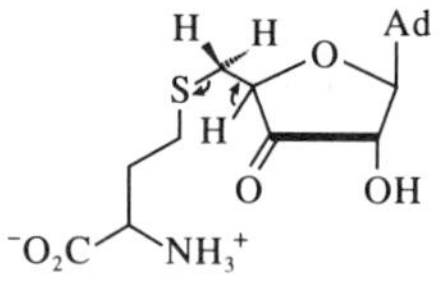 H, H, O, Ad; S, H; O, OH; ^-O_2C, NH_3^+	*286*
Dehydroquinate synthase (EC 4.6.1.3)	*E. coli*	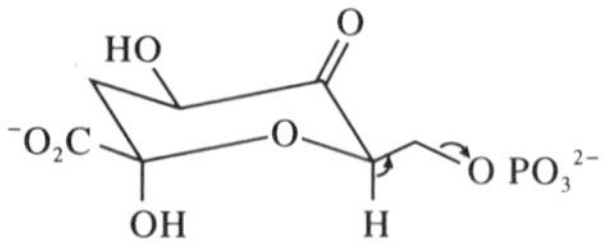 HO, O; ^-O_2C, O, O PO_3^{2-}; OH, H	*287*
Muconate cycloisomerase (EC 5.5.1.1)	*P. putida*	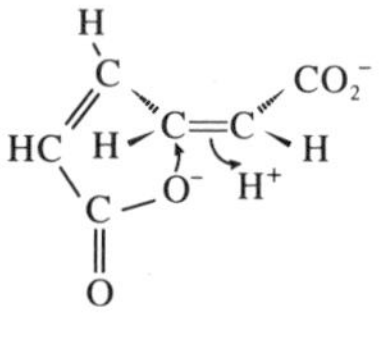 H; C, CO_2^-; HC, H, C=C, H; C, O^-, H^+; O	*288, 289*
3-Carboxy-cis-cis-muconate cyclase (EC 5.5.1.5)	*Neurospora crassa*	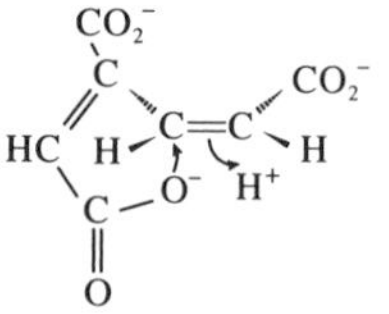 CO_2^-; C, CO_2^-; HC, H, C=C, H; C, O^-, H^+; O	*290*

TABLE XI

ENZYME-CATALYZED 1,2-ELIMINATION–ADDITION REACTIONS INVOLVING anti-STEREOCHEMISTRY

Enzyme	Source	Chemistry of elimination step	Ref.
Fumarase (EC 4.2.1.2; fumarate hydratase)	Heart, *B. cadaveris*	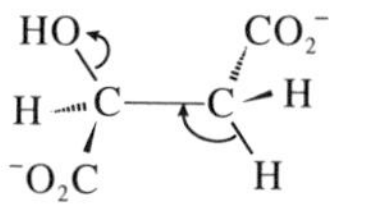 HO; CO_2^-; H—C—C—H; ^-O_2C; H	*291–296*
Maleate hydratase (EC 4.2.1.31)	Kidney	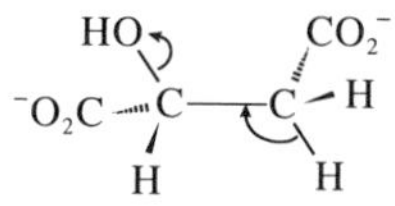 HO; CO_2^-; ^-O_2C—C—C—H; H; H	*295*
Mesaconate hydratase (EC 4.2.1.34)	*C. tetanomorphum*	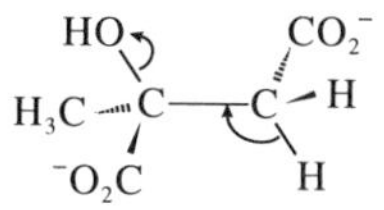 HO; CO_2^-; H_3C—C—C—H; ^-O_2C; H	*296*
Citraconate hydratase (EC 4.2.1.35)	*C. tetanomorphum*	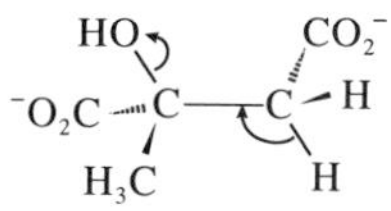 HO; CO_2^-; ^-O_2C—C—C—H; H_3C; H	*296*
Aconitase (EC 4.2.1.3; aconitate hydratase)	Heart	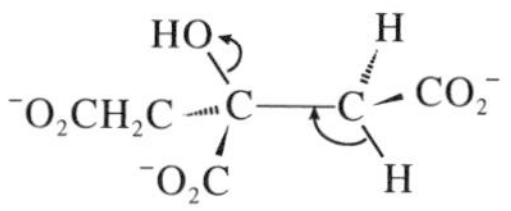 HO; H; $^-O_2CH_2C$—C—C—CO_2^-; ^-O_2C; H	*284*
Enolase (EC 4.2.1.11)	Muscle	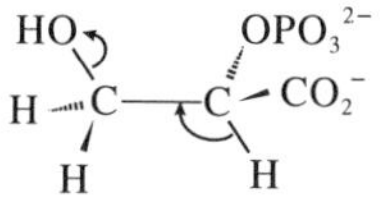 HO; OPO_3^{2-}; H—C—C—CO_2^-; H; H	*297*
2-Isopropylmalate dehydratase (EC 4.2.1.33)	*N. crassa*	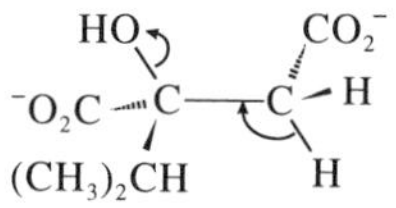 HO; CO_2^-; ^-O_2C—C—C—H; $(CH_3)_2CH$; H	*298*
Aspartate ammonia-lyase (EC 4.3.1.1)	*B. cadaveris, Proteus vulgaris*	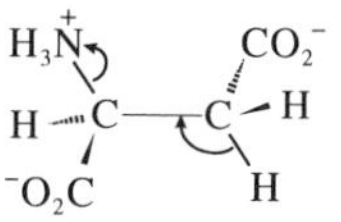 $H_3\overset{+}{N}$; CO_2^-; H—C—C—H; ^-O_2C; H	*292, 293, 299*
Histidine ammonia-lyase (EC 4.3.1.3)	*P. putida*	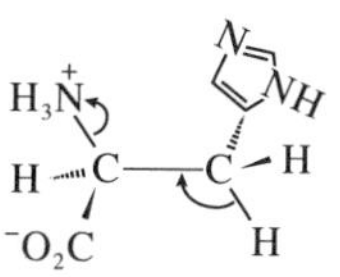 $H_3\overset{+}{N}$; imidazolyl (N, NH); H—C—C—H; ^-O_2C; H	*300, 301*
Phenylalanine ammonia-lyase (EC 4.3.1.5)	Potato, *Rhodotorula texenis, Colchicum*	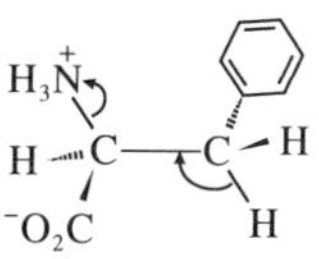 $H_3\overset{+}{N}$; phenyl; H—C—C—H; ^-O_2C; H	*302–304*

(*continued*)

TABLE XI (*continued*)

Enzyme	Source	Chemistry of elimination step	Ref.
Tyrosine ammonia-lyase	Maize	$H_3\overset{+}{N}$, H, ^-O_2C, C—C, H, H, OH	*305, 306*
Argininosuccinate lyase (EC 4.3.2.1)	Liver	$H_2\overset{+}{N}$, (Arg)—NH—C—NH, H, ^-O_2C, C—C, CO_2^-, H, H	*307*
Adenylosuccinate lyase (EC 4.3.2.2)	Yeast	(Ad)—NH, H, ^-O_2C, C—C, CO_2^-, H, H	*308*
Pyrophosphomevalonate decarboxylase (EC 4.1.1.33; diphosphomevalonate decarboxylase)	Liver	$^{2-}O_3PO$, $^{3-}O_6P_2OH_2CH_2C$, H_3C, C—C, H, H, CO_2^-	*309*
3-Carboxy-cis-cis-muconate cycloisomerase (EC 5.5.1.2)	*P. putida*, *Acinetobacter calcoaceticus*	O=C, HC, HC, O^-, ^-O_2C, C=C, CO_2^-, H, H^+	*310*

from ancestral enzymes, the stereochemistries of which were selected at random. Others have noted empirical correlations between substrate structure and reaction stereochemistry that could serve as a basis for a mechanistic hypothesis (*16, 281, 287*). For example, syn-eliminations are usually observed in cases where the abstracted proton is α to the carbonyl of a ketone, aldehyde, or thioester; anti-eliminations are generally observed when the abstracted proton is α to a carboxyl group (*281, 287*). These trends imply a relationship between stereochemistry and the relative kinetic stabilities of carbanion intermediates that might form along the reaction pathways for these enzymes. There is substantial precedent in the chemical literature for E1cB mechanisms of this type with substrates that are "activated" for elimination owing to the presence of electron-withdrawing groups (*311–318*), such as an acyl function (*315, 316*).

A. syn-ELIMINATIONS

The mechanistic features of several enzyme-catalyzed syn-elimination reactions either clearly demonstrate or strongly imply the formation of a stable carbanion intermediate. These include the previously discussed PLP-dependent syn-eliminations as well as several of the enzymes listed in Table X.

S-Adenosylhomocysteine hydrolase contains tightly bound NAD^+ that functions to oxidize the C-3′ hydroxyl of substrate prior to the elimination of homocysteine to yield bound keto-4′,5′-dehydro-5′-deoxyadenosine (*319, 320*). The latter then adds water to produce 3′-ketoadenosine, which is reduced to adenosine by bound NADH to complete the catalytic cycle. Presumably, the cyclic oxidation of the C-3′ hydroxyl of substrate serves to stabilize an incipient carbanion formed as an obligatory intermediate during catalysis.

Dehydroquinate synthase catalyzes the β-elimination of phosphate from 3-deoxy-D-arabinoheptulosonate 7-phosphate (**23**) to form (**26**), as a partial reaction during the formation of dehydroquinate (**28**) [Eq. (60)]:

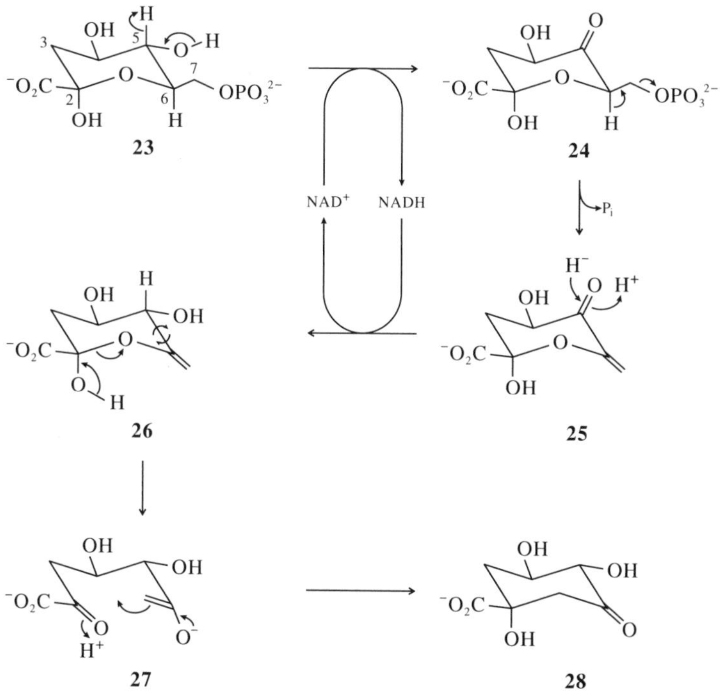

(60)

Early work primarily from the laboratory of Sprinson (*321, 322*) and subsequent studies by Knowles and co-workers (*323–325*) suggest that tightly bound NAD^+ functions to reversibly oxidize the C-5 hydroxyl of substrate in a fashion analogous to that for *S*-adenosylhomocysteine hydrolase.syn-Stereochemistry for the elimination step (**24** → **25**) was indirectly established on the basis of the observation that the 2-deoxy analog of substrate (**29**), containing deuterium in the *pro*-(*S*) position at C-7, is converted by the enzyme to the enol ether **30,** established to have the *E*-configuration by the application of NOE difference spectroscopy to a bicyclic lactone derivative of **30** (*325*) [Eq. (61)]:

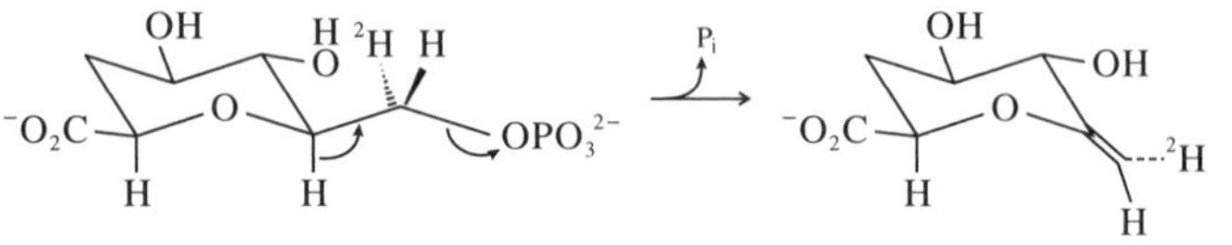

29 → **30** (61)

In support of an E1cB mechanism, dehydroquinate synthase is capable of catalyzing solvent deuterium exchange at C-6 of a phosphonate analog of substrate incapable of undergoing the elimination reaction (*324*).

The subsequent metabolic reaction catalyzed by dehydroquinate dehydratase presumably also involves a stable carbanion intermediate, judging from the demonstrated formation of a Schiff base during catalysis (*326*). The likely conformation of the intermediate is that of a skew-boat in which the C-4 and C-5 hydroxyls are di-equitorial (**31**) (*327*).

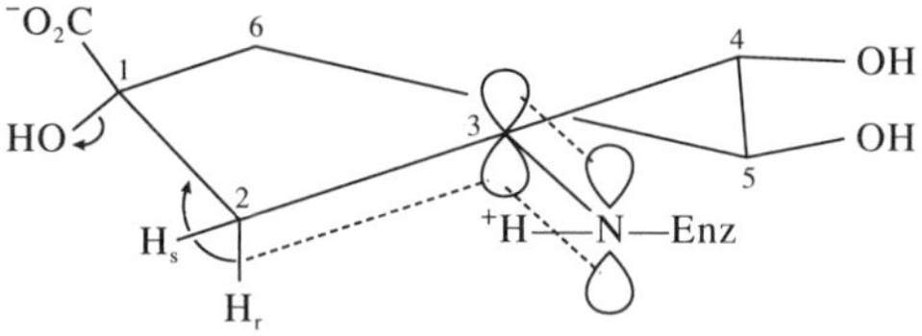

31

This is supported by the observation that the isopropylidene derivative of the substrate, in which the hydroxyls are constrained to be di-equitorial, is also a good substrate for the enzyme.

Thus, syn-stereochemistry coupled with the evidence for stable carbanion intermediates for at least some of the reactions listed in Table X might be interpreted to reflect the operation of a single active site base alternately functioning in a general base and general acid capacity during catalysis [e.g., Eq. (55)]. [In the case of dehydroquinate synthase, the base may actually be the phosphate group of substrate (*324*).] Clearly, this interpretation of syn-stereochemistry is analogous to the single-base interpretation of retention of configuration observed among the carbon–carbon lyase reactions that appear to involve stable carbanion intermediates. However, it must be emphasized that an E1cB mechanism may not be operative for all of the reactions listed in Table X. The substrates for the muconate and 3-carboxymuconate cycloisomerases, contrary to the aforementioned empirical generalization, do not contain functional groups that would readily stabilize a carbanion intermediate. A stepwise E1 mechanism (involving an allylic carbocation intermediate) is a reasonable mechanistic alternative for this enzyme, for which there is precedent in the chemical literature (*328*). A single-base mechanism could also operate in the case of an E1 mechanism.

B. anti-ELIMINATIONS

In comparison with the syn-eliminations of Table X, the substrates for enzyme-catalyzed anti-eliminations (Table XI) are generally not so well suited for stabilization of the carbanion intermediates associated with E1cB mechanisms. Therefore, it is tempting to interpret anti-stereochemistry in support of the stereoelectronically allowed concerted (E2) mechanism that is usually associated with the conversion of unactivated alkyl compounds to alkenes in simple chemical systems (*311–313*). However, there may not be a clear mechanistic distinction between enzyme-catalyzed syn- and anti-eliminations. Carbanion intermediates have been implicated in the reactions catalyzed by fumarase (*329*) and phenylalanine ammonia-lyase (*330*) (Table XI) on the basis of kinetic isotope effect measurements. Moreover, carbanion intermediates, in their *aci*-carboxylate forms [$-CH{=}C(O^-)O^-$], have been proposed to form during catalysis by fumarase (*331*), aspartase (aspartate ammonia-lyase) (*331*), aconitase (aconitate hydratase) (*332*), and enolase (*333*) in order to explain the observation that nitronate analogs of substrates are tight-binding inhibitors of these enzymes. Nevertheless, these observations do not exclude a mechanistic interpretation of reaction stereochemistry provided that the carbanions formed during anti-eliminations are kinetically less stable than those formed during syn-eliminations. Thus, the potential mechanistic advantage of the anti-mode is that it may allow the concerted or tightly coupled action of two or more active site functional groups to trap kinetically unstable carbanion intermediates, perhaps in a fashion analogous to that envisioned for the carbon–carbon lyases involving inversion of configuration.

Finally, the stereochemical divergence between the 3-carboxymuconate cycloisomerase (the last entry in Table XI) and the two corresponding muconate cycloisomerases catalyzing syn-eliminations (the last two entries in Table X) is intriguing. Whether this difference has a basis in mechanism or in convergent evolution from different ancestral proteins (or both) is unclear (*310, 334*). Additional mechanism studies on these enzymes would be most welcome.

VIII. Methyl Group Transfer Reactions: One- versus Two-Step Mechanisms

The chiral [$^{1}H,^{2}H,^{3}H$]methyl group has been an essential experimental tool for evaluating stereochemical trends among enzyme-catalyzed reactions involving the interconversion of methyl and methylene functions ($-CH_3 \rightleftharpoons -CH_2-$) (Tables VI–VIII). Therefore, it seems appropriate to end this chapter with a review of the most recent general application of chiral labeled methyl groups, namely, that involving the elucidation of stereochemical trends among the *C*-, *N*-, *O*-, and *S*-methyltransferases ($-YCH_3 + X- \rightleftharpoons H_3CX- + Y-$) (Table XII) (*335–345*). Both retention and inversion of configuration have been observed among these reactions. There appears to be a general correlation between the structure of the methyl group donor and the stereochemical fate of the methyl group.

A. Methods

The laboratories of Floss and Arigoni laid the experimental foundation for the stereochemical analysis of methyl group transfer reactions, on the basis of the synthesis from chiral [$^{1}H,^{2}H,^{3}H$]acetate of methionine and *S*-adenosylmethionine carrying a chiral [$^{1}H,^{2}H,^{3}H$]methyl group (*335, 341*), for example,

$$Na^+\ {}^-O_2C{-}C(^{1}H)(^{2}H)(^{3}H)\ (S) \xrightarrow{NaN_3,\ H_2SO_4} H_2N{-}C(^{1}H)(^{2}H)(^{3}H)\ (S) \xrightarrow{TsCl,\ NaOH} TsNH{-}C(^{1}H)(^{2}H)(^{3}H)\ (S)$$

$$\xrightarrow[DMF]{TsCl,\ NaH} (Ts)_2N{-}C(^{1}H)(^{2}H)(^{3}H)\ (S) \xrightarrow[Na^0,\ HMPA]{S\text{-benzyl-L-homocysteine}} (^{1}H)(^{2}H)(^{3}H)C{-}S(CH_2)_2C(\overset{+}{N}H_3)(H)CO_2^-\ (R)$$

$$\xrightarrow[ATP,\ Mg^{2+}]{S\text{-Adenosylmethionine synthetase}} (^{1}H)(^{2}H)(^{3}H)C{-}\overset{Ad}{S^+}(CH_2)_2C(\overset{+}{N}H_3)(H)CO_2^-\ (R) \tag{62}$$

Floss and co-workers first employed (2*S*)-(*S*-methyl)- and (2*S*)-(*R*-methyl)-[*methyl*-$^{14}C,^{1}H,^{2}H,^{3}H$]methionine to demonstrate that both the C- and N-methylation reactions associated with the *in vivo* biosynthesis (*Streptomyces griseus*) of indolmycin involve inversion of configuration (Table XII). Configurational analysis of the labeled methyl groups of the isolated indolmycin involved stereospecific chemical degradation of the indolmycin to labeled acetic acid, followed by chiral analysis using malate synthase/fumarase (Fig. 15). Most of the subsequent studies summarized in Table XII have employed the common strategy of transferring the labeled methyl group of product to CN^-, followed by conversion of the labeled acetonitrile to labeled acetic acid for chiral analysis.

TABLE XII

STEREOCHEMISTRY OF ENZYME-CATALYZED METHYL GROUP TRANSFER REACTIONS

System	Stereochemical analysis			
	Active donor species	Product species	Reaction stereo-chemistry	Ref.
Purified catechol *O*-methyltransferase (liver) (EC 2.1.1.6)	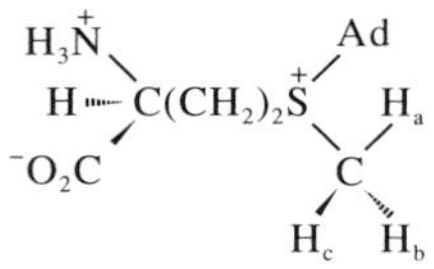 $H_3\overset{+}{N}$, H—C(CH_2)$_2\overset{+}{S}$(Ad)—C(H_a)(H_b)(H_c), $^{-}O_2C$	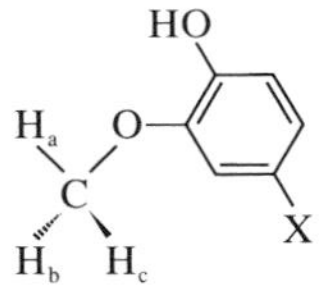 HO; H_a, H_b, H_c—C—O; X X = —CH(OH)CH_2NHCH_3; —CO_2^-	Inversion	*335*
Purified histamine *N*-methyltransferase (brain) (EC 2.1.1.8)	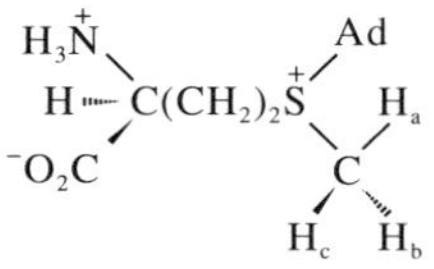 $H_3\overset{+}{N}$, H—C(CH_2)$_2\overset{+}{S}$(Ad)—C(H_a)(H_b)(H_c), $^{-}O_2C$	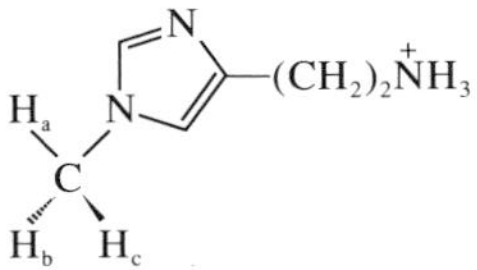 N; H_a, H_b, H_c—C—N; —(CH_2)$_2\overset{+}{N}H_3$	Inversion	*336*
Purified norreticuline *N*-methyltransferase from *Berberis koetineana*	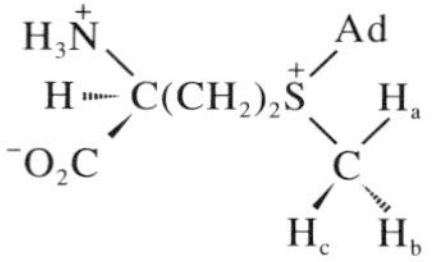 $H_3\overset{+}{N}$, H—C(CH_2)$_2\overset{+}{S}$(Ad)—C(H_a)(H_b)(H_c), $^{-}O_2C$	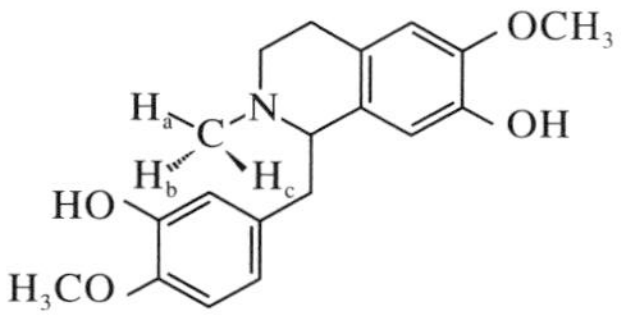 OCH_3; OH; H_a, H_b, H_c—C—N; HO; H_3CO	Inversion	*337*

(continued)

TABLE XII (*continued*)

System	Stereochemical analysis: Active donor species	Stereochemical analysis: Product species	Stereochemical analysis: Reaction stereo-chemistry	Ref.
Purified *O*-methyl-transferase from *Berberis koetineana*	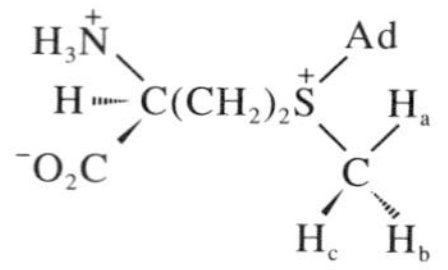 $H_3\overset{+}{N}$, H, $C(CH_2)_2\overset{+}{S}$, Ad, $^{-}O_2C$, H_a, C, H_c, H_b	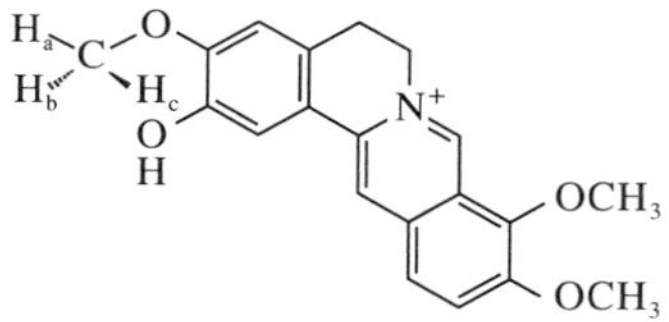 H_a, H_b, C, H_c, O, OH, N^+, OCH_3, OCH_3	Inversion	*338*
Purified *O*-methyl-transferase from *Phaseolus aureus* (mung beans)	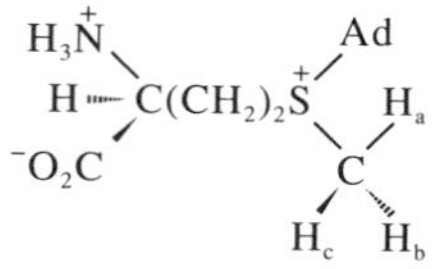 $H_3\overset{+}{N}$, H, $C(CH_2)_2\overset{+}{S}$, Ad, $^{-}O_2C$, H_a, C, H_c, H_b	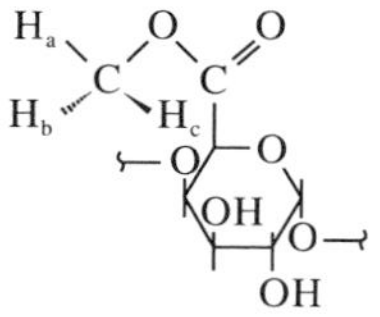 H_a, H_b, C, H_c, O, C, O, O, O, OH, O, OH	Inversion	*339*
Streptomyces griseus (indolmycin biosynthesis)	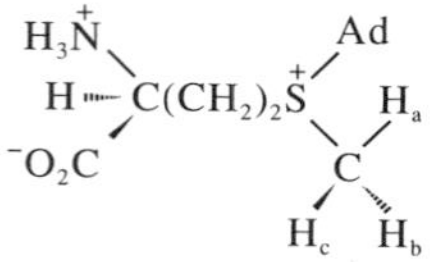 $H_3\overset{+}{N}$, H, $C(CH_2)_2\overset{+}{S}$, Ad, $^{-}O_2C$, H_a, C, H_c, H_b	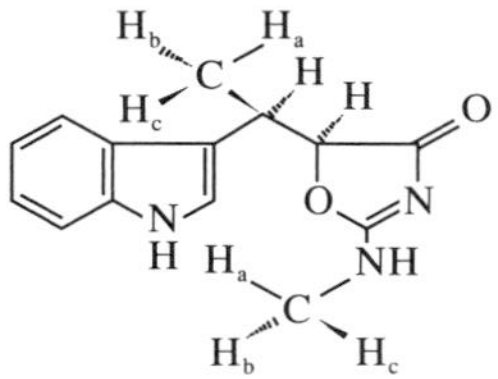 H_b, H_a, C, H_c, H, H, O, N, H, O, N, NH, H_a, C, H_b, H_c	Inversion	*340*

Purified homocysteine methyltransferase from jack beans	$H_3\overset{+}{N}$, H, ^-O_2C–$C(CH_2)_2\overset{+}{S}$–[$C(H_a)(H_b)(H_c)$]–[$C(H_a)(H_c)(H_b)$]	$C(H_a)(H_b)(H_c)$–$S(CH_2)_2C(CO_2^-)(H)(\overset{+}{N}H_3)$ from homocysteine	Inversion	*341*
Menyanthes trifoliata (loganin biosynthesis)	$H_3\overset{+}{N}$, H, ^-O_2C–$C(CH_2)_2\overset{+}{S}$(Ad)–$C(H_a)(H_c)(H_b)$	$C(H_a)(H_b)(H_c)$–O–C(=O)– [ring: O, OH, CH_3, $O_{glucose}$]	Inversion	*341*
P. shermanii (vitamin B12 biosynthesis)	$H_3\overset{+}{N}$, H, ^-O_2C–$C(CH_2)_2\overset{+}{S}$(Ad)–$C(H_a)(H_c)(H_b)$	$\{(H_a)(H_b)(H_c)CC\}_7$ cyanocobalamin	Inversion	*341*
Methanosarcina barkeri (cell-free extract)	HO–$C(H_a)(H_c)(H_b)$	$C(H_a)(H_c)(H_b)$–$S(CH_2)_2SO_3^-$ methyl-coenzyme M	Retention	*342*
Streptomyces cattleya (thienamycin biosynthesis)	$H_3\overset{+}{N}$, H, ^-O_2C–$C(CH_2)_2S$–$C(H_a)(H_c)(H_b)$?	$C(H_a)(H_c)(H_b)$–CH(OH)– [H H, O=, N, CO_2^-]–$S(CH_2)_2\overset{+}{N}H_3$	Retention	*343*

(*continued*)

TABLE XII (*continued*)

System	Stereochemical analysis			Ref.
	Active donor species	Product species	Reaction stereo- chemistry	
Purified methionine syn- thase from *E. coli*	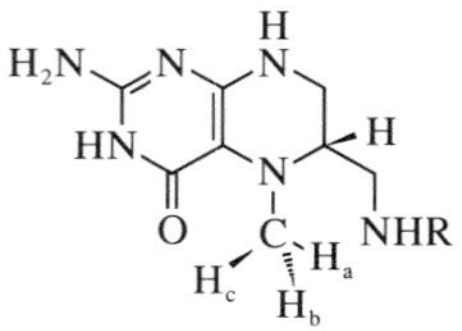 H_2N, N, H N, H, HN, O, N, C, H_a, H_b, H_c, NHR	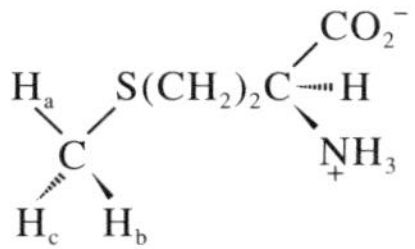 H_a, $S(CH_2)_2C$, CO_2^-, H, $\overset{+}{N}H_3$, C, H_c, H_b	Retention	*344*
Clostridium thermo- aceticum (cell-free extract)	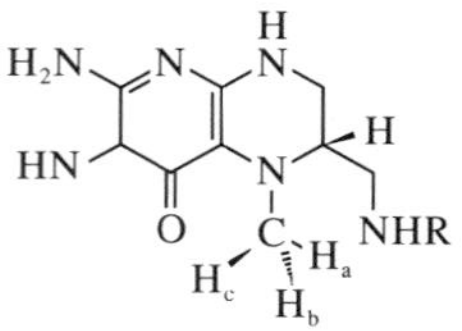 H_2N, N, H N, H, HN, O, N, C, H_a, H_b, H_c, NHR	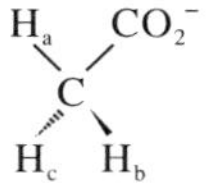 H_a, CO_2^-, C, H_c, H_b	Retention	*345*

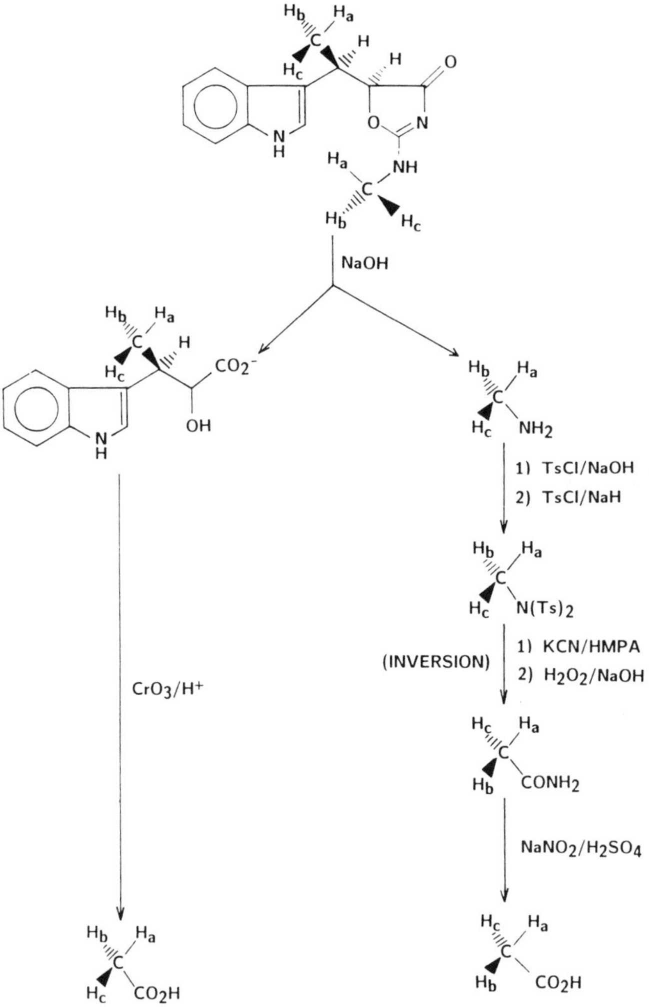

FIG. 15. Chemical methods for converting labeled indolmycin to samples of chiral [^{3}H,^{2}H]acetic acid (*340*).

A particularly elegant method has recently been devised for the stereoselective synthesis of (*R*-methyl)- and (*S*-methyl)-N^5-[*methyl*-^{1}H,^{2}H,^{3}H]tetrahydrofolate (N^5-CH_3-THF) for examining the stereochemistry of THF-dependent enzymes (*344*). For example, synthesis of the former substrate first involves reduction of N^5,N^{10}-methenyl-THF with NaB^2H_4, demonstrated to be approximately 80% stereoselective on the basis of proton nuclear Overhauser effect measurements on the product (*346*) [Eq. (63)]:

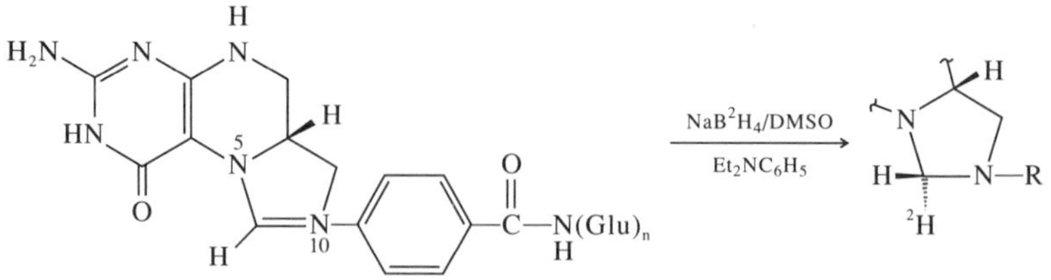

$$\xrightarrow[\text{Et}_2\text{NC}_6\text{H}_5]{\text{NaB}^2\text{H}_4/\text{DMSO}} \qquad (63)$$

Further reduction using NaB^3H_4 resulted in a 44% enantiomeric excess of (*R*-methyl)-[*methyl*-^{1}H,^{2}H,^{3}H]THF [Eq. (64)]:

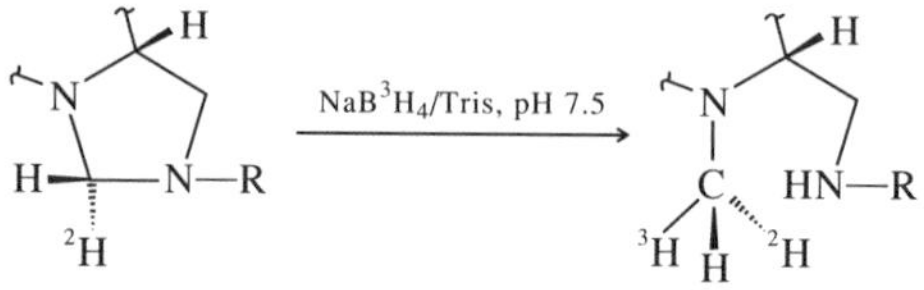

$$\xrightarrow{\text{NaB}^3\text{H}_4/\text{Tris, pH 7.5}} \qquad (64)$$

as judged by chiral analysis of the labeled acetic acid resulting from chemical degradation of the product mixture. Use of this material as a substrate for the cobalamin-dependent methionine synthase from *E. coli* resulted in a 21% enantiomeric excess of (2*S*)-(*R*-methyl)-[*methyl*-^{1}H,^{2}H,^{3}H]methionine, indicating substantial retention of configuration during methyl transfer (Table XII).

B. Empirical Generalization

Inspection of Table XII suggests a tentative correlation between the structure of the methyl group donor and net reaction stereochemistry. In cases where the donor species has been clearly identified as an *active* methylating agent, owing to the presence of a sulfonium ion, inversion of configuration is observed. This is consistent with a one-step S_N2 mechanism of the same type that is normally observed with *S*-adenosylmethionine in nonenzymic displacement reactions [Eq. (65)]:

$$-\text{YC}(\text{H}_a)(\text{H}_b)(\text{H}_c) + :\text{X}- \rightleftharpoons (\text{H}_a)(\text{H}_b)(\text{H}_c)\text{CX}- + :\text{Y}- \qquad (65)$$

In principle, inversion of configuration can be explained by any odd number of S_N2 displacement reactions. However, there is no clear evidence for the formation of methyl–enzyme intermediates among any of the enzyme-catalyzed methyl transfer reactions that exhibit net inversion of configuration. In the case of catechol *O*-methyltransferase, the observation of inversion of configuration argues against a previously proposed Ping-Pong mechanism presumably involving a single methyl–enzyme intermediate (*347*). Such a mechanism would involve two nucleophilic displacements resulting in net retention of configuration. Kinetic α-^{2}H and ^{13}C isotope effects on the methyl transfer reaction catalyzed by catechol *O*-methyltransferase support an S_N2-like transition state in which the transferring methyl is "symmetrically" positioned between the donor and acceptor atoms (*348*).

In those cases where the methyl group donor is chemically less reactive, as with N^5-methyl-THF and methanol, retention of configuration is observed (Table XII). For the cobalamin-dependent methionine synthase, retention of configuration is consistent with a postulated mechanism for this enzyme involving two sequential transfers of the methyl group, one from N^5-methyl-THF to cobalt to generate methylcobalamin and a second from cobalt to the sulfur of homocysteine (*349*) [Eq. (66)]:

Enz $\xrightarrow{N^5\text{—}CH_3\text{—}THF}$ [N^5-methyl-THF bound; H_a, H_b, H_c methyl C attacked by Co^I; N, N, N, N; DBC] $\xrightarrow{\text{THF}\quad\text{homocysteine}}$ [$^-S(CH_2)_2CH(NH_3^+)CO_2^-$; H_a, H_c, C—H_b on $Co^{(III)}$; N, N, N, N; DBC]

$\xrightarrow{\text{Enz}}$ H_a, H_b, H_cC—$S(CH_2)_2CH(NH_3^+)CO_2^-$ (66)

where DBC is dimethylbenzimidazolyl cobalamide.

Similar explanations can be envisioned for net retention of configuration during the biosynthesis of methyl–coenzyme M from methanol and the formation of acetic acid from N^5-methyl-THF, as these systems also depend on corrinoid enzymes (*350, 351*). Thus, retention of configuration in these systems may reflect a fundamental mechanistic advantage of using $Co^{(I)}$, a powerful nucleophile (*352*), to overcome the problem of cleaving the relatively inert C–O and C–N bonds of methanol and N^5-methyl-THF, respectively. Whether the biosynthesis of thienamycin from methionine conforms to the empirical generalization suggested here is unclear. *S*-Adenosylmethionine may be the active methylating agent *in vivo* (predicting inversion of configuration). However, the involvement of a corrin cannot be excluded, since the biosynthesis of this antibiotic appears to have a requirement for cobalt (*353*).

IX. Concluding Remarks

Since 1970 there has been a dramatic increase in the number of enzymes of known stereochemistry, owing to the broad application of numerous elegant methods of stereochemical analysis. An important challenge facing bioorganic chemists is to distinguish between those cases in which the stereochemical features of an enzymic reaction are functionally neutral and those that reflect underlying mechanistic advantages unavailable through alternative stereochemical modes. Nowhere is the potential difficulty of making such distinctions more clearly illustrated than by the competing functional versus nonfunctional explanations for stereochemical trends among the $NAD(P)^+$-dependent dehydrogenases. Nevertheless, several of the empirical generalizations noted in this chapter, when combined with structural and mechanistic data, *imply* that the kinetic stabilities of reaction intermediates are an important factor controlling the stereochemical properties of many enzymic reactions: The occurrence of stable intermediates along the reaction pathway of an enzyme appears to allow the deployment and reuse of a minimum number of catalytic groups [the "minimum number rule" of Hanson and Rose (*16*)], accounting for suprafacial proton transfers among the isomerases, retention of configuration among many Schiff base- and metal ion-dependent carbon–carbon lyases, and syn-stereochemistry for enzyme-catalyzed β-eliminations involving substrates that are structurally suited for stabilizing carbanion intermediates. In contrast, the occurrence of kinetically unstable intermediates requires the deployment of multiple catalytic groups and their tightly coupled action in order to trap the intermediate, accounting for inversion of configuration among selected carbon–carbon lyases and anti-stereochemistry for enzyme-catalyzed β-elimination reactions whose substrates are poorly suited for stabilizing carbanion intermediates. Nevertheless, the empirical generalizations on which these interpretations rest are not without exceptions. Perhaps the next 20 years will lead to a more quantitative assessment of the relative importance of functional versus ancestral explanations for enzyme reaction stereochemistry.

Acknowledgments

We thank Professors Ralph M. Pollack, Joel F. Liebman, Dale L. Whalen, and Paul E. Dietze for critical comments, and Ms. Patricia L. Gagné for preparing the manuscript.

References

1. Popják, G. (1970). "The Enzymes," 3rd Ed., Vol. 2, p. 115.
2. Westheimer, F. H. (1987). *In* "Pyridine Nucleotide Coenzymes" (D. Dolphin, R. Poulson, and O. Avramorić, eds.), Part A, p. 253. Wiley, New York.
3. Cornforth, J. W., Redmond, J. W., Eggerer, H., Buckel, W., and Gutschow, C. (1969). *Nature (London)* **221,** 1212.
4. Luthy, J., Retey, J., and Arigoni, D. (1969). *Nature (London)* **221,** 1213.
5. Cahn, R. S., and Ingold, C. K. (1951). *J. Chem. Soc.*, p. 612.
6. Cahn, R. S., Ingold, C. K., and Prelog, V. (1956). *Experientia* **12,** 81.

7. Cahn, R. S., Ingold, C. K., and Prelog, V. (1966). *Angew. Chem. Int. Ed. Engl.* **5,** 385.
8. Hanson, K. R. (1966). *J. Am. Chem. Soc.* **88,** 2731.
9. Bentley, R. (1969, 1970). "Molecular Asymmetry in Biology." Vols. 1 and 2. Academic Press, New York.
10. Alworth, W. L. (1972). "Stereochemistry and Its Application in Biochemistry." Wiley (Interscience), New York.
11. Cornforth, J. W. (1969). *Q. Rev. Chem. Soc.* **23,** 125.
12. Rose, I. A. (1970). *Annu. Rev. Biochem.* **35,** 23.
13. Rose, I. A. (1972). *Crit. Rev. Biochem.* **1,** 33.
14. Hanson, K. R. (1972). *Annu. Rev. Plant Physiol.* **23,** 335.
15. Goodwin, T. W. (1973). *Essays Biochem.* **9,** 103.
16. Hanson, K. R., and Rose, I. A. (1975). *Acc. Chem. Res.* **8,** 1.
17. Hanson, K. R. (1976). *Annu. Rev. Biochem.* **45,** 307.
18. Floss, H. G., and Tsai, M.-D. (1979). *Adv. in Enzymol.* **50,** 243.
19. Floss, H. G. (1982). "Methods in Enzymology," Vol. 87, p. 126.
20. Floss, H. G. (1986). *In* "Mechanisms of Enzymatic Reactions: Stereochemistry" (P. A. Frey, ed.), p. 71. Elsevier, New York.
21. Knowles, J. R. (1980). *Annu. Rev. Biochem.* **49,** 877.
22. Eckstein, F. (1983). *Angew. Chem. Int. Ed. Engl.* **22,** 423.
23. Frey, P. A. (1982). *Tetrahedron* **38,** 1541.
24. Buchwald, S. L., Hansen, D. E., Hassett, A., and Knowles, J. R. (1982). "Methods in Enzymology," Vol. 87, p. 279.
25. Eckstein, F., Romaniuk, P. J., and Connolly, B. A. (1982). "Methods in Enzymology," Vol. 87, p. 197.
26. Benner, S. A., and Ellington, A. D. (1988). *Crit. Rev. Biochem.* **23,** 369.
27. Anderson, V. E., and LaReau, R. D. (1988). *J. Am. Chem. Soc.* **110,** 3695.
28. Popják, G., and Cornforth, J. W. (1966). *BJ* **101,** 553.
29. Grau, U. M., Trommer, W. E., and Rossman, M. G. (1981). *JMB* **151,** 289.
30. Eklund, H., Plapp, B. V., Samama, J.-P., and Branden, C.-I. (1982). *JBC* **257,** 14349.
31. Shapiro, S., Arunachalam, T., and Caspi, E. (1983). *J. Am. Chem. Soc.* **105,** 1642.
32. Jornvall, H., Eklund, H., and Branden, C.-I. (1978). *JBC* **253,** 8414.
33. Plapp, B. V. (1986). *In* "Mechanisms of Enzymatic Reactions: Stereochemistry" (P. A. Frey, ed.), p. 343. Elsevier, New York.
34. You, K.-S. (1985). *Crit. Rev. Biochem.* **17,** 313.
35. Westheimer, F. H., Fisher, H. F., Conn, E. E., and Vennesland, B. J. (1951). *J. Am. Chem. Soc.* **73,** 2403.
36. Talalay, P., Loewus, F. A., and Vennesland, B. (1955). *JBC* **212,** 801.
37. Vennesland, B. (1956). *J. Cell. Comp. Physiol. Suppl.* **1,** 201.
38. Nakamoto, T., and Vennesland, B. (1960). *JBC* **235,** 202.
39. Levy, H. R., Talalay, P., and Vennesland, B. (1962). *Prog. Stereochem.* **3,** 299.
40. Levy, H. R., and Vennesland, B. (1957). *JBC* **228,** 85.
41. Krakow, G., Ludoweig, J., Mather, J. H., Normore, W. M., Tosi, L., Udaka, S., and Vennesland, B. (1963). *Biochemistry* **2,** 1009.
42. Cornforth, J. W., Ryback, G., Popják, G., Donninger, C., and Schroepfer, G., Jr. (1962). *BBRC* **9,** 371.
43. Weber, H., Seibl, J., and Arigoni, D. (1966). *Helv. Chim. Acta* **49,** 741.
44. Arigoni, D., and Eliel, E. L. (1969). *In* "Topics in Stereochemistry" (E. L. Eliel and N. L. Allinger, eds.), Vol. 4, p. 127. Wiley (Interscience), New York.
45. You, K., Arnold, L. J., Jr., and Kaplan, N. O. (1977). *ABB* **100,** 550.
46. Ehmke, A., Flossdorf, J., Habicht, W., Schiebel, H. M., and Schulten, H. R. (1980). *Anal. Biochem.* **101,** 413.

47. Bentley, R. (1970). "Molecular Asymmetry in Biology," Vol. 2. Academic Press, New York.
48. Davies, D. D., Teixeria, A., and Kenworthy, P. (1972). *BJ* **127,** 335.
49. Davies, D. D., and Kenworth, P. (1982). *BJ* **205,** 581.
50. Mansell, R. L., Gross, G. G., Stockigt, J., Franke, H., and Zenk, M. H. (1974). *Phytochemistry* **13,** 2427.
51. You, K.-S., Arnold, L. J., Allison, W. S., and Kaplan, N. O. (1978). *TIBS* **3,** 265.
52. Srivastava, D. K., and Bernhard, S. A. (1985). *Biochemistry* **24,** 623.
53. Srivastava, D. K., and Bernhard, S. A. (1986). *Curr. Top. Cell. Regul.* **28,** 1.
54. Srivastava, D. K., and Bernhard, S. A. (1986). *Science* **234,** 1081.
55. Srivastava, D. K., and Bernhard, S. A. (1987). *Annu. Rev. Biophys. Chem.* **16,** 175.
56. Srere, P. A. (1987). *Annu. Rev. Biochem.* **56,** 89.
57. Srivastava, D. K., Bernhard, S. A., Langridge, R., and McClarin, J. A. (1985). *Biochemistry* **24,** 629.
58. Chock, P. B., and Gutfreund, H. (1988). *PNAS* **85,** 8870.
59. Srivastava, D. K., and Bernhard, S. A. (1987). *Biochemistry* **26,** 1240.
60. Benner, S. (1982). *Experientia* **32,** 634.
61. Benner, S. A., Nambiar, K., and Chambers, G. K. (1985). *J. Am. Chem. Soc.* **107,** 5513.
62. Oppenheimer, N. J. (1984). *J. Am. Chem. Soc.* **106,** 3032.
63. Nambiar, K. P., Stauffer, D. M., Kolodziej, P. A., and Benner, S. A. (1983). *J. Am. Chem. Soc.* **105,** 5886.
64. Albery, W. J., and Knowles, J. R. (1976). *Biochemistry* **15,** 5631.
65. Moras, D., Olsen, K. W., Sabesan, M. N., Buehner, M., Ford, G. C., and Rossman, M. G. (1975). *JBC* **250,** 9137.
66. Walsh, C. (1979). "Enzymatic Reaction Mechanisms." Freeman, San Francisco, California.
67. Rossman, M. G., Liljas, A., Branden, C.-I., and Banaszak, L. J. (1975). "The Enzymes," 3rd Ed., Vol. 11, p. 61.
68. Gronenborn, A. M., and Clore, G. M. (1982). *JMB* **157,** 155.
69. Levy, H. R., Ejchart, A., and Levy, G. C. (1983). *Biochemistry* **22,** 2792.
70. Matthews, D. A., Alden, R. A., Freer, S. T., Xuong, N., and Kraut, J. (1979). *JBC* **254,** 4144.
71. Feeney, J., Birdsall, B., Roberts, G. C. K., and Burgen, A. S. V. (1983). *Biochemistry* **22,** 628.
72. Gronenborn, A. M., Clore, G. M., and Jeffery, J. (1984). *JMB* **172,** 559.
73. Biesecker, G., Harris, J. I., Thierry, J. C., Walker, J. E., and Wonacott, A. J. (1977). *Nature (London)* **266,** 328.
74. Pai, E. F., and Schulz, G. E. (1983). *JBC* **258,** 1752.
75. Sarma, R., and Mynott, R. J. (1973). *J. Am. Chem. Soc.* **95,** 1641.
76. Birdsall, B., Birdsall, N. J. M., Feeney, J., and Thornton, J. (1975). *J. Am. Chem. Soc.* **97,** 2845.
77. Wu, Y.-D., and Houk, K. N. (1987). *J. Am. Chem. Soc.* **109,** 2226.
78. Walsh, C. (1980). *Acc. Chem. Res.* **13,** 148.
79. Bruice, T. C. (1980). *Acc. Chem. Res.* **13,** 256.
80. Hersh, L., and Jorns, M. (1975). *JBC* **250,** 8728.
81. Jorns, M., and Hersh, L. (1974). *J. Am. Chem. Soc.* **96,** 4012.
82. Fisher, J., and Walsh, C. (1974). *J. Am. Chem. Soc.* **96,** 4345.
83. Spencer, R., Fisher, J., and Walsh, C. (1976). *Biochemistry* **15,** 1043.
84. Fisher, J., Spencer, R., and Walsh, C. (1976). *Biochemistry* **15,** 1054.
85. Yamazaki, S., Tsai, L., Stadtman, T. C., Teshima, T., Nakaji, A., and Shiba, T. (1985). *PNAS* **82,** 1364.
86. Jacobson, F., and Walsh, C. (1984). *Biochemistry* **23,** 979.
87. Manstein, D. J., Pai, E. F., Schopter, L. M., and Massey, V. (1986). *Biochemistry* **25,** 6807.

88. Manstein, D. J., Massey, V., Ghisla, S., and Pai, E. F. (1988). *Biochemistry* **27,** 2300.
89. Lindqvist, Y., and Branden, C.-I. (1985). *PNAS* **82,** 6855.
90. Williams, C. H., Jr. (1976). "The Enzymes," 3rd Ed., Vol. 13, p. 89.
91. Farber, G. K., Glasfeld, A., Tirby, G., Ringe, D., and Petsko, G. A. (1989). *Biochemistry* **28,** 7289.
92. Lin, H.-C., Lei, S.-P., and Wilcox, G. (1985). *Gene* **34,** 123.
93. Miles, J. S., and Guest, J. R. (1984). *Gene* **32,** 41.
94. Raines, R. T., Sutton, E. L., Straus, D. R., Gilbert, W., and Knowles, J. R. (1986). *Biochemistry* **25,** 7145.
95. Hermes, J. D., Blacklow, S. C., and Knowles, J. R. (1987). *Cold Spring Harbor Symp. Quant. Biol.* **52,** 597.
96. Banner, D. W., Bloomer, A. C., Petsko, G. A., Phillips, D. C., Pogson, C. I., Wilson, I. A., Corran, P. H., Furth, A. J., Milman, J. D., Offord, R. E., Priddle, J. D., and Waley, S. G. (1975). *Nature (London)* **255,** 609.
97. Alber, T. C., Davenport, R. C., Jr., Giammona, D. A., Lolis, F., Petsko, G. A., and Ringe, D. (1987). *Cold Spring Harbor Symp. Quant. Biol.* **52,** 603.
98. O'Connell, E. L., and Rose, I. A. (1973). *JBC* **248,** 2225.
99. Dyson, J. E. D., and Noltmann, E. A. (1968). *JBC* **243,** 1401.
100. Benkovic, S. J., and Shray, K. J. (1976). *Adv. Enzymol. Relat. Areas Mol. Biol.* **44,** 139.
101. Shaw, P. J., and Muirhead, H. (1976). *FEBS Lett.* **65,** 50.
102. Ekwall, K., and Mannervik, B. (1973). *BBA* **297,** 297.
103. Vander Jagt, D. L., Daub, E., Krohn, J. A., and Han, L.-P. B. (1975). *Biochemistry* **14,** 3669.
104. Creighton, D. J., and Pourmotabbed, T. (1988). *In* "Mechanistic Principles of Enzyme Activity," Volume 9: Molecular Structure and Energetics" (J. F. Liebman and A. Greenberg, eds.), p. 353. VCH, New York.
105. Griffis, C. E. F., Ong, L. H., Buettner, L., and Creighton, D. J. (1983). *Biochemistry* **22,** 2945.
106. Creighton, D. J., Migliorini, M., Pourmotabbed, T., and Guha, M. K. (1988). *Biochemistry* **27,** 7376.
107. Hall, S. S., Doweyko, A. M., and Jordan, F. (1978). *J. Am. Chem. Soc.* **98,** 7460.
108. Shinkai, S., Yamashita, T., Kusano, Y., and Manabe, O. (1981). *J. Am. Chem. Soc.* **103,** 2070.
109. Kozarich, J. W., Chari, R. V. J., Wu, J. C., and Lawrence, T. L. (1981). *J. Am. Chem. Soc.* **103,** 4593.
110. Chari, R. V. J., and Kozarich, J. W. (1981). *JBC* **256,** 9785.
111. Kozarich, J. W., and Chari, R. V. J. (1982). *J. Am. Chem. Soc.* **104,** 2655.
112. Chari, R. V. J., and Kozarich, J. W. (1983). *J. Am. Chem. Soc.* **105,** 7169.
113. Sellin, S., Rosevear, P. R., Mannervik, B., and Mildvan, A. S. (1982). *JBC* **257,** 10023.
114. Rosevear, P. R., Chari, R. V. J., Kozarich, J. W., Sellin, S., Mannervik, B., and Mildvan, A. S. (1983). *JBC* **258,** 6823.
115. Talalay, P., and Wang, V. S. (1955). *BBA* **18,** 300.
116. Kawahara, F. S., and Talalay, P. (1960). *JBC* **235,** PC1.
117. Wang, S.-F., Kawahara, F. S., and Talalay, P. (1963). *JBC* **238,** 576.
118. Malhotra, S. K., and Ringold, H. J. (1965). *J. Am. Chem. Soc.* **87,** 3228.
119. Klinman, J. P., and Rose, I. A. (1971). *Biochemistry* **10,** 2259.
120. Gunther, H., and Simon, H. (1973). *FEBS Lett.* **33,** 81.
121. Rilling, H. C., and Coon, M. J. (1960). *JBC* **235,** 3087.
122. Pollack, R. M., Bounds, P. L., and Bevins, C. L. (1989). *In* "The Chemistry of Enones" (S. Patai and Z. Rappoport, eds.), p. 559. Wiley, New York.
123. Martyr, R. J., and Benisek, W. F. (1973). *Biochemistry* **12,** 2172.
124. Martyr, R. J., and Benisek, W. F. (1975). *JBC* **250,** 1218.

125. Orgez, J. R., Tivol, W. F., and Benisek, W. F. (1977). *JBC* **252,** 6151.
126. Kuliopulos, A., Mildvan, A. S., Shortle, D., and Talalay, P. (1989). *Biochemistry* **28,** 149.
127. Kuliopulos, A., Westbrook, E. M., Talalay, P., and Mildvan, A. S. (1987). *Biochemistry* **26,** 3927.
128. Westbrook, E. M., Piro, O. E., and Sigler, P. B. (1984). *JBC* **259,** 9096.
129. Kayser, R. H., Bounds, P. L., Bevins, C. L., and Pollack, R. M. (1983). *JBC* **258,** 909.
130. Bounds, P. L., and Pollack, R. M. (1987). *Biochemistry* **26,** 2263.
131. Bevins, C. L., Bantia, S., Pollack, R. M., Bounds, P. L., and Kayser, R. H. (1984). *J. Am. Chem. Soc.* **106,** 4957.
132. Bantia, S., Bevins, C. L., and Pollack, R. M. (1985). *Biochemistry* **24,** 2606.
133. Eames, T. C. M., Pollack, R. M., and Steiner, R. F. (1989). *Biochemistry* **28,** 6269.
134. Viger, A., Coustal, S., and Marquet, A. (1981). *J. Am. Chem. Soc.* **103,** 451.
135. Viger, A., and Marquet, A. (1977). *BBA* **485,** 482.
136. Floss, H. G., and Tsai, M.-D. (1979). *Adv. Enzymol.* **50,** 243.
137. Floss, H. G. (1982). "Methods in Enzymology," Vol. 87, p. 126.
138. Floss, H. G. (1984). *Top. Stereochem.* **15,** 253.
139. Lenz, H., and Eggerer, H. (1976). *EJB* **65,** 237.
140. Lenz, H., Wunderwald, P., Buschmeier, V., and Eggerer, H. (1971). *Hoppe-Seyler's Z. Physiol. Chem.* **352,** 517.
141. Townsend, C. A., Scholl, T., and Arigoni, D. (1975). *J. Chem. Soc., Chem. Commun.* p. 921.
142. Kajiwara, M., Lee, S.-F., Scott, A. I., Akhtar, M., Jones, C. R., and Jordan, P. M. (1978). *J. Chem. Soc., Chem. Commun.* p. 967.
143. Fryzuk, M. D., and Bosnich, B. (1979). *J. Am. Chem. Soc.* **101,** 3043.
144. Caspi, E., Piper, J., and Shapiro, S. (1981). *J. Chem. Soc., Chem. Commun.* p. 76.
145. Townsend, C. A., Neese, A. S., and Theis, A. B. (1982). *J. Chem. Soc., Chem. Commun.* p. 116.
146. Kobayashi, K., Jadhav, P. K., Zydowsky, T. M., and Floss, H. G. (1983). *J. Org. Chem.* **48,** 3510.
147. Caspi, E., Aranachalam, T., and Nelson, P. A. (1983). *J. Am. Chem. Soc.* **105,** 6987.
148. Kobayashi, K., Kakinuma, K., and Floss, H. G. (1984). *J. Org. Chem.* **49,** 1290.
149. Coates, R. M., Kock, S. C., and Hegde, S. (1986). *J. Am. Chem. Soc.* **108,** 2762.
150. Zydowsky, T. M., Courtney, L. F., Frasca, V., Kobayashi, K., Shimizu, H., Yuen, L.-D., Matthews, R. G., Benkovic, S. J., and Floss, H. G. (1986). *J. Am. Chem. Soc.* **108,** 3152.
151. O'Connor, E. J., Kobayashi, K., Floss, H. G., and Gladysz, J. A. (1987). *J. Am. Chem. Soc.* **109,** 4837.
152. Rose, I. A. (1970). *JBC* **245,** 6052.
153. Creighton, D. J., and Rose, I. A. (1976). *JBC* **251,** 61.
154. Altman, L. J., Han, C. Y., Bertolino, A., Handy, G., Laungaini, D., Muller, W., Schwartz, S., Shanker, D., de Wolf, W. H., and Yang, F. (1978). *J. Am. Chem. Soc.* **100,** 3235.
155. Rozell, J. D., Jr., and Benner, S. A. (1983). *J. Org. Chem.* **48,** 1190.
156. Rose, I. A. (1958). *J. Am. Chem. Soc.* **80,** 5835.
157. Chiu, T. H., and Feingold, D. S. (1967). *Fed. Proc., Fed. Am. Soc. Exp. Biol.* **26,** 835.
158. Meloche, H. P., Mehler, L., and Wurstur, J. M. (1975). *JBC* **250,** 6870.
159. Meloche, H. P., and Monti, C. T. (1975). *JBC* **250,** 6875.
160. Meloche, H. P., and Mehler, L. (1973). *JBC* **248,** 6333.
161. Loewus, M. W., Loewus, F. A., Brillinger, G.-U., Otsuka, H., and Floss, H. G. (1980). *JBC* **255,** 11710.
162. Grazi, E., Cheng, T., and Horecker, B. L. (1962). *BBRC* **7,** 250.
163. Horecker, B. L., Rowley, P. T., Grazi, E., Chang, T., and Tchola, O. (1963). *Biochem. Z.* **330,** 36.
164. Model, P., Ponticorvo, L., and Rittenberg, D. (1968). *Biochemistry* **7,** 1339.

165. Wurstur, B., and Hess, B. (1974). *FEBS Lett.* **38,** 257.
166. Schray, K. J., Fishbein, R., Bullard, W. P., and Benkovic, S. J. (1975). *JBC* **250,** 4883.
167. Bloom, B., and Topper, Y. J. (1956). *Science* **124,** 982.
168. Rose, I. A., and Rieder, S. V. (1958). *JBC* **231,** 315.
169. Ferran, H. E., Jr., Roberts, R. D., Jacob, J. N., and Spencer, T. A. (1978). *J. Chem. Soc., Chem. Commun.*, p. 49.
170. Periana, R. A., Motiu-DeGrood, R., Chiang, Y., and Hupe, D. J. (1980). *J. Am. Chem. Soc.* **102,** 3923.
171. Grazi, E., Sivieri-Pecorari, C., Gagliano, R., and Trombetta, G. (1973). *Biochemistry* **12,** 2583.
172. Lowe, G., and Pratt, R. F. (1976). *EJB* **66,** 95.
173. Sygusch, J., Beaudry, D., and Allaire, M. (1987). *PNAS* **84,** 7846.
174. Mavridis, I. M., Hatada, M. H., Tulinsky, A., and Lebioda, L. (1982). *JMB* **162,** 419.
175. Meloche, H. P., Monti, C. T., and Hogue-Angeletti, R. A. (1978). *BBRC* **84,** 589.
176. Suzuki, N., and Wood, W. A. (1980). *JBC* **255,** 3427.
177. DiIasio, A., Trombetta, G., and Grazi, E. (1977). *FEBS Lett.* **73,** 244.
178. Meloche, H. P., Sparks, G. R., Monti, C. T., Waterbor, J. W., and Lademan, T. H. (1980). *ABB* **203,** 702.
179. Harris, C. E., Kobes, R. D., Teller, D. C., and Rutter, W. J. (1969). *Biochemistry* **8,** 2442.
180. Rutter, W. J. (1964). *Fed. Proc., Fed. Am. Soc. Exp. Biol.* **23,** 1248.
181. Smith, G. M., Harper, E. T., and Mildvan, A. S. (1980). *Biochemistry* **19,** 1248.
182. Kadonaga, J. T., and Knowles, J. R. (1983). *Biochemistry* **22,** 130.
183. Belasco, J. G., and Knowles, J. R. (1983). *Biochemistry* **22,** 122.
184. Loewus, F. A., and Kelly, S. (1962). *BBRC* **7,** 204.
185. Barnett, J. E. G., Rasheed, A., and Corina, D. L. (1973). *BJ* **131,** 21.
186. Chen, C. J.-J., and Eisenberg, F., Jr. (1975). *JBC* **250,** 2963.
187. Loewus, M. W. (1977). *JBC* **252,** 7221.
188. Loewus, M. W., Loewus, F. A., Brillinger, G.-U., Otsuka, H., and Floss, H. G. (1980). *JBC* **255,** 11710.
189. Byun, S. M., Jenness, R., Ridley, W. P., and Kirkwood, S. (1973). *BBRC* **54,** 961.
190. Benkovic, S. J., and Schray, K. J. (1976). *Adv. Enzymol. Relat. Areas Mol. Biol.* **44,** 139.
191. Schray, K. J., and Benkovic, S. J. (1978). *Acc. Chem. Res.* **11,** 136.
192. Wurster, B., and Hess, B. (1973). *BBRC* **55,** 985.
193. Koerner, T. A. W., Jr., Voll, R. J., and Younathan, E. S. (1977). *FEBS Lett.* **84,** 207.
194. Schray, K. J., Howell, E. E., Waud, J. M., Benkovic, S. J., and Cunningham, B. A. (1980). *Biochemistry* **19,** 2593.
195. Midelfort, C. F., Gupta, R. J., and Meloche, H. P. (1977). *JBC* **252,** 3486.
196. Lenz, H., and Eggerer, H. (1976). *EJB* **65,** 237.
197. Eggerer, H., Buckel, W., Lenz, H., Wunderwald, P., Gottschalk, G., Cornforth, J. W., Donninger, C., Mallaby, R., and Redmond, J. W. (1970). *Nature (London)* **226,** 517.
198. Klinman, J. P., and Rose, I. A. (1971). *Biochemistry* **10,** 2267.
199. Buckel, W., Lenz, H., Wunderwald, P., Buschmeier, V., Eggerer, H., and Gottschalk, G. (1971). *EJB* **24,** 201.
200. Rétey, J., Luthy, J., and Arigoni, D. (1970). *Nature (London)* **226,** 519.
201. Lenz, H., Buckel, W., Wunderwald, P., Biedermann, G., Buschmeier, V., Eggerer, H., Cornforth, J. W., Redmond, J. W., and Mallaby, R. (1971). *J. Biochem. (Tokyo)* **24,** 207.
202. Rétey, J., Luthy, J., and Arigoni, D. (1970). *Chimia* **24,** 34.
203. Wunderwald, P., Buckel, W., Lenz, H., Buschmeier, V., Eggerer, H., Gottschalk, G., Cornforth, J. W., Redmond, J. W., and Mallaby, R. (1971). *EJB* **24,** 216.
204. Hanson, K. R. (1965). *Fed. Proc., Fed. Am. Soc. Exp. Biol.* **24,** 229.
205. Sprecher, M., Berger, R., and Sprinson, D. B. (1964). *JBC* **239,** 4268.

206. Cornforth, J. W., Phillips, G. T., Messner, B., and Eggerer, H. (1974). *EJB* **42,** 591.
207. Messner, B., Eggerer, H., Cornforth, J. W., and Mallaby, R. (1975). *EJB* **53,** 255.
208. Martinoni, B., and Arigoni, D. (1975). *Chimia* **29,** 26.
209. Buckel, W., and Bobi, A. (1975). *Biochem. Soc. Trans.* **3,** 924.
210. Lenz, H., Wunderwald, P., and Eggerer, H. (1976). *EJB* **65,** 225.
211. Willadsen, P., and Eggerer, H. (1975). *EJB* **54,** 253.
212. Eggerer, H. (1965). *Biochem. Z.* **343,** 111.
213. Kurtz, L. C., Ackerman, J. J. H., and Drysdale, G. R. (1985). *Biochemistry* **24,** 452.
214. Karpusas, M., Branchaud, B., and Remington, S. J. (1990). *Biochemistry* **29,** 2213.
215. Englard, S., and Listowsky, I. (1963). *BBRC* **12,** 356.
216. Lienhard, G. E., and Rose, I. A. (1964). *Biochemistry* **3,** 185.
217. Rose, Z. B. (1966). *JBC* **241,** 2311.
218. Rose, I. A. (1970). *JBC* **245,** 6052.
219. Lienhard, G. E., and Rose, I. A. (1964). *Biochemistry* **3,** 190.
220. Hill, R. K., Sawada, S., and Arfin, S. M. (1979). *Bioorg. Chem.* **8,** 175.
221. Schutzbach, J. S., and Feingold, D. S. (1970). *JBC* **245,** 2476.
222. Rozzell, J. D., Jr., and Benner, S. A. (1984). *J. Am. Chem. Soc.* **106,** 4937.
223. Piccirilli, J. A., Rozzell, J. D., Jr., and Benner, S. A. (1987). *J. Am. Chem. Soc.* **109,** 8084.
224. Hsu, R. Y., Mildvan, A. S., Chang, G.-G., and Fung, C.-H. (1976). *JBC* **251,** 6574.
225. Grissom, C. B., and Cleland, W. W. (1988). *Biochemistry* **27,** 2927.
226. Bratcher, S. C., and Hsu, R. Y. (1982). *BBA* **702,** 54.
227. Steinberger, R., and Westheimer, F. H. (1949). *J. Am. Chem. Soc.* **71,** 4158.
228. Steinberger, R., and Westheimer, F. H. (1951). *J. Am. Chem. Soc.* **73,** 429.
229. Thibblin, A., and Jencks, W. P. (1979). *J. Am. Chem. Soc.* **101,** 4963.
230. Depuy, C. H., Breitbeil, F. W., and DeBruin, K. R. (1966). *J. Am. Chem. Soc.* **88,** 3347.
231. Floss, H. G., and Vederas, J. C. (1982). *In* "Stereochemistry" (C. Tamm, ed.), pp. 161–199. Elsevier Biomedical Press, Amsterdam.
232. Metzler, D. E. (1979). *Adv. Enzymol.* **50,** 1.
233. Walsh, C. (1979). "Enzymatic Reaction Mechanisms," pp. 777–833. Freeman, San Francisco, California.
234. Dunathan, H. C. (1966). *PNAS* **55,** 713.
235. Tsai, M.-D., Weintraub, H. J. R., Byrn, S. R., Chang, C., and Floss, H. G. (1978). *Biochemistry* **17,** 3183.
236. Besmer, P., and Arigoni, D. (1969). *Chimia* **23,** 190.
237. Austermuhle-Bertola, E. (1973). Ph.D. Dissertation No. 5009, Eidgenössische Technische Hochschule, Zurich.
238. Ayling, J. E., Dunathan, H. C., and Snell, E. E. (1968). *Biochemistry* **7,** 4537.
239. Bailey, G. B., Kusamvarn, T., and Vuttivej, K. (1970). *Fed. Proc., Fed. Am. Soc. Exp. Biol.* **29,** 857.
240. Eichele, G., Ford, G. C., Glor, M., Jansonius, J. N., Mavrides, C., and Christen, P. (1979). *JMB* **133,** 161.
241. Ford, G. C., Eichele, G., and Jansonius, J. N. (1980). *PNAS* **77,** 2559.
242. Kirsch, J. F., Eichele, G., Ford, G. C., Vincent, M. G., and Jansonius, J. N. (1984). *JMB* **174,** 497.
243. Ivanov, V. I., and Karpeisky, M. Y. (1969). *Adv. Enzymol.* **32,** 21.
244. Metzler, C. M., Metzler, D. E., Martin, D. S., Newman, R., Arnne, A., and Rogers, P. (1978). *JBC* **253,** 525.
245. Vederas, J. C., Schleicher, E., Tsai, M.-D., and Floss, H. G. (1978). *JBC* **253,** 5350.
246. Belleau, B., and Burba, J. (1960). *J. Am. Chem. Soc.* **82,** 5751.
247. Battersby, A. R., Chrystal, E. J. T., and Staunton, J. (1980). *J. Chem. Soc., Perkin Trans. I*, p. 31.

248. Marshall, K. S., and Castagnoli, N., Jr. (1973). *J. Med. Chem.* **16,** 266.
249. Leistner, E., and Spenser, I. D. (1975). *J. Chem. Soc. Chem. Commun.*, p. 378.
250. Gerdes, H. J., and Leistner, E. (1979). *Phytochemistry* **18,** 771.
251. Orr, G. R., and Gould, S. J. (1982). *Tetrahedron Lett.* **23,** 3139.
252. Yamada, H., and O'Leary, M. H. (1978). *Biochemistry* **17,** 669.
253. Santaniello, E., Kienle, M. G., Manzocchi, A., and Bosisio, E. (1979). *J. Chem. Soc., Perkin Trans. 1*, p. 1677.
254. Bouclier, M., Jung, M. J., and Lippert, B. (1979). *EJB* **98,** 363.
255. Battersby, A. R., Joyeau, R., and Staunton, J. (1979). *FEBS Lett.* **107,** 231.
256. Orr, G. R., Gould, S. J., Pegg, A. E., Seely, J. E., and Coward, J. K. (1984). *Bioorg. Chem.* **12,** 252.
257. Robins, D. J. (1983). *Phytochemistry* **22,** 1133.
258. Asada, Y., Tanizawa, K., Nakamura, K., Moriguchi, M., and Soda, K. (1984). *J. Biochem. (Tokyo)* **95,** 277.
259. Asada, Y., Tanizawa, K., Sawada, S., Suzuki, T., Misono, H., and Soda, K. (1981). *Biochemistry* **20,** 6881.
260. Kelland, J. G., Palcic, M. M., Pickard, M. A., and Vederas, J. C. (1985). *Biochemistry* **24,** 3263.
261. Sukhareva, B. S., Dunathan, H. C., and Braunstein, A. E. (1971). *FEBS Lett.* **15,** 241.
262. Vederas, J. C., Reingold, I. D., and Sellers, H. W. (1979). *JBC* **254,** 5053.
263. Dunathan, H. C., and Voet, J. G. (1974). *PNAS* **71,** 3888.
264. Asada, Y., Tanizawa, K., Kawabata, Y., Misono, H., and Soda, K. (1981). *Agric. Biol. Chem.* **45,** 1218.
265. Riley, W. D., and Dixon, E. E. (1968). *Biochemistry* **7,** 3520.
266. Chang, G. W., and Snell, E. E. (1968). *Biochemistry* **7,** 2005.
267. Battersby, A. R., Nicoletti, M., Staunton, J., and Vleggaar, R. (1980). *J. Chem. Soc., Perkin Trans. 1*, p. 43.
268. Santaniello, E., Manzocchi, A., and Biondi, P. A. (1981). *J. Chem. Soc., Perkin Trans. 1*, p. 307.
269. Allen, R. R., and Klinman, J. P. (1981). *JBC* **256,** 3233.
270. Gallagher, T., Snell, E. E., and Hackert, M. L. (1989). *JBC* **264,** 12737.
271. Vederas, J. C., Schleicher, E., Tsai, M. D., and Floss, H. G. (1978). *JBC* **253,** 5350.
272. Palcic, M. M., Antoun, M., Tanizawa, K., Soda, K., and Floss, H. G. (1985). *JBC* **260,** 5248.
273. Chang, C.-C., Laghai, A., O'Leary, M. H., and Floss, H. G. (1982). *JBC* **257,** 3564.
274. Guggenheim, S., and Flavin, M. (1968). *BBA* **151,** 664.
275. Posner, B. I., and Flavin, M. (1972). *JBC* **247,** 6402.
276. Posner, B. I., and Flavin, M. (1972). *JBC* **247,** 6412.
277. Guggenheim, S., and Flavin, M. (1969). *JBC* **244,** 6217.
278. Hansen, P. E., Feeney, J., and Roberts, G. C. K. (1975). *J. Magn. Reson.* **17,** 249.
279. Fuganti, C., and Coggiola, D. (1977). *Experientia* **33,** 847.
280. Chang, M. N. T., and Walsh, C. (1980). *J. Am. Chem. Soc.* **102,** 7370.
281. Schwab, J. M., Klassen, J. B., and Habib, A. (1986). *J. Chem. Soc., Chem. Commun.*, p. 357.
282. Sedgwick, B., Morris, C., and French, S. J. (1978). *J. Chem. Soc., Chem. Commun.*, p. 193.
283. Messner, B., Eggerer, H., Cornforth, J. W., and Mallaby, R. (1975). *EJB* **53,** 255.
284. Hanson, K. R., and Rose, I. A. (1963). *PNAS* **50,** 981.
285. Mazumder, A., Gerlt, J. A., Rabow, L., Absalon, M. J., Stubbe, J., and Bolton, P. H. (1989). *J. Am. Chem. Soc.* **111,** 8029.
286. Parry, R. J., and Askonas, L. J. (1985). *J. Am. Chem. Soc.* **107,** 1417.
287. Widlanski, T. S., Bender, S. L., and Knowles, J. R. (1987). *J. Am. Chem. Soc.* **109,** 1873.

288. Avigad, G., and Englard, S. (1969). *Fed. Proc., Fed. Am. Soc. Exp. Biol.* **28,** 345.
289. Chari, R. V. J., Whitman, C. P., Kozarich, J. W., Ngai, K.-L., and Ornston, L. N. (1987). *J. Am. Chem. Soc.* **109,** 5514.
290. Kirby, G. W., O'Laughlin, G. J., and Robins, D. J. (1975). *J. Chem. Soc., Chem. Commun.,* p. 402.
291. Farrer, T. C., Gutowsky, H. S., Albery, R. A., and Miller, W. G. (1957). *J. Am. Chem. Soc.* **79,** 3978.
292. Englard, S. (1958). *JBC* **233,** 1003.
293. Anet, F. A. L. (1960). *J. Am. Chem. Soc.* **82,** 994.
294. Gawron, O., Glaid, A. J., and Fondy, T. P. (1961). *J. Am. Chem. Soc.* **83,** 3634.
295. Englard, S., Britten, J. S., and Litowsky, I. (1967). *JBC* **242,** 2255.
296. Subramanian, S. S., Raghavendra, R., and Bhattacharyya, P. K. (1966). *Indian J. Biochem.* **3,** 260.
297. Cohn, M., Pearson, J. E., O'Connell, E. L., and Rose, I. A. (1970). *J. Am. Chem. Soc.* **92,** 4095.
298. Cole, F. E., Kalyanpur, M. G., and Stevens, C. M. (1973). *Biochemistry* **12,** 3346.
299. Krasna, A. I. (1958). *JBC* **233,** 1010.
300. Givot, I. L., Smith, T. A., and Ables, R. H. (1969). *JBC* **244,** 6341.
301. Rétey, J., Fierz, H., and Zeylemaker, W. P. (1970). *FEBS Lett.* **6,** 203.
302. Wightman, R. H., Staunton, J., Battersby, A. R., and Hanson, K. R. (1972). *J. Chem. Soc., Perkin Trans. 1,* p. 2355.
303. Sawada, S., Kumagai, H., Yamada, H., Hill, R. K., Mugibayashi, Y., and Ogata, K. (1973). *BBA* **315,** 204.
304. Ife, R., and Haslam, E. (1971). *J. Chem. Soc. C,* p. 2818.
305. Strange, P. G., Staunton, J., Wiltshire, H. R., Battersby, A. R., Hanson, K. R., and Havir, E. A. (1972). *J. Chem. Soc., Perkin Trans. 1,* p. 2364.
306. Ellis, B. E., Zenk, M. H., Kirby, G. W., Michael, J., and Floss, H. G. (1973). *Phytochemistry* **12,** 1057.
307. Hoberman, H. D., Havir, E. A., Rochovansky, O., and Ratner, S. (1964). *JBC* **239,** 3818.
308. Miller, R. W., and Buchanan, J. M. (1962). *JBC* **237,** 491.
309. Cornforth, J. W., Cornforth, R. H., Popják, G., and Yengoyan, L. (1966). *JBC* **241,** 3970.
310. Chari, R. V. J., Whitman, C. P., Kozarich, J. W., Ngai, K.-L., and Ornston, L. N. (1987). *J. Am. Chem. Soc.* **109,** 5514.
311. Bordwell, F. G. (1970). *Acc. Chem. Res.* **3,** 281.
312. Bordwell, F. G. (1972). *Acc. Chem. Res.* **5,** 374.
313. Bordwell, F. G., Vestling, M. M., and Yee, K. C. (1970). *J. Am. Chem. Soc.* **92,** 5950.
314. Hine, J., Wiesboeck, R., and Ghirardellia, R. G. (1961). *J. Am. Chem. Soc.* **83,** 1219.
315. Fedor, L. R., and Glave, W. R. (1971). *J. Am. Chem. Soc.* **93,** 985.
316. Fedor, L. R. (1969). *J. Am. Chem. Soc.* **91,** 908.
317. More O'Ferrall, R. A., and Slae, S. J. (1969). *Chem. Soc., Chem. Commun.,* p. 486.
318. More O'Ferrall, R. A., and Slae, S. J. (1970). *J. Chem. Soc. B,* p. 260.
319. Palmer, J. L., and Ables, R. H. (1979). *JBC* **254,** 1217.
320. Palmer, J. L., and Ables, R. H. (1976). *JBC* **251,** 5817.
321. Srinivasan, P. R., Rothschild, J., and Sprinson, D. B. (1963). *JBC* **238,** 3176.
322. Maitra, U. S., and Sprinson, D. B. (1978). *JBC* **253,** 5426.
323. Bender, S. L., Mehdi, S., and Knowles, J. R. (1989). *Biochemistry* **28,** 7555.
324. Bender, S. L., Widlanski, T., and Knowles, J. R. (1989). *Biochemistry* **28,** 7560.
325. Widlanski, T., Bender, S. L., and Knowles, J. R. (1989). *Biochemistry* **28,** 7572.
326. Butler, J. R., Alworth, W. L., and Nugent, M. J. (1974). *J. Am. Chem. Soc.* **96,** 1617.
327. Varz, A. D. M., Butler, J. R., and Nugent, M. J. (1975). *J. Am. Chem. Soc.* **97,** 5914.

328. Goering, H. L., and Pombo, M. M. (1960). *J. Am. Chem. Soc.* **82,** 2515.
329. Blanchard, J. S., and Cleland, W. W. (1980). *Biochemistry* **19,** 4506.
330. Hermes, J. D., Weiss, P. M., and Cleland, W. W. (1985). *Biochemistry* **24,** 2959.
331. Porter, D. J. T., and Bright, H. J. (1980). *JBC* **255,** 4772.
332. Schloss, J. V., Porter, D. J. T., Bright, H. J., and Cleland, W. W. (1980). *Biochemistry* **19,** 2358.
333. Anderson, V. E., Weiss, P. M., and Cleland, W. W. (1984). *Biochemistry* **23,** 2779.
334. Benner, S. A. (1988). *In* "Mechanistic Principles of Enzyme Activity, Volume 9: Molecular Structure and Energetics" (J. F. Liebman and A. Greenberg, eds.), p. 27. VCH, New York.
335. Woodard, R. W., Tsai, M.-D., Floss, H. G., Crooks, P. A., and Coward, J. K. (1980). *JBC* **255,** 9124.
336. Asano, Y., Woodard, R. W., Houck, D. R., and Floss, H. G. (1984). *ABB* **231,** 253.
337. Frenzel, T., Beale, J. M., Kobayashi, M., Zenk, M. H., and Floss, H. G. (1988). *J. Am. Chem. Soc.* **110,** 7878.
338. Kobayashi, M., Frenzel, T., Lee, J. P., Zenck, M. H., and Floss, H. G. (1987). *J. Am. Chem. Soc.* **109,** 6184.
339. Woodard, R. W., Weaver, J., and Floss, H. G. (1981). *ABB* **207,** 51.
340. Woodard, R. W., Mascaro, L., Hörhammer, R., Jr., Eisenstein, S., and Floss, H. G. (1980). *J. Am. Chem. Soc.* **102,** 6314.
341. Arigoni, D. (1978). *Ciba Found. Symp.* **60** (new series), 243.
342. Zydowsky, L. D., Zydowsky, T. M., Hass, E. S., Brown, J. W., Reeve, J. N., and Floss, H. G. (1987). *J. Am. Chem. Soc.* **109,** 7922.
343. Houck, D. R., Kobayashi, K., Williamson, J. M., and Floss, H. G. (1986). *J. Am. Chem. Soc.* **108,** 5365.
344. Zydowsky, T. M., Courtney, L. F., Frasca, V., Kobayashi, K., Shimizu, H., Yuen, L.-D., Matthews, R. G., Benkovic, S. J., and Floss, H. G. (1986). *J. Am. Chem. Soc.* **108,** 3152.
345. Lebertz, H., Simon, H., Courtney, L. F., Benkovic, S. J., Zydowsky, L. D., Lee, K., and Floss, H. G. (1987). *J. Am. Chem. Soc.* **109,** 3173.
346. Slieker, L. J., and Benkovic, S. J. (1984). *J. Am. Chem. Soc.* **106,** 1833.
347. Borchardt, R. T. (1973). *J. Med. Chem.* **16,** 377.
348. Hegazi, M. F., Borchardt, R. T., and Schowen, R. L. (1979). *J. Am. Chem. Soc.* **101,** 4359.
349. Matthews, R. G. (1984). *In* "Folates and Pterins, Volume 1: Chemistry and Biochemistry of Folates" (R. L. Blakley and S. J. Benkovic, eds.), p. 497. Wiley, New York.
350. Van der Meijden, P., Heythuysen, H. J., Pouwels, A., Houwen, F., Van der Drift, C., and Vagels, G. D. (1983). *Arch. Microbiol.* **134,** 238.
351. Wood, H. G., Ragsdale, S. W., and Pezacka, E. (1986). *Biochem. Int.* **12,** 421.
352. Hogenkamp, H. (1975). *In* "Cobalamins" (B. Babior, ed.), p. 21. Wiley, New York.
353. Williamson, J. M., Inamine, E., Wilson, K. E., Douglas, A. W., Liesch, J. M., and Albers-Schönberg, G. J. (1985). *JBC* **260,** 4637.

Author Index

Numbers is parentheses are reference numbers and indicate that an author's work is referred to although the name is not cited in the text. Numbers in italics refer to the page numbers on which the complete reference appears.

G

Subject Index